Modern Insights in Biotechnology

Modern Insights in Biotechnology

Editor: Joy Adam

R CALLISTO REFERENCE

www.callistoreference.com

Callisto Reference,
118-35 Queens Blvd., Suite 400,
Forest Hills, NY 11375, USA

Visit us on the World Wide Web at:
www.callistoreference.com

ISBN: 978-1-64116-156-5 (Hardback)

Cataloging-in-Publication Data

Modern insights in biotechnology / edited by Joy Adam.
    p. cm.
Includes bibliographical references and index.
ISBN 978-1-64116-156-5
1. Biotechnology. 2. Genetic engineering. I. Adam, Joy.
TP248.2 .M63 2019
660.6--dc23

Table of Contents

Preface

Biotechnology is an interdisciplinary field that uses biological systems, organisms and their derivatives to develop products and processes for specific use. It was traditionally used for the domestication of animals and cultivation of plants, as well as for their enhancement by using artificial selection and hybridization. Modern biotechnology integrates techniques of genetic engineering, cell and tissue culture for developing products such as biodegradable plastics, biofuels, vegetable oil, as well as for the manufacture of beer and milk products. Besides these, bioleaching, bioremediation and biological weapons production are upcoming applications of biotechnology. The major branches of biotechnology are bioinformatics, blue biotechnology, green biotechnology, red biotechnology and white biotechnology. This book elucidates the concepts and innovative models around prospective developments with respect to biotechnology. It presents researches and studies performed by experts across the globe. It will help new researchers by foregrounding their knowledge in this field. This book is a resource guide for experts as well as students.

Various studies have approached the subject by analyzing it with a single perspective, but the present book provides diverse methodologies and techniques to address this field. This book contains theories and applications needed for understanding the subject from different perspectives. The aim is to keep the readers informed about the progresses in the field; therefore, the contributions were carefully examined to compile novel researches by specialists from across the globe.

Indeed, the job of the editor is the most crucial and challenging in compiling all chapters into a single book. In the end, I would extend my sincere thanks to the chapter authors for their profound work. I am also thankful for the support provided by my family and colleagues during the compilation of this book.

Editor

Spindle assembly checkpoint is sufficient for complete Cdc20 sequestering in mitotic control

Bashar Ibrahim *

Bio System Analysis Group, Friedrich-Schiller-University Jena, and Jena Centre for Bioinformatics (JCB), 07743 Jena, Germany
Umm Al-Qura University, 1109 Makkah, Saudi Arabia
Al-Qunfudah Center for Scientific Research (QCSR), 21912 Al-Qunfudah, Saudi Arabia

ARTICLE INFO

Keywords:
Spindle assembly checkpoint
Anaphase promoting complex
MCC
Cdc20
Systems biology

ABSTRACT

The spindle checkpoint assembly (SAC) ensures genome fidelity by temporarily delaying anaphase onset, until all chromosomes are properly attached to the mitotic spindle. The SAC delays mitotic progression by preventing activation of the ubiquitin ligase anaphase-promoting complex (APC/C) or cyclosome; whose activation by Cdc20 is required for sister-chromatid separation marking the transition into anaphase. The mitotic checkpoint complex (MCC), which contains Cdc20 as a subunit, binds stably to the APC/C. Compelling evidence by Izawa and Pines (Nature 2014; 10.1038/nature13911) indicates that the MCC can inhibit a second Cdc20 that has already bound and activated the APC/C. Whether or not MCC per se is sufficient to fully sequester Cdc20 and inhibit APC/C remains unclear. Here, a dynamic model for SAC regulation in which the MCC binds a second Cdc20 was constructed. This model is compared to the MCC, and the MCC-and-BubR1 (dual inhibition of APC) core model variants and subsequently validated with experimental data from the literature. By using ordinary nonlinear differential equations and spatial simulations, it is shown that the SAC works sufficiently to fully sequester Cdc20 and completely inhibit APC/C activity. This study highlights the principle that a systems biology approach is vital for molecular biology and could also be used for creating hypotheses to design future experiments.

1. Introduction

Faithful DNA segregation, prior to cell division at mitosis, is vital for maintaining genomic integrity. Eukaryotic cells have evolved a conserved surveillance control mechanism for DNA segregation called the Spindle Assembly Checkpoint (SAC; [1]). The SAC monitors the existence of chromatids that are not yet attached correctly to the mitotic spindle and delays the onset of anaphase until all chromosomes have made amphitelic tight bipolar attachments to the mitotic spindle. A dysfunction in the SAC can lead to aneuploidy [2] and furthermore its reliable function is important for tumor suppression [3,4].

SAC acts by inhibiting the anaphase-promoting complex (APC/C or APC), a ubiquitin ligase, presumably through sequestering the ACP-activator Cdc20 (cf. Fig. 1A). APC activity is inhibited by the Mitotic Checkpoint Complex (MCC), which consists of the four checkpoint proteins Mad2, BubR1, Bub3, and Cdc20 [5]. A key MCC component is Mad2, a small protein that can adopt two conformations: 'open' inactive

form (O-Mad2) and 'closed' active form (C-Mad2) [6,7]. C-Mad2 only forms when Mad2 binds to its kinetochore receptor, Mad1, or its checkpoint target Cdc20. The resulting C-Mad2–Cdc20 then binds to the BubR1–Bub3 complex, forming the MCC, which can then stably bind to the APC [5,8,9].

Furthermore, BubR1 has been suggested to interact with APC [10]. The complex Cdc20:C-Mad2 can also bind to the APC and form an inactive complex [11]. Another inhibitor, called the mitotic checkpoint factor 2 (MCF2), is associated with APC merely in the checkpoint arrested state but its composition is not known [12]. Recently and based on computational modeling, it has been shown that MCC alone is insufficient for fully inhibiting Cdc20 and APC. The same study has shown that cooperation between MCC and BubR1 is required to fully inhibit APC activity [13]. Very recent compelling evidence indicates that the MCC can inhibit a second Cdc20 that has already bound and activated the APC [14]. This data can enhance and elaborate on potential predictions from an integrative systems biology prospective.

So far, modeling of the SAC has helped to pinpoint advantages and problems of putative regulatory mechanisms [15–30]. These models can serve as a basis to integrate further findings and evaluate novel

E-mail address: bashar.ibrahim@uni-jena.de.

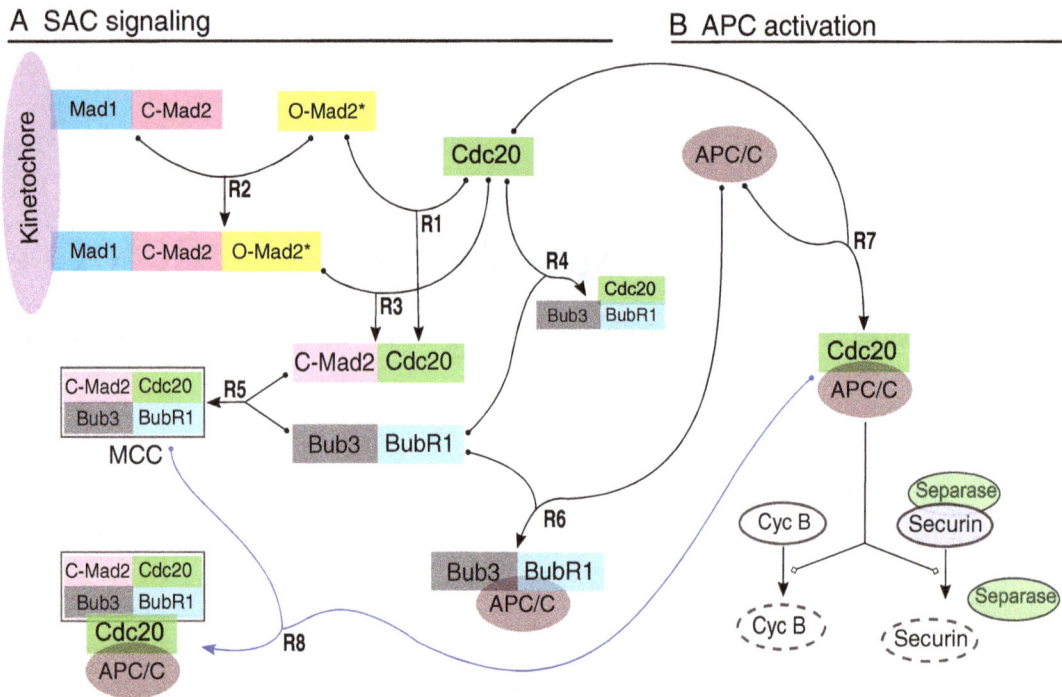

Fig. 1. Schematic representation of the core mechanism of SAC. (A) The SAC acts mainly through sequestration of the APC/C-activator Cdc20 by Mad2. Mad2 in closed conformation (C-Mad2) anchored at the kinetochore via Mad1 recruits cytosolic Mad2 in open conformation (O-Mad2). The so recruited Mad2 is stabilized in an intermediate conformation (Mad2*), which in turn is able to bind Cdc20 efficiently. The resulting C-Mad2–Cdc20 dimers are released from the kinetochore and form the mitotic checkpoint complex (MCC) together with Bub3 and BubR1. The Cdc20-containing complexes are not stable and dissociate with a certain rate, thus Cdc20 becomes available for APC/C activation soon after the last signaling kinetochore is silenced by proper microtubule attachment. (B) When SAC signaling is turned off, Cdc20 binds to and thereby activates the APC/C. Active APC/C:Cdc20 promotes degradation of securin, which leads to cohesin cleavage by now active separase. The resulting separation of sister-chromatids is the hallmark of anaphase. Simultaneously, APC/C:Cdc20 promotes degradation of cyclin B, a requirement for mitotic exit.

hypothesis related to checkpoint architecture and regulation. SAC models either consider few interacting elements using ordinary differential equations [18,21] or partial differential equations [15–17,28,29]; or conceive many interacting elements [20,22]. Other models use unconventional modeling approaches like Rule-Based modeling in space [25,26,28,31].

In this study a dynamical model for SAC activation and maintenance was constructed. This model considered all components of APC regulation in human cells in three variants: the MCC basic model variant, the MCC–BubR1 and the MCC that binds a second Cdc20 model variant. These models are validated with experimental data from the literature. A wide range of parameter values have been tested to find critical values of the APC binding rate. Simple mathematical analysis and computer simulations have helped to show that the MCC model variant in which MCC binds a second Cdc20 is sufficient to fully sequester Cdc20 and eventually completely inhibit APC activity.

2. Materials and methods

2.1. Model assumptions

Some reactions can depend on the attachment status of the kinetochores, so all reactions can be classified by whether they are unaffected ("uncontrolled"), turned off ("off-controlled") or turned on ("on-controlled") upon microtubule attachment. Only reactions involving kinetochore localized species can be controlled. For example, formation of Mad1:C-Mad2:O-Mad2* (Reaction 2) can only take place as long as the kinetochores are unattached. In this model, if the kinetochore is unattached, u is set to u = 1, otherwise u = 0 [22,23]. Note that mass-action-kinetics is used for all reactions. Mad1:Mad2 is considered to be a preformed complex and the complex formation is not considered. It should be noted that this complex is a tetrameric 2:2

Mad1:Mad2 and not a monomer complex. From a mathematical point of view, considering the complex as a species would not make any difference in this case as long as there is one model. All previous mathematical models have considered the similar assumption to the template model (e.g., [18,20,23], see R1–R3).

For the spatial simulations, the mitotic cell is assumed as a 3D-ball with radius r. The last unattached kinetochore is a 2-sphere with radius r in the center of the cell (Table 1). A lattice based model was used, which implies that the reaction volume of the mitotic cell is segmented into equal compartments. The initial concentrations of all freely diffuse species like Cdc20 and O-Mad2 are distributed randomly over all compartments of the mitotic cell. Localized species like Mad1:C-Mad2 and Mad1:C-Mad2:Mad2* are present at the kinetochore, their initial amount is located on the surface of the modeled 2-sphere. In order to observe a more accurate spatial behavior of the model variants, any symmetrical restrictions were not considered. All boundary conditions are reflective in order that the amount of particles is conserved.

2.2. Numerical simulation of ODEs system

The reaction rules are converted into sets of time dependent nonlinear ordinary differential equations (ODEs) by computing $dS/dt = \mathbf{N}v(S)$ with state vector S, flux vector v(S) and stoichiometric matrix $\mathbf{N}$. The actual initial amounts for reaction species are taken from literature (cf. Table 1). The kinetic rate constants (k_{on} and k_{off}) are also taken from literature as far as they are known. In the other cases, representative values that exemplified a whole physiologically possible range were selected. A summary of all simulation parameters is given in Table 1. Also parameter scans were used to determine the critical and ideal rate values. In a typical simulation run, all reaction partners were initialized according to Table 1 and the ODEs were numerically solved

Table 1
Model parameters.

Parameters		Remarks
Rate constants		
k_1	$1 \times 10^3 \, M^{-1} \, s^{-1}$	[21,61]
k_2	$2 \times 10^5 \, M^{-1} \, s^{-1}$	[34,59]
k_3	$1 \times 10^7 \, M^{-1} \, s^{-1}$	[21]
k_4	$2 \times 10^4 \, M^{-1} \, s^{-1}$	[21,28]
k_5	$10^3 – 10^9 \, M^{-1} \, s^{-1}$	[20,28]
k_6	$10^3 – 10^9 \, M^{-1} \, s^{-1}$	This study
k_7	$5 \times 10^6 \, M^{-1} \, s^{-1}$	[20,22]
k_8	$10^3 – 10^9 \, M^{-1} \, s^{-1}$	This study
k_{-1}	$1 \times 10^{-2} \, s^{-1}$	[21]
k_{-2}	$2 \times 10^{-1} \, s^{-1}$	[59]
k_{-3}	$0 \, s^{-1}$	[21]
k_{-4}	$2 \times 10^{-2} \, s^{-1}$	[21,28]
k_{-5}	$1 \times 10^{-1} \, s^{-1}$	[28]
k_{-6}	$1 \times 10^{-2} \, s^{-1}$	This study
k_{-7}	$1 \times 10^{-1} \, s^{-1}$	[20,22]
k_{-8}	$8 \times 10^{-2} \, s^{-1}$	This study
Initial amount		
Cdc20	$0.22 \, \mu M$	[39,62]
O-Mad2	$0.15 \, \mu M$	[59]
Mad1:C-Mad2	$0.05 \, \mu M$	[33]
BubR1:Bub3	$0.13 \, \mu M$	[20,39,63]
APC	$0.09 \, \mu M$	[62]
Other species start from zero		
Diffusion constants		
Cdc20	$19.5 \, \mu m^2 \, s^{-1}$	[64]
O-Mad2	$5 \, \mu m^2 \, s^{-1}$	[28]
Mad1:C-Mad2	$0 \, \mu m^2 \, s^{-1}$	[28]
Mad1:C-Mad2:Mad2*	$0 \, \mu m^2 \, s^{-1}$	[28]
Bub3:BubR1	$4 \, \mu m^2 \, s^{-1}$	[16,65]
APC	$1.8 \, \mu m^2 \, s^{-1}$	[64]
Other species diffusion coefficients are calculated from $D_{AB} = \frac{D_A \cdot D_B}{D_A + D_B}$, where DA and DB are the diffusion coefficient for A and B, respectively.		This study
Convection constant		
O-Mad2	$10 \, \mu m \, s^{-1}$	[28]
Environment		
Radius of the kinetochore	$0.1 \, \mu m$	[66]
Radius of the cell	$10 \, \mu m$	[28]

until steady state was reached before attachment (using u = 1). After attachment, switching u to 0, the equations are again simulated, until steady state is reached. The implementation and simulation code are written based on MATLAB (Mathworks, Natick, MA).

2.3. Spatial simulation of PDEs system

Adding a second spatial-derivative as a diffusion term and a first-derivative as a convection term transforms the system of ODEs in coupled partial differential equations (PDEs) known as a *reaction-diffusion-convection system* (see for details [28]).

Partial differential equations resulting from the reaction-diffusion-convection system were solved numerically using the open access Virtual Cell software [32]. The simulations are conducted using 3D geometries. Each dimension is divided into 51 parts, which results in 132.651 compartments in total. All parameters are set up consistent with the model assumptions. The system of PDEs with boundary and initial conditions is solved using the "fully implicit finite volume with variable time-step" method. This method employs Sundials stiff solver CVODE for time stepping (method of lines) [32]. The derivations, necessary for diffusion and convection, are computed numerically. The human system is simulated for 1000 s which is sufficient to reach steady

state, with a maximum time-step of 0.1 s and an absolute and relative tolerance of 1.0×10^{-7}. One simulation run takes between 1 and 10 h, dependent on the parameter-set. The time dependent concentration plots add up the amount of every species over all compartments and are generated with MATLAB (Mathworks, Natick, MA).

3. Results

3.1. Biochemical background of the model

The reaction network of the SAC activation and maintenance mechanism (Fig. 1) can be divided into three main parts: Mad2-activation template, MCC formation, and APC inhibition.

The essential component of the SAC-network is a kinetochore-bound template complex made up from Mad1 and C-Mad2. This template complex recruits O-Mad2 and stabilizes an intermediate conformation (O-Mad2*) which can bind Cdc20 efficiently and switches to closed conformation upon Cdc20-binding [6,33,34] (the biochemical equations of are described by reaction (R1–R3), Fig. 1, and reaction scheme). The C-Mad2-Cdc20 complexes formed by this mechanism, which has been given the name "template-model" [33], can further associate with the two proteins BubR1 (homologue of budding yeast Mad3) and Bub3 to form the tetrameric mitotic checkpoint complex (MCC; [5,35,36]). Another trimeric complex Bub3:BubR1:Cdc20 can form faster in the presence of unattached chromosomes [35] and it may be that MCC forms as an intermediate complex from which O-Mad2 rapidly dissociates [35,37,38]. The MCC and Bub3:BubR1:Cdc20 formations are described by the reaction (R4–R5, see chemical reaction scheme, below).

The APC is believed to be inhibited in multiple ways. Complexes of APC together with either Cdc20:C-Mad2 [11,39], Bub3:BubR1 [10], Bub3:BubR1:Cdc20 [10,39], MCC [5,8,9] or MCF2 [12] have been found to be inactive [5,8,12,35,37,38]. Recent work based on a systems biology approach, has shown that the MCC–BubR1 alone is able to reproduce both wild-type as well as mutation experiments of SAC mechanism. Hence these reactions, described by the reaction (R6–R7) (see chemical reaction scheme, below), are included. Free Cdc20 binds to and thereby activates the APC (R8) which promotes degradation of securin, which leads to cohesion cleavage by now active separase [40–42] (cf. Fig. 1B).

The following four model variants are considered: First is the core MCC model which consists of reactions (R1–R5, and R7, see also [20]). The second variant is the MCC–BubR1 model which consists of reactions (R1–R7, see also [13]). These model variants serve as the basic and reference models to compare with. The third and fourth model variants are the extension of the basic model variants with the addition of the MCC's ability to bind a second Cdc20 (R9, see also [14]).

3.2. Chemical reaction scheme

The SAC mechanism consists of 9 biochemical reaction equations describing the dynamics of the following 14 species: Mad1:C-Mad2, O-Mad2, Mad1:C-Mad2:O-Mad2*, Cdc20, Cdc20:C-Mad2, Bub3:BubR1, MCC, Bub3:BubR1:Cdc20, APC, MCC:APC, APC:BubR1:Bub3, APC:Cdc20: BubR1:Bub3, APC:Cdc20:MCC, and APC:Cdc20 (see also Fig. 1A).

3.3. Time dependant dynamics of SAC regulation

The SAC models, where either MCC is the exclusive inhibitor of APC [20] and both MCC together with BubR1, have been previously analyzed [13]. These models (see R1–R6, and R8; and R1–R8, respectively) are used in this study to build upon and to compare with the other two new model variants that mainly involve the ability of the MCC to bind a second Cdc20 that is already bound to APC [14]. In these two new model variants, in addition to the basic model variants (above), MCC binds a second Cdc20 that is already bound to APC (R9 is added).

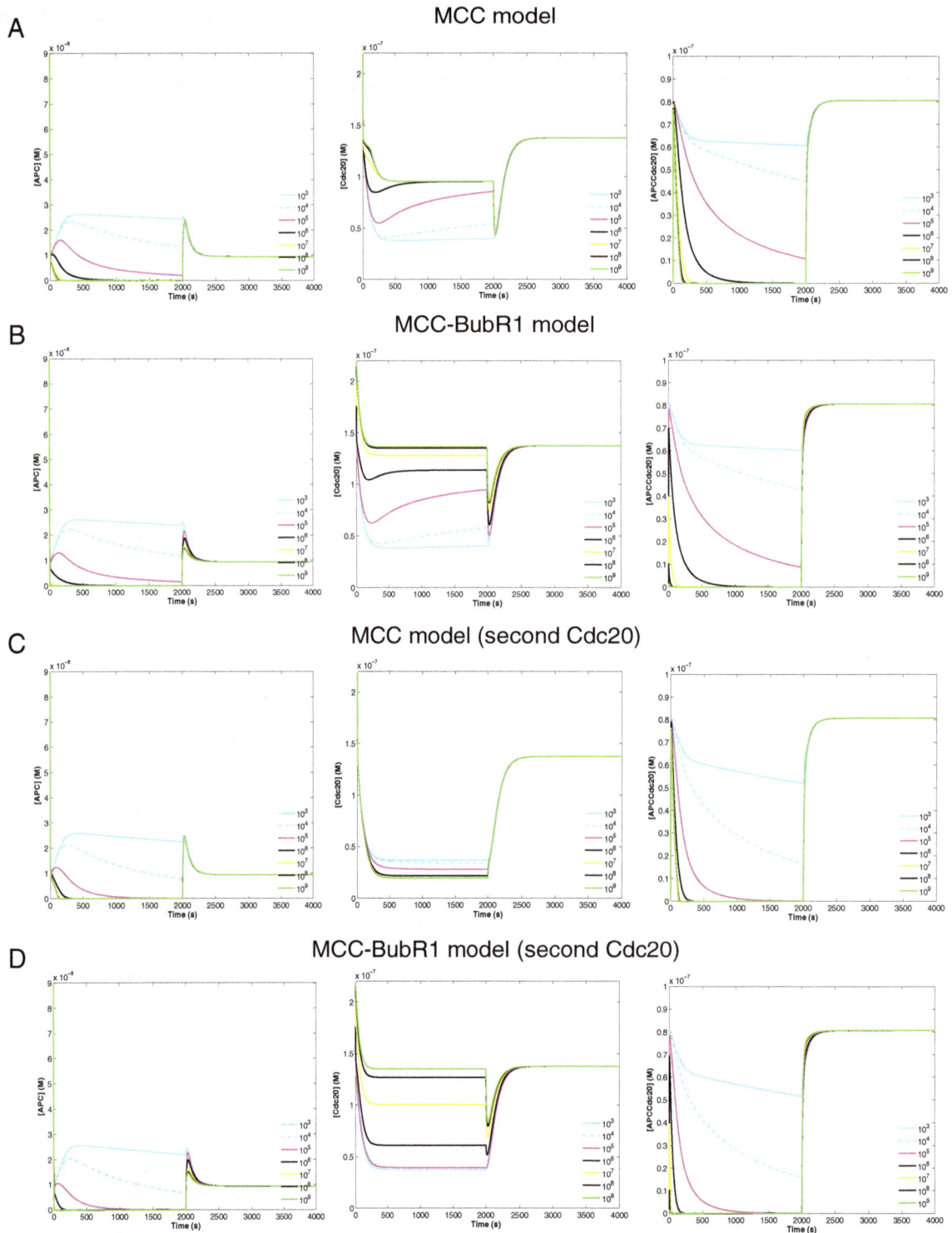

Fig. 2. Dynamical behavior of core SAC component concentration versus time. The columns from left to right show the APC, Cdc20, and APC:Cdc20 concentration (spindle attachment occurs at $t = 2000$ s). All results are presented for different values of the rate k_5 (MCC binding to APC). Parameters setting are according to Table 1. Free APC concentration (left column) in all model variants is similar where its value at any given time is less than 30% of its initial concentration. The APC:Cdc20 dynamics (right column) in all model variants is also very similar, shows fast recovery and only with high MCC binding rate to APC shows fast inhibition for APC:Cdc20 activity. Cdc20 sequestration is depicted in the middle column. All model variants except the MCC-model variant that binds second Cdc20 (c.f. Panels A, B and D) are able to sequester about 80% of the free Cdc20 only with low MCC–APC binding rate. The MCC model variant that binds second Cdc20 (c.f. Panel C) is able to sequester around 95% of the free Cdc20 and independent of MCC–APC binding rate.

Simulation results of these four model variants as non-linear ODEs are shown in Fig. 2. As for the dynamics of free APC, all variants behave qualitatively similarly (Fig. 2 left column). For the two new variants where MCC is able to bind a second Cdc20, slow MCC–APC binding rate is sufficient for fully APC:Cdc20 inhibition. Cdc20 sequestration reached about 95% with the MCC model variant that binds a second Cdc20 (Fig. 2C middle column).

Table 2
In-silico mutation experiments for validation.

Species	Exp.	Experimental effects	Effects in the model variants			
			MCC core	MCC–BubR1 core	MCC extended	MCC–BubR1 extended
BubR1	D	SAC dysfunction [53–56]	Failed to arrest	Failed to arrest	Failed to arrest	Failed to arrest
BubR1	O	Chromosomal instability [57]	Arrested $k_5 = 10^5$	Arrested	Arrested $k_5 \geq 10^4$	Arrested
Mad2	D	Cells are unable to arrest and impaired SAC (e.g., [11,47–50])	Failed to arrest	Failed to arrest $k_5 = 10^6$	Failed to arrest $k_5 = 10^4$	Failed to arrest $k_5 = 10^4$
Mad2	O	Activates the SAC and blocks mitosis and stabilizes microtubule attachment [34,51–52]	Arrested	Arrested	Arrested	Arrested
Cdc20	D	Cells arrested in metaphase [58–60]	Arrested $k_5 = 10^5$	Arrested	Arrested	Arrested
Cdc20	O	Impairment SAC and allows cells with a depolymerized spindle or damaged DNA to leave mitosis [61–62].	Failed to arrest	Failed to arrest	Failed to arrest $k_5 \leq 10^4$	Failed to arrest $k_5 \leq 10^4$

D refers to deletion or knockdown experiment, and O refers to an over-expression experiment. Failed to arrest means very high level of [APC:Cdc20] and low sequestration level of Cdc20. Arrested means very low level of [APC:Cdc20] and fully sequestration of Cdc20. Green means fully consistent with experiments and capture the desire behavior. Yellow means consistent with experiments but required specific MCC–APC binding rate (see text for details).

MCC dominated

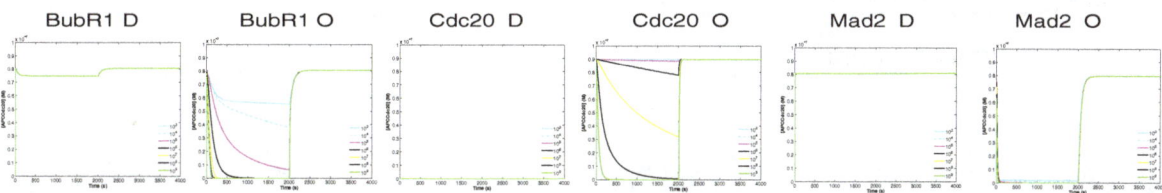

MCC-BubR1 dominated

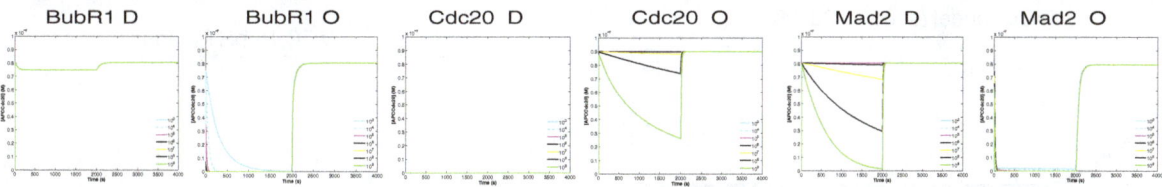

MCC dominated (second Cdc20)

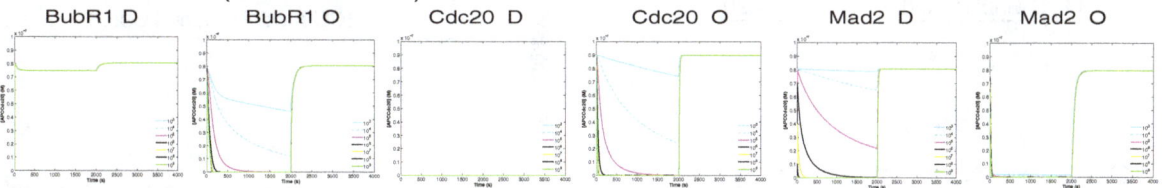

MCC-BubR1 dominated (second Cdc20)

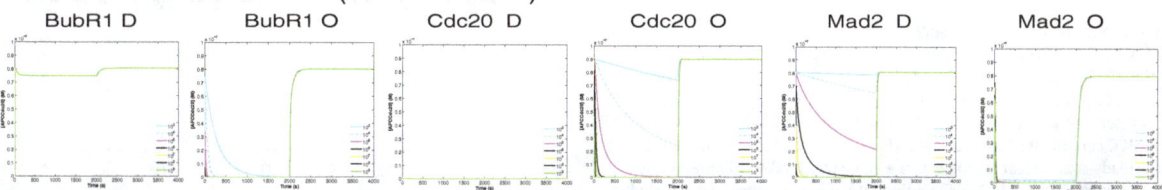

Fig. 3. Simulation of Mad2, BubR1, and Cdc20 mutations for each model variant. For deletion we set the respective initial concentration to zero, and for over-expression 100 folds higher. Towards proper wild type functioning, APC:Cdc20 concentration should be very low (zero) before the attachment, and should increase quickly after attachment. Deletion of Mad2 or BubR1 or an overexpression of Cdc20 leads to inability of the cell to arrest, that is in the simulation, the concentration of APC:Cdc20 keeps high. Overexpression of Mad2 or BubR1 or deleting Cdc20 results in arresting the cell, that is, the concentration of APC:Cdc20 is very low or zero. Each row represents the mutation simulations of a model variant and a range of parameter rate for APC binding. Spindle attachment occurs at t = 2000 s (switching parameter u from 1 to 0). All parameter settings are according to Table 1. See text for more details.

To validate all model variants, different mutations (deletion and over-expression) of the species involved were tested (Table 2). There are many experimental studies reported in the literature where deletion and also overexpression in different organisms of any of the core components, Mad2 [11,33,43–48], BubR1 [49–53], and Cdc20 [54–58], resulted in SAC failures, such as failed or successful mitotic arrest. These experiments may help in validating all model variants and additionally discriminating between them. The experiments from literature are listed in Table 2.

In the simulations, the respective initial concentration was set to zero for the deletions, and 100 fold higher concentrations for over-expression. The desired proper wild type functioning, APC:Cdc20 concentration should be very low (zero) before the attachment, and should increase quickly after attachment. Cells failing to arrest meant a very high level of APC:Cdc20 and low sequestration level of Cdc20. Arrested cells meant a very low level of APC:Cdc20 and full sequestration of Cdc20. The simulations show that all model variants are able to fully reproduce all known experimental findings for specific MCC–APC binding rate range (Fig. 3, Table 2). The simulations additionally indicate that the ideal MCC–APC binding rate for mutant type is 10^4–10^5 M^{-1} s^{-1} (Table 2).

Together, WT simulations of the four model variants were qualitatively similar. However, the variants where MCC binds a second Cdc20 did not require a high MCC–APC binding rate. Additionally, the Cdc20 sequestration level is higher for the variant where MCC-binds a second Cdc20.

All model variants provided an ideal SAC functioning and were able to reproduce all experimental findings based on ODEs where only time but no space is included.

3.4. Spatial dynamics of SAC regulation

Mathematical studies have recently shown that spatial properties such as diffusion and active transportation can play important roles in SAC activity and maintenance [17,23,28,29]. Lohel et al. [23] extended previously existing models of the SAC, to enable a detailed analysis of the kinetic consequences of localization. They found that the binding kinetics and stoichiometry are limiting factors for the overall dynamics of the SAC. Therefore, 3D space simulation was considered and spatial properties like diffusion and active transportation was tested with a range of kinetic reaction rates for APC binding. The environmental, diffusion and convection parameters are listed in Table 1. For numerical simulation and geometry details see Materials and methods.

The spatial simulation was run for each model variants four times. Three times as a reaction–diffusion system (Materials and methods) for different MCC–APC binding rates; low rate 10^6 M^{-1} s^{-1} (Fig. 4, blue line), moderate rate 10^8 M^{-1} s^{-1} (Fig. 4, black lines) or high rate 10^{10} M^{-1} s^{-1} (Fig. 4, red lines). Additionally, the simulation was run a fourth time to consider an active transportation for Mad2 as suggested by [28] as a reaction–diffusion–convection systems with a moderate

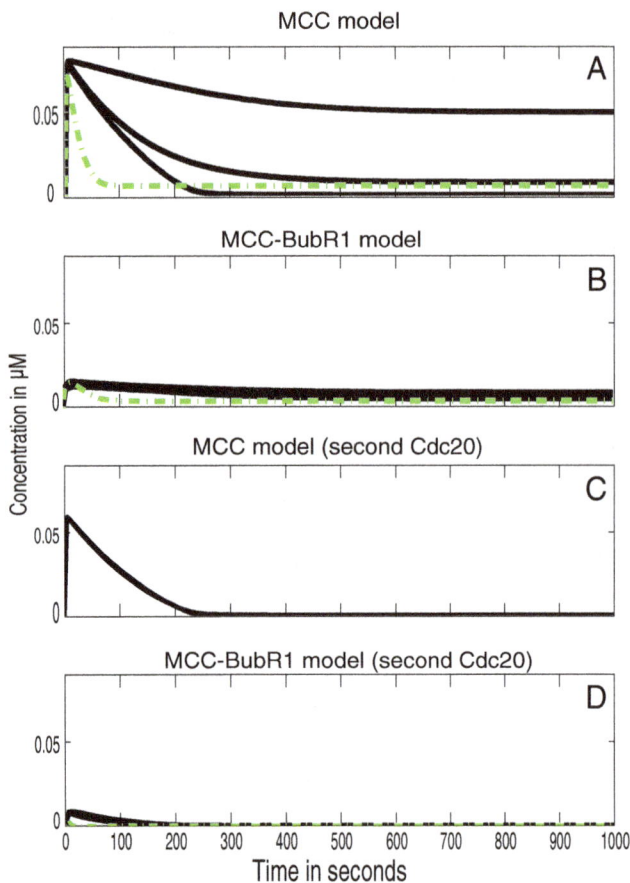

Fig. 4. Spatial simulation of SAC model variants. The figures show the total concentrations over time for APC:Cdc20 with different parameter sets. All results are presented for different values of the APC/C binding rates (k_5, k_6 and k_8). Blue, black and red lines refer to the different APC/C binding rates, 10^6 M^{-1} s^{-1}, 10^8 M^{-1} s^{-1} and 10^{10} M^{-1} s^{-1} respectively. Dotted lines represent the simulations when Mad2 convection is included. (A) Outcome of the simulated MCC core model (Reactions (1)–(5), and (7); cf. Table 1). It takes about 5 min to reach steady state except for the low rate value which takes 10 min. APC:Cdc20 is 90% inhibited only with high MCC–APC binding rate or when convection is included. (B) Outcome of the simulated MCC–BubR1 core model (Reactions (1)–(7); cf. Table 1). It takes 3 min to reach steady state for any parameter set. (C) Outcome of the simulated MCC model that binds second Cdc20 (Reactions (1)–(5), (7),and (8); cf. Table 1). It takes about 3 min to reach steady state for any parameter set. (D) Outcome of the simulated MCC–BubR1 model that binds second Cdc20 (Reactions (1)–(8); cf. Table 1). It takes about 1.5 min to reach steady state for any parameter set.

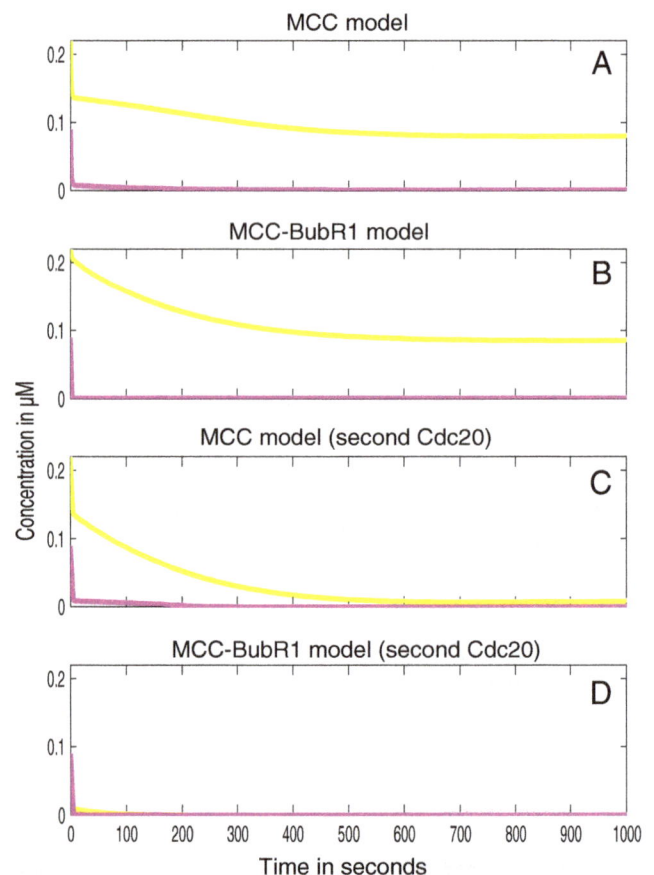

Fig. 5. Spatial simulation of APC and Cdc20 dynamics. The figures show the total concentrations over time for free APC and free Cdc20. All results are presented for 10^8 M^{-1} s^{-1} value of the APC/C binding rates (k_5, k_6 and k_8). We can clearly see that only for the MCC–BubR1 model variant that binds second Cdc20, complete APC and Cdc20 sequestration is achieved.

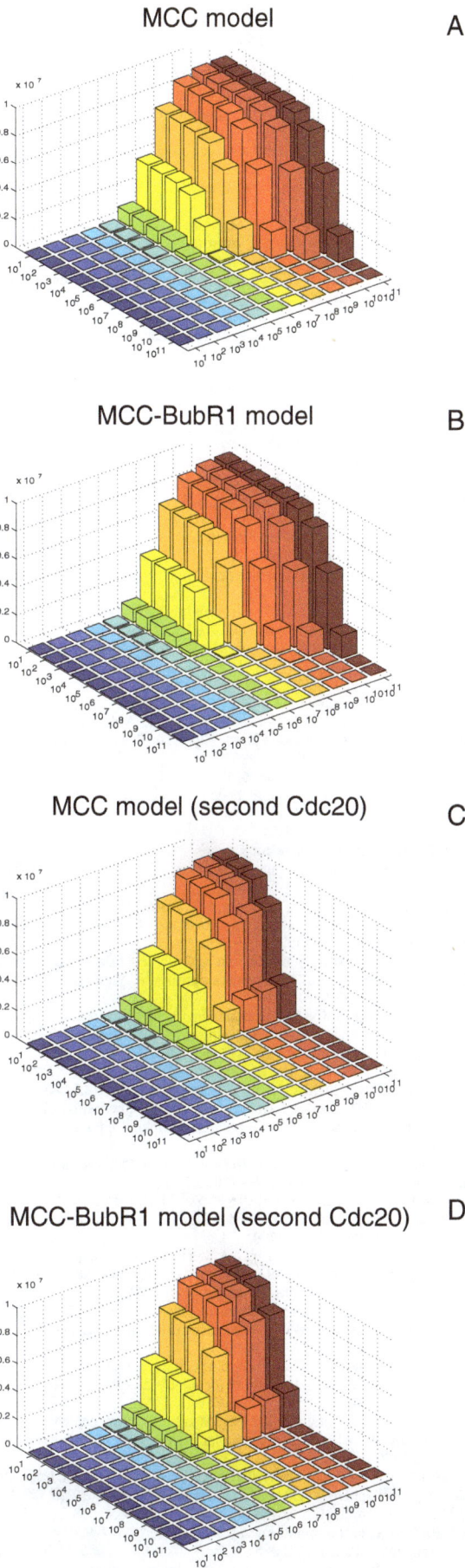

MCC model A

MCC-BubR1 model B

MCC model (second Cdc20) C

MCC-BubR1 model (second Cdc20) D

MCC binding rate (see green dot lines in Fig. 4). Fig. 4 depicts the wild type behavior of the average APC:Cdc20 concentration over time. All models should, in principle, be able to reproduce the desired behavior that is a very low level of APC:Cdc20. The MCC core model was able to reproduce the desired behavior only with a high MCC–APC binding rate or when convection is presented (Fig. 4A). These rates however are very high compared to the known SAC binding rate (e.g. Mad2–Cdc20 or Mad1–Mad2 [34,59]). The MCC–BubR1 core model variant was able to reproduce the desired behavior with any parameter set. However, the APC:Cdc20 was not fully inhibited (reached 90% of APC level, Fig. 4B). The new variants, in which MCC is able to bind a second Cdc20, were able to reproduce the desired behavior for any given parameter set and additionally were able to fully inhibit APC:Cdc20 activity (Fig. 4C–D). Only the MCC–BubR1 model variants that included additional Cdc20 binding, were able to reach steady state very fast (in about minute) while all other variants needed at least 3 min (Fig. 4A–D).

The level of free APC as well as free Cdc20 in each of the model variants was examined and these results are shown in Fig. 5. The core models (MCC and MCC with BubR1) were able to sequester only 50% of Cdc20 amount (Fig. 5A–B yellow lines). The new variants that included a second binding of Cdc20 were able to fully sequester Cdc20 together with the APC (Fig. 5C–D, yellow and rose lines). However, the MCC–BubR1 variant that has additional Cdc20 binding was able to fully sequester both APC and Cdc20 after few seconds of the simulation (Fig. 5D).

Taken together, the MCC–BubR1 model variant is able to capture ideal SAC behavior while not requiring very high binding rates or convection properties. Secondary Cdc20 binding [14] enhances SAC functioning.

4. Discussion

Building on the investigation [20] of different models for Cdc20:Mad2 complex formation, the mathematical description of the SAC model has been enhanced by those reaction equations which describe additional Cdc20 sequestration by MCC, as reported recently [14]. A major role is played by the MCC and the BubR1, which in turn blocks APC activity. Four SAC model variants were analyzed; distinguishing the APC binding partners MCC or MCC and BubR1, and additionally the MCC's ability to bind a second Cdc20 that is already bound to APC. The latter succeeded to describe the correct metaphase to anaphase switching and also the ability to complete Cdc20 sequestering and APC inhibition. The calculations are in full agreement with the recent findings [14]. The model also indicated the value for the MCC–APC binding via a parameter scan (Fig. 6) and additionally favored the variant where both the MCC and BubR1 bind APC and additionally where the MCC binds a second Cdc20.

Computational modeling is a very important tool to elucidate how elaborate systems work. So far, mathematical models have helped to elucidate the kinetochore structure and with that the mitotic checkpoint mechanism [15,16,18,19,21,23,24,25,26,28,60]. These models mostly focus on either a minimal spatial model of SAC [15,16,23,28, 29], namely the template model, or a detailed model excluding spatial effects [18–22]; consequently, previous models ignore the spatial and temporal regulation of multiple APC inhibition for SAC activity. In this work, both the approaches of using ODEs and PDEs were combined and this has enhanced the most detailed model available in the literature

Fig. 6. Sensitivities of the steady state concentrations of APC:Cdc20 associated with rate coefficients (k_5) and (k_7). Both parameters were varied in a range from 10^0 M^{-1} s^{-1} to 10^{11} M^{-1} s^{-1}. This analysis has been repeated for each model variants (Panels A, B, C, and D, respectively). Each model variant was simulated 121 times, and each simulation run until steady state reached. Panels A and B (core model variants) are very similar. The same is true for Panels C and D (where MCC binds a second Cdc20). The scan of the new model variants, Panels C and D, indicates that the MCC–APC binding rate must be at least 10^5 M^{-1} s^{-1} and meanwhile APC–Cdc20 binding rate must not exceed 10^6 M^{-1} s^{-1}.

[22]. This work has also been confirmed by the very recent experimental findings that the MCC binds a second Cdc20 [14].

In order to accelerate the pace of cell biology knowledge, systems analysis should be developed to link computational models of biological networks to experimental data in tight rounds of analysis and synthesis in an integrative systems biology framework. It is anticipated that such an approach for the SAC mechanism will serve as a basis to design experiments and evaluate novel hypotheses related to mitotic checkpoint control.

Acknowledgment

The author would like to thank Fouzia Ahmad for proofreading the manuscript. This work was supported by the European Commission HIERATIC Grant 062098/14.

References

[1] Minshull J, Sun H, Tonks NK, Murray AW. A MAP kinase-dependent spindle assembly checkpoint in Xenopus egg extracts. Cell 1994;79:475–86.

[2] Suijkerbuijk SJ, Kops GJ. Preventing aneuploidy: the contribution of mitotic checkpoint proteins. Biochim Biophys Acta 2008;1786:24–31.

[3] Holland AJ, Cleveland DW. Boveri revisited: chromosomal instability, aneuploidy and tumorigenesis. Nat Rev Mol Cell Biol 2009;10:478–87.

[4] Morais da Silva S, Moutinho-Santos T, Sunkel CE. A tumor suppressor role of the Bub3 spindle checkpoint protein after apoptosis inhibition. J Cell Biol 2013;201:385–93.

[5] Sudakin V, Chan GK, Yen TJ. Checkpoint inhibition of the APC/C in HeLa cells is mediated by a complex of BUBR1, BUB3, CDC20, and MAD2. J Cell Biol 2001;154:925–36.

[6] Luo X, Tang Z, Xia G, Wassmann K, Matsumoto T, Rizo J, et al. The Mad2 spindle checkpoint protein has two distinct natively folded states. Nat Struct Mol Biol 2004;11:338–45.

[7] Mapelli M, Musacchio A. MAD contortions: conformational dimerization boosts spindle checkpoint signaling. Curr Opin Struct Biol 2007;17:716–25.

[8] Herzog F, Primorac I, Dube P, Lenart P, Sander B, Mechtler K, et al. Structure of the anaphase-promoting complex/cyclosome interacting with a mitotic checkpoint complex. Science 2009;323:1477–81.

[9] Hein JB, Nilsson J. Stable MCC binding to the APC/C is required for a functional spindle assembly checkpoint. EMBO Rep 2014;15:264–72.

[10] Han JS, Holland AJ, Fachinetti D, Kulukian A, Cetin B, Cleveland DW. Catalytic assembly of the mitotic checkpoint inhibitor BubR1–Cdc20 by a Mad2-induced functional switch in Cdc20. Mol Cell 2013;51:92–104.

[11] Fang G, Yu H, Kirschner MW. The checkpoint protein MAD2 and the mitotic regulator CDC20 form a ternary complex with the anaphase-promoting complex to control anaphase initiation. Genes Dev 1998;12:1871–83.

[12] Eytan E, Braunstein I, Ganoth D, Teichner A, Hittle JC, Yen TJ, et al. Two different mitotic checkpoint inhibitors of the anaphase-promoting complex/cyclosome antagonize the action of the activator Cdc20. Proc Natl Acad Sci U S A 2008;105:9181–5.

[13] Ibrahim B. Systems biology modeling of five pathways for regulation and potent inhibition of the anaphase-promoting complex (APC/C): pivotal roles for MCC and BubR1. OMICS 2015;19:294–305.

[14] Izawa D, Pines J. The mitotic checkpoint complex binds a second CDC20 to inhibit active APC/C. Nature 2015;517:631–4.

[15] Doncic A, Ben-Jacob E, Barkai N. Evaluating putative mechanisms of the mitotic spindle checkpoint. Proc Natl Acad Sci U S A 2005;102:6332–7.

[16] Sear RP, Howard M. Modeling dual pathways for the metazoan spindle assembly checkpoint. Proc Natl Acad Sci U S A 2006;103:16758–63.

[17] Mistry HB, MacCallum DE, Jackson RC, Chaplain MA, Davidson FA. Modeling the temporal evolution of the spindle assembly checkpoint and role of Aurora B kinase. Proc Natl Acad Sci U S A 2008;105:20215–20.

[18] Simonetta M, Manzoni R, Mosca R, Mapelli M, Massimiliano L, Vink M, et al. The influence of catalysis on mad2 activation dynamics. PLoS Biol 2009;7:e10.

[19] Ibrahim B, Dittrich P, Diekmann S, Schmitt E. Stochastic effects in a compartmental model for mitotic checkpoint regulation. J Integr Bioinform 2007;4.

[20] Ibrahim B, Diekmann S, Schmitt E, Dittrich P. In-silico modeling of the mitotic spindle assembly checkpoint. PLoS One 2008;3:e1555.

[21] Ibrahim B, Dittrich P, Diekmann S, Schmitt E. Mad2 binding is not sufficient for complete Cdc20 sequestering in mitotic transition control (an in silico study). Biophys Chem 2008;134:93–100.

[22] Ibrahim B, Schmitt E, Dittrich P, Diekmann S. In silico study of kinetochore control, amplification, and inhibition effects in MCC assembly. Biosystems 2009;95:35–50.

[23] Lohel M, Ibrahim B, Diekmann S, Dittrich P. The role of localization in the operation of the mitotic spindle assembly checkpoint. Cell Cycle 2009;8:2650–60.

[24] Kreyssig P, Escuela G, Reynaert B, Veloz T, Ibrahim B, Dittrich P. Cycles and the qualitative evolution of chemical systems. PLoS One 2012;7:e45772.

[25] Ibrahim B, Henze R, Gruenert G, Egbert M, Huwald J, Dittrich P. Spatial rule-based modeling: a method and its application to the human mitotic kinetochore. Cells 2013;2:506–44.

[26] Tschernyschkow S, Herda S, Gruenert G, Doring V, Gorlich D, Hofmeister A, et al. Rule-based modeling and simulations of the inner kinetochore structure. Prog Biophys Mol Biol 2013;113:33–45.

[27] Ibrahim B. In silico spatial simulations reveal that MCC formation and excess BubR1 are required for tight inhibition of APC/C. Mol BioSyst 2015 (under review).

[28] Ibrahim B, Henze R. Active transport can greatly enhance Cdc20:Mad2 formation. Int J Mol Sci 2014;15:19074–91.

[29] Chen J, Liu J. Spatial-temporal model for silencing of the mitotic spindle assembly checkpoint. Nat Commun 2014;5:4795.

[30] Ibrahim B. Toward a systems-level view of mitotic checkpoints. Prog Biophys Mol Biol 2015;117:217–24.

[31] Gruenert G, Ibrahim B, Lenser T, Lohel M, Hinze T, Dittrich P. Rule-based spatial modeling with diffusing, geometrically constrained molecules. BMC Bioinf 2010;11:307.

[32] Loew LM, Schaff JC. The Virtual Cell: a software environment for computational cell biology. Trends Biotechnol 2001;19:401–6.

[33] De Antoni A, Pearson CG, Cimini D, Canman JC, Sala V, Nezi L, et al. The Mad1/Mad2 complex as a template for Mad2 activation in the spindle assembly checkpoint. Curr Biol 2005;15:214–25.

[34] Vink M, Simonetta M, Transidico P, Ferrari K, Mapelli M, De Antoni A, et al. In vitro FRAP identifies the minimal requirements for Mad2 kinetochore dynamics. Curr Biol 2006;16:755–66.

[35] Kulukian A, Han JS, Cleveland DW. Unattached kinetochores catalyze production of an anaphase inhibitor that requires a Mad2 template to prime Cdc20 for BubR1 binding. Dev Cell 2009;16:105–17.

[36] Chao WC, Kulkarni K, Zhang Z, Kong EH, Barford D. Structure of the mitotic checkpoint complex. Nature 2012;484:208–13.

[37] Malureanu LA, Jeganathan KB, Hamada M, Wasilewski L, Davenport J, van Deursen JM. BubR1 N terminus acts as a soluble inhibitor of cyclin B degradation by APC/C-Cdc20 in interphase. Dev Cell 2009;16:118–31.

[38] Medema RH. Relaying the checkpoint signal from kinetochore to APC/C. Dev Cell 2009;16:6–8.

[39] Fang G. Checkpoint protein BubR1 acts synergistically with Mad2 to inhibit anaphase-promoting complex. Mol Biol Cell 2002;13:755–66.

[40] Izawa D, Pines J. How APC/C-Cdc20 changes its substrate specificity in mitosis. Nat Cell Biol 2011;13:223–33.

[41] Izawa D, Pines J. Mad2 and the APC/C compete for the same site on Cdc20 to ensure proper chromosome segregation. J Cell Biol 2012;199:27–37.

[42] Rattani A, Vinod PK, Godwin J, Tachibana-Konwalski K, Wolna M, Malumbres M, et al. Dependency of the spindle assembly checkpoint on Cdk1 renders the anaphase transition irreversible. Curr Biol 2014;24:630–7.

[43] Nezi L, Rancati G, De Antoni A, Pasqualato S, Piatti S, Musacchio A. Accumulation of Mad2–Cdc20 complex during spindle checkpoint activation requires binding of open and closed conformers of Mad2 in Saccharomyces cerevisiae. J Cell Biol 2006;174:39–51.

[44] Dobles M, Liberal V, Scott ML, Benezra R, Sorger PK. Chromosome missegregation and apoptosis in mice lacking the mitotic checkpoint protein Mad2. Cell 2000;101:635–45.

[45] Michel LS, Liberal V, Chatterjee A, Kirchwegger R, Pasche B, Gerald W, et al. MAD2 haplo-insufficiency causes premature anaphase and chromosome instability in mammalian cells. Nature 2001;409:355–9.

[46] Nath S, Moghe M, Chowdhury A, Godbole K, Godbole G, Doiphode M, et al. Is germline transmission of MAD2 gene deletion associated with human fetal loss? Mol Hum Reprod 2012;18:554–62.

[47] He X, Patterson TE, Sazer S. The Schizosaccharomyces pombe spindle checkpoint protein mad2p blocks anaphase and genetically interacts with the anaphase-promoting complex. Proc Natl Acad Sci U S A 1997;94:7965–70.

[48] Kabeche L, Compton DA. Checkpoint-independent stabilization of kinetochore-microtubule attachments by Mad2 in human cells. Curr Biol 2012;22:638–44.

[49] Davenport J, Harris LD, Goorha R. Spindle checkpoint function requires Mad2-dependent Cdc20 binding to the Mad3 homology domain of BubR1. Exp Cell Res 2006;312:1831–42.

[50] Harris L, Davenport J, Neale G, Goorha R. The mitotic checkpoint gene BubR1 has two distinct functions in mitosis. Exp Cell Res 2005;308:85–100.

[51] Chan GK, Jablonski SA, Sudakin V, Hittle JC, Yen TJ. Human BUBR1 is a mitotic checkpoint kinase that monitors CENP-E functions at kinetochores and binds the cyclosome/APC. J Cell Biol 1999;146:941–54.

[52] Ouyang B, Knauf JA, Ain K, Nacev B, Fagin JA. Mechanisms of aneuploidy in thyroid cancer cell lines and tissues: evidence for mitotic checkpoint dysfunction without mutations in BUB1 and BUBR1. Clin Endocrinol (Oxf) 2002;56:341–50.

[53] Yamamoto Y, Matsuyama H, Chochi Y, Okuda M, Kawauchi S, Inoue R, et al. Overexpression of BUBR1 is associated with chromosomal instability in bladder cancer. Cancer Genet Cytogenet 2007;174:42–7.

[54] Zhang Y, Lees E. Identification of an overlapping binding domain on Cdc20 for Mad2 and anaphase-promoting complex: model for spindle checkpoint regulation. Mol Cell Biol 2001;21:5190–9.

[55] Mondal G, Baral RN, Roychoudhury S. A new Mad2-interacting domain of Cdc20 is critical for the function of Mad2–Cdc20 complex in the spindle assembly checkpoint. Biochem J 2006;396:243–53.

[56] Shirayama M, Toth A, Galova M, Nasmyth K. APC(Cdc20) promotes exit from mitosis by destroying the anaphase inhibitor Pds1 and cyclin Clb5. Nature 1999;402:203–7.

[57] Hwang LH, Lau LF, Smith DL, Mistrot CA, Hardwick KG, Hwang ES, et al. Budding yeast Cdc20: a target of the spindle checkpoint. Science 1998;279:1041–4.

[58] Mondal G, Sengupta S, Panda CK, Gollin SM, Saunders WS, Roychoudhury S. Overexpression of Cdc20 leads to impairment of the spindle assembly checkpoint and aneuploidization in oral cancer. Carcinogenesis 2007;28:81–92.

[59] Howell BJ, Hoffman DB, Fang G, Murray AW, Salmon ED. Visualization of Mad2 dynamics at kinetochores, along spindle fibers, and at spindle poles in living cells. J Cell Biol 2000;150:1233–50.

[60] Kreyssig P, Wozar C, Peter S, Veloz T, Ibrahim B, et al. Effects of small particle numbers on long-term behaviour in discrete biochemical systems. Bioinformatics 2014;30:i475–81.

[61] Musacchio A, Salmon ED. The spindle-assembly checkpoint in space and time. Nat Rev Mol Cell Biol 2007;8:379–93.

[62] Stegmeier F, Rape M, Draviam VM, Nalepa G, Sowa ME, Ang XL, et al. Anaphase initiation is regulated by antagonistic ubiquitination and deubiquitination activities. Nature 2007;446:876–81.

[63] Burton JL, Solomon MJ. Mad3p, a pseudosubstrate inhibitor of APCCdc20 in the spindle assembly checkpoint. Genes Dev 2007;21:655–67.

[64] Wang Z, Shah JV, Berns MW, Cleveland DW. In vivo quantitative studies of dynamic intracellular processes using fluorescence correlation spectroscopy. Biophys J 2006; 91:343–51.

[65] Howell BJ, Moree B, Farrar EM, Stewart S, Fang G, Salmon ED. Spindle checkpoint protein dynamics at kinetochores in living cells. Curr Biol 2004;14:953–64.

[66] Cherry LM, Faulkner AJ, Grossberg LA, Balczon R. Kinetochore size variation in mammalian chromosomes: an image analysis study with evolutionary implications. J Cell Sci 1989;92(Pt 2):281–9.

2

Novel Cell-Ess ® supplement used as a feed or as an initial boost to CHO serum free media results in a significant increase in protein yield and production

Adam Elhofy

Essential Pharmaceuticals, Ewing, New Jersey, United States

ARTICLE INFO

Keywords:
Bioprocessing
Titer
Lipid
Cell-Ess
CHO
Monoclonal antibody
Serum free media
Animal component free
Viable cell density
Bioreactor
Single use bioreactor

ABSTRACT

Many metrics, including metabolic profiles, have been used to analyze cell health and optimize productivity. In this study, we investigated the ability of a lipid supplement to increase protein yield. At a concentration of 1% (v/v) the lipid supplement caused a significant increase in protein titer (1118 ± 65.4 ng 10^5 cells^{-1} days^{-1}) when compared to cultures grown in the absence of supplementation (819.3 ± 38.1 ng 10^5 cells^{-1} days^{-1}; $p < 0.05$). This equated to a 37% increase in productivity. Furthermore, metabolic profiles of ammonia, glutamate, lactate, and glucose were not significantly altered by the polar lipid supplement. In a separate set of experiments, using the supplement as a feed resulted in 2 notable effects. The first was a 25% increase in protein titer. The second was an extension of peak protein production from 1 day to 2 days. These results suggest that lipid supplementation is a promising avenue for enhancing protein production. In addition, our results also suggest that an increase in protein production may not necessarily require a change in the metabolic state of the cells.

1. Introduction

Using proteins as therapies was ushered in during the early 1980's with the introduction of EPO. Since then, the biologics market has seen a significant increase in the number of products and the total revenue generated each year. The sales of biologics have surpassed $125 billion [1]. As the market has grown, there has been increased competition as numerous companies target therapeutics for the same indication as well as new indications. In addition, biosimilars, already popular in Europe, are growing rapidly in the United States and elsewhere. This creates competitive price pressure and the need for biotech companies to optimize their protein production processes.

Enhancing protein production of cultured cells is a common goal throughout the biomanufacturing industry. This commonly consists of an ongoing process of optimizing cell lines in serum free conditions in order to balance cell proliferation and protein production.

Serum free media have been optimized based on certain principles. The original serum free media was made by Ham in the 1960s [2]. It was based on adding the amino acids, vitamins, and salts needed for cells to grow. The same concept is generally used as a base media today but as technology has improved so have the methods of analysis. Assays testing metabolic byproducts and investigating proteomics are

used today to better optimize a media for CHO cell lines [3–7]. The use rate of amino acids or glucose can be determined to identify the constituents rapidly used during a run and may need replenishment, usually in the form of a feed and or feeds. None of these assays screen for lipids or cholesterol. In fact, the typical metabolic analysis looking at the TCA cycle would not be able to predict a benefit or show the harm of not having lipids and cholesterol. It has been reported that essential fats can be beneficial for cell health and protein production [6,8–10].

Here we investigate the utility of a novel approach to optimization using Cell-Ess® as a feed or an initial supplement to show proof of principle of adding lipids to culture. Furthermore Cell-Ess is tested as a feed supplement and in a small bioreactor as a model for use in large scale bio-production and use in a single use bioreactor. In this report we hypothesize that adding lipids will benefit protein production and then set out a set of experiments to test the hypothesis.

2. Materials and methods

2.1. Cell culture

Suspension Chinese Hamster Ovary (CHO) cells expressing several different monoclonal antibodies (mAb) from multiple individual clones were grown in 50 mL suspension cultures for 7 days. The lipid supplement was administered using a dosing paradigm as indicated for each experiment. The effect on protein titer and common metabolic profiles were determined and analyzed.

For these experiments two different stably transfected mAb single isolated CHO cell clones were used. The CHO cells were either grown in SAFC (Sigma, St. Louis, MO) serum free media or FortiCHO (Thermo Life). CHO cells were thawed and expanded prior to seeding in shake flasks.

2.2. Supplement with Cell-Ess

Cells were seeded in shake flasks. Cell-Ess was given as a supplement independent of other supplements and in addition to normal feeds at increasing concentrations. Cell-Ess (Essential Pharma, Ewing, NJ) was either added at the beginning of culture for protein expression or added as a feed as indicated.

2.3. Protein analysis

The mAb titers were calculated using two different methods. For the GS0 cells grown in SAFC media the cells were purified on a protein A column and then quantified by in a spectrophotometer at a wavelength of 280. The mAb from the DG44 cells grown in FortiCHO were quantified by an antibody specific ELISA.

2.4. Metabolic analysis

Samples were taken at the indicated time points and frozen. At the end of the run the samples were thawed and run on the BioProfile Flex analyzer (Nova Biomedical, Waltham, MA).

3. Results

3.1. Addition of Cell-Ess® as a supplement significantly increases titer

Large-scale protein production has predominantly moved to serum free systems developed for CHO cell lines. Serum free media systems are made by multiple companies in the life science industry, like Thermo Fisher Scientific, one of the industry leaders. Thermo provides several brands of CHO cell media including but not limited to FortiCHO and OptiCHO. In theory the complete media systems are ready for use. Interestingly, even though there are readily available products, many biomanufacturers will make internal modifications to optimize protein production for a specific cell line suggesting genetic differences were introduced to each CHO cell clone. In this study we investigated if the addition of a lipid supplement, Cell-Ess®, would lead to an increase in protein production across different clones in two different genetic backgrounds. A multi-step approach was utilized where proof of principle was researched to test the hypothesis of the benefit of adding lipids in a simplified controlled model. The second step was to test a complex model for fed batch in shake flasks where there are multiple variables.

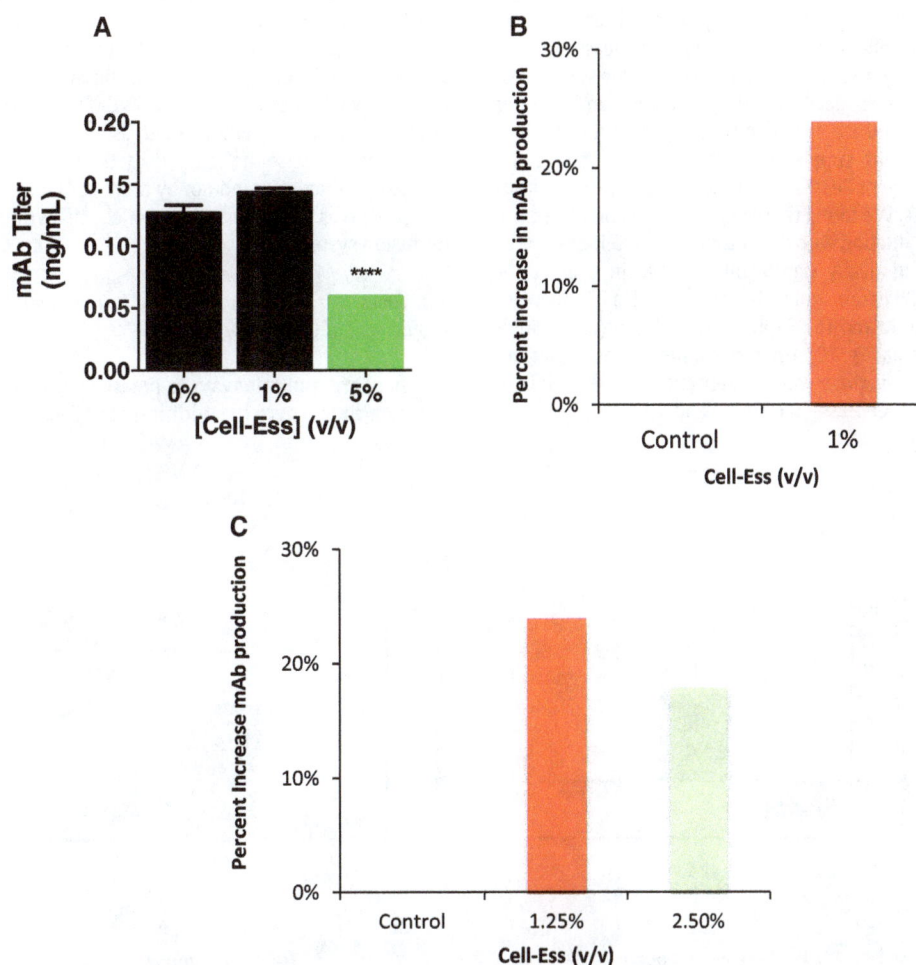

Fig. 1. Increased mAb titer with initial lipid supplement. The addition of Cell-Ess as an initial supplement in a short 7 day culture model was done in multiple media systems investigating the ability of the lipids supplement to increase mAb titers over control. In this figure the increase in titer due to the addition of Cell-Ess to SAFC media is shown in 1A and in 1B the increase is shown as a percent over control. In a similar experiment with addition of lipid at a different site with a different CHO clone the cells were grown in the Thermo media, Forti-CHO and shown as a percent increase over control.

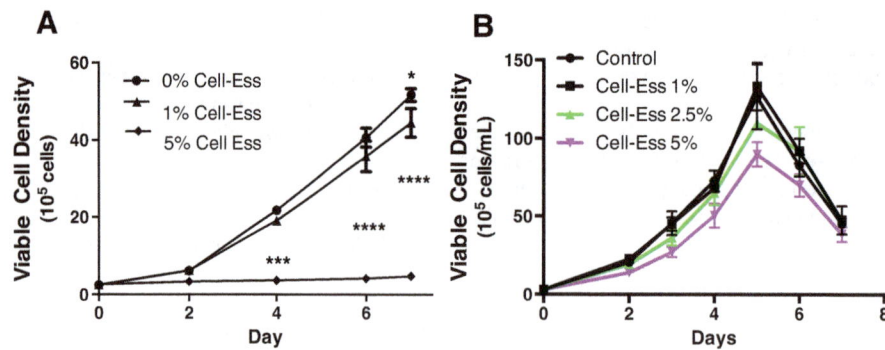

Fig. 2. Viable cell density in mAb producing CHO cells with lipid supplement. Cell-Ess was added at increasing amounts in two independent experiments done with two different stably transfected CHO cell clones. In SAFC SFM (A) 5% Cell-Ess reduced growth while in FortiCHO media (B) 5% Cell-Ess had a similar growth pattern to control.

The last part of the investigation was the addition of lipids to a scalable small bioreactor utilizing a single use bioreactor (SUB) where lipid additives previously.

The first study on proof of principle was designed to be open-ended allowing for a variety of outcomes. Four theoretical outcomes were possible; an advantage of using the lipid supplement at a defined percentage; an advantage of using Cell-Ess regardless of percentage of supplement used; the addition of the lipid supplement provided no effect; and lastly Cell-Ess ® would have a negative effect on protein titer.

As mentioned previously, there are multiple manufacturers of serum free media where variability in performance has been observed. In an effort to understand how Cell-Ess ® can affect titer, we measured Cell-Ess's effect on titer output in two independent systems. One system utilized a clone designed for use specifically with FortiCHO and another clone was designed to produce specifically in Sigma's media. In a shake flask study the lipid supplement was added once at the initiation of culture. A range of Cell-Ess® was added to the shake flasks starting at 0.5%, up to 15%. We found there was a significant increase in monoclonal antibody production with the addition of 1% Cell-Ess ® (see Fig. 1). There was a statistically significant increase in titer at 1%, increasing output from 126 μgrams/ml to 143 μgrams/ml, a 25% increase using the SAFC media (see 1a and 1b). Similarly, the addition of the lipids to LifeTech's FortiCHO media(Fig. 1C) resulted in a similar and significant increase in titer. Interestingly, there was a significant increase with both the addition of 1% and 2.5% Cell-Ess. (See Figs. 2 and 4.)

3.2. Viable cell density is not altered by the addition of Cell-Ess ®

An increase in titer can be attributed to either an increase in cell density, an increase in the number of cells making the same amount of protein per cell, or increased protein production on a per cell basis. To better understand if the increase in titer was a result of increased cell density we investigated the viable cell density (VCD) and viability of the two different clones treated with Cell-Ess ®. At 0.5% and 1% additions of Cell-Ess, there was no alteration in VCD. At increasing amounts of Cell-Ess®, the VCD began to display an effect starting at the 2.5% concentration in SAFC media. In the Thermo Fisher media the VCD was also unchanged at 1% Cell-Ess supplementation but with the addition of 2.5% Cell-Ess there was no effect of the VCD. Adding Cell-Ess in a specified ranged of 1%–2.5% did not have any effect on proliferation or cell viability when added as an initial supplement. Since Cell-Ess did not increase the cell density the increase in titer was due to increased productivity, and not by more cells making the same or slightly less protein as in the perfusion model. This is often the case in constant perfusion systems.

3.3. There is a significant increase in the yield per cell using an initial supplement of Cell-Ess ®

There are multiple ways to measure protein output. One way is to measure titer where the addition of Cell-Ess increased titer. Another

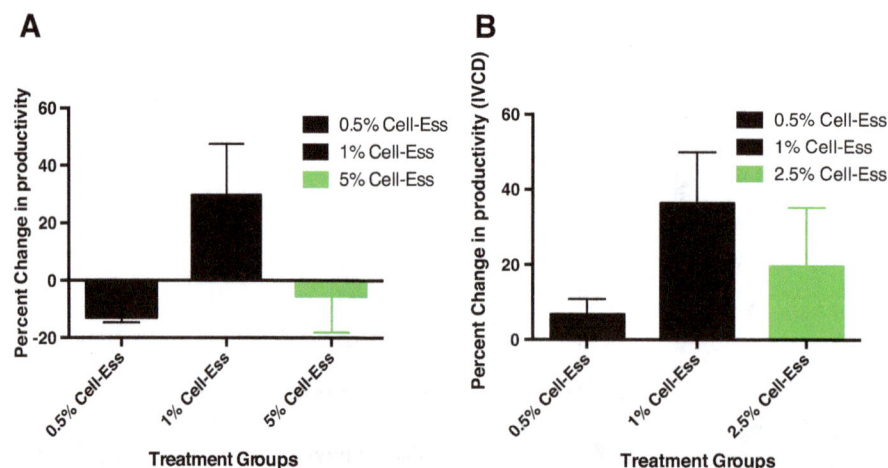

Fig. 3. Increased CHO cell productivity with initial lipid supplement. The addition of Cell-Ess as an initial supplement in a short 7 day culture model was analyzed for the ability to increase productivity of CHO cells. In this figure the increase in productivity in SAFC media is measured by two different methods. In (A) the calculation is the final titer in g/l divided by the final VCD cells/l resulting in g/cell and then increase is shown as a percentage. In (B) the integral under the VCD curve is calculated and used as the denominator to divide into the final titer and shown as a percent increase over control.

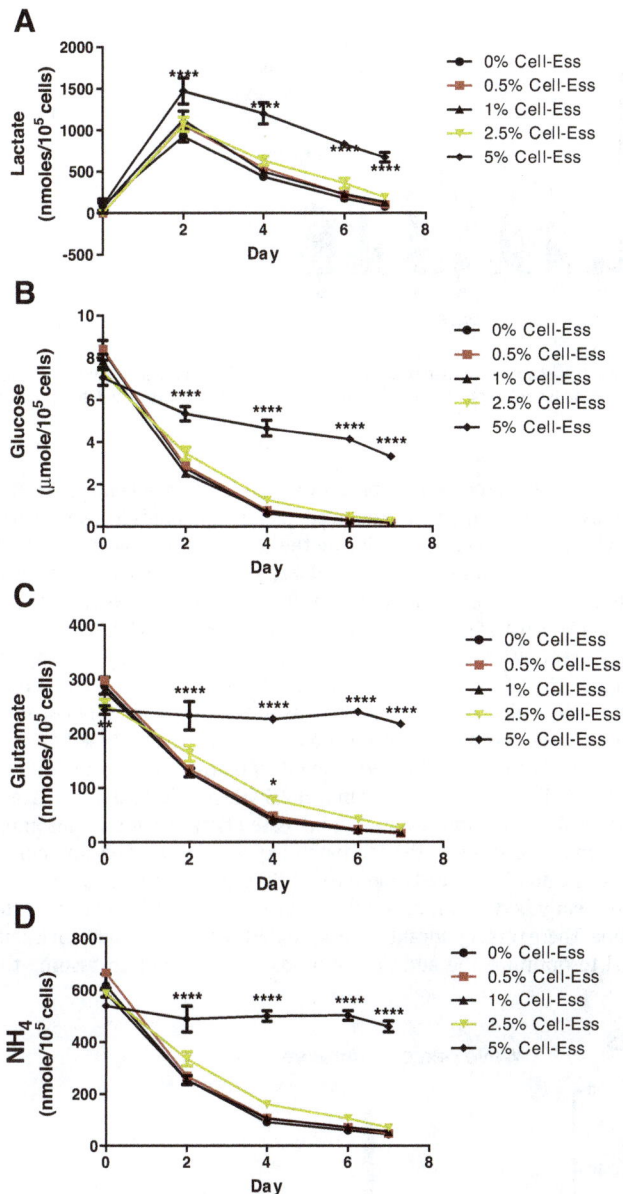

Fig. 4. Addition of lipids does not affect the metabolic profile. Cell-Ess was added at the indicated levels to SAFC media at the indicated concentrations. Samples were collected from the associated flasks and frozen until the final collection. Samples were thawed and metabolic analysis for lactate, glucose, glutamate, and ammonia (A, B, C, and D respectively) was done using a BioProfile Nova Flex.

measure of output is productivity. In this assay we measured if there was an increase in the amount of productivity on a per cell basis. Productivity can be measured by one of several different methods. The most common methodology is to divide the titer from a set time point and then divide this by the number of cells at that same time point. The result is an indication of the productivity or amount of protein secreted on a per cell basis. The addition of Cell-Ess resulted in a significant increase in the amount of protein secreted per cell. This increase in productivity was not dependent on cell proliferation as there was no significant increase in total cell growth ($129.2 \pm 9.0 \ 10^5$ cells day) when compared to control ($145.2 \pm 5.3 \ 10^5$ cells day). At 1% addition of Cell-Ess, there was a greater than 40% increase in productivity. The addition of lipids increased the amount of total mAb titer by increasing the productivity of CHO cells. Another more conservative measure of productivity is to divide the titer of a single time point by the area under the VCD curve or IVCD. That value describes the production of the aggregate production over the entire time frame. The addition of

Cell-Ess increased productivity of cells using the IVCD method regardless of Cell-Ess concentration. These data suggest the addition of lipids allows CHO cells to more efficiently produce protein.

3.4. There is no alteration in the metabolites or salts despite a significant increase in mAb production

Since the addition of lipids increases titer primarily by increasing the efficiency of CHO cells, an experiment was then conducted to determine if the increased protein secretion was causing the CHO cells to build up toxic by- products. As mentioned in the introduction, metabolic analysis has been used to optimize CHO cell antibody production. We examined the entire metabolic profile (Fig. 3) and identified no significant differences in any of the salts or metabolic by-products. These are common indicators for potential harmful by-products. Lactate is one factor that can increase the pH causing cell necrosis. In no fewer than 6 independent runs in both models, addition of Cell-Ess as an initial supplement or as a fed batch supplement did not result in an accumulation of lactate. In fact after the initial peak, lactate was controlled while simultaneously there was a constant demand for glucose. The addition of lipids did not feed into the TCA cycle which would result in an increase in lactate as occurs with the addition of glucose. Since there was an increase in the amount of protein, there is a potential for the increase the amount of ammonia, a toxic by-product of additional protein.

3.5. The addition of lipids in a fed batch system results in increased titers

The most common method to extend culture times and increase titer is to use a feed during the batch run. The method, called fed batch, uses a variety of feeds to add back nutrients which were utilized by the cells in the bioreactor. There is a way to model the fed batch system in shake flasks. This study sought to understand - in a system where feeds are added to an optimized serum free culture condition - if the use of Cell-Ess as a feed can increase titer or have other beneficial effects, such as improving cell viability or extending potential run times.

To test this hypothesis we modeled a fed batch system in shake flasks, with the study variable being feeds of Cell-Ess® at varying concentrations. Feeds already utilized in these previously optimized systems were retained to control for the effect of adding Cell-Ess. There was a benefit of Cell-Ess at all concentrations tested in the fed batch model unlike the use of the lipid supplement Cell-Ess when used as a one-time initial supplement. The optimal result was seen with multiple feedings of Cell-Ess at a concentration of 5%. As seen in Fig. 5, there were multiple benefits. The first is an increase in titer over the controls. The increase in titer was greater than 30%. The second observation was the addition of Cell-Ess led to an increase in titer early, and throughout the run. Finally, the viability improved with the use of Cell-Ess. The addition of lipids as a feed behaved similarly to when the lipids were used only as an initial supplement. The titer was significantly increased without increasing the cell proliferation. In both models there was a productivity increase through the use of Cell-Ess.

3.6. A novel liposome can be used in a single use bioreactor

Shake flasks experiments are very useful in modeling behavior and optimizing conditions prior to entering a bioreactor. The behavior of the culture in bioreactors can be different than that seen in shake flasks. We then wanted to determine if the lipid supplement, could be used to increase titer as shown in the shake flask model Figs. 1 and 5, would be repeatable in a bioreactor. This experiment involved testing Cell-Ess as both an initial supplement at 1% and also adding it as a feed at 5% as determined in the shake flask model (Fig. 5). Furthermore, we wanted to test the lipid supplement in a single use system where lipid use has been shown to be toxic or not beneficial.

Cell-Ess was tested in a 10 l Wave bag with EVA film. Bioreactor runs done at the 5 L size have been shown to be representative for large-scale

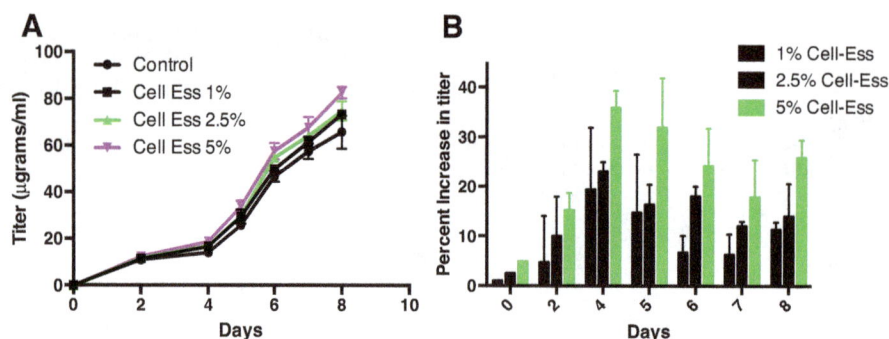

Fig. 5. Increased mAb titer with a lipid as feed in a fed batch model. Cell-Ess was used as an initial supplement and as a feed in shake flask model for fed-batch protein production. Cell-Ess was added as an initial supplement at 1% as determined in Fig. 1 and then increasing feed amounts were added as indicated. Protein was quantified by ELISA shown in A and the increase in titer as a percent is shown in B.

bioreactor runs up to the 10,000 L scale. The hypothesis would be the use of Cell-Ess in a SUB would be as efficacious as the run done in the smaller volume shake flasks.

4. Discussion

The use of lipids has been shown to be beneficial but to date there has not been an optimized method to deliver the lipids into a long term culture system. Long term culture systems are the standard for growing and expanding CHO cells for protein production. Many methods have been employed to determine the optimal media for each CHO cell and CHO cell clone [7,11,12]. The methods include, but are not limited to, metabolic analysis [3,11,12], proteomic analysis [11, 13], and network analysis. In this paper we demonstrated a novel lipid mixture using a proprietary delivery system called Cell-Ess® can deliver lipids and cholesterol resulting in increased titer by increasing cell productivity.

During the process of empirically testing Cell-Ess, several interesting discoveries were made related to supplementing lipids in a serum free environment. It was found during the proof of principle experiment, where Cell-Ess was only added at the beginning of a short protein run, the amount of lipids might have different effects between different media When 1% Cell-Ess was added to FortiCHO and SAFC media there was an equivalent increase in mAb titer. But the FortiCHO SFM supported growth and increased titer with concentration of lipids greater than 1% while with more than 1% Cell-Ess in SAFC media did not sustain growth, demonstrating there was a different sensitivity based on constituents in the SFM. There was an effect on the proliferation of CHO cells at 2.5% in the SAFC media but, at 2.5% the FortiCHO did not change the proliferation and resulted in increased titer. We found this effect was much broader in the fed batch experiment. In data not shown here, the number of feeds and the Cell-Ess® concentration per feed varied greatly between media but the result - once optimized - was the same. There was significant increase in the titer in both models. It is critical to optimize the addition of lipids. While we often assume the

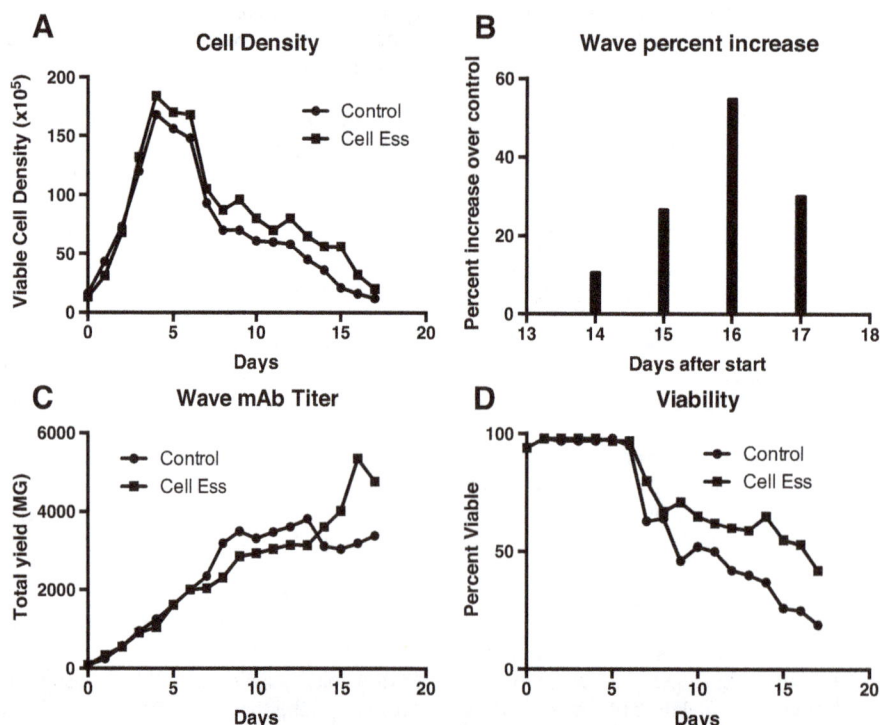

Fig. 6. Increased mAb titer in single use bioreactor. Cell-Ess was used as an initial supplement and as a feed in a single use bioreactor (SUB). A 5 l batch run was done in 10 l Wave bags. Cell-Ess was added as an initial supplement at 1% as determined in Fig. 1 and then 5% feed was added every 3 days. Protein was quantified by ELISA shown in C and the increase in titer as a percent is shown in B. The VCD and viability rates are shown in A and D respectively.

difference in formulations explain why there are different optimization parameters, it is not clear what specific ingredient or ingredients are contributing to this variance.

There is a movement toward single use systems [11,14]. The single use systems take advantage of a smaller bioreactor and utilize a constant perfusion to replenish the media and maintain cell health. The result is increased cell density and increased titer. The increase in cell density is in the range of 3–5 fold while the increase in titer is usually 1.5 to 2 fold. From a downstream processing perspective, this approach increases the amount of cell bioburden and host cell protein on a per gram purification. An increase in productivity (seen in figure 2) is more desirable as the amount of protein per cell is increased while the purification burden downstream is not affected. The use of the novel lipid supplement Cell-Ess® increases the productivity on a per cell basis. Furthermore Cell-Ess functioned equally well in glass and on a single use film.

There are numerous supplements on the market provided by a variety of companies, but in many cases the supplement is designed to work with a specific media formulation. Cell-Ess is designed to be robust so the results are media agnostic. To date, Cell-Ess has been added to SAFC, LifeTech (Figs. 1, 5 and 6) and Lonza media (data not shown) with an increase in titer observed all systems. This was done at difference sites with different clones and reproduced multiple times with the same result. As mentioned earlier, the design of an experiment is crucial to define the optimal amount of lipid supplement for establishing reproducible results.

The methods used to optimize CHO cell production, such as metabolic analysis, and proteomic analysis will not predict or demonstrate the benefit of adding lipids. The output is focused on the TCA cycle or what proteins are being produced. Here it was demonstrated that there were no alterations in the metabolic profile even while there was a significant increase in the protein titer. This suggests the increase in titer does not alter the TCA cycle since there is no increased lactate. This can mean there is more efficient energy usage which the cell can then utilize for increased protein production. There have been reports recently suggesting a metabolic profile in the oxidative state is conducive to increased titer [3,4]. Another explanation is the lipid's stabilizing effect on the lipid membranes throughout the entire protein synthesis and secretion pathway.

This study showed the addition of lipids increased mAb titer by increasing productivity. This increase did not affect the metabolic profile of the cell. The next level of investigation will be on protein quality.

While preliminary data does exist that is not shown here, the glycolytic pattern is not significantly altered by the addition of Cell-Ess. Further exploration of the role of adding lipids on protein quality is planned.

Conflicts

I am employed by Essential Pharmaceuticals, LLC the maker of Cell-Ess. The work done in the manuscript was done by third parties who did not have any financial interest in Essential Pharmaceuticals, LLC or receive any incentives based on the outcome of experiments done.

References

[1] Emerton DA. Profitability in the biosimilars market. Can you translate scientific excellence into a healthy commercial return? BioProcess Int 2013;11(6)s (Supplement).
[2] Ham RG. Clonal growth of mammalian cells in a chemically defined, synthetic medium. Proc Natl Acad Sci U S A 1965;53:288–93.
[3] Dean J, Reddy P. Metabolic analysis of antibody producing CHO cells in fed-batch production. Biotechnol Bioeng 2013;110:1735–47.
[4] Templeton N, Dean J, Reddy P, Young JD. Peak antibody production is associated with increased oxidative metabolism in an industrially relevant fed-batch CHO cell culture. Biotechnol Bioeng 2013;110:2013–24.
[5] Bai Y, et al. Role of iron and sodium citrate in animal protein-free CHO cell culture medium on cell growth and monoclonal antibody production. Biotechnol Prog 2011;27:209–19.
[6] Li F, Vijayasankaran N, Shen A, Kiss R, Amanullah A. Cell culture processes for monoclonal antibody production. MAbs 2010;2:466–79.
[7] Legmann R, et al. A predictive high-throughput scale-down model of monoclonal antibody production in CHO cells. Biotechnol Bioeng 2009;104:1107–20.
[8] Jeon MK, Lim J-BB, Lee GM. Development of a serum-free medium for in vitro expansion of human cytotoxic T lymphocytes using a statistical design. BMC Biotechnol 2010;10:70.
[9] Van der Valk J, et al. Optimization of chemically defined cell culture media— replacing fetal bovine serum in mammalian in vitro methods. Toxicol In Vitro 2010;24:1053–63.
[10] Sato JD, Kawamoto T, McClure DB, Sato GH. Cholesterol requirement of NS-1 mouse myeloma cells for growth in serum-free medium. Mol Biol Med 1984;2:121–34.
[11] Meleady P, et al. Sustained productivity in recombinant Chinese Hamster Ovary (CHO) cell lines: proteome analysis of the molecular basis for a process-related phenotype. BMC Biotechnol 2011;11:78.
[12] Unknown. CHO_cell_metabolomics.pdf.
[13] Zheng X, et al. Proteomic analysis for the assessment of different lots of fetal bovine serum as a raw material for cell culture. Part IV application of proteomics to the manufacture of biological drugs. Biotechnol Prog 2008;22:12941300.
[14] Diekmann S, Dürr C, Herrmann A, Lindner I, Jozic D. Single use bioreactors for the clinical production of monoclonal antibodies – a study to analyze the performance of a CHO cell line and the quality of the produced monoclonal antibody. BMC Proc 2011;5:P103.

3

A random forest classifier for detecting rare variants in NGS data from viral populations

Raunaq Malhotra [a],[*], Manjari Jha [a], Mary Poss [b], Raj Acharya [c]

[a] The School of Electrical Engineering and Computer Science, The Pennsylvania State University, University Park, PA, 16802, USA
[b] Department of Biology, The Pennsylvania State University, University Park, PA 16802, USA
[c] School of Informatics and Computing, Indiana University, Bloomington, IN 47405, USA

ARTICLE INFO

Keywords:
Sequencing error detection
Reference free methods
Next-generation sequencing
Viral populations
Multi-resolution frames
Random forest classifier

ABSTRACT

We propose a random forest classifier for detecting rare variants from sequencing errors in Next Generation Sequencing (NGS) data from viral populations. The method utilizes counts of varying length of k-mers from the reads of a viral population to train a Random forest classifier, called MultiRes, that classifies k-mers as erroneous or rare variants. Our algorithm is rooted in concepts from signal processing and uses a frame-based representation of k-mers. Frames are sets of non-orthogonal basis functions that were traditionally used in signal processing for noise removal. We define discrete spatial signals for genomes and sequenced reads, and show that k-mers of a given size constitute a frame.

We evaluate MultiRes on simulated and real viral population datasets, which consist of many low frequency variants, and compare it to the error detection methods used in correction tools known in the literature. MultiRes has 4 to 500 times less false positives k-mer predictions compared to other methods, essential for accurate estimation of viral population diversity and their *de-novo* assembly. It has high recall of the true k-mers, comparable to other error correction methods. MultiRes also has greater than 95% recall for detecting single nucleotide polymorphisms (SNPs) and fewer false positive SNPs, while detecting higher number of rare variants compared to other variant calling methods for viral populations. The software is available freely from the GitHub link https://github.com/raunaq-m/MultiRes.

1. Introduction

The sequence diversity present in a population of closely related genomes is important for their survival under environmental pressures. Viral population within a host is an example of such population of closely related genomes, where some viral strains survive even when large segments of their genome are deleted. The sequence variants that occur at low frequency in the population, also known as rare variants, have been known to impact the population's survival and understanding their prevalence is important for drug design and in therapeutics [1].

However, detection of rare variants from Next Generation Sequencing (NGS) data is still a challenge as the rare variants are tangled with errors in sequencing technologies due to their similar prevalence [2,3]. The NGS data technologies are error prone and even though their error profiles are well studied [4,5], removing sequencing errors is essential before downstream processing of NGS data such as assembly of haplotypes in viral populations [2,6-9] and variant calling for viral populations [7,9,10].

In order to remove sequencing errors from NGS data, the first step is detecting the errors from true biological sequences and then correcting the errors to the true sequence. For NGS data obtained from a viral population, the reads are mapped to a reference genome to detect true variants from sequencing errors based on a probabilistic model [6,7,9-11], and then the sequencing errors are corrected to the sequence of the reference genome. However, as virus population contains a large diversity of true sequences, accurate mapping of reads to any one reference may not be possible.

Alternatively, sampled reads are broken into small fixed length sub-strings called k-mers and their counts are used for error detection (e.g. [12-16]). The erroneous k-mers are corrected by changing minimum number of bases in the reads using the detected true

* Corresponding author.
  E-mail address: raunaq.123@gmail.com (R. Malhotra).

k-mers. These methods use a generative model for *k*-mer counts to determine if an observed *k*-mer is erroneous or a true *k*-mer [12] based on a counts threshold [12-14,17].

For *k*-mer based error detection, the length of the *k*-mer and the frequency threshold are important parameters. The size of a *k*-mer can effect the performance of error detection method, as it either decreases the evidence for a segment of the genome for a large *k*, or combines evidences from multiple segments for a small *k*. However, a single appropriate *k*-mer size for error detection in viral populations is restrictive in nature, as a combination of different sized overlapping *k*-mers, although redundant, can provide richer information.

The error detection part in most *k*-mer based error correction tools [12–15] has been designed assuming the reads are sampled from a single diploid genome and rely on a single counts threshold. However, for viral populations a single threshold is not suitable as viral strains occur at different relative frequencies. Currently, a number of time and memory efficient *k*-mer counting algorithms are available [18,19]. Thus, choosing an appropriate size of *k*-mer is possible by performing *k*-mer counts at multiple sizes [20].

With the availability of large amounts of data from NGS technologies, data driven classifiers have also been used for detection of sequencing errors [21] and for variant calling [5,12,22,23]. However, identifying the features for classification of rare variants and sequencing errors is still a challenge, due to their similar characteristics in the NGS data.

We propose, MultiRes, a reference-free *k*-mer based error detection algorithm for a viral population. The algorithm uses *k*-mer counts of different sizes to train a Random Forest Classifier that classifies *k*-mers as erroneous or rare variant *k*-mers. We also propose a mechanism for selecting the optimal combination of *k*-mer sizes. The rare variant *k*-mers along with high frequency *k*-mers can be used as is in downstream tools for variant calling and for de novo assembly of viral populations.

MultiRes uses a collection of sizes of *k*-mers as features for detecting sequencing errors and rare variants. Our rationale to choose a combination of sizes for *k*-mers is rooted in signals processing, where analysis of signals at different resolutions has been used for noise removal [24]. Signals are projected onto a series of non-orthonormal basis functions, known as a **frame**, [25–27]; these projections are used for error removal and signal recovery [28,29].

The classifier in MultiRes is trained on a simulated dataset that models NGS data generated from a replicating viral population. We evaluate the performance of MultiRes on simulated and real datasets, and compare it to the error detection algorithms of error correction tools BLESS [13], Quake [12], BFC [14], and Musket [15]. We also compare our results to BayesHammer [30] and Seecer [31], which can handle variable sequencing coverage across the genome and polymorphisms in the RNA sequencing data respectively.

MultiRes has a high recall of the true *k*-mers, comparable to other methods and has 5 to 500 times better removal of erroneous *k*-mers compared to other methods. Our results demonstrate that the classifier in MultiRes performs well for error detection on real sequencing data obtained from the same sequencing technology. Thus, the classifier in MultiRes is generalizable to viral population data from the same sequencing technology.

As MultiRes detects the rare variant *k*-mers in an NGS data, its output can be directly used for identifying rare variants in a viral population. Variant calling for viral populations typically relies on a single reference genome or on a consensus genome generated from the population being studied [7,9,10]. We compare the rare variants detected by MultiRes to variant calling methods VPhaser-2 [9], LoFreq [10] and the outputs from haplotype reconstruction method ShoRAH [7]. MultiRes has the higher recall of true SNPs compared to the SNPs called by VPhaser-2, LoFreq and ShoRAH on both simulated and real datasets, and misses the least number of true SNPs

amongst all methods. This demonstrates its applicability for rare variant detection in viral populations.

2. Methods

MultiRes is a classifier for detecting sequencing errors from rare variants. The counts of the *k*-mers along with the counts of their sub-sequences (sub *k*-mers within a *k*-mer) are used as features for training a classifier. The true *k*-mers observed in the viral haplotypes with counts in the reads less than a threshold T_{High} are defined to be rare variant *k*-mers, while the rest of *k*-mers with counts less than T_{High} are erroneous *k*-mers. The *k*-mers that occur at counts greater than T_{High} are known as common *k*-mers, as they occur frequently in the viral haplotypes. The common *k*-mers are assumed to be error-free and the classifier is trained only for the erroneous and rare variant *k*-mers.

The premise of our method is that reads sequenced from a population of genomes can be modeled as discrete spatial signals. Discrete spatial signals can be projected on to a **frame** [25-27,32] for their representation (See Supplementary Material for details), where the coefficients of projections characterize the discrete spatial signals. Similarly, we show that *k*-mers (of a given size *k*) form a **frame** and the maximal projection of *k*-mers correspond to their counts in a sequencing run. Additionally, a *k*-mer can be projected on to a collection of **frames**, where each **frame** represents counts of *k'*-mers ($k' < k$) that are sub-strings of the given *k*-mer.

The choice of *k* for a **frame** is important and should be large enough such that a *k*-mer only occurs once in the haplotypes. On the other hand, it should be smaller than the read lengths so that *k*-mer counting is still meaningful.

The minimum *k* can be approximated by ensuring that the probability of picking a string of length equal to the genome length (say |*H*|) where all *k*-mers in it occur only once is low [12]. Thus the probability of picking approximately |*H*| unique *k*-mers out of a set of $4^{k/2}$ (considering reverse complements) should be low. We set $2 \cdot |H|/4^k \approx \epsilon$, where ϵ is a small number, to determine the smallest possible choice of *k* (k_{min}) for the **frame**.

As an example, a *k*-mer *u* occurring $c(u)$ times in the reads when projected on a **frame** of size *k* is in-fact represented by its maximal projection $c(u)$. The same *k*-mer *u*, can also be represented on **frames** of sizes (k', k'' in the range [k_{min}, k]. Now the maximal projections for *u* in these **frames** are the counts of *k'*-mers and *k''*-mers present within *u*. This representation of *k*-mer *u* can be used to train a classifier for identifying erroneous versus rare variant *k*-mers.

2.1. MultiRes: Classification algorithm for detecting sequencing errors and rare variants

We define a classifier, *EC*, for classifying a *k*-mer as erroneous, a rare variant, or a common *k*-mer in the dataset. Algorithm 1 describes MultiRes, the proposed algorithm for detecting rare variants and sequencing errors. The algorithm takes as input the sampled reads, the classifier *EC*, an ordered array (k, k', k''), and a threshold parameter T_{High}. It outputs for every *k*-mer observed in the sampled reads a status: whether the *k*-mer is erroneous or a rare variant.

It first computes the counts of *k*-mers, *k'*-mers, and *k''*-mers using the dsk *k*-mer counting software [18]. The *k*-mers *u* that have counts greater than T_{High} are marked as true *k*-mers while the rest of the *k*-mers are classified using the classifier *EC* based on their counts on *k'*-mers, and *k''*-mers.

The classifier *EC* captures the profile of erroneous versus rare variant *k*-mers from Illumina sequencing of viral populations. We used the software dsk (version 1.6066) [18] for *k*-mer counting, which can perform the *k*-mer counts in a limited memory and disk space machine quickly. The run time of MultiRes is linearly dependent on the number of unique k-mers in a dataset, as once the classifier EC is trained, it can be used for all datasets, and it can be easily parallelized.

Algorithm 1. MultiRes: Error detection in the sampled reads by **frame**-based classification of k-mers

---

Input: Sampled reads, classifier EC, (k, k', k''), T_{High}
Output: $y = EC(u)$ for all k-mers u in the sampled reads, where $y = \{0, 1\}$ (True k-mer, Erroneous k-mer)

1: Compute counts $c(\cdot)$ of k-mers, k'-mers and k''-mers from the sampled reads.
2: **for** each k-mer $u = (u_1, u_2, \ldots, u_k)$ **do**
3:    **if** $c(u) < T_{high}$ **then**
4:       $F(u) \qquad = \qquad [c(u), c(u_1, u_2, \ldots, u_{k'}),$
      $c(u_2, u_3, \ldots, u_{k'+1}), \ldots, c(u_{k-k'+1}, \ldots, u_k),$
      $c(u_1, u_2, \ldots, u_{k''}), \qquad\qquad c(u_2, u_3, \ldots, u_{k''+1}),$
      $\ldots, c(u_{k-k''+1}, \ldots, u_k)]$
5:       Use classifier EC to classify the k-mer u using its features $F(u)$
6:       Output $\{0, 1\}$ for k-mer u based on the classifier output
7:    **else**
8:       Output $\{1\}$ for the k-mer u // Mark the k-mer as true
9:    **end if**
10: **end for**

2.1.1. Simulated data for classifier training

MultiRes assumes the availability of a classifier EC which can distinguish between the erroneous and rare variant k-mers. We use simulated datasets to train a series of classifiers and set EC to the classifier which has the highest accuracy. The simulated viral population consists of 11 haplotypes and is generated by mutating 10% of positions on a known HIV-1 reference sequence of length 9.18 kb (NC_001802). These mutations also model the evolution of a viral population under a high mutation rate. The mutations introduced are randomly and uniformly distributed across the length of the genome so that the classifier is not biased towards the distribution of true variants. This introduces a total of 195,000 ground truth unique 35-mers in the simulated HIV-1 dataset.

We next simulate Illumina paired-end sequencing reads using the software dwgsim (https://github.com/nh13/DWGSIM) at 400x sequencing coverage from this viral population. The status of each k-mer in this dataset is known as being erroneous, rare variant or a common k-mer. We use close to 100,000 k-mers in the training dataset. Thus, there is a test dataset of k-mers left for evaluating the efficacy of the classifiers.

In order to train a classifier, we need to choose the size of the k-mer, the sizes of k'-mers for computing the projections of k-mer signals, and the number of such projections needed. The choice of the smallest of $\{k, k', \ldots\}$ should be above the minimum length k_{min} to ensure that each k-mer still corresponds to a unique location on a viral genome.

For HIV populations, with genome length 9180 base pairs (9.1 kbp) and taking $\epsilon = 0.001$ (a small value, as mentioned before), the minimum length of k-mer is $k_{min} = \lceil \log_4 2 \cdot G / \epsilon \rceil = \lceil 12.06 \rceil = 13$. As in signal processing domain, we choose $k \approx 3 \cdot k_{min} = 35$ (an integral multiple of k_{min}) as the largest k-mer, and consider its projections on **frames** of sizes ranging from 13 to 35.

MultiRes assumes that k-mers above the threshold count T_{high} are error-free, and only classifies the k-mers with counts less than T_{high}.

The choice of T_{high} should ensure that the probability of erroneous k-mers with counts above T_{high} is negligible. We use the gamma distribution model mentioned in the Quake error correction paper [12] for modeling erroneous k-mers, as it approximates the observed distribution of errors. Based on this gamma distribution, we set $T_{high} = 30$ for the simulated HIV population data. The classifiers are therefore trained on 35-mers with counts less than 30.

Three training datasets consisting of both erroneous and rare variant 35-mers are generated. The features in the three datasets are the projection of the 35-mers onto (i) the frame of size 23, (ii) the frame of size 13, and (iii) a combination of both frames (Fig. 1 a). The features in the three settings translate to the counts of the 13-mers and 23-mers observed within the 35-mer along with the counts of the 35-mer. We observed 11.9 million unique 35-mers in the simulated HIV-1 population, from which features from 76,000 erroneous 35-mers and 32,000 true variant 35-mers distributed uniformly over counts 1–30 were used for training the classifiers.

2.1.2. Classifier selection

Classifiers Nearest Neighbor, Decision Tree, Random Forest, Adaboost, Naive Bayes, Linear Discriminant Analysis (LDA), and Quadratic Discriminant Analysis (QDA) are trained on the three training datasets and evaluated based on their test data accuracy over a 5-fold cross validation dataset. The classifiers are implemented in the scikit-learn library (version is 0.16.1) in python programming language (version 2.7.6). For all the classifiers, the accuracy improves as the 35-mers are projected onto 13-mers rather than 23-mers (higher resolution, lower size of k'-mers), and improves even further when 35-mers are resolved onto both 13-mers and 23-mers (Fig. 1). No further feature selection was performed when 35-mers were resolved onto 13-mers and 23-mers. The Random Forest Classifier performs the best on all three datasets, where the accuracy for dataset (iii) is 98.12%. The accuracy for Naive Bayes and QDA classifiers are lower for all datasets, and also decreases when the projections in both 13-mers and 23-mers are considered, indicating that inadequacy of their models for the classification of 35-mers in these projections. The performance of other classifiers are comparable and follows similar trends.

2.1.3. Exploring additional feature spaces

Additionally, we generate a series of 4 projections of the 35-mers onto frames of sizes a) 15, b) {15 + 20}, c) {15 + 20 + 25}, and d) {15 + 20 + 25 + 30} to evaluate the effect of number of frames used for projection on the performance (Fig. 1 b). Increasing the number of projections has no visible effect on increasing the accuracy of performance, although it increases the memory requirements and time complexity for computing counts of all five different values of k. Based on this, we chose the Random Forest classifier with a resolution of 35-mers decomposed into a combination of 13-mers and 23-mers for other simulated and real datasets.

3. Results

3.1. Error detection for reconstruction of haplotypes

We evaluate MultiRes on simulated HIV and HCV datasets and a laboratory mixture of HIV-1 strains. MultiRes is compared to the detection algorithms in the error correction tools Quake (last checked version Feb 2012) [12], BLESS (version 0.15) [13], Musket (last downloaded October 2015) [15], BFC (last downloaded October 2015) [14], BayesHammer (version 3.6.2) [30] and Seecer (version 0.1.3) [31]. As these tools are traditionally designed for error correction, the error corrected reads or k-mers from these methods were used for comparison with the rare variant k-mers and common k-mers predicted by MultiRes. ShoRAH [7] reconstructs a set of haplotypes as its final output rather than error corrected reads

(a) 2Frames

(b) 4Frames

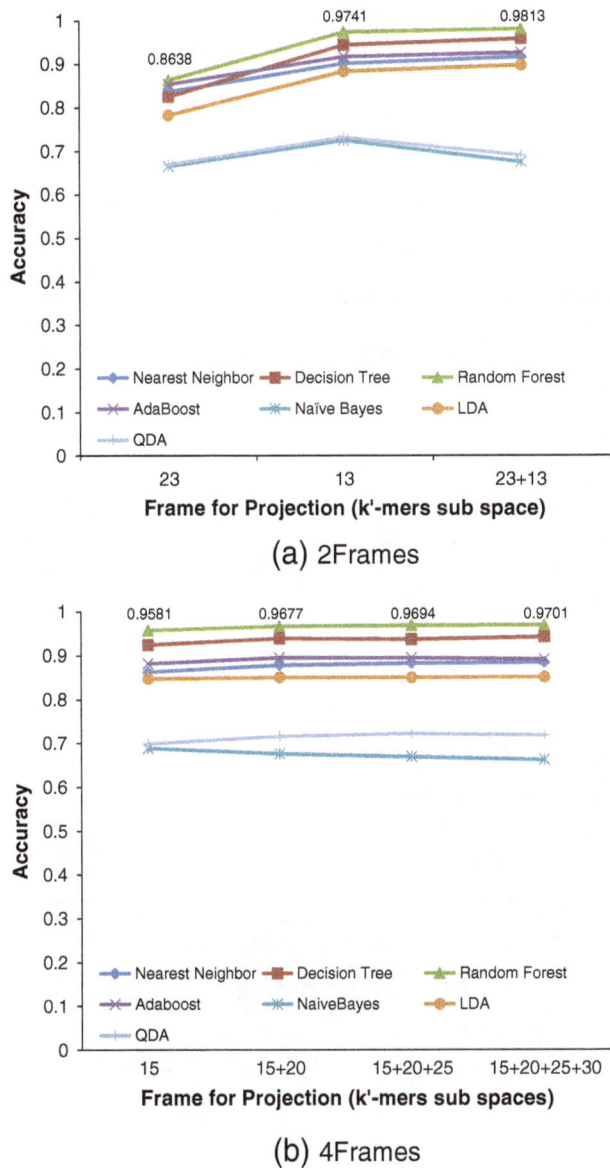

Fig. 1. Performance of classification algorithms for erroneous versus rare variant k-mer classification. The performance of mentioned classification algorithms for classifying 35-mers are compared over two sets of features. 35-mers are either projected onto a family of (a) 23-mers, 13-mers, and a 13 + 23-mers, and (b) projections onto 15-mers, 15 + 20-mers, 15 + 20 + 25-mers, and 15 + 20 + 25 + 30-mers. The accuracy reported is over fivefold cross validation on 35-mers extracted from HIV viral population. Accuracy improves when 35-mers are projected onto smaller sized k'- mers and as the number of projections increases. Random Forest Classifier has the best accuracy across different classification algorithms.

and thus was not evaluated for error correction. The error corrected reads, although available as an intermediate output, are not reported due to their low precision numbers, but ShoRAH is used for single nucleotide variant calling and comparison later in the text. Other recent error correction methods available for viral populations such as PredictHaplo http://bmda.cs.unibas.ch/software.html, HaploClique [11], and Viral Genome Assembler (VGA) [8] were not evaluated in this study.

Three measures, defined in terms of the true and erroneous k-mers, are used for comparing the detection algorithms in all methods. *Precision* is defined as the ratio of the known true k-mers identified to the total number of k-mers predicted as true variants by an algorithm. *Recall* is defined as the ratio of the true variant

k-mers identified to the total number of true k-mers by an algorithm and measures the goodness of a method to retain true k-mers for a dataset. *False Positives to True Positives Ratio* (FP/TP ratio) is the ratio of the erroneous k-mers predicted as true variants to the true variant k-mers identified by the algorithm. FP/TP ratio measures the number of erroneous k-mers identified by an algorithm to detect a single true variant k-mer and is a measure of the overall volume of k-mers predicted by an algorithm.

3.1.1. HIV simulated datasets

We first assess the performance of MultiRes on the reads simulated from the HIV-1 population containing 11 haplotypes, generated from a single HIV-1 reference sequence (NC_001802) as mentioned before. Two datasets are generated from the simulated reads: one with average haplotype coverage of 100x (denoted as HIV 100x), and second where the average coverage is 400x (denoted as HIV 400x) as increasing sequencing depth increases the absolute number of erroneous k-mers introduced in the data.

The recall of MultiRes is 95% and 98% on HIV 100x and HIV 400x datasets, respectively, indicating that performance of MultiRes improves with increasing sequencing depth as expected. The recall numbers are comparable to around 98% recall of other methods on the HIV 100x dataset and 94% to 99% for HIV 400x dataset (Table 1).

The precision of MultiRes is 89% in the HIV 100x while all other methods have low precisions for HIV 100x. While precision in all other methods is less than 5% for HIV 400x dataset, the precision of MultiRes is 95%, suggesting that precision decreases for other methods with increasing sequencing depth. As higher depth samples also have higher sequencing errors, the detection algorithms in these methods are not able to differentiate between rare variants and sequencing errors. Seecer and BayesHammer, methods which can handle variability in sequencing coverage, also have very low precision values compared to the proposed method.

The FP/TP ratio obtained by MultiRes are 4 to 500 times better than other methods and the number of k-mers retained is close to the true set of k-mers in the two datasets (FP/TP ratio is close to zero & recall close to 95–98 %).Thus, while all methods retain the true k-mers to the same extent, only MultiRes reduces the number of false positive k-mers. This is important as the memory requirements for de novo assembly tools linearly increases based on the number of k-mers. Thus the k-mers predicted by MultiRes would have a 500 times reduction in memory consumption for downstream de novo assembly tools as compared to current error correction methods.

Table 1

Comparison of performance metrics of error detection on simulated HIV datasets. FP/TP ratio is the measure of false positive to true positive ratio, Recall measures the percentage of true k-mers out of all true k-mer predicted by an algorithm, Precision measures the percentage of predicted k-mers by an algorithm that are true k-mers.

Algorithm	FP/TP ratio		Recall		Precision	
	HIV 100x	HIV 400x	HIV 100x	HIV 400x	HIV 100x	HIV 400x
Uncorrected	53	121	98.91	99.67	1.85	0.82
Quake	9.26	29.5	**98.63**	94.84	9.74	3.27
BLESS	0.71	76.7	98.38	99.36	58.48	1.28
Musket	0.46	121	98.46	**99.67**	68.48	0.82
BFC	2.12	112	98.47	99.57	32.01	0.89
BayesHammer	0.37	69.1	98.47	98.59	73.04	1.42
Seecer	12.1	110	98.49	98.31	7.65	0.90
MultiRes	**0.11**	**0.048**	95.01	98.17	**89.34**	**95.39**

The False positive/True Positive ratios (FP/TP ratios), Recall, and Precision are compared on two HIV datasets for the methods: Quake, BLESS, Musket, BFC, BayesHammer, Seecer, and the proposed method MultiRes. The error corrected reads from each method are broken into k-mers and compared to the true k-mers in the HIV-1 viral populations. Uncorrected denotes the statistics when no error correction is performed. Bold in each column indicates the best method for the dataset and the metric evaluated.

3.1.2. Generalizability: Testing MultiRes on a Hepatitis C virus dataset

We also evaluate our method on reads simulated from viral populations consisting of the E1/E2 gene of Hepatitis C virus (HCV). The purpose of using HCV strains is to understand the generalization of the MultiRes classifier on other viral population datasets. Two HCV populations observed in patients in previous studies are used as simulated viral populations. The first, denoted as HCV 1, consists of 36 HCV strains from E1/E2 region and are of length 1672 bps [33]. The second, denoted as HCV 2, consists of 44 HCV strains from the E1/E2 regions of the HCV genome with lengths 1734 bps [8,17]. We simulate 500 K Illumina paired end reads from both datasets under a power law (with ratio 2) of reads distribution amongst the strains [34]. The two simulated datasets are denoted as HCV1P and HCV2P respectively. The power law distribution of reads also helps in evaluating the performance of MultiRes when more than 50% of the haplotypes are present at less than 5% relative abundances.

All methods have recall greater than 90% on both datasets (Table 2). Again, the difference between MultiRes and other methods is evident from the FP/TP ratios and precision. The false positive to true positive ratios for MultiRes are less than other methods at least by a factor of 5 (Table 2). MultiRes still outperforms all other methods on predicting the smallest set of predicted k-mers while maintaining high recall levels of true k-mers.

The recall for MultiRes is respectively 96% and 94% on HCV1P and HCV2P datasets, which is less than the method Seecer that has recall values around 99%. Seecer marks more than 90% of the observed k-mers as true, which explains the high recall values. However, this also leads to a large number of false positive k-mers being predicted as true k-mers in Seecer, leading to low precision values. All other methods also achieve high recall by retention of all large fraction of observed k-mers, as indicated by their precision values being less than 1% and false positive to true positive ratios being greater than 100.

The similar performance of MultiRes on a dataset, such as the HCV population, which is diverse in genome composition from the simulated HIV-1 sequences used in simulation indicates the generalizability of the Random Forest Classifier in MultiRes. The classifier is capturing properties of the Illumina sequencing platform and the fact that both datasets contain a large number of rare variants occurring at k-mer counts close to the sequencing errors. Thus, MultiRes can be used as it is for error and rare variant detection in diverse datasets.

As the performance of MultiRes on HCV population is not as impressive as on the HIV simulated populations, it is also important to understand the cause for this decrease in performance. It is possible that the decrease in performance is correlated to the large number of low-frequency variants that are being misclassified by MultiRes. In order to test this, we investigate MultiRes' classification as a function of the count of the 35-mer which is being classified. MultiRes predicts about one-fourth of the observed k-mers as rare

variants for k-mer counts less than 15, and predicts more than 99% all of the observed k-mers as true for counts greater 20 (Fig. 2 (a)). This suggests that MultiRes predicts rare variant k-mers for all observed counts and detects more rare variant k-mers than a method based on a single threshold.

Most of the k-mer based error correction methods use a single threshold over the k-mer counts, which will clearly lose true rare variant k-mers (Fig. 2 (b)). On the other hand, MultiRes has a recall of 50% for k-mers observed 3 times, while still correctly identifying more than 75% of the k-mers as erroneous. The recall of MultiRes increases to 100% as the counts of the observed k-mers increases to 35. This indicates the importance of not having a single threshold for distinguishing between sequencing errors and rare variants in viral

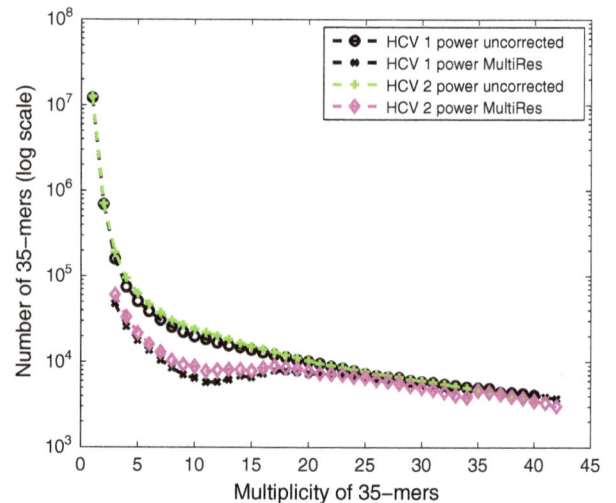

(a) Total k-mer Multiplicity Plots

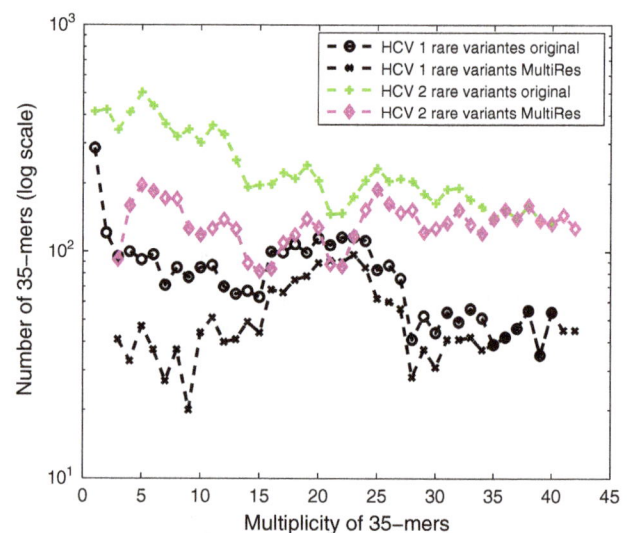

(b) True rare Variants Multiplicity plots

Fig. 2. Performance of MultiRes on HCV datasets under power law distributions of viral haplotypes with respect to count of k-mer. 35-mer multiplicity plots for HCV1P and HCV2P datasets are shown. x-axis indicates the number of times a 35-mer was observed while y-axis indicates the number of 35-mers at a count. (a) The predicted true 35-mers from MultiRes (HCV1P red, HCV2P pink) compared to the uncorrected data (HCV1P blue, HCV2P green), and (b) The true positive rare variants 35-mers from MultiRes (HCV1P red, HCV2P pink) versus the ground truth 35-mers (HCV1P red, HCV2P pink). MultiRes predicts rare variants k-mers at all counts greater than 3, with its accuracy improving as counts of k-mer increases.

Table 2

Comparison of performance metrics of different methods on HCV population datasets.

Algorithm	FP/TP ratio		Recall		Precision	
	HCV1P	HCV2P	HCV1P	HCV2P	HCV1P	HCV2P
Uncorrected	1201	571	99.51	99.88	0.08	0.17
Quake	303.3	149	96.41	97.23	0.32	0.66
BLESS	202	112	98.35	97.18	0.49	0.88
Musket	938	463	93.53	89.17	0.10	0.21
BFC	352	161	99.32	99.84	0.28	0.61
BayesHammer	699	340	98.12	97.1	0.14	0.29
Seecer	1095	528	**99.48**	**99.85**	0.09	0.19
MultiRes	**37.4**	**19.54**	96.5	94.25	**2.6**	**4.87**

The false positive to true positive ratios, Recall, and Precision of error correction methods on the two simulated HCV datasets are shown. Uncorrected refers to the statistics when no error correction is performed. Bold font in each column indicates the best method for each dataset on the evaluated measure.

population datasets, and our MultiRes bypasses a single threshold by training a Random Forest classifier.

3.1.3. Evaluation on population of 5 HIV-1 sequences

We also evaluate MultiRes on a laboratory mixture of five known HIV-1 strains [35], which captures the variability occurring during sample preparation, errors introduced in a real sequencing project, and mutations occurring during reverse transcription of RNA samples. Five HIV-1 strains (named YU2, HXB2, NL43, 89.6, and JRCSF) of lengths 9.1 kb were pooled and sequenced using Illumina paired end sequencing technology (Refer to [35] for details). Each HIV strain was also sequenced separately in their study and aligned to their known reference sequence (from Genbank) to generate a consensus sequence for each HIV-1 strain [35]. This provides us with a dataset of actual sequence reads where the ground truth is known allowing us to assess the performance of MultiRes and other methods. We extracted 35-mers from the paired end sequencing data and classify them using the Random Forest classifier of MultiRes trained on the simulated HIV sequencing data.

All the error correction methods and MultiRes have recall values around 97%, indicating that the performance for recovery of true k-mers is comparable across all methods (Table 3). The false positive to true positive ratio for MultiRes is 13 while all other methods have ratios more than 120. MultiRes predicts 359 thousand unique k-mers in the set of true k-mers while all other methods predict more than 5 million unique k-mers. Even methods that take variance in sequencing depths while performing error correction, such as BayesHammer and Seecer, predict 11.3 million and 6.3 million unique k-mers which is two orders more than the ground truth number of k-mers in the consensus sequence of the 5 HIV-1 strains (53 thousand unique k-mers). Thus, even considering the artifacts introduced in sequencing, MultiRes has by far the most compact set of predicted error free k-mers amongst all methods while retaining high number of true k-mers. As mentioned earlier, as the number of k-mers linearly affects the memory requirements for downstream *de novo* assembly methods, the error detection from MultiRes would translate to a 10-fold reduction in memory.

3.1.4. Runtime and memory

MultiRes has comparable running times to BayesHammer on the five-viral mix dataset (Fig. 3) on a Dell system with 8 GB main memory, and 2X Dual Core AMD Opteron 2216 CPU type. The

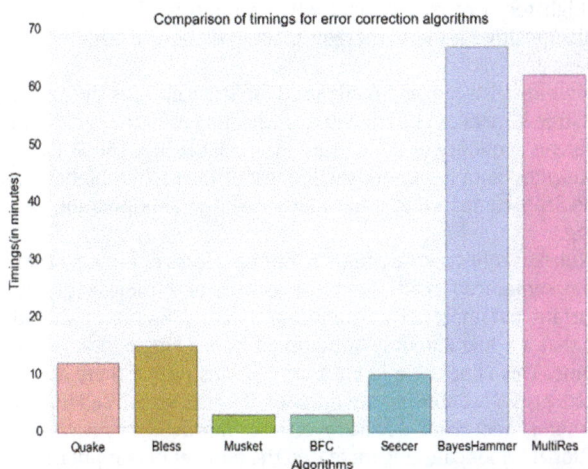

Fig. 3. Runtime comparison on five-viral mix dataset. Comparison of running times for different algorithms on 5-viral mix dataset on 8GB memory nodes of 2X Dual Core AMD Opteron 2216 systems from Dell. The time noted for BayesHammer is only the time reported for BayesHammer error correction step in SPADES (version 3.6.2). The time reported for MultiRes is the combined time for k-mer counting, predicting k-mers as erroneous and rare variants and generating the final output.

performance on all other datasets was similar indicating that the timings are comparable. Additionally, while other methods have parallel implementations, the error detection classifier step in MultiRes is a single thread serial implementation. As the random forest classifier used by MultiRes is already trained and independent of the input k-mers for classification, the runtime of MultiRes can be significantly improved via parallelization of the k-mer classification step.

3.2. Comparison of MultiRes to variant calling methods for viral populations

As one of the objectives in NGS studies of viruses is to identify the single nucleotide polymorphisms (SNPs) in a population [2,7,9] which is sensitive to erroneous reads, we evaluate the inference of SNPs from the k-mers predicted by MultiRes, and compare it to known SNP profiling methods for viral populations. We first align the predicted k-mers from MultiRes to a reference sequence of the viral population using the bwa mem aligner and a base is called as a SNP when its relative fraction amongst the k-mers aligned at that position is greater than 0.01. All the variants that occur at a frequency greater than the error threshold at that position are reported as SNPs. The choice of the reference sequence is based on the viral population data being evaluated, and the same reference sequence is used for calling true SNPs and the SNPs predicted by a method.

Each SNP detected at a base position of the reference and detection of the reference base itself are treated as *true positives* for a method; thus the number of true positives can be greater than the length of the reference sequence. All the SNPs predicted by a method and the number of bases mapped to the reference sequence are known as the *total SNP predictions* of a method. We use three measures for evaluating the SNPs called by any method. *Precision* is defined as the ratio of the number of *true positives* to the *total SNP predictions* made by a method, while *recall* is defined as the ratio of the *true positives* to the total number of SNPs and reference bases in the viral population. Finally, *false positive to true positive ratio* is a ratio of the number of false SNP predictions to the number of *true positives* detected by a method.

We compare our results to state-of-the-art variant calling methods for viral populations VPhaser-2 [9], a rare variant calling method LoFreq [10], and viral haplotype reconstruction algorithm ShoRAH [7] using the above three measures. The reference sequence used by variant calling methods VPhaser-2 and LoFreq is the same as that used by *samtools* to determine the true SNPs, while the SNPs predicted by ShoRAH at default parameters are compared directly to the true SNPs. We only used the SNP calls from VPhaser-2 for evaluation, as length polymorphisms are not generated by the other methods, but the results from VPhaser-2 were not penalized when comparing the SNPs.

We report results for LoFreq [10], VPhaser [9], ShoRAH [7] and our method MultiRes on all datasets (Table 4). Overall, MultiRes has greater than 94% recall and precision values greater than 83% in

Table 3
Comparison of performance metrics on 5-viral mix HIV-1 dataset.

Algorithm	Recall	Precision	FP/TP ratio	# of unique 35-mers
Uncorrected	98.01	0.2	439	11.4 M
BLESS	97.31	0.4	227	5.89 M
Musket	**97.91**	0.3	366	11.2 M
BFC	97.55	0.3	316	9.6 M
BayesHammer	97.49	0.8	122	6.3M
Seecer	97.84	0.5	220	11.3M
MultiRes	96.64	7.1	**13**	**359 K**

The recall, precision, and FP/TP ratios of each method are evaluated on the 5-viral mix HIV-1 dataset. The number of unique 35-mers indicates the number of unique 35-mers predicted by a method. There are 53 thousand true unique 35-mers in the consensus sequences of the 5 viral strains. Bold indicates the best method for the measure in each column.

Table 4
Comparison with Variant Calling methods on all datasets.

Dataset	Method	Recall (%)	FP/TP ratio	Precision (%)	# of False negatives	Mapped reads (%)
HIV 100x	LoFreq	97.33	0.004	99.60	444	89.51
	Vphaser	98.90	0.007	99.26	183	89.51
	ShoRAH	55.21	**0**	**100**	7746	**98.04**
	MultiRes	**99.69**	0.011	98.88	**51**	97.89
HIV 400x	LoFreq	84.83	0	99.99	2522	99.55
	Vphaser	**95.92**	0.292	77.37	**678**	99.55
	ShoRAH	55.21	**0**	**100**	7746	**99.95**
	MultiRes	95.57	0.007	99.33	736	97.34
HCV1P	LoFreq	**98.30**	1.282	43.82	**31**	**99.99**
	Vphaser	93.51	1.628	38.05	118	**99.99**
	ShoRAH	91.92	**0**	**100**	147	**99.99**
	MultiRes	98.24	0.597	62.64	32	97.32
HCV2P	LoFreq	97.10	1.046	48.87	60	**100**
	Vphaser	95.65	1.492	40.13	90	**100**
	ShoRAH	83.73	**0**	**100**	337	99.95
	MultiRes	**98.79**	0.201	83.27	**25**	85.14
5-viral mix	LoFreq	99.06	0.085	92.15	101	98.59
	Vphaser	92.68	0.039	96.25	789	98.59
	ShoRAH	98.66	**0.014**	**98.99**	109	**99.3**
	MultiRes	**99.39**	0.077	92.82	**66**	96.29

The Recall, false positive to true positive ratios (FP/TP), Precision, number of false negatives, and % of mapped reads by methods LoFreq, VPhaser-2, ShoRAH, and MultiRes are computed for listed datasets. All reads from a sample were aligned using bwa-mem tool for LoFreq and VPhaser-2 under default settings. ShoRAH uses its own aligner for read alignment and variant calling, while k-mers detected by MultiRes were aligned using bwa-mem for MultiRes. Outputs from LoFreq (version 2.1.2), VPhaser-2 (last downloaded version October 2015), and ShoRAH (last downloaded version from November 2013) are compared against known variants for simulated datasets. For 5-viral mix, the consensus reference provided by [35] was used to determine ground truth variants. MultiRes variants are determined by aligning 35-mers to a reference sequence and bases occurring at more than 0.01 frequency as variants. Bold for each dataset indicates the best method for the performance measures.

all datasets. LoFreq and VPhaser have comparable recall but lower precision values and an increase in the FP/TP ratios on the HCV population datasets, indicating a decrease in performance. ShoRAH overall has lower recall values, nevertheless a 100% precision in all but the 5-viral mix dataset, suggesting that it misses true SNPs but is very accurate when it calls a base as SNP. Overall all methods have low values for FP/TP ratio as compared to before, indicating that the number of false positive SNP predictions are low. The metric where MultiRes outperforms others is the lowest number of true SNPs missed. This shows that even with a simplistic SNP prediction method used in MultiRes, it is able to capture the true variation of the sampled viral population and has the lowest false negatives of well established methods. This demonstrates that using error-free set of k-mers can vastly increase the variant detection in viral populations.

The number of reads or k-mers aligned to the reference sequence are comparable across the methods, except for HCV2P dataset where MultiRes has 85% k-mers mapped compared to 100% read mapping (Table 4). It is possible that the unmapped k-mers correspond to the length variants and could be verified by haplotype reconstruction using the predicted k-mers, but that was not the focus in this paper.

4. Discussion and conclusions

We have proposed a classifier MultiRes for detecting rare variant and erroneous k-mers obtained from Illumina sequencing of viral populations. Our method does not rely on a reference sequence and uses concepts from signal processing to justify using the counts of sets of k-mers of different sizes. We utilize the projections of sampled reads signals onto multiple **frames** as features for our classifier MultiRes.

We demonstrated the performance of MultiRes on simulated HIV and HCV viral populations and real HIV viral populations containing viral haplotypes at varying relative frequencies, where it outperformed the error detection algorithms used in error correction methods in terms of recall and the total number of predicted k-mers. Though, the error detection algorithms in the error correction methods evaluated assumed that sequenced reads originated from a single genome sequenced at uniform coverage, our method also works better than the method BayesHammer, which can tackle

non-uniform sequencing coverage, and the method Seecer, which additionally incorporates methods for detecting alternative splicing and polymorphisms.

The error-free k-mers predicted by MultiRes enable the usage of *de novo* assembly methods for viral genomes. A major challenge for using De Bruijn graph based methods for viral populations has been the increased complexity of the graph due to the presence of large number of sequencing errors [36]. Moreover, the memory footprint of a De Bruijn assembly graph increases linearly with the number of k-mers in the NGS data. Thus the low false positives along with high recall of k-mers predicted by MultiRes drastically reduce the memory requirements for De Bruijn graphs. An edge-centric De Bruijn graph of size $k - 1$ can be directly generated from error free k-mers, such as in *de novo* assembly tools SPADES, Cortex [37,38] for reconstruction of viral haplotypes in a viral population. The graph can be used for calling structural variants in the viral population data. MultiRes has high recall of true k-mers while outputting the least number of false positive k-mers, thereby making *de novo* assembly graphs manageable.

MultiRes also can be directly used for SNP calling as the predicted error-free k-mers can be aligned to an existing reference genome or a consensus sequence of the current viral population. The SNPs called by MultiRes' data has either the highest or the second highest recall of the SNPs compared to other methods for viral population variant calling.

MultiRes relies on the counts of multiple sizes of k-mers observed in the sequenced reads, and the choice of k-mer length is an important parameter. The minimum value of k chosen should be such that a k-mer can only be sampled from a single location in the genome. This is possible in viral populations where there are small repeats present. Choosing the number of k-mer sizes used is another parameter, and while accuracy can be improved by increasing it, additional k-mer counting increased the number of computations. As demonstrated by our experiments, choosing three different values of k, namely $(k, 2 \cdot k, 3 \cdot k)$ was sufficient for accurate results.

MultiRes also has applications for studying the large scale variation in closely related genomes, including as viral populations. The complexity of De Bruijn graphs, useful for studying structural variants and rearrangements in the population, increases because of sequencing errors. Our method can provide a compact set of k-mers while still retaining high recall of the true k-mers, which can be utilized for constructing the graph. Additionally, the error-free k-mers predicted by MultiRes can be directly used for understanding the SNPs observed in the viral population to a high degree of accuracy.

MultiRes' classifier also has its limitations. The model, although trained to model the features of an Illumina sequencing machine, does have a decreased performance on different viral populations with a large number of rare variants, as is evident from its 50% accuracy for HCV2P population for k-mers observed only 3 times. Although it is able to eliminate a large number of false positive k-mers (more than 75% of k-mers at counts of 3), the classifier model can be improved with additional training data and an ensemble of classifier models.

MultiRes was primarily developed for detection of sequencing errors and rare variants in viral populations, which have small genomes. Extending our method for larger genomes may require additional tuning of the parameters via re-training of the classifier, but the concepts developed here are applicable to studying variation in closely related genomes such as cancer cell lines. It is also applicable for understanding somatic variation in sequences as their variation frequency is close to the sequencing error rates. The technique can also be explored for newer sequencing machines, such as PacBio sequences and Oxford Nanopore long read sequencing, where the type of sequencing errors are different, but the concepts of projections of signals are still applicable. The software is available for download from the github link (https://github.com/raunaq-m/MultiRes).

Acknowledgments

This research was funded by the National Science Foundation (NSF) Awards #1421908, 1724008 and 1720635.

References

[1] Nguyen DX, Massagué J. Genetic determinants of cancer metastasis. Nat Rev Genet 2007;8(5):341–52.

[2] McElroy K, Thomas T, Luciani F. Deep sequencing of evolving pathogen populations: applications, errors, and bioinformatic solutions. Microb Inform Exp 2014;4(1).

[3] Beerenwinkel N, Gunthard HF, Roth V, Metzner KJ. Challenges and opportunities in estimating viral genetic diversity from next-generation sequencing data. Front Microbiol 2012;3(329). https://doi.org/10.3389/fmicb.2012.00329.

[4] Schirmer M, Ijaz UZ, D'Amore R, Hall N, Sloan WT, Quince C. Insight into biases and sequencing errors for amplicon sequencing with the illumina miseq platform. Nucleic Acids Res 2015;43(6):e37. https://doi.org/10.1093/nar/gku1341.

[5] Meacham F, Boffelli D, Dhahbi J, Martin DI, Singer M, Pachter L. Identification and correction of systematic error in high-throughput sequence data. BMC Bioinf 2011;12(1):451.

[6] Töpfer A, Zagordi O, Prabhakaran S, Roth V, Halperin E, Beerenwinkel N. Probabilistic inference of viral quasispecies subject to recombination. J Comput Biol 2013;20(2):113–23.

[7] Zagordi O, Bhattacharya A, Eriksson N, Beerenwinkel N. Shorah: estimating the genetic diversity of a mixed sample from next-generation sequencing data. BMC Bioinf 2011;12(1):119.

[8] Mangul S, Wu NC, Mancuso N, Zelikovsky A, Sun R, Eskin E. Accurate viral population assembly from ultra-deep sequencing data. Bioinformatics 2014;30(12):i329–i337. https://doi.org/10.1093/bioinformatics/btu295.

[9] Yang X, Charlebois P, Macalalad A, Henn M, Zody M. V-phaser 2: variant inference for viral populations. BMC Genomics 2013;14(1):674. https://doi.org/10.1186/1471-2164-14-674.

[10] Wilm A, Aw PPK, Bertrand D, Yeo GHT, Ong SH, Wong CH. et al. Lofreq: a sequence-quality aware, ultra-sensitive variant caller for uncovering cell-population heterogeneity from high-throughput sequencing datasets. Nucleic Acids Res 2012.

[11] Töpfer A, Marschall T, Bull RA, Luciani F, Schönhuth A, Beerenwinkel N. Viral quasispecies assembly via maximal clique enumeration. PLoS Comput Biol 2014;10(3). https://doi.org/10.1371/journal.pcbi.1003515.

[12] Kelley DR, Schatz MC, Salzberg SL. et al. Quake: quality-aware detection and correction of sequencing errors. Genome Biol 2010;11(11):R116.

[13] Heo Y, Wu X-L, Chen D, Ma J, Hwu W-M. Bless: bloom filter-based error correction solution for high-throughput sequencing reads. Bioinformatics 2014.

[14] Li H. Bfc: correcting illumina sequencing errors. Bioinformatics 2015; https://doi.org/10.1093/bioinformatics/btv290.

[15] Liu Y, Schröder J, Schmidt B. Musket: a multistage k-mer spectrum-based error corrector for illumina sequence data. Bioinformatics 2013;29(3):308–15. https://doi.org/10.1093/bioinformatics/bts690.

[16] Medvedev P, Scott E, Kakaradov B, Pevzner PA. Error correction of high-throughput sequencing datasets with non-uniform coverage. Bioinformatics [ISMB/ECCB] 2011;27(13):137–41.

[17] Skums P, Dimitrova Z, Campo DS, Vaughan G, Rossi L, Forbi JC. et al. Efficient error correction for next-generation sequencing of viral amplicons. BMC Bioinf 2012;13(Suppl 10):S6.

[18] Rizk G, Lavenier D, Chikhi R. Dsk: k-mer counting with very low memory usage. Bioinformatics 2013;29(5):652–3. https://doi.org/10.1093/bioinformatics/btt020.

[19] Deorowicz S, Kokot M, Grabowski S, Debudaj-Grabysz A. Kmc 2: fast and resource-frugal k-mer counting. Bioinformatics 2015;31(10):1569–76.

[20] Chikhi R, Medvedev P. Informed and automated k-mer size selection for genome assembly. Bioinformatics 2013.

[21] Feng S, Lo C-C, Li P-E, Chain PS. Adept, a dynamic next generation sequencing data error-detection program with trimming. BMC Bioinf 2016;17(1):109.

[22] Ding J, Bashashati A, Roth A, Oloumi A, Tse K, Zeng T. et al. Feature-based classifiers for somatic mutation detection in tumour-normal paired sequencing data. Bioinformatics 2011;28(2):167–75.

[23] Poplin R, Newburger D, Dijamco J, Nguyen N, Loy D, Gross SS. et al. Creating a universal snp and small indel variant caller with deep neural networks. BioRxiv 2016;092890.

[24] Ferreira P. Mathematics for multimedia signal processing II: discrete finite frames and signal reconstruction. Nato ASI Series of Computer and Systems Sciences 1999;174:35–54.

[25] Duffin RJ, Schaeffer AC. A class of nonharmonic fourier series. Trans Am Math Soc 1952;341–66.

[26] Daubechies I, Grossmann A, Meyer Y. Painless nonorthogonal expansions. J Math Phys 1986;27(5):1271–83.

[27] Daubechies I, Han B, Ron A, Shen Z. Framelets: MRA-based constructions of wavelet frames. Appl Comput Harmon Anal 2003;14(1):1–46.

[28] Unser M. Texture classification and segmentation using wavelet frames. IEEE Trans Image Process 1995;4(11):1549–60.

[29] Ron A, Shen Z. Frames and stable bases for shift-invariant subspaces of $l^2(r^d)$. Can J Math 1995;47(5):1051–94.

[30] Nikolenko SI, Korobeynikov AI, Alekseyev MA. Bayeshammer: Bayesian clustering for error correction in single-cell sequencing. BMC Genomics 2013;14(Suppl 1):S7.

[31] Le H-S, Schulz MH, McCauley BM, Hinman VF, Bar-Joseph Z. Probabilistic error correction for rna sequencing. Nucleic Acids Res 2013.

[32] Kaiser G. A friendly guide to wavelets. Springer Science & Business Media; 2010.

[33] Hussein N, Zekri A-RN, Abouelhoda M, El-din HMA, Ghamry AA, Amer MA. et al. New insight into hcv e1/e2 region of genotype 4a. Virol J 2014;11(1):2512.

[34] Angly FE, Willner D, Rohwer F, Hugenholtz P, Tyson GW. Grinder: a versatile amplicon and shotgun sequence simulator. Nucleic Acids Res 2012; http://nar.oxfordjournals.org/content/early/2012/03/19/nar.gks251.abstract. https://doi.org/10.1093/nar/gks251.

[35] Giallonardo FD, Töpfer A, Rey M, Prabhakaran S, Duport Y, Leemann C. et al. Full-length haplotype reconstruction to infer the structure of heterogeneous virus populations. Nucleic Acids Res 2014;42(14):e115. arXiv:http://nar.oxfordjournals.org/content/42/14/e115.abstract. https://doi.org/10.1093/nar/gku537.

[36] Yang X, Charlebois P, Gnerre S, Coole MG, Lennon NJ, Levin JZ. et al. De novo assembly of highly diverse viral populations. BMC Genomics 2012;13(1):475.

[37] Bankevich A, Nurk S, Antipov D, Gurevich AA, Dvorkin M, Kulikov AS. et al. Spades: a new genome assembly algorithm and its applications to single-cell sequencing. J Comput Biol 2012;19(5):455–77.

[38] Iqbal Z, Caccamo M, Turner I, Flicek P, McVean G. De novo assembly and genotyping of variants using colored de Bruijn graphs. Nat Genet 2012.

4

Meta-analysis of Liver and Heart Transcriptomic Data for Functional Annotation Transfer in Mammalian Orthologs

Pía Francesca Loren Reyes, Tom Michoel, Anagha Joshi*, Guillaume Devailly*

The Roslin Institute, The University of Edinburgh, Easter Bush, Midlothian, EH25 9RG, Scotland, UK

ARTICLE INFO

Keywords:
Gene function
Transcriptomics
Liver
Heart
Orthologs
Paralogs
Co-expression
Gene networks

ABSTRACT

Functional annotation transfer across multi-gene family orthologs can lead to functional misannotations. We hypothesised that co-expression network will help predict functional orthologs amongst complex homologous gene families. To explore the use of transcriptomic data available in public domain to identify functionally equivalent ones from all predicted orthologs, we collected genome wide expression data in mouse and rat liver from over 1500 experiments with varied treatments. We used a hyper-graph clustering method to identify clusters of orthologous genes co-expressed in both mouse and rat. We validated these clusters by analysing expression profiles in each species separately, and demonstrating a high overlap. We then focused on genes in 18 homology groups with one-to-many or many-to-many relationships between two species, to discriminate between functionally equivalent and non-equivalent orthologs. Finally, we further applied our method by collecting heart transcriptomic data (over 1400 experiments) in rat and mouse to validate the method in an independent tissue.

1. Introduction

Annotation of gene function is a crucial step to understand the DNA sequencing data currently generated at an unprecedented rate. The lack of functional annotation forms a major bottleneck in analyses across diverse fields, including de novo genome sequencing [1], Genome Wide Association Studies (GWAS) in model and non-model organisms [2], and metagenomics [3]. An experimental validation of each gene is impractical to this end as it demands high financial and time cost. It is estimated that only 1% of proteins have experimental functional annotations [4]. Bioinformatic approaches therefore provide an attractive alternative [5]. The most widely used and successful gene annotation strategy has been the annotation transfer between homologous genes. Automated annotation pipelines from sequence alone are widely used, including GOtcha [6] and BlastGO [7]. They allow fast annotation of thousands of genes for newly sequenced genomes [8]. This approach can be used within a species, where gene families (paralogs), might share common functions, or across species, where known function(s) of a gene in one species are used to infer functions of the homologous gene(s) in another species.

Despite being widely used, fast computational annotation comes at a cost of misannotation, which is present at high levels (over 10%) and is believed to be increasing [9] due to misannotation transfer. The most common misannotation is over-annotation, where a gene is assigned a specific but incorrect function [10]. This is partly because one of the major challenges in functional annotation transfer across species is that the orthology relationships are not always one-to-one. Specifically, a single gene in one species can be homologous to multiple paralogs in another (one-to-many homologies), after gene duplication or gene loss event(s). After a gene duplication, the two paralogs can have redundant functions, and thus should share similar functional annotations, or one copy might diverge (lose functionality, or gain new functionalities, or change cellular localisation or tissue specificity), and thus paralogs should have different functional annotations despite their homology. Similarly, multigene families (with many-to-many homologies) are highly prone to over-annotation errors.

Protein structure information can act as source for functional distinction within multigene family proteins [4]. Protein-protein interaction networks have also been successfully used to identify functional orthologs [11]; two orthologs interacting with the same proteins in each species are likely to share similar functions.

* Corresponding authors.
  E-mail addresses: Anagha.Joshi@roslin.ed.ac.uk (A. Joshi), Guillaume.Devailly@roslin.ed.ac.uk (G. Devailly).

Similar strategy has been applied to biochemical pathway information [12]. Co-expression gene networks have also been used in this context [13–15], as they offer two main advantages over protein-protein interactions and biochemical pathways. First, they can be inferred from transcriptomic datasets, which are more abundant than protein-protein interaction datasets. Second, they allow functional annotation of the various classes of RNA genes. We have previously shown that multi-species information improves gene network reconstruction [16].

In order to further explore the potential of co-expressed gene networks to identify functional equivalents in complex homologous families, we collected transcriptomic data from mouse and rat liver samples. To minimise technical variation, we collected datasets generated using a single microarray platform in each species, resulting into 920 experiments in mouse and 620 experiments in rat. We firstly identified clusters of co-expressed genes using hierarchical clustering and found biologically relevant clusters. We applied an hyper-graph clustering method, SCHype [17] to simultaneously cluster co-expressed orthologous genes between species. We then focussed on 18 complex (one-to-many or many-to-many) homology groups, where at least one member in mouse and in rat where present in similar co-regulated gene clusters providing an independent source of evidence for shared functionality amongst orthologous genes in complex homologous families. We successfully applied the same method on heart transcriptomic data from mouse and rat, and investigated functional relevance of 11 other orthologous groups. Our results show the potential of this method to use co-expression as an independent measure to evaluate shared functionality amongst orthologs and limit over-zealous annotation transfers.

2. Methods

2.1. Data Collection and Normalisation

Microarray data for liver and heart samples in mouse and rat were collected from GEO, where data for mouse was generated using Affymetrix Mouse Genome 430 2.0 Array, and data for rat was generated using Affymetrix Rat Genome 230 2.0 Array as they were the platforms with a large number of experiments available for each species. Liver experiments came from 62 (mouse) and 28 (rat) independent studies or GEO series. Heart experiments came from 20 (mouse) and 19 (rat) independent studies or GEO series. The GEO accession numbers for individual studies are provided in Supplementary Table 1. Processed data was not directly comparable between studies, as different studies used different normalisation methods, leading to different distribution of values (Supplementary Fig. 1, A and B, Supplementary Fig. 3, A and B). As some datasets had a trimmed lower quartile for reduction in noise by limiting the variability of lowly expressed genes, we applied lower quartile trimming on all datasets (Supplementary Fig. 1, C and D, Supplementary Fig. 3, C and D). Specifically, we set the expression value of all probes belonging to the lower quartile to the value of the 25 percentile. We then applied quantile normalisation resulting into a uniform distribution of values for each experiment. To facilitate the comparison between mouse and rat data, we used liver mouse data as a target for quantile normalisation of heart mouse data and liver and heart rat data, using preprocessCore functions `normalize.quantiles.determine.target` and `normalize.quantiles.use.target` [18]. Liver mouse data was selected as the target because it contained more experiments than the liver rat dataset. Thus, after our normalisation steps, the distribution of values was identical for each experiment in both species.

2.2. Data Clustering

We selected genes with variable expression across experiments by selecting probes with a standard deviation greater than one across experiments. As shown in Fig. 1, such probes included genes of low as well as high expression levels, and largely excluded probes showing very low expression in all experiments. Microarray data being already log-transformed, log fold change over the average values were obtained by subtracting the mean expression of each probes.

Hierarchical clustering was done on the log fold change matrices using R functions `dist` ad `hclust` with default parameters (euclidean distance, complete linkage). Dendrogram branches were reordered using the function `order.optimal` from the cba package [19]. Both rows (probes) and columns (experiments) were clustered using this approach.

Gene homology information was retrieved from the Homologen database [20], and probe orthology information was obtained using the R package annotationTools [21]. Due to one-to-many homologs, rat probes and mouse probes intersections resulted into slightly different numbers for each species. Average of the two numbers was used to obtain Jaccard indexes. Jaccard index significance was obtained using the hypergeometric test, and P-values were corrected for multiple testing using Bonferroni correction.

SCHype takes as input a list of conserved interactions which was generated as follows. First Spearman correlation coefficient between each pair of probes was obtained independently for both Mouse and Rat expression data. Pairs of probes with a correlation coefficient ≥ 0.5 were selected. Then if orthologs of two connected probes were connected in the other species, they were kept as an SCHype input. SCHype was run using default parameters. In liver, SCHype identified 132 clusters of homologous genes co-expressed both in mouse and in rat, which included 825 nodes in mouse and 778 nodes in rat. SCHype allows probes to be included in multiple clusters. The different number of probes in mouse and rat is due to the presence of one-to-many and many-to-many orthologs, as well as the presence of gene measured by multiple probes on the array.

2.3. Gene Ontology Analysis

Gene ontology analysis was performed using PantherDB [22], using as a control gene set the genes analysed by the microarray, or only the variable gene sets previously defined.

2.4. Scripts and Data Availability

R scripts used for this analysis are available in a Github repository https://github.com/gdevailly/liver_mouse_rat. Normalised expression matrices, fold change matrices, as well as probe clusters (hierarchical clustering and SCHype clustering) are available through

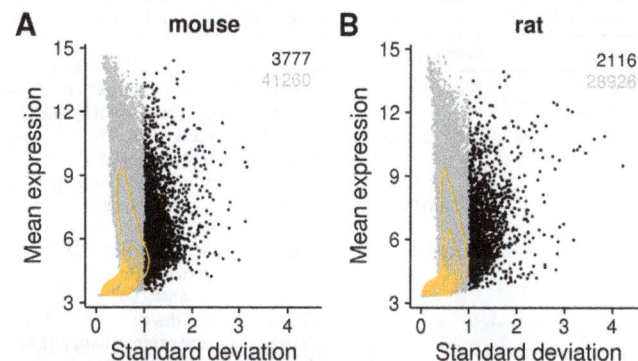

Fig. 1. Identification of variable probes in mouse (A) and rat (B) datasets. Each dot represents a single probe. X axis: standard deviation across experiments. Y-axis: mean expression values across experiments (in arbitrary units). In black the probes with a standard deviation ≥ 1, in grey the probes with a standard deviation < 1. Orange lines: 2D kernel density.

two Zenodo collections: https://zenodo.org/record/439483 (liver data) and https://zenodo.org/record/839015 (heart data).

3. Results

3.1. Identification of Variable Genes Across Datasets

We downloaded 920 and 620 experiments for gene expression data in rat and mouse liver from the GEO database. We firstly normalised the data using lower quartile trimming (Supplementary Fig. 1, C and D) and quantile normalisation (Supplementary Fig. 1, E and F) independently for each species. We then selected the probes with dynamic expression across samples (standard deviation ≥ 1). This resulted into 3777 probes in mouse (8.4%) and 2116 probes in rat (6.8%), with a wide range of expression values (Fig. 1). 735 mouse variable probes out of 3777 had a homologue in rat variable probes, and 624 rat variable probes out of 2116 had a homologue in mouse variable probes. Variable genes were enriched for pathways and functions related to liver biology (Table 1), including metabolism of lipid an protein (rat, adjusted P value $\leq 10^{-4}$), regulation of cholesterol biosynthesis by SREBP (mouse and rat, respectively adjusted P value ≤ 0.01 and ≤ 0.03), synthesis of bile acid and salt via 24-hydroxycholesterol (rat, adjusted P value ≤ 0.03), and fatty acid metabolic process (mouse and rat, respectively adjusted P value $\leq 10^{-4}$ and ≤ 0.03). As the biological processes enriched in variable genes reflected functions associated with liver, we concluded that the expression variability across samples was due to biological variability, and not only technical variations, and therefore was of significance for further investigation.

3.2. Independent Hierarchical Clustering of Mouse and Rat Data

Hierarchical clustering was applied to the mouse and rat expression matrices independently (Fig. 2, A and B). We defined 7 major clusters of variable probes, while the experiments were grouped in 4 clusters. The two major clusters of experiments in mouse showed broadly opposite expression patterns (Fig. 2A). Two major experimental groups were also noted in rat, albeit to a lesser extent compared to mouse (Fig. 2B). Experiments were annotated according to their series of origin (Fig. 2A and B, bottom of the heatmap), revealing that most experiments from the same series grouped together (including cases and controls). Notably, no series of experiments were split in the two main experiment clusters.

We characterised the main experiment clusters by looking at the most different non-trivial terms in the element-term matrix

build from the metadata retrieved from GEO (characteristic field, Fig. 2, C and D). No clear difference between experiment clusters was observed in mouse. Experiment cluster 3 in rat seems to be composed mostly of F344 strains of rat and/or of rat treated with the microcystinlr toxin. To note, this cluster is dominated by experiments from a single experiment series (Fig. 2B). Since experiment clustering matched series of origin of the data, this hinders correction for batch effects to get biological differences.

Given that mouse and rat probes formed two major clusters anti-correlated with each other despite diverse experimental set ups in each species, we investigated whether the mouse and rat probe clusters were composed of probes measuring similar genes (Fig. 2E). We calculated the overlap between genes in each cluster in mouse with genes in each cluster in rat. Cluster 2 in mouse (golden colour, Fig. 2A) and cluster 2 in rat (golden colour, Fig. 2B) showed a very high overlap with the highest Jaccard index across all clusters. Neither mouse cluster 2 nor rat cluster 2 were enriched for any gene ontology term or reactome pathway terms, when using the set of variable probes as background. Most clusters did not show a very high genes overlap across species. This might be due to the fact that the experiments carried out in each species were different, resulting in distinct set of genes perturbed in each species, resulting into little overlap of co-expression clusters across species. Functional enrichment analyses of other clusters were suggestive that observed gene variations reflected differences in the liver physiology. Specifically, cluster 1 in mouse (claret red colour, Fig. 2A) was enriched for generation of precursor metabolites and energy (adjusted P value $\leq 10^{-6}$), steroid metabolic process (adjusted P value ≤ 0.001), fatty acid metabolic process (adjusted P value ≤ 0.001), and Cytochrome P450 - arranged by substrate type (adjusted P value $\leq 10^{-6}$). Cluster 3 in mouse (green colour) was enriched for arachidonic acid metabolic process (adjusted P value ≤ 0.01), icosanoid metabolic process (adjusted P value ≤ 0.05), fatty acid derivative metabolic process (adjusted P value ≤ 0.05), and Cytochrome P450 - arranged by substrate type (adjusted P value ≤ 0.05). Cluster 6 in rat (blue) was enriched for proteolysis (FE 10, adjusted P value ≤ 0.01). More terms related to the liver metabolism were enriched when the same analysis was performed using all genes as a background (Supplementary Table 2).

3.3. Co-clustering of Mouse and Rat Expression Data

To identify clusters of homologous probes between mouse and rat, we used the hyper-graph clustering tool SCHype [17]. SCHype uses a recursive spectral clustering algorithm to identify sets of

Table 1

Variable genes are enriched for categories and pathways related to liver functions. FE: fold enrichment between actual over expected number of genes. GO: gene ontology. BP: biological process. Only categories with a fold enrichment > 2 are shown. All P-values were corrected for multiple testing with the Bonferroni method.

Species	Category	Term	Gene	FE	P-value
Mouse	Reactome	Synthesis of (16-20)-hydroxyeicosatetraenoic acids (HETE)	11	4.78	4.29E−02
		Activation of gene expression by SREBF (SREBP)	15	4.34	5.18E−03
		Regulation of cholesterol biosynthesis by SREBP (SREBF)	17	3.94	4.36E−03
		Cytochrome P450 - arranged by substrate type	27	2.72	7.78E−03
		Phase 1 - Ff unctionalization of compounds	37	2.55	7.58E−04
	GO slim BP	Fatty acid metabolic process	52	2.26	2.95E−05
		Steroid metabolic process	50	2.18	1.31E−04
Rat	Reactome	Synthesis of bile acids and bile salts via 24-hydroxycholesterol	7	8.63	2.95E−02
		Endosomal/vacuolar pathway	10	7.93	1.15E−03
		Striated muscle contraction	11	6.78	1.48E−03
		ER-phagosome pathway	10	6.53	6.29E−03
		Activation of gene expression by SREBF (SREBP)	10	6.53	6.29E−03
		Antigen presentation: folding, assembly and peptide loading of class I MHC	13	6.27	3.84E−04
		Regulation of cholesterol biosynthesis by SREBP (SREBF)	10	5.55	2.51E−02
		Biological oxidations	25	2.95	3.58E−03
		Metabolism of lipids and lipoproteins	68	2.13	1.15E−05
	GO slim BP	Response to biotic stimulus	12	4.16	1.12E−02
		Fatty acid metabolic process	22	2.52	2.52E−02

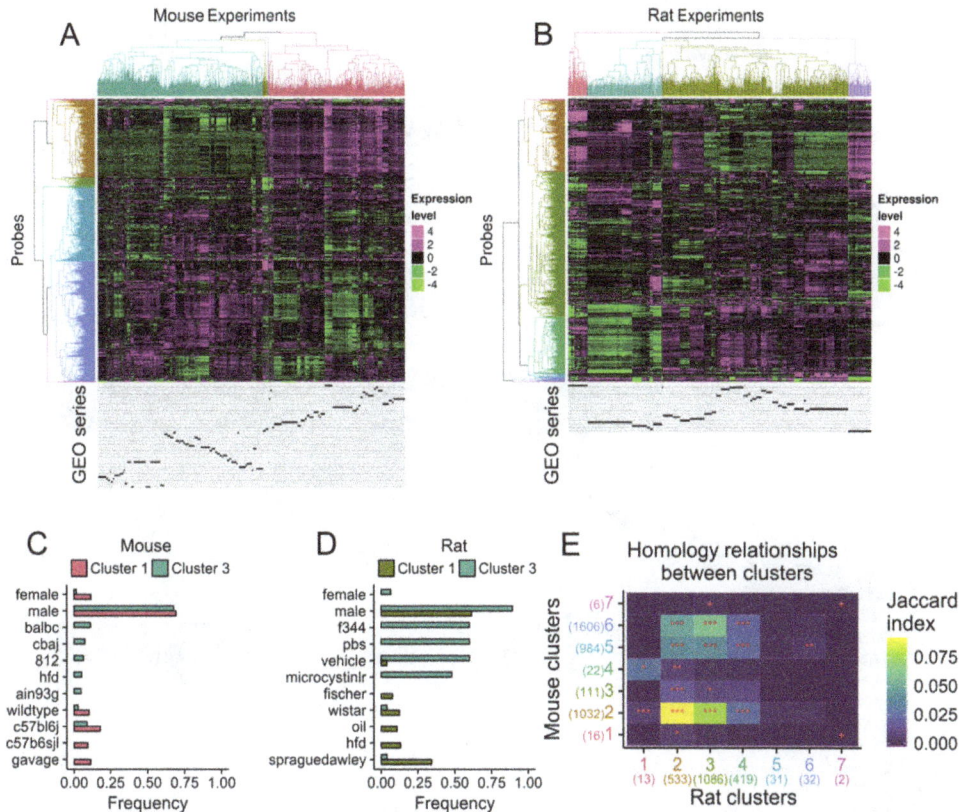

Fig. 2. Hierarchical clustering of variable probes in mouse (A) and in rat (B). Four clusters were defined for experiments and seven for probes as reflected by the dendrogram colours. Below the heatmaps, localisation of experiments from each series are shown in black, one line per series. FC: fold change. C and D. Metadata term frequencies of the two biggest experiment clusters were compared for mouse (C) and rat (D). Colour-code matches the experiments trees in panels A and B. E. Homology relationships between probe clusters between rat (X-axis) and mouse (Y-axis). Cell colour: Jaccard index. Cell label: Bonferroni adjusted P-values: *** $\leq$ 0.0001, ** $\leq$ 0.001, * $\leq$ 0.01, + $\leq$ 0.05. Cluster number colours match the probes dendrogram colours in panels A and B. The numbers in parenthesis denote the number of probes in each cluster.

nodes in each species with a greater than expected number of conserved interactions (based on co-expression in this case) between them (Fig. 3A). Input data for SCHype was built using three graphs: a mouse probe graph built from pairs of probes with a Spearman correlation coefficient $\geq$ 0.5 (Supplementary Fig. 2A), a rat probe graph with pairs of probes with a Spearman correlation coefficient $\geq$ 0.5 (Supplementary Fig. 2B), and a probe to probe homology graph between rat and mouse built using the Homologene database [20] and the annotationTools package [21]. SCHype identified 132 clusters of homologous genes co-expressed both in mouse and in rat, which included 825 nodes in mouse and 778 nodes in rat (Fig. 3B). SCHype allows probes to be included in multiple clusters resulting into 474 unique probes in mouse and 425 unique probes in rat. It identified four clusters with over 30 homologous genes in each species, eighteen clusters with over 10 probes in each species, thirty-five clusters with only 2 co-expressed probes in each species (Fig. 3B). We further focussed on the first four (c1–c4) SCHype clusters (Fig. 3C). We firstly compared SCHype clusters with results obtained by clustering data from each species independently. SCHype cluster c3 highly overlapped with the previous cluster 2 in mouse (golden colour, Fig. 2A) and the cluster 2 in rat (golden colour, Fig. 2B). These two clusters were shown to share a high number of homologous probes (Fig. 2E). Gene ontology analysis of the four biggest SCHype clusters, both over the set of variable probes or over the full set of probes, did not lead to any significant results, most likely due to small number of genes in each cluster. Importantly, the experiments in each series no longer clustered together after restricting the data to

each of the four biggest SCHype clusters (Fig. 3C). Individual experiments from each series nevertheless belonged to the same large experiment cluster (Fig. 3C) highlighting the need for building an expression compendium to obtain these results.

3.4. Co-clustering Across Species as Source of Information for Inferring Shared Functionality Amongst Orthologs

SCHype clustering successfully identified clusters of homologous genes co-expressed in both mouse and rat datasets. This information adds an independent evidence in support of a functional annotation transfer for pairs of orthologous genes across species found in the same SCHype cluster(s), as functionally equivalent orthologs would be co-expressed with the same set of genes in both species, and therefore would be included in the same SCHype cluster(s). We investigated if SCHype clusters could help identify functionally equivalent orthologs amongst complex homology groups. Eighteen homology groups of three members or more had at least one member of each species in the same SCHype cluster(s). For example, for homology group 137299 (Table 2), *Anp32a* in mouse and *Anp32a* in rat were in the same SCHype cluster 69, while *LOC100909983*, another homologue of rat *Anp32a*, was not. This suggests that indeed *Anp32a* in rat is the functional equivalent of *Anp32a* in mouse, but *LOC100909983* is not. In this case, our method found back a functional equivalent already known [23]. Similar observations were made for homology groups 68982 (*Ccnb1*), 10699 (*Cdc248*), 3938

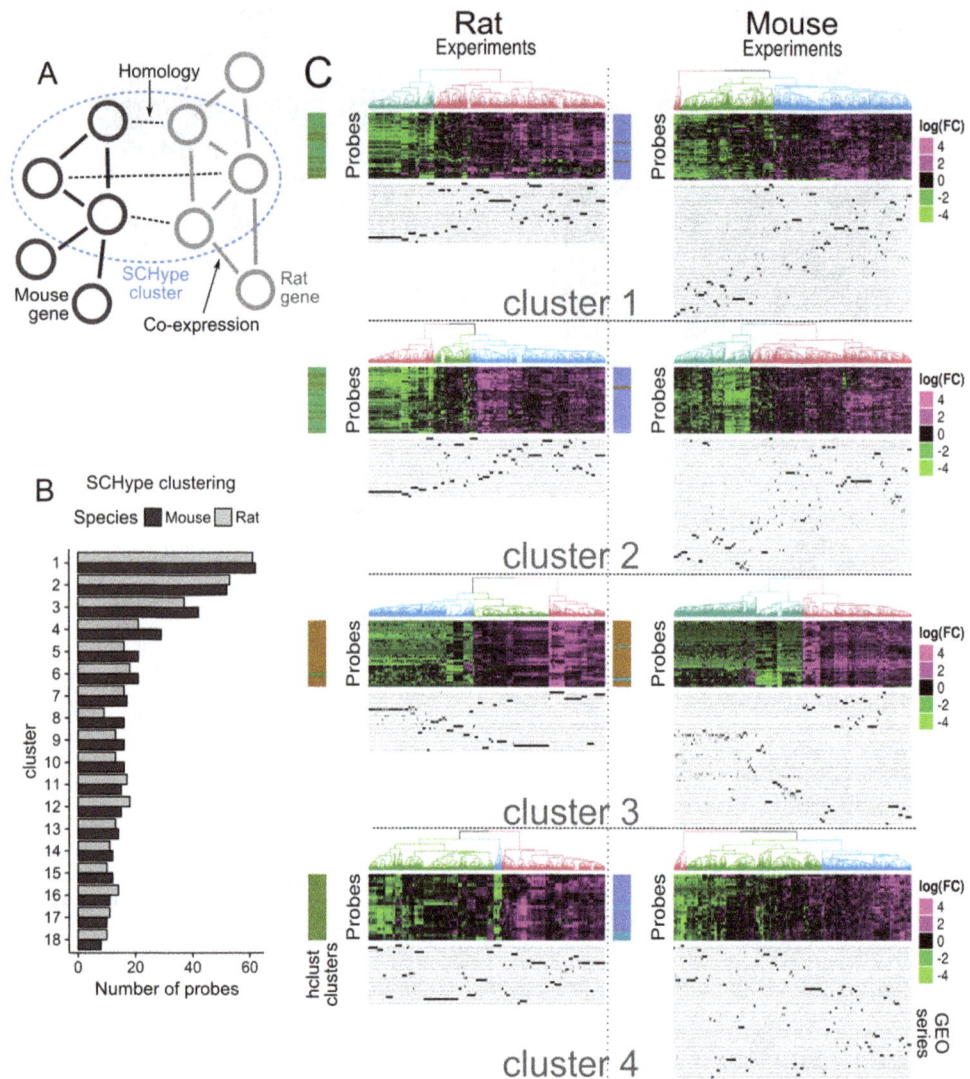

Fig. 3. Co-clustering of rat (middle) and mouse (right) liver data using SCHype. A. SCHype is a clustering tool for hypergraphs, built here from two co-expression graphs and an homology graph. B. Number of mouse (dark grey) and rat (light grey) probes for the SCHype clusters with more than 10 probes for each species. X-axis: number of probes included in each SCHype cluster. Y-axis: SCHype predicted clusters, numbered according to the number of probes per cluster in decreasing order. C. The biggest four SCHype clusters are shown. Genes in mouse and rat in each cluster are homologous to each other. The results of hierarchical clustering for each species is shown as a colour bar on the left. Colour-code matches the experiments trees in Fig. 1. Under the heatmap, clustering localisation of experiments from each series is shown in black, one line per series.

(*Ppp1r3c*), and 14108 (*Rasl10b*) (Table 2). In five cases, all members of the homology groups were included in the same SCHype clusters (Table 3), suggesting that all orthologs are likely to share the same function(s). Finally, eight homology groups showed more complex situations, where neither only one nor all the homologs where present in the same groups (Table 4 and Supplementary Table 3). For example, in homology group 117945, *Cyp2c7* in rat had three homologous genes in mouse but only *Cyp2c38* in mouse belonged to the same SCHype cluster (Table 4) predicting that mouse *Cyp2c38* (and not mouse *Cyp2c29* or mouse *Cyp2c39*) is a functional ortholog of rat *Cyp2c7*. We further explored the impact of the correlation threshold used to build the hypergraph (0.5, Supplementary Fig. 2) on the functional transfer evidence generated by assessing the predictions made using a higher correlation threshold of 0.75 (Supplementary Table 3). As expected, this resulted in reduction of co-expression edges, and thus reduction in identified clusters. Of the 18 groups described, 6 retained at the threshold of 0.75, with no major changes on the predictions of shared functionality.

Altogether, hypergraph clustering of co-expression network from rat and mouse liver microarray data was able to provide new evidence for functional annotation transfer between orthologous groups.

3.5. Functional Annotation Transfer Across Rat and Mouse Using Heart Transcriptomic Data

To test whether the approach described above was extendible to other tissues, we collected expression data in heart for mouse and rat. Data from heart samples processed using the same microarray platform were downloaded from GEO, for a total of 248 experiments from 20 studies in mouse and 1202 experiments from 19 studies in rat. Same data processing pipeline as for the liver samples (Supplementary Fig. 3) resulted into selection of 7371 (mouse) and 917 (rat) variable genes for clustering (Supplementary Fig. 4). The large difference between the selected number of genes might be due to the large difference in the number of samples available

Table 2

SCHype clustering of homologous groups: SCHype gene clustering reflects gene names. Homology groups were obtained from the Homologene database. Tick mark indicates the inclusion of the gene in the corresponding SCHype cluster.

Homology group	Species	Gene name	SCHype cluster	
137229			Cluster 69	
	Mouse	Anp32a	✓	
	Rat	Anp32a	✓	
	Rat	LOC100909983		
68982			Cluster 7	Cluster 30
	Mouse	Ccnb1	✓	✓
	Mouse	Gm5593		
	Rat	Ccnb1	✓	✓
10699			Cluster 2	Cluster 118
	Mouse	Cd248	✓	✓
	Rat	Cd248	✓	✓
	Rat	LOC100911932		
	Rat	LOC100911882		
3938			Cluster 1	
	Mouse	Ppp1r3c	✓	
	Rat	Ppp1r3c	✓	
	Rat	LOC100910671		
14108			Cluster 2	
	Mouse	Rasl10b	✓	
	Rat	Rasl10b	✓	
	Rat	LOC100912246		

Table 4

SCHype clustering of homologous groups: new predictions for functional orthologous relations. Homology groups are obtained from the Homologene database. Tick mark indicates the inclusion of the gene in the corresponding SCHype cluster. Four additional, more complex, homology groups are shown in supplementary Table 3.

Homology group	Species	Gene name	SCHype cluster
117948			Cluster 102
	Mouse	Cyp2c38	✓
	Mouse	Cyp2c29	
	Mouse	Cyp2c39	
	Rat	Cyp2c7	✓
104115			Cluster 33
	Mouse	Hsd3b5	✓
	Mouse	Gm10681	
	Mouse	Hsd3b4	
	Mouse	Gm4450	
	Rat	Hsd3b5	✓
	Rat	LOC100911116	✓
137425			Cluster 2
	Mouse	Lce3c	✓
	Rat	LOC100361951	✓
	Rat	LOC100911982	✓
	Rat	Lce3d	
129514			Cluster 17
	Mouse	Rdh9	✓
	Mouse	Rdh1	
	Mouse	Rdh16	
	Mouse	Rdh19	
	Mouse	BC089597	
	Rat	Rdh16	✓
	Rat	LOC100365958	✓

(and therefore used in analysis) for each species. Functional enrichment analysis of the variable genes revealed pathways related to the heart functions (Supplementary Table 4), confirming that at least part of the gene expression variability was reflecting biological differences. Hierarchical clustering of the mouse and rat dataset separately revealed clusters of probes and clusters of experiments (Supplementary Fig. 5). Notably, experiments from a same series (GEO) were split in distinct clusters.

Co-clustering of the mouse and rat co-expression network in heart, using Homologene homology information and the SCHype hypergraph clustering tool, identified clusters of homologous genes co-expressed in both species (Supplementary Fig. 6). Notably, this provided an independent evidence in favour or against shared functionality for 12 complex orthology groups (Supplementary Table 5). In details, for 6 groups all homologous genes where found in the same SCHype cluster (i.e. were in homologous co-expressed gene networks in both species). For 3 one-to-two homology groups (*Ifit3*, *Ogn* and *Ppp1r3c*), only one of the two paralogs was included in

Table 3

SCHype clustering of homologous groups: all members of the homology groups share predicted functionalities. Homology groups are obtained from the Homologene database. Tick mark indicates the inclusion of the gene in the corresponding SCHype cluster.

Homology group	Species	Gene name	SCHype cluster		
128630			Cluster 9	Cluster 12	Cluster 45
	Mouse	Ceacam1	✓		
	Mouse	Ceacam2	✓	✓	✓
	Rat	Ceacam1	✓	✓	✓
11456			Cluster 5		
	Mouse	Elovl6	✓		
	Rat	Elovl6	✓		
	Rat	LOC102549542	✓		
20277			Cluster 35		
	Mouse	Rrm2	✓		
	Rat	Rrm2	✓		
	Rat	LOC100359539	✓		
55991			Cluster 1	Cluster 119	
	Mouse	Tmed2	✓	✓	
	Mouse	Gm21540	✓	✓	
	Rat	Tmed2	✓	✓	
11890			Cluster 10	Cluster 43	Cluster 81
	Mouse	Tnks2	✓	✓	✓
	Rat	LOC100910717	✓	✓	✓
	Rat	Tnks2	✓	✓	✓

the same SCHype cluster(s) as the ortholog copy, suggesting a loss or change of functionality for the second paralog. The remaining 3 groups presented complex scenarios where more than one, but not all paralogs were included in the same SCHype cluster(s). One of these complex groups, Homologene group 128352, was found both in heart and liver data. Altogether, the proposed method was able to provide evidence to support annotation transfer from transcriptomic data not only in liver, but also in heart, suggesting that the approach is applicable to different tissues.

4. Discussion

Here we have shown that transcriptomic data can be used to provide evidence for functional annotation transfer between orthologs (see Fig. 4), using co-expression networks built from mouse and rat liver and heart samples. In liver, we identified 18 complex homologous groups (i.e. with paralogs in at least one of the species), including 54 genes in mouse and 46 genes in rat, with at least one gene in mouse and one gene in rat in the same SCHype cluster(s). Twelve more groups (of which 11 groups non overlapping with the liver analysis) were found when applying the same method to heart transcriptomic data. Increasing the correlation thresholds resulted in loss of total number of predictions, as expected. Lowering the correlation threshold and the standard deviation threshold, on the other hand, will likely increase the number of homologous groups, potentially with a higher false positive rate. The use of co-expression network to provide evidence for functional annotation transfer has been previously demonstrated [13–15]. These studies combined samples from various tissues, while we analysed two tissues (liver and heart) independently. Both approaches are complementary. While mixing tissues might result in broader co-expression network (many edges), it might also lack the fine resolution needed to improve functional annotation inference in a tissue specific context. We used microarray data in this study as it is by far the most abundant dataset. However, consortia like GTEx [24] have generated large amount of RNA sequencing data, and we envisage application of the method described here to RNA sequencing data in the future.

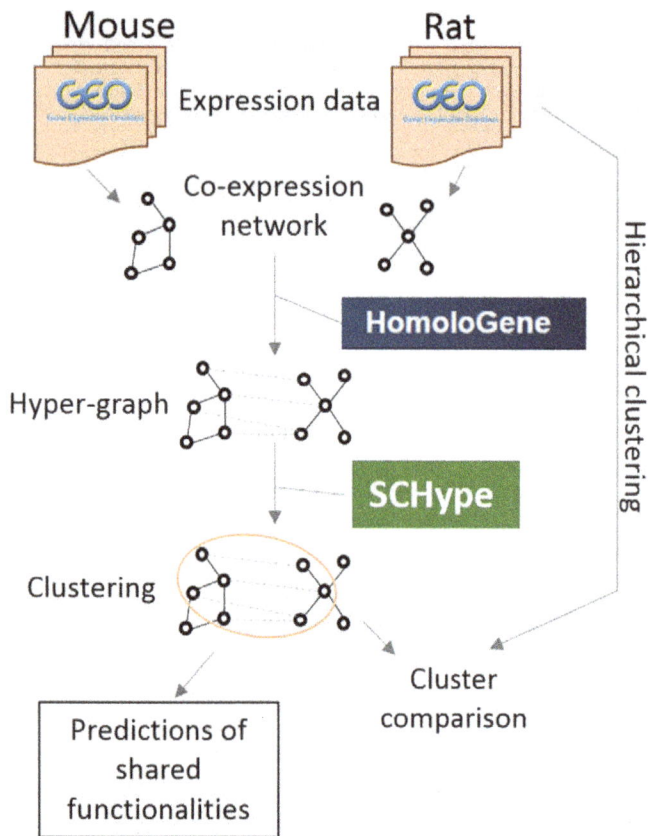

Fig. 4. Work-flow diagram. Transcriptomic data (microarray) was gathered from GEO to build species-specific co-expression network. Homology information from the Homologene database together with co-expression networks were used to extract hypergraph clusters using the SCHype software. Resulting clusters were firstly compared to species specific hierarchical clusters, and were used to infer shared functionality links in complex homology groups.

The greater sensitivity of RNA sequencing over microarray [25] might allow the identification of more co-expressed genes.

Despite rigorous data normalisation, liver experiments from the same series tended to cluster together, cases and controls included. While this could be a sign of technical biases, gene ontology analysis of the variable genes demonstrated that they are related to liver functions. Thus it appears that the gene expression variability we observed is, at least partially, reflecting biological variations impacting the liver physiology. Importantly, individual experiments from series did not cluster together in SCHype clusters. We applied various approaches but could not identify the biological origin(s) of the observed variations. This is in part due to the lack of standardised experiment metadata fields in GEO (not all datasets even had a strain or a sex information, for example), and the lack of controlled vocabulary used to describe experiments. It is a possibility that better annotation of the metadata would have allowed the identification of critical confounding factors. Noteworthy, heart experiments were not clustered by series of experiments. It could be that the heart tissue is less sensitive than the liver to differences in the animal environment.

SCHype clustering was able to find some known as well as some yet to be experimentally validated ortholog functional relationships. For example, only mouse and rat *Ccnb1* were in the same SCHype cluster, and not *Gm5593*. While mouse *Ccnb1* and rat *Ccnb1* are annotated as protein coding genes, *Gm5593* in mouse is annotated as a processed pseudogene [23].

Finally, we note that conserved co-expression of orthologous genes is not a direct proof of shared functionality, but it can be used as an additional layer of evidence. While protein-protein interaction networks could be used for the same aim, transcriptomic data are more easily generated and therefore more likely to be widely available for many species. Thus the method described here shows a promise to enhance functional gene annotation transfer across species. It can provide an experimental support for one-to-one ortholog annotation transfer, and can help identify functionally similar and non similar orthologs in one-to-many and many-to-many orthology groups.

Funding

AJ is a Chancellor's fellow and AJ and TM labs are supported by institute strategic funding from Biotechnology and Biological Sciences Research Council (BBSRC, BBSRC-BB/P013732/1-ISPG 2017/22 and BBSRC-BB/P013740/1-ISPG 2017/22). GD is funded by the People Programme (Marie Curie Actions FP7/2007-2013) under REA grant agreement No PCOFUND-GA-2012-600181.

References

[1] Salavati R, Najafabadi HS. Sequence-based functional annotation: what if most of the genes are unique to a genome? Trends Parasitol 2010;26:225–9.
[2] Edwards SL, Beesley J, French JD, Dunning AM. Beyond GWASs: illuminating the dark road from association to function. Am J Hum Genet 2013;93:779–97.
[3] Rinke C, Schwientek P, Sczyrba A, Ivanova NN, Anderson IJ, Cheng J-F. et al. Insights into the phylogeny and coding potential of microbial dark matter. Nature 2013;499:431–7.
[4] Das S, Orengo CA. Protein function annotation using protein domain family resources. Methods (San Diego, Calif) 2016;93:24–34.
[5] Davidson D, Baldock R. Bioinformatics beyond sequence: mapping gene function in the embryo. Nat Rev Genet 2001;2:409–17.
[6] Martin DMA, Berriman M, Barton GJ. GOtcha: a new method for prediction of protein function assessed by the annotation of seven genomes. BMC Bioinf 2004;5:178.
[7] Götz S, García-Gómez JM, Terol J, Williams TD, Nagaraj SH, Nueda MJ. et al. High-throughput functional annotation and data mining with the Blast2GO suite. Nucleic Acids Res 2008;36:3420–35.
[8] Cozzetto D, Jones DT. Computational methods for annotation transfers from sequence. Methods Mol Biol (Clifton, NJ) 2017;1446:55–67.
[9] Schnoes AM, Brown SD, Dodevski I, Babbitt PC. Annotation error in public databases: misannotation of molecular function in enzyme superfamilies. PLoS Comput Biol 2009;5:e1000605.
[10] Zallot R, Harrison K, Kolaczkowski B, de Crécy-Lagard V. Functional annotations of paralogs: a blessing and a curse. Life 2016;6:39.
[11] Kolár M, Lässig M, Berg J. From protein interactions to functional annotation: graph alignment in Herpes. BMC Syst Biol 2008;2:90.
[12] Mao F, Su Z, Olman V, Dam P, Liu Z, Xu Y. Mapping of orthologous genes in the context of biological pathways: an application of integer programming. Proc Natl Acad Sci U S A 2006;103:129–34.
[13] Towfic F, VanderPlas S, Oliver CA, Couture O, Tuggle CK, Greenlee MHW. et al. Detection of gene orthology from gene co-expression and protein interaction networks. BMC Bioinf 2010;11:S7.
[14] Chikina MD, Troyanskaya OG. Accurate quantification of functional analogy among close homologs. PLoS Comput Biol 2011;7:e1001074.
[15] Yan K-K, Wang D, Rozowsky J, Zheng H, Cheng C, Gerstein M. OrthoClust: an orthology-based network framework for clustering data across multiple species. Genome Biol 2014;15:R100.
[16] Joshi A, Beck Y, Michoel T. Multi-species network inference improves gene regulatory network reconstruction for early embryonic development in Drosophila. J Comput Biol 2015;22:253–65.
[17] Michoel T, Nachtergaele B. Alignment and integration of complex networks by hypergraph-based spectral clustering. Phys Rev E 2012;86:056111.
[18] Bolstad B. preprocessCore: a collection of pre-processing functions.2016.
[19] Buchta C, Hahsler M. cba: clustering for business analytics.2017.R package version 0.2-18.
[20] Database resources of the National Center for Biotechnology Information. Nucleic Acids Res 2016;44:D7–D19.
[21] Kuhn A, Luthi-Carter R, Delorenzi M. Cross-species and cross-platform gene expression studies with the bioconductor-compliant R package 'annotationTools'. BMC Bioinf 2008;9:26.
[22] Mi H, Huang X, Muruganujan A, Tang H, Mills C, Kang D. PANTHER version 11: expanded annotation data from gene ontology and reactome pathways, and data analysis tool enhancements. Nucleic Acids Res 2017;45:D183–D189.
[23] Yates A, Akanni W, Amode MR, Barrell D, Billis K, Carvalho-Silva D. et al. Ensembl 2016. Nucleic Acids Res 2016;44:D710–6.
[24] Melé M, Ferreira PG, Reverter F, DeLuca DS, Monlong J, Sammeth M. et al. The human transcriptome across tissues and individuals. Science 2015;348:660–5.

5

Genetic Code Optimization for Cotranslational Protein Folding: Codon Directional Asymmetry Correlates with Antiparallel Betasheets, tRNA Synthetase Classes

Hervé Seligmann [a,b,*], Ganesh Warthi [a]

[a] Aix-Marseille Univ, Unité de Recherche sur les Maladies Infectieuses et Tropicales Emergentes, UM 63, CNRS UMR7278, IRD 198, INSERM U1095, Institut Hospitalo-Universitaire Méditerranée-Infection, Marseille, Postal code 13385, France
[b] Dept. Ecol Evol Behav, Alexander Silberman Inst Life Sci, The Hebrew University of Jerusalem, IL-91904 Jerusalem, Israel

ARTICLE INFO

Keywords:
Secondary structure
Codon-amino acid assignment
Mitochondrial genetic code
Synonymous codon
Alpha helix
Beta turn

ABSTRACT

A new codon property, codon directional asymmetry in nucleotide content (CDA), reveals a biologically meaningful genetic code dimension: palindromic codons (first and last nucleotides identical, codon structure XZX) are symmetric (CDA = 0), codons with structures ZXX/XXZ are 5'/3' asymmetric (CDA = −1/1; CDA = −0.5/0.5 if Z and X are both purines or both pyrimidines, assigning negative/positive (−/+) signs is an arbitrary convention). Negative/positive CDAs associate with (a) Fujimoto's tetrahedral codon stereo-table; (b) tRNA synthetase class I/II (aminoacylate the 2'/3' hydroxyl group of the tRNA's last ribose, respectively); and (c) high/low antiparallel (not parallel) betasheet conformation parameters. Preliminary results suggest CDA-whole organism associations (body temperature, developmental stability, lifespan). Presumably, CDA impacts spatial kinetics of codon-anticodon interactions, affecting cotranslational protein folding. Some synonymous codons have opposite CDA sign (alanine, leucine, serine, and valine), putatively explaining how synonymous mutations sometimes affect protein function. Correlations between CDA and tRNA synthetase classes are weaker than between CDA and antiparallel betasheet conformation parameters. This effect is stronger for mitochondrial genetic codes, and potentially drives mitochondrial codon-amino acid reassignments. CDA reveals information ruling nucleotide-protein relations embedded in reversed (not reverse-complement) sequences (5'-ZXX-3'/5'-XXZ-3').

1. Introduction

The genetic code is optimised along several dimensions. Correlations between codon and amino acid properties have frequently been interpreted as resulting from evolutionary optimizations of the genetic code's codon-amino acid assignments. These minimise effects of: replicational/transcriptional nucleotide substitutions on amino acid hydrophobicity [1–11] and along multiple properties [12]. The genetic code is also optimised in relation to other processes, such as tRNA misloading with non-cognate amino acids [13–16]; ribosomal frame-shifts [17–23]; and protein folding kinetics [24–26].

Another approach assumes that the genetic code coevolved with codon/amino acid metabolic pathways [27–31]. It remains unclear whether genetic code optimizations are circumstantial byproducts of the metabolic coevolution hypothesis [32–36], or whether some combination of both processes produced the genetic code [34,37–42].

Here we present a previously unknown dimension of the genetic code. Analyses suggest that the genetic code is optimised in relation to this new property. The property reflects differences between nucleotides at first versus second codon positions, as compared to differences between nucleotides at third versus second codon positions. In this context, previous analyses [43] showed that the subtraction of dipole moments of nucleotides at first and second codon positions correlate with hydrophobicities of corresponding amino acids, after accounting for another, previously reported, correlation between codon and amino acid hydrophobicities [44,45]. Here analyses generalise the principle to all codon positions and nucleotide properties.

2. Codon Directional Asymmetry

The new codon property is derived from comparing two differences in nucleotide contents, the difference between nucleotides at first and second codon positions, and the difference between nucleotides at second and third codon positions. This defines a codon's directional

* Corresponding author at: Aix-Marseille Univ, Unité de Recherche sur les Maladies Infectieuses et Tropicales Emergentes, UM 63, CNRS UMR7278, IRD 198, INSERM U1095, Institut Hospitalo-Universitaire Méditerranée-Infection, Marseille, Postal code 13385, France.
E-mail address: varanuseremius@gmail.com (H. Seligmann).

asymmetry in nucleotide content, CDA. CDA reflects semi-quantitatively extents by which a nucleotide at either 5′ or 3′ codon extremity differs from the codon's two remaining nucleotides. Along this principle, palindromic codons with the same nucleotide at 5′ and 3′ extremities (at first and third positions, XZX (including codons with X = Z)) are symmetric, CDA = 0. When the nucleotide at the 5′ extremity belongs to a different nucleotide group (purine/pyrimidine) than the two other positions and the latter are identical (ZXX), CDA = −1. When the nucleotide at the 3′ extremity differs from other positions (XXZ), CDA = +1. Signs for 5′-and 3′-dominant CDAs are arbitrary, but necessarily opposite (positive versus negative).

2.1. Purines and Pyrimidines

For codons of types ZXX/XXZ, CDA = −0.5/+0.5, when both X and Z are purines, or both pyrimidines. This reflects lesser purine-purine and pyrimidine-pyrimidine structural differences than for purine-pyrimidine comparisons. This principle assigns a CDA score also for some codons of type XZW, where all three nucleotides differ, and Z belongs to the same chemical group (purine or pyrimidine) as the nucleotide at either codon extremity. For codons where nucleotides Z and W are both purines/pyrimidines, X is the most different nucleotide (CDA = −0.5), because chemical structural differences between X and Z are greater than between W and Z. According to that rationale, for codons where nucleotides X and Z are both purines (or both pyrimidines), W is the most different nucleotide (CDA = +0.5).

2.2. Complementarity Between Nucleotides at Different Codon Positions

For some codons with structure XZW, Z does not belong to the same group (in terms of purines/pyrimidines) as any nucleotide at the other positions. In these cases, an additional rule determines which of the nucleotides among X or W, differs more from the two others. We propose that complementarity between canonical base pairs (C:G and A:T/U) defines that complementary nucleotide pairs are the most different pairs. Hence for codons with structure XZW, CDA = −0.5 and CDA = +0.5 when X is the canonical complement of Z, and when W is the complementary of Z, respectively. This rule set defines CDA for all 64 codons (Table 1).

Table 1

The genetic code's 64 codons and their codon directional asymmetry, CDA. Shaded nucleotides indicate the nucleotide at one of the codon's extremities that is the most different from nucleotides at other positions, along rules described in text, and which determines the dominant side of codon directional asymmetry: negative CDA when the first (5′) codon position has the most different nucleotide, and positive CDA when the third (3′) position has the most different nucleotide. Codons assigned to amino acids aminoacylated by class I tRNA synthetases are framed, remaining amino acids are aminoacylated by class II tRNA synthetases.

TTT	F	0	TCT	S	0	TAT	Y	0	TGT	C	0
TTC	F	0.5	TCC	S	-0.5	TAC	Y	-0.5	TGC	C	0.5
TTA	L	1	TCA	S	0.5	TAA	*	-1	TGA	*	-0.5
TTG	L	1	TCG	S	0.5	TAG	*	-0.5	TGG	W	-1
CTT	L	-0.5	CCT	P	0.5	CAT	H	0.5	CGT	R	-0.5
CTC	L	0	CCC	P	0	CAC	H	0	CGC	R	0
CTA	L	0.5	CCA	P	1	CAA	Q	-1	CGA	R	-0.5
CTG	L	0.5	CCG	P	1	CAG	Q	-0.5	CGG	R	-1
ATT	I	-1	ACT	T	-0.5	AAT	N	1	AGT	S	0.5
ATC	I	-0.5	ACC	T	-1	AAC	N	1	AGC	S	0.5
ATA	I	0	ACA	T	0	AAA	K	0	AGA	R	0
ATG	M	-0.5	ACG	T	0.5	AAG	K	0.5	AGG	R	-0.5
GTT	V	-1	GCT	A	-0.5	GAT	D	0.5	GGT	G	1
GTC	V	-0.5	GCC	A	-1	GAC	D	0.5	GGC	G	1
GTA	V	0.5	GCA	A	-0.5	GAA	E	-0.5	GGA	G	0.5
GTG	V	0	GCG	A	0	GAG	E	0	GGG	G	0

3. A New Dimension of the Genetic Code

The distribution of CDA in Table 1 is symmetric. Therefore, the genetic code table could probably be reordered so as to reveal graphically this symmetry, as done for other symmetry properties of the genetic code [46].

To what extent does CDA represent a dimension of the genetic code that is independent of other dimensions? In this respect, we compare Table 1 with the binary representation of the genetic code [47, therein figure 6], a rather complete 6-bit representation of each codon. It assigns to each codon position two binary values, the first representing the purine-pyrimidine divide, the second value represents whether the nucleotide forms two or three hydrogen interactions when in duplex conformation with an inverse-complementary strand. This defines two binary variables for each codon position, hence six binary variables for each codon.

Pearson correlation coefficients r of CDA with any of these six binary codon properties are 'zero', indicating that CDA is independent of each of these properties. Correlations with sums and subtractions between any pairs of these six binary values also yield r = 0. Results are identical if one pairs nucleotides according to keto versus amino nucleotides as previously reported [47,48]. This means that CDA catches a genetic code dimension that differs from classically recognised codon properties.

3.1. Tetrahedral Representations and CDA

The genetic code can also be presented as a tetrahedron, with four equal triangular faces each subdivided into 16 equilateral, smaller triangles, representing the 64 codons. Castro-Chavez [49] reviews these representations, and proposes a tetrahedral representation, placing codons so that hydrophobic amino acids are central to each tetrahedral face, named faces A–D. Applying CDA to Castro-Chavez's tetrahedral representation, faces A and D tend to have CDA < 0, and faces C and B CDA > 0. Within each face, in total 19 triangle vertices (over all 4 faces) with CDA < 0 are common with vertices belonging to triangles with CDA > 0. This is very close to the 18 vertices expected if codons were randomly distributed in relation to CDA (P > 0.5, chi-square test), considering that 24 codons have CDA < 0, 16 have CDA = 0, and 24 have CDA > 0. Eleven among 24 vertices common between triangles from different faces of the tetrahedron are for triangles/codons with opposite CDA. This is slightly more than the 6.75 expected by random CDA distribution (P = 0.054, chi-square test). Hence the tetrahedral representation of Castro-Chavez [49] is random in relation to CDA within tetrahedral faces, and probably also between faces.

Fujimoto's tetrahedral codon stereo-table [50] is much more ordered in relation to CDA's distribution among and within tetrahedral faces (Fig. 1): Faces A–D each have six codons with CDA < 0, six codons with CDA > 0, and four codons with CDA = 0. Within each face, there are exactly two contacts between codons/triangles with opposite CDA. This total of eight contacts between triangles with opposite CDA is significantly less than the expected 18 contacts for randomly distributed CDA within faces of the tetrahedron (P = 0.018, chi-square test). There are no contacts between tetrahedron faces for codons/triangles with opposite CDA (P = 0.0096, chi-square test). Hence Fujimoto's tetrahedral representation is most compatible with the genetic code's symmetries implied by CDA in Table 1.

The specific examples used here illustrate randomness versus CDA, and close to perfect reorganisation of the genetic code in relation to CDA, respectively. Other representations might reorganise the genetic code more optimally in relation to CDA. However, these representations may not relate to interpretable phenomena in the real world.

3.2. Codon Directional Asymmetry and Codon Participation in Error Correcting Codes

Genetic codes include a subjacent punctuation code called the natural circular code that enables retrieving the ribosomal translation frame

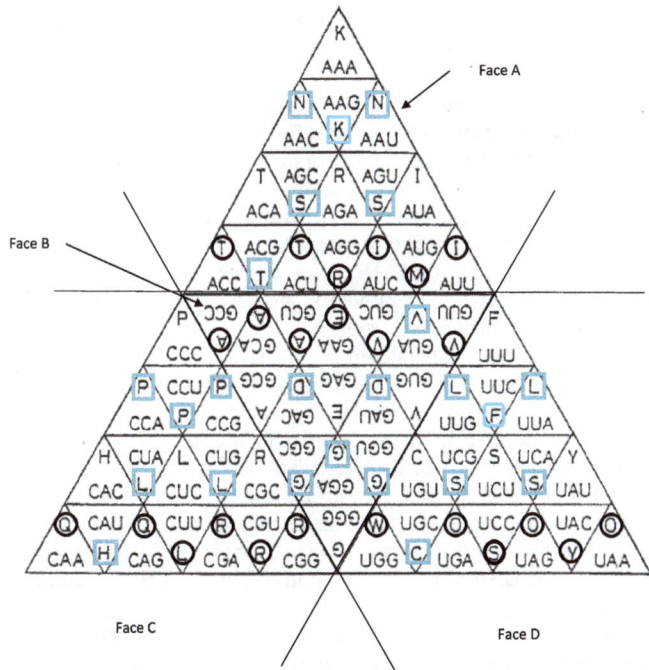

Fig. 1. Fujimoto's tetrahedral codon stereo-table, a genetic code's representation that seems non-random in relation to codon directional asymmetry. The tetrahedron has four equal, equilateral faces (A–D), and consist each of 16 equilateral triangles representing each one codon. Red circles: CDA < 0; blue squares: CDA > 0.
Adapted from Fig. 1 at http://www.google.com/patents/US4702704.

[51–55]. Mechanisms for coding frame retrieval remain unknown, but are probably associated with circular code motifs conserved in tRNAs and ribosomal RNAs [56–59]. Codon symmetry is particularly informative in relation to frame retrieval, as codons of type XZX (CDA = 0) have maximal capacity for reading frame retrieval [55,60,61], and have highest occurrences within various types of error-correcting codes [62]. Absolute values of CDA are lower for codons belonging to the natural circular code than for the remaining codons ($P = 0.016$, two tailed Mann-Whitney test). This principle is confirmed also when comparisons imply only codons belonging to the natural circular code: their absolute CDA increases with codon-specific reading frame retrieval (r = −0.615, $P = 0.002$; rs = 0.44, $P = 0.026$, one tailed tests). Hence processes determining the near-universal natural circular code probably contributed biological functions to CDA.

4. Codon Directional Asymmetry and tRNA Synthetase Classes

CDA in Table 1 reflects a genetic code symmetry that does not follow the purine-pyrimidine, keto-amino, nor the weak-strong base-pairing patterns. A little known symmetry within the genetic code relates to Rumer's transformation [63–65], which replaces systematically all adenine (A) with cytosine (C) and vice versa, and also all guanine (G) with thymine (T) and vice versa. It is one among 23 bijective transformations [60], also called systematic nucleotide exchanges [66,67] or 'swinger' transformations [68–71]. RNA and DNA sequenced by several different methods and published in GenBank by various groups match these transformations. Hence while a priori, transformations such as Rumer's seem theoretical processes, they reflect biological realities, such as actual nucleotide sequences that were presumably produced by replication or transcription that systematically inserts a specific nucleotide instead of another specific nucleotide. This phenomenon of systematic nucleotide exchanges has similarities with isolated nucleotide misinsertions [60,66,67].

Rumer's transformation also correlates with a notable biological property, tRNA synthetase classes [72] of amino acids assigned to codons. The tRNA synthetases are enzymes that load amino acids to their cognate tRNA. The twenty tRNA synthetases form two groups of equal size, tRNA synthetase classes I and II based on structural homology [73,74]. tRNA synthetases class I covalently link cognates to the 2′ hydroxyl group of the tRNA's last ribose, and class II to its 3′ hydroxyl group [75,76].

The symmetry in the genetic code that correlates with tRNA synthetase classes exchanges nucleotides at the first and third codon positions along rule A↔C + G↔T (Rumer's transformation), and A↔G + C↔T at the second codon position. If instead of applying the nucleotide exchange rule A↔C + G↔T to the third codon position, one applies the exchange rule A↔T + C↔G, the symmetry between codons whose corresponding tRNA is aminoacylated by tRNA synthetase class I or class II is also recovered [77]. These symmetries by nucleotide exchanges are not mere theoretical considerations. Homologies of some DNA and RNA sequences in GenBank were detected after accounting for systematic nucleotide exchanges for the mitogenome [66–71,78–80]. In addition, the regular human mitogenome includes numerous repeats that can only be detected when assuming systematic exchanges [81], including palindromes [82].

CDA associates with tRNA synthetase classes. On average, codons assigned to amino acids aminoacylated by tRNA synthetases class I have CDA < 0 (15 among 21 codons (stops excluded), $P = 0.039$, two tailed sign test). For tRNA synthetases class II, the situation is opposite: most codons have CDA > 0 (17 among 24, CDA = 0, $P = 0.032$, two-tailed sign test). Sign tests are inadequate to handle codons with CDA = 0, therefore codons with CDA = 0 are excluded from these calculations. Mean CDA for tRNA synthetase classes differ significantly (two-tailed $P = 0.002$ for each t-test and Mann-Whitney test). These comparisons between means include codons with CDA = 0.

CDAs are averaged for codons assigned to specific amino acids. Mean CDA < 0 for 8 among 10 amino acids for class I; and CDA > 0 for 8 among 10 amino acids for class II ($P = 0.006$, two-tailed sign test for each tRNA synthetase class). Exceptions are Cys and Leu for class I, and Ala, and Thr for class II. Overall, the sign of mean CDA for codons assigned to an amino acid follows expected patterns (class I, CDA < 0; class II, CDA > 0) for 16 among 20 amino acids/tRNA synthetases ($P = 0.00296$, one tailed sign test).

Note that stop codons have CDA < 0, predicting tRNA synthetase class I. However, the tRNA synthetase of pyrrolysine, which is inserted at some stop codons, belongs to tRNA synthetase class II [83]. Exceptions might reflect historical constraints on the genetic code's genesis [77].

Hence the rationale defining CDA reveals a symmetry that is close to that of the combination of nucleotide exchanges that reveal the genetic code's symmetry in relation to tRNA synthetase classes. However, the rationale behind CDA is simpler and perhaps more amenable to mechanistic reduction.

4.1. Alternative Scores for Codons with CDA = |0.5|

Three different types of codons get CDA = |0.5|, based on different rationales: (a) codons with structures ZXX/XXZ where both X and Z are purines/pyrimidines; (b) codons with structure XZW where Z belongs to the same nucleotide family (purine/pyrimidine) as either X or W; and (c) codons with structure XZW where Z belongs to a different nucleotide family than X and Z. This scoring is somewhat arbitrary, and might not be optimal to reflect biological properties. Keeping signs, we rescore each of these three codon types with values |0.25| and |0.75|, resulting in different scoring systems for these three codon groups: alternative CDAs of groups (a, b, c) are (0.5, 0.25, 0.75), (0.5, 0.75, 0.25), (0.25, 0.5, 0.75), (0.25, 0.75, 0.5), (0.75, 0.5, 0.25), and (0.75, 0.25, 0.5). CDA of codons with CDA = 0 and CDA = |1| remain unchanged. These different scoring systems do not alter the strength of the CDA-tRNA synthetase class association: according to all these scoring systems, the same 8 among 10 codon families in class I have CDA < 0, and 8 among

10 amino acids in class II have CDA > 0. Excluding palindromic codons (CDA = 0) from calculations does not change results.

This heuristic approach suggests that associations between tRNA synthetase classes (an ancient property of the translational apparatus) and CDA are robust in relation to CDA's semi-quantitative scoring.

5. Translation Kinetics

The tRNA synthetase classes differ in the position of aminoacylation of the amino acid on the tRNA's acceptor stem. This probably affects the spatial kinetics of peptide elongation. We suggest that CDA also affects the spatial kinetics of codon-anticodon interactions in the ribosome's translational core (site P [84]; site A [85]). Hence both tRNA synthetase class and CDA would affect cotranslational protein folding, meaning folding during the process of peptide extension by ribosomal translation [86–97]. Tentatively, we consider that associations between CDA and tRNA synthetase classes suggest synergistic effects on cotranslational protein folding by each CDA and tRNA synthetase class.

Note that cotranslational protein folding does not occur for all proteins [98]. Cotranslational protein folding frequently increases the yield of proper folds, but is not always an absolute requirement [99–103]; yet decreases misfolding probabilities [104–106]. Among others, at least in some cases, cotranslational folding requires complete protein structural subdomains [107,108]. Cotranslational protein folding following the sense of translation (from the N terminal) predicts more accurately protein structures than when proceeding in the opposite sense (from the C terminal) [109,110], indicating that cotranslational protein folding is a reality for most proteins. Nevertheless, cell free protein folding shows that cotranslational folding is not always required [111].

mRNA properties affecting translation speed and ribosomal pausing [112–114], also affect protein folding independently of that protein's amino acid sequence. Synonymous codons associate with different types of protein secondary structures [115,116], in particular for clusters of rare codons on mRNAs [117–119]. These associations might explain effects of synonymous single nucleotide polymorphisms on protein function [120–123] and are in line with selection at amino acid level that affects synonymous codon choice [124,125].

More specifically, rare codons concentrate in mRNA regions that code for transmembrane helical structures [116]. Optimization of codon usage means that organisms match codon usage frequencies with anticodons of common tRNAs [126–133], speeding translation, affecting cotranslational protein folding [134]. Lopez and Pazos [135] suggest that proper folding into transmembrane structures requires specific spatial kinetics and particular accuracy in the process. Cotranslational protein folding is most apparent on alpha helices and betasheet secondary structures [136–140]. Hence one expects associations between CDA and these conformational indices of amino acids. Chemical kinetics of the transfer of the amino acid loaded on the tRNA's acceptor stem to the elongating peptide (kinetic estimates from [141]) also constrain codon-anticodon interactions [43].

Following these rationales, CDA might reflect (a) indirectly tRNA synthetase classes and their effects on amino acid positioning during peptide elongation; and (b) directly the spatial kinetics of codon-anticodon interactions, such as tRNA-mRNA approach angles during codon-anticodon duplex formation in the ribosomal translational core(s). These two components should affect according to the cotranslational protein folding hypothesis folding patterns of elongating peptides. Hence CDA is predicted to correlate with amino acid secondary structure conformational parameters for alpha helices, beta turns and/or betasheets (conformational indices are from [142–145]). The main candidates are the conformational parameters associated with transmembrane foldings (beta turns, and/or parallel and antiparallel betasheets, from references [146,147]).

6. Antiparallel Betasheet Formation and Codon Directional Asymmetry

The hypothesis that CDA associates with cotranslational protein folding predicts correlations between CDA and secondary structure conformation parameters. Betasheets are the major secondary structures found in transmembrane proteins, antiparallel betasheets are more frequent than parallel betasheets [147]. Biases in tRNA synthetase amino acid contents correlate with the amino acid's antiparallel betasheet conformation parameter [148]. Hence, we predict correlations between CDA and conformation parameters, and in particular antiparallel betasheet conformation parameters.

Indeed, antiparallel betasheet conformation parameters correlate negatively with mean CDA of codons assigned to the amino acid according to the standard genetic code (Pearson correlation coefficient $r = -0.642$, two-tailed $P = 0.0023$; non-parametric Spearman rank correlation coefficient $rs = -0.564$, two-tailed $P = 0.01$; Fig. 2). In contrast, and functioning as a negative control, the correlation between mean CDA and parallel betasheet conformation parameters is not statistically significant ($r = -0.28$, two-tailed $P = 0.23$, not shown). The presumed effect of CDA is specific for formation of antiparallel, not parallel, betasheets.

The variation around the regression line is similar for negative and positive CDA ranges (Fig. 2). Hence the determinism of CDA on conformation is comparable for 5′ versus 3′ CDA dominance: effects are independent of coding importance of codon positions. In other words, the 'information' in CDA that is relevant to protein secondary structure is similar for asymmetry at first and third codon positions. Alternative scores (Section 4.1) do not change qualitatively the results (P values for rs remain above 0.05).

The correlation between mean CDA of codons assigned to amino acids and these amino acids' antiparallel betasheet conformational indices might be due to transitivity, due to associations between CDA and tRNA synthetase classes (see above section) and the association between tRNA synthetase class and conformational indices. In order to control for effects of tRNA synthetase classes, we calculate mean CDA and mean antiparallel betasheet index separately for each tRNA synthetase class. These means are subtracted from CDA and conformational indices of each amino acid in that respective class. These values are residual CDA and conformational indices after excluding effects of tRNA synthetase classes. Residual CDA and residual antiparallel betasheet indices correlate negatively ($r = -0.435$, $P = 0.0275$; $rs = -0.461$, $P = 0.0205$, one tailed tests). Hence the correlation between CDA and antiparallel betasheet indices is not indirect, through colinearity with tRNA synthetase classes.

The association between CDA and antiparallel betasheet indices has rs with $P < 0.05$ for eight among ten alternative scores (as in Section 4.1) after controlling for tRNA synthetase class. The genetic code seems structured so as to enable synergistic effects of CDA and tRNA synthetase classes on antiparallel betasheet formation, presumably by cotranslational protein folding.

Independently of the correlation between CDA and antiparallel betasheet conformation parameters, a weaker correlation exists between CDA and alpha-helix conformation parameters ($r = -0.556$, $P = 0.011$; $rs = -0.499$, $P = 0.05$, two-tailed test, not shown). This further correlation confirms that CDA affects protein folding. To our knowledge, these are the first described correlations between a codon property and secondary structure conformational parameters of assigned amino acids. CDA < 0 associates independently with each alpha and antiparallel beta conformational indices, in line with the literature on cotranslational protein folding [136–140]. Hence according to the working hypothesis, similar kinetic conditions favor each of these two very different secondary structures. Presumably, factors other than CDA (for example chain polarity) determine whether an alpha helix rather than an antiparallel betasheet is initiated during peptide elongation.

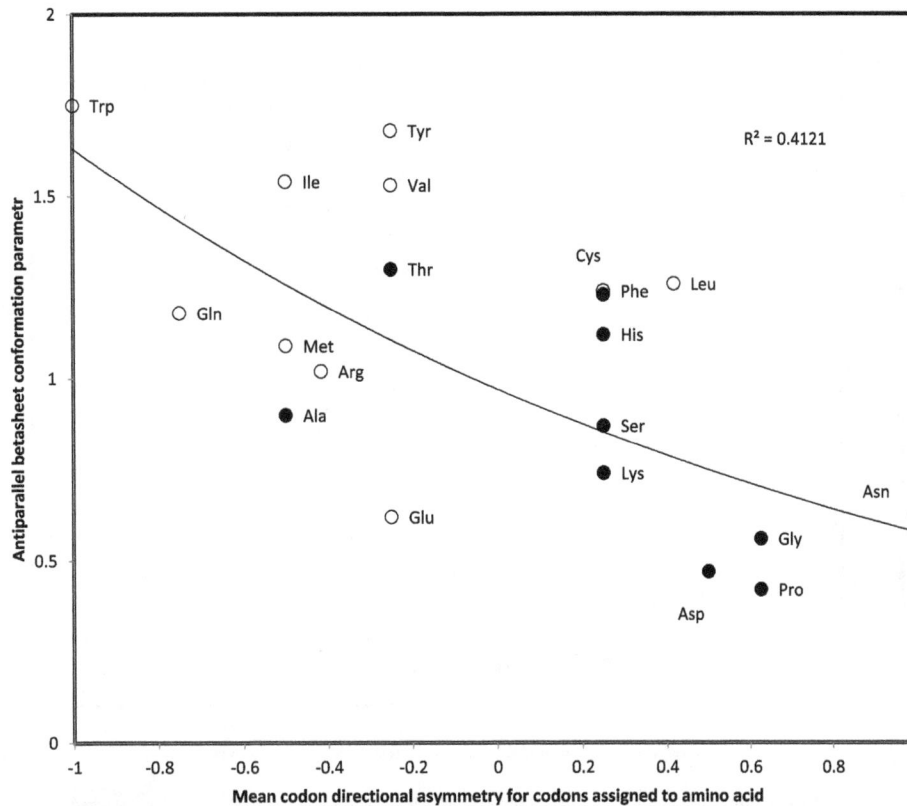

Fig. 2. Antiparallel betasheet conformation parameter of amino acids as a function of the mean codon directional asymmetry (CDA) of codons assigned to that amino acid, for the standard genetic code. Amino acids aminoacylated by tRNA synthetases from class I have open circles, filled circles are for tRNA synthetases from class II.

7. Codon Directional Asymmetry and Prediction of Protein Secondary Structure

The correlation between CDA and conformation parameters might have two causes. First, it could be intrinsic to the genesis of the genetic code, but relatively inconsequent to modern organisms. Secondly, CDA still affects protein folding. In the latter case, correlations between CDA and secondary structure conformation parameters could explain that some synonymous mutations perturb protein function. Indeed, several amino acids have some synonymous codons with opposite CDA, such as for alanine, leucine, serine and valine. Putatively, this would indicate that for these amino acids, synonymous codons with CDA < 0 occur preferentially for mRNA regions coding for antiparallel betasheets, and those with CDA > 0 in other mRNA regions.

Codon usage frequencies are adapted to minimise effects of mutations and translation errors [149–152]. Hence weighing mean CDA for a given amino acid according to observed synonymous codon usages might increase correlations between CDA and conformation parameters. However, this is not the case for the pool of genes encoded by the human nucleus, nor those coded by the human mitogenome: correlations become in both cases weaker (not shown).

CDAs of stop codons are negative, suggesting a bias for amino acids with high tendencies to participate in antiparallel betasheets when amino acids are inserted at stop codons. Indeed, the evolution of mitochondrial genetic codes seems best reconstructed when assuming insertion of amino acids at stops [153], in line with coevolution between predicted suppressor tRNAs [154–156] and protein alignment analyses [16,78,79,157–161]. However, frequencies of amino acids inserted at stops [71,162–166] do not significantly correlate with antiparallel betasheet conformation parameters.

This does not mean that associations between synonymous codons in modern mRNAs and secondary structures of modern proteins do not exist. However, this suggests that testing these predictions is not as straightforward as it seems. Among others, secondary structure annotations available in GenBank don't indicate whether a betasheet is parallel or antiparallel. Hence these tests will require involvement of more adequately equipped specialised proteomics teams (for example Caudron and Jestin [147]). Until then, the contribution of CDA for improving secondary structure predictions [26,167], especially such based on optimization of multiple approaches [168], will remain speculative.

8. Mitochondrial Genetic Codes Optimise Codon Directional Asymmetry

Many variant genetic codes are from mitochondria [169]. The reduced mitogenomes almost exclusively encode for mitochondrial transmembrane proteins, which include mainly antiparallel betasheets. In contrast, nuclear genomes encode also for large proportions of cytosolic proteins, which include much fewer betasheets. Hence, we predict that the correlation between CDA and antiparallel betasheet conformation parameters is weaker for genetic codes associated with nucleus-encoded proteomes than for mitochondrial genetic codes. The correlation in Fig. 2 (for the standard genetic code) is calculated for the remaining genetic codes listed by Elzanowski and Ostell [169], after recalculating mean amino acid CDA, considering codon-amino acid reassignments. The correlation's strength for each genetic code is estimated by the Pearson correlation coefficient r.

The correlation between tRNA synthetase classes and CDA is also calculated, by assigning to tRNA synthetase classes I and II values '1' and '2', respectively, and calculating the Pearson correlation coefficients r between this dummy variable representing tRNA synthetase classes and the mean CDA of codons assigned to the corresponding amino acid, for each variant genetic code. The CDA-antiparallel betasheet correlation coefficients are plotted as a function of the CDA-tRNA synthetase class correlation coefficients for the various genetic codes (Fig. 3).

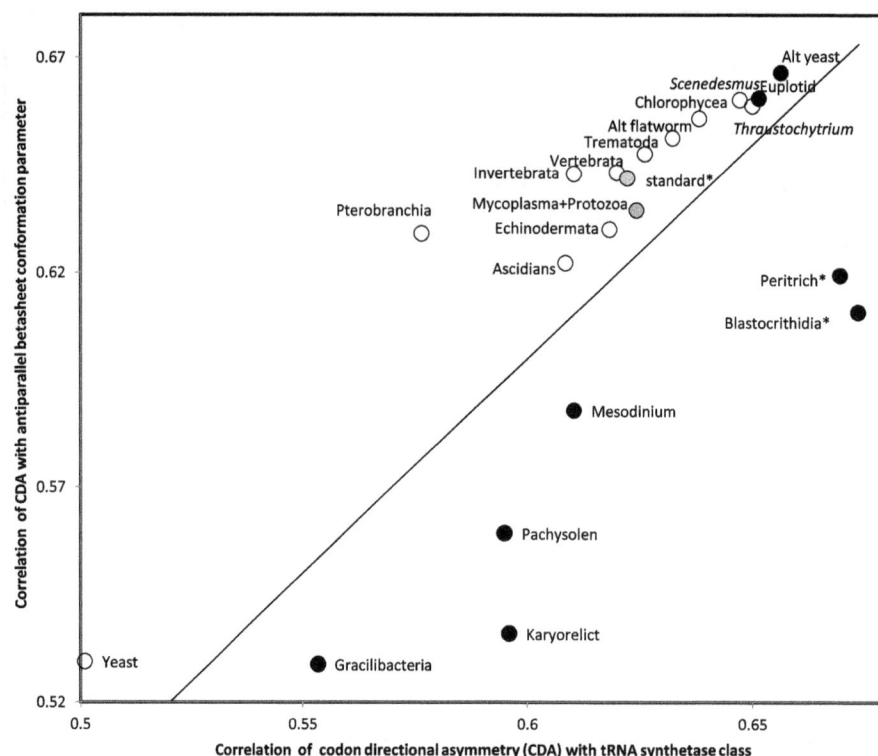

Fig. 3. Correlation between antiparallel betasheet conformation parameter of amino acids and mean directional asymmetry (CDA) of codons assigned to that amino acid as a function of the correlation between CDA and the tRNA synthetase class for the corresponding amino acid for different genetic codes. Correlations are Pearson correlation coefficients. Filled/open circles are nuclear/mitochondrial genetic codes, shaded circles are for genetic codes existing in nuclei and mitochondria. The line indicates y = x. Nuclear genetic codes tend to optimise the association between CDA and tRNA synthetase classes, mitochondrial genetic codes tend to optimise the association between CDA and the antiparallel betasheet conformation parameter. Most mitogenome-encoded proteins are transmembrane proteins, hence antiparallel betasheets are particularly frequent in these proteins. Hence genetic code evolution optimises the CDA-antiparallel betasheet association in mitochondria. Open circles: mitochondrial genetic codes; filled circles: nuclear genetic codes; shaded circles: genetic codes used in nuclei and mitochondria.

The line in Fig. 3 indicates y = x, meaning that both correlations have equal strengths. Note that in context of this particular section, Pearson correlation coefficients are used as quantitative estimates of the strength of a correlation, not as test statistics to infer that a correlation exists.

All eleven mitochondrial genetic codes have stronger correlations between CDA and antiparallel betasheet conformation parameters than between CDA and tRNA synthetase classes. Obtaining this result for all eleven mitochondrial genetic codes has $P = 0.00049$ (two-tailed sign test). Two additional genetic codes occur in nuclear and mitochondrial genomes: the standard genetic code, and the *Mycoplasma/Spiroplasma* genetic code that also occurs in mold, protozoan and coelenterate mitochondria. These two genetic codes follow the pattern observed for the eleven genetic codes only found in mitochondria.

Six among eight genetic codes associated only with nuclear genomes are below the line y = x in Fig. 2, indicating that the CDA-tRNA synthetase class correlation is frequently a greater constraint for nuclear genetic codes than mitochondrial ones. This qualitative difference between nuclear and mitochondrial genetic codes has $P = 0.001$ (two-tailed Fisher exact test). This divide might reflect different constraints on protein folding for populations of mitochondrion-encoded versus nucleus-encoded proteins. This pattern might indicate stronger synergy between effects of CDA and tRNA synthetase class on cotranslational folding for nucleus-encoded proteins translated in the cytosol than mitogenome-encoded ones.

These results indicate that associations between CDA and conformation parameters, and between CDA and tRNA synthetase classes, drive differentially evolutions of mitochondrial versus nuclear genetic codes. Tentatively, amino acid positioning on the tRNA acceptor stem is less relevant for mitochondrial translation than CDA, the opposite is true for cytosolic translations.

9. Whole Organism Properties and Codon Directional Asymmetry

Whole organism properties correlate sometimes with molecular properties [170,171]: morphological versus molecular rates of evolution [172–174]; growth rates and genome sizes [175–178]; and metabolic costs of protein synthesis [179]; body temperatures and predicted expanded codons [180–183]; developmental stability estimated by lateral differences between bilateral morphological traits and accuracy of various aspects of molecular processes, such as replication [184], ribosomal translation [20,185], and tRNA loading [14]. CDA might also correlate with whole organism properties.

9.1. Lepidosaurian Body Temperature and Codon Directional Asymmetry

Temperature reflects noise in molecular movements, potentially affecting contranslational protein folding, which indeed depends on optimal temperatures [186]. Hence, formation of antiparallel betasheets might be impeded by high temperatures. Therefore, we expect that negative CDAs promote betasheet formation despite high temperature. Hence when comparing the mean CDA calculated across all 13 membrane-embedded mitogenome-encoded proteins of different organisms, we expect that organisms with high temperatures have low mean CDA for the same homologous genes. Indeed, the mean CDA of lepidosaurian mitochondrion-encoded proteins decreases with their body temperature (ro = -0.283, one tailed $P = 0.018$, Fig. 4, temperature data compiled for species with complete mitogenome available in GenBank by Seligmann and Labra [183], therein Table 1).

This correlation is also statistically significant within the family Lacertidae (ro = -0.842, one-tailed $P = 0.001$). It is negative for Agamidae (ro = -0.255, one-tailed $P = 0.238$), Gekkota (ro = -0.25, one-tailed $P = 0.258$), iguanid lizards (ro = -0.333, one-

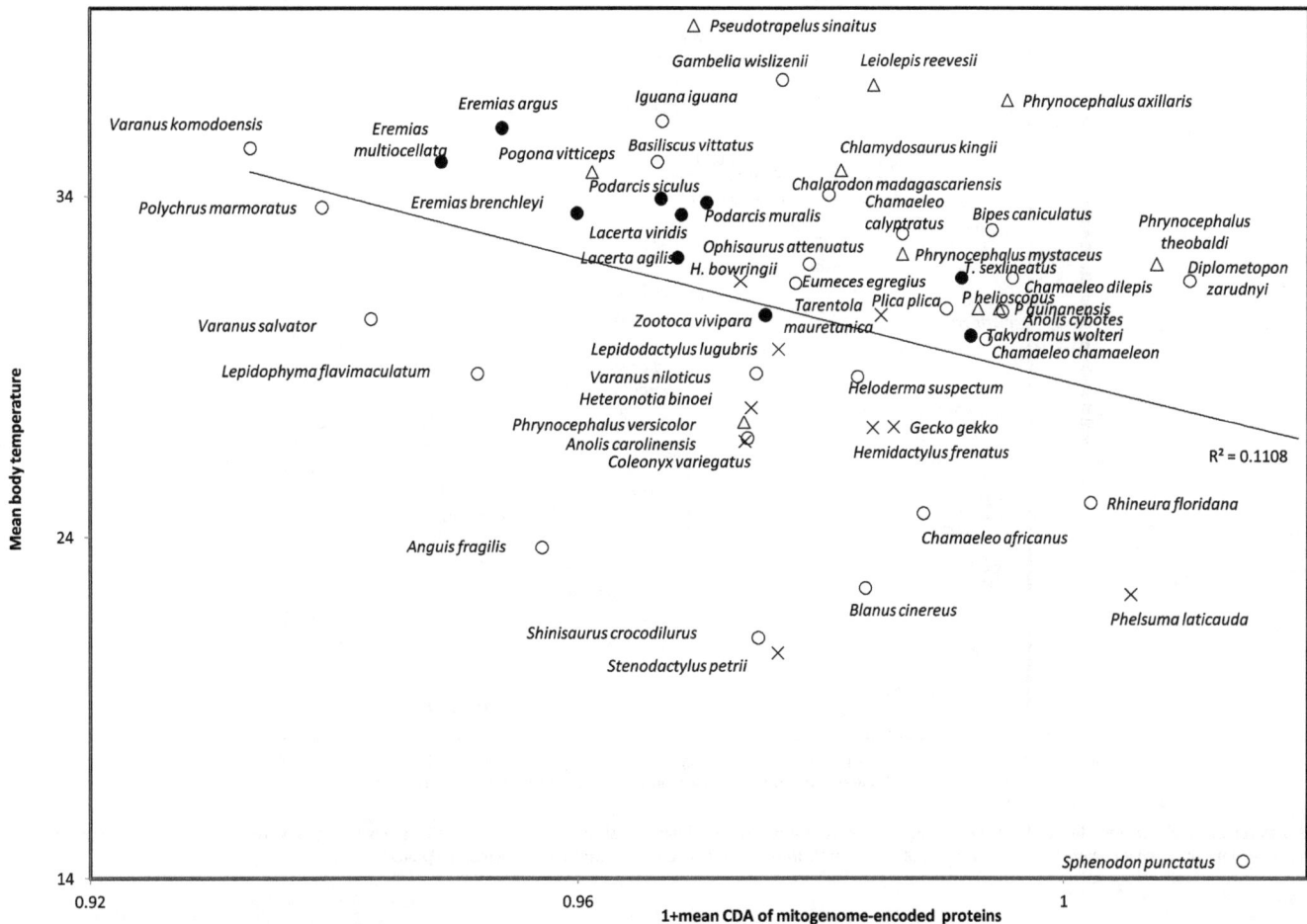

Fig. 4. Lepidosaurian body temperature as a function of mean codon directional asymmetry of codons in protein coding genes encoded by complete mitogenomes available in GenBank. Compilation of body temperatures and mitochondrial genomes as in Table 1 of Seligmann and Labra [183]. Agamidae are indicated by triangles, Gekkota by crosses and Lacertidae by filled circles. Species from various other families have open circles.

tailed $P = 0.21$), Varanidae (ro $= -1.00$, one-tailed $P = 0.005$) and *Chamaeleo* (ro $= -0.40$, one-tailed $P = 0.30$) and for the pool of remaining isolated species from various families (*Heloderma, Shinisaurus, Lepidophyma, Sphenodon* (ro $= -0.238$, one tailed $P = 0.285$). The correlation is positive for Amphisbaenia (ro $= 0.40$, one tailed $P = 0.30$). Hence seven among eight phylogenetically independent samples yield negative correlations, which is a significant majority according to a sign test ($P = 0.0176$). Considering the qualitative direction of correlations for phylogenetically independent species samples follows the principle of phylogenetically independent contrasts [187]. This confirms that positive results are not confounded by phylogenetic inertia among species. Results of this sign test are valid independently of P value adjustments for multiple tests.

GC contents could confound this correlation, because G:C base pairs are linked by three hydrogen interactions, while A:T and A:U base pairs by only two hydrogen bridges. Hence GC contents usually increases with temperature, as it confers higher stability to structures formed by nucleotide chains [188,189,190]. However, GC codon content does not correlate with body temperature for mitochondria of the above mentioned lepidosaurian species ($r = -0.0425$, one-tailed $P = 0.379$). This is in line with results from various analyses [191–193]; that didn't detect the expected GC-temperature correlation. This negative control stresses that the association in Fig. 4 is not trivial.

9.2. Developmental Stability and CDA

Molecular noise (in terms of erratic molecular movements) affecting mitochondrial transmembrane protein folding might cause developmental inaccuracies at the whole organism level. Hence, we explore the correlation between mean CDA of mitogenome-encoded proteins and developmental stability of the 4th toe of Lepidosauria, estimated by the Pearson correlation coefficient r between subdigital lamellae counts on left and right sides (data from [194–198]). Developmental stability/accuracy decreases with mean CDA of mitogenome-encoded proteins (ro $= -0.316$, one-tailed $P = 0.0235$), as expected by the working hypothesis. However, analyzing separately species grouped according to phylogenetic groups (as in previous section) yields negative correlations only in five among eight groups, which is not statistically significant at $P < 0.05$ according to a one sided sign test. Hence this preliminary result on CDA and developmental stability is at best tentative.

9.3. Lifespan and CDA

Patterns between CDA and temperature, and CDA and developmental stability (Figs. 4 and 5) suggest that CDA < 0 for mitogenome-encoded proteins associates with longevity. For this purpose, we compared codon contents in mitogenomes of 112 semi-supercentenarians and 96 centenarians versus those of 97 healthy young controls [199,200] (Table 3). Codons with CDA $= -1$ are more frequent in supercentenarians than in controls for seven among eight comparisons, which is a significant majority according to a one tailed sign test ($P = 0.0176$). No tendencies are observed for other CDA values ($-0.5, 0, 0.5, 1$), nor for comparisons between centenarians and controls. The result is suggestive that CDA < 0 could contribute to extreme longevity, but the high number of tests and the small differences in codon frequencies stress cautious interpretation.

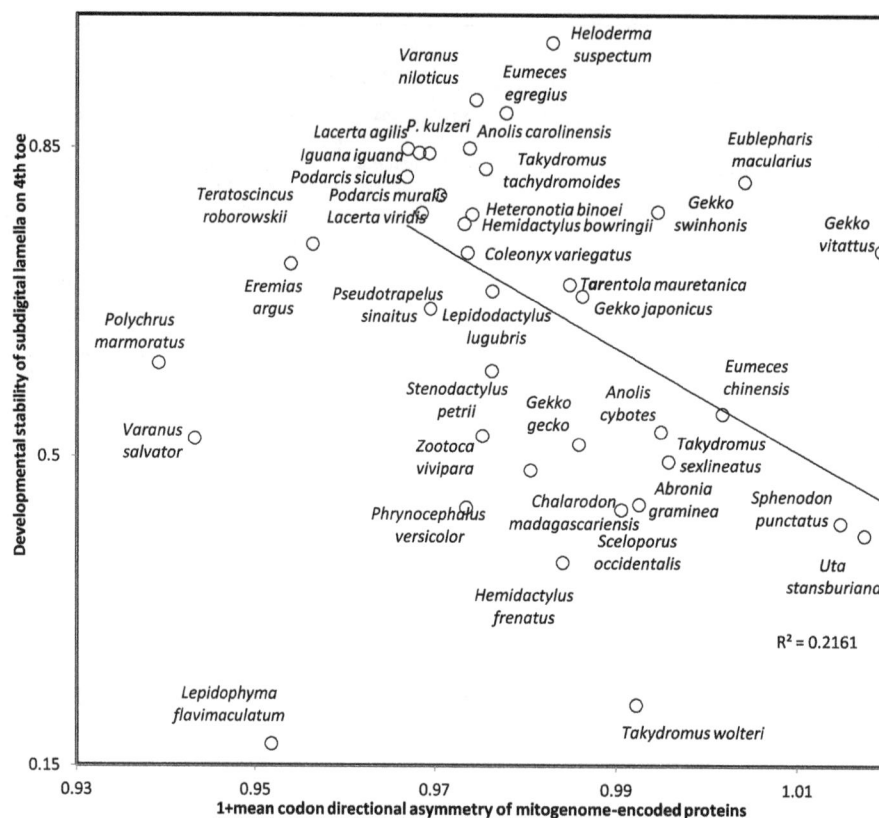

Fig. 5. Developmental stability of bilateral counts of subdigital lamellae on 4th toe of Lepidosauria (estimated by Pearson correlation coefficients r between counts on left and right sides) as a function of mean codon directional asymmetry (CDA) of codons in all 13 genes of mitogenome-encoded transmembrane proteins.

Overall, analyses weakly confirm predictions for correlations between CDA and whole organism properties (body temperature, developmental stability, longevity). These suggest that analyses considering additional information, such as residue-specific location in three dimensional protein structures, might yield positive results. More up-to-date methods for including phylogenetic information in relation to evolutionary adaptive optima might also alter conclusions [201].

9.4. Replicational Deamination Gradients

Mitochondrial DNA replication differs from nuclear chromosome replication [202] and is usually strand asymmetric resulting in replicational deamination gradients where $C \rightarrow T$ and $A \rightarrow G$ substitutions exceed reversed mutations proportionally to the time spent single stranded during replication [203–208]. Inverting the direction of the light strand replication origin also inverts the direction of the replicational deamination gradients [209–217]. These physico-chemical mutation pressures could affect CDA according to gene locations on the mitogenome, independently of protein properties.

Mean CDA of the 13 human mitogenome-encoded proteins does not correlate with time spent single stranded by that gene during replication, assuming light strand replication initiates at the OL, the light strand replication origin (ro = 0.033, P = 0.92, two tailed test). DNA templating for tRNA genes presumably also functions sometimes as replication origins [184,218,219]. Integrating the possibility of these multiple replication origins yields gene-wise single-strand durations that converge with transcriptional singlestrandedness [220]. The correlation between transcriptional duration of singlestrandedness and mean gene CDA is also not statistically significant (ro = 0.418, P = 0.156, two tailed test). Hence, we do not detect statistically significant

effects of mutation pressures on mean CDA of human mitochondrial genes.

9.5. Adjusting Statistical Significances for Multiple Tests

Analyses that include several tests have to adjust P values according to the number of tests. This is because, when deciding that a result is positive at $P < 0.05$, when k tests are performed, on average, $k \times 0.05$ tests are false positives. Bonferroni's correction considers that when performing k tests, results are statistically significant at P = 0.05 for any specific test among k tests if $P < 0.05/k$. This correction is reputedly overconservative [221,222]. Unadjusted Ps minimise risks of false negative results, Bonferroni's method minimises risks of false positives. The Benjamini-Hochberg adjustment for false discovery rates [223] optimises between these two risks and seems most adequate [224]. This method ranks all k P values from highest to lowest (best), adjusted Ps are the product of P with k divided by the rank i, where i ranges from 1 to k. This means that the 'best' (lowest) P is unchanged, and that the 'worst' (highest) P value after adjustment follows Bonferroni's adjustment. Ps with intermediate rank are intermediate between these extremes.

Here we consider only P values from non-parametric tests, when also parametric tests were done. For some of the associations described, more than one test was done, but these are then summarised by a test that integrates the previous tests. Adjustments consider in these cases only the latter P value. Along this approach a total of 29 hypothesis tests were done, as detailed in Table 2. Control analyses (such as with GC contents, and mutational gradients, in total 29 tests) are also included in the list of multiple tests. These are not related to the main CDA hypothesis and could arguably be excluded. Excluding controls does not alter qualitatively results of the adjustments of P values.

Table 2

Benjamini-Hochberg adjustment of P values of non-redundant hypothesis tests. The total number of tests including controls is 58 (Rank 1 and adjusted P1), excluding controls and considering only tests pertaining directly to CDA, there are 29 tests (rank 2 and adjusted P2). Only tests with unadjusted $P < 0.05$ are shown, all these tests pertain to CDA directly.

Test	P	Rank 1	Adj P1	Rank 2	Adj P2
Number of mitochondrial genetic codes above line in Fig. 3	0.00050	58	0.00050	29	0.00050
Number of mitochondrial genetic codes above line vs number of nuclear codes below line in Fig. 3	0.00100	57	0.00102	28	0.00100
tRNA synthetase classes and CDA	0.00200	56	0.00207	27	0.00207
tRNA synthetase classes and mean CDA of codons assigned to amino acids	0.00295	55	0.00312	26	0.00319
Contacts between Fujimoto's tetrahedron faces	0.00960	54	0.01031	25	0.01075
Correlation CDA-antiparallel betasheet indices	0.01000	53	0.01094	24	0.01167
Absolute CDA and circular code	0.01600	52	0.01785	23	0.01948
Temperature and mean CDA of 13 lepidosaurian mitogenome-encoded proteins	0.01760	51	0.02002	22	0.02240
Human lifespan and mitochondrial codon usages-CDA	0.01760	50	0.02042	21	0.02347
Contacts within Fujimoto's tetrahedron faces	0.01800	49	0.02131	20	0.02520
Partial correlation CDA-antiparallel betasheet indices	0.02050	48	0.02477	19	0.03021
CDA-developmental stability	0.02350	47	0.02900	18	0.03656
Absolute CDA-reading frame retrieval capacity	0.02600	46	0.03278	17	0.04435
Alpha helix-CDA	0.05000	45	0.06444	16	0.08750

The analysis for codon usage associated with lifespan includes 10 tests (for CDA values -1, -0.5, 0, 0.5, and 1, and this for comparisons between controls and centenarians, and between controls and supercentenarians). Among unadjusted P values with $P < 0.05$, only the adjusted P value for the correlation between mean CDA of codons assigned to amino acids and the amino acids' alpha helix conformational indices is above 0.05. This occurs when considering all 58 tests, and when considering only the 29 tests directly pertaining to the working hypothesis about CDA. Qualitatively, results of P adjustments are robust in relation to numbers of tests included in this analysis: for example, for P with rank 17 to get $P > 0.05$ after adjustment, one requires k = 89 when including negative controls and k = 33 when excluding negative controls. Hence even if one was to increase numbers of tests included in the analyses, the relevant cutoff property of the distribution of adjusted

Table 3

Mean codon frequencies (promil) in the 13 mitogenome-encoded genes of three groups of Japanese males: 97 healthy controls, 112 semi-supercentenarians and 96 centenarians from references [199,200].

Codon	CDA	Control	Super	Cent	Codon	CDA	Control	Super	Cent
UUU	0	20.34	20.32	20.34	UAU	0.00	12.11	12.13	12.15
UUC	0.5	36.59	36.61	36.60	UAC	−0.50	23.39	23.35	23.35
UUA	1	19.07	19.04	19.01	UAA	−1.00	2.09	2.09	2.10
UUG	1	4.59	4.57	4.57	UAG	−0.50	0.81	0.81	0.80
CUU	−0.5	17.02	17.03	17.00	CAU	0.50	4.71	4.71	4.71
CUC	0	43.89	43.88	43.89	CAC	0.00	20.80	20.80	20.79
CUA	0.5	72.88	72.87	72.90	CAA	−1.00	21.61	21.61	21.59
CUG	0.5	11.74	11.78	11.79	CAG	−0.50	2.08	2.09	2.10
AUU	−1	33.06	33.03	33.07	AAU	1.00	8.38	8.39	8.41
AUC	−0.5	51.34	51.30	51.25	AAC	1.00	34.72	34.74	34.72
AUA	0	43.77	43.81	43.82	AAA	0.00	22.37	22.40	22.39
AUG	−0.5	10.72	10.74	10.73	AAG	0.50	2.63	2.61	2.61
GUU	−1	8.19	8.19	8.18	GAU	0.50	3.89	3.92	3.93
GUC	−0.5	12.61	12.64	12.68	GAC	0.50	13.43	13.38	13.38
GUA	0.5	18.39	18.40	18.36	GAA	−0.50	16.87	16.85	16.86
GUG	0	4.71	4.71	4.73	GAG	0.00	6.23	6.25	6.24
UCU	0	8.42	8.42	8.41	UGU	0.00	1.30	1.30	1.31
UCC	−0.5	25.99	25.99	26.00	UGC	0.50	4.50	4.50	4.50
UCA	0.5	21.81	21.81	21.80	UGA	−0.50	24.48	24.46	24.42
UCG	0.5	1.80	1.79	1.80	UGG	−1.00	2.91	2.92	2.97
CCU	0.5	10.70	10.66	10.68	CGU	−0.50	1.80	1.80	1.80
CCC	0	31.39	31.43	31.42	CGC	0.00	6.80	6.80	6.80
CCA	1	13.71	13.71	13.72	CGA	−0.50	7.40	7.40	7.40
CCG	1	1.78	1.78	1.78	CGG	−1.00	0.50	0.50	0.50
ACU	−0.5	13.41	13.39	13.43	AGU	0.50	3.70	3.69	3.69
ACC	−1	40.44	40.45	40.36	AGC	0.50	10.26	10.27	10.29
ACA	0	34.80	34.78	34.77	AGA	0.00	0.30	0.29	0.30
ACG	0.5	2.57	2.58	2.58	AGG	−0.50	0.30	0.30	0.30
GCU	−0.5	11.73	11.73	11.74	GGU	1.00	6.30	6.32	6.33
GCC	−1	32.55	32.56	32.60	GGC	1.00	22.88	22.87	22.85
GCA	−0.5	21.60	21.58	21.60	GGA	0.50	18.00	18.02	18.03
GCG	0	2.08	2.09	2.08	GGG	0.00	8.55	8.52	8.51

Ps is relatively robust, so that issues related to multiple tests are unlikely to alter conclusions.

10. A New Directional Codon Dimension

Intuitively, it seems conceivable that CDA, via its plausible effects on codon-anticodon interactions, affects cotranslational protein folding. However, developing a mechanistic scenario that explains why this effect should occur for antiparallel betasheets rather than parallel ones, or for alpha helices, is more difficult. We propose that some (unspecified) conformations depend on translational speed. Other conformations might be favored by random movements of the tRNA's loaded acceptor stem in relation to the elongating peptide, versus more directed movements of that stem, hence some ratio between kinetic noise and direction. Our educated guess (but nothing beyond that) is that CDA relates more to the latter type of mechanisms. We also lack clues on why CDA < 0 promotes antiparallel betasheets, and CDA > 0 prevents them. Alpha helices might be more simple structures that require less order than antiparallel betasheets. A similar rationale might function for parallel and antiparallel betasheets. In addition, the ratio between parallel and antiparallel betasheets is about 1:7 [26]: the genetic code might be optimised towards 'coding' for the most frequent protein conformation.

The genetic code can be characterised as a hypercomplex mathematical multidimensional symmetry structure [225]. In other terms, the genetic code reminds spontaneously self-organizing structures such as crystals [226, 227]. Crystals result from specific rules organizing relations between atoms. Similarly, but at a much higher level of molecular complexity, the genetic code organises relations between nucleic and amino acid sequences. The genetic code might be thought as an imaginary polyhedron with 64 triangular faces (64 codons with three nucleotide positions). The geometrical form of this structure remains unknown, but several symmetries implied by RNA/DNA structure and chemistry are known, such as reverse-complementarity (implied by the double helix structure), and the purine-pyrimidine as well as the alpha-keto groupings of nucleic acids. Formulation of a generalised description of this complex structure is a difficult task. It is simplified by projections of the complex structure on specific scales/planes of probable biological interest.

Here learned intuition detects a new symmetry property, based on codon content directionality. Analyses here can be seen as projecting that complex genetic code structure on the CDA scale, enabling to detect some new properties of the genetic code. The details of the scale of CDA scores as presented here is probably inaccurate and will hopefully be amended. CDA implies that a directional dimension that had not been apprehended links codons and amino acids: biologically meaningful information relating to protein structure is embedded in the comparison between codons and their reversed (not reverse-complemented) sequence. This palindrome-minded approach to codons probably reflects error-correcting properties of primitive genetic code(s) [228].

11. Conclusions

A property of codons, codon directional asymmetry (CDA), is defined for the genetic code. Codons are classified into symmetric (CDA = 0), 5′- and 3′-asymmetric (negative and positive CDA). CDA maps non-randomly on Fujimoto's tetrahedral representation of the genetic code. Symmetric codons are the most common codons in frame-error-correcting codes, such as comma-free and circular codes. Most codons assigned to amino acids aminoacylated to cognate tRNAs by tRNA synthetases class I have CDA < 0, those assigned to cognates of tRNA synthetases class II have usually CDA > 0.

Amino acid tendencies to participate in antiparallel betasheets decrease with CDA. Results suggest that CDA and tRNA synthetase class affect spatial kinetics of peptide elongation. These spatial kinetics affect local peptide elongation rates, which determine cotranslational peptide folding during peptide synthesis. Hence CDA, a property of gene sequences, bears useful information to predict protein folding. Some synonymous codons have CDA with opposite signs, potentially explaining how some synonymous mutations alter protein function.

CDA probably played a role in the evolution of genetic codes. Mitochondrial genetic codes optimise associations between CDA and antiparallel betasheet formation, nuclear genetic codes tend to optimise associations between CDA and tRNA synthetase class. This difference might mean that synergistic effects of CDA and tRNA synthetase class on cotranslational protein folding are stronger for nuclear than mitochondrial genetic codes. CDA affects codon-amino acid (re)assignments, hence plays an important role in genetic code evolution.

Preliminary analyses suggest that average CDA of mitochondrion-encoded proteins decreases with body temperature, increases developmental stability and lifespan, but further controlled analyses are required to confirm these potential whole organism effects of codon directional asymmetry (CDA).

Acknowledgments

This study was supported by Méditerranée Infection and the National Research Agency under the program "Investissements d'avenir", reference ANR-10-IAHU-03 and the A*MIDEX project (no ANR-11-IDEX-0001-02).

References

[1] Woese CR. Order in the genetic code. Proc Natl Acad Sci 1965;54:71–5.
[2] Di Giulio M. The extension reached by the minimization of the polarity distances during the evolution of the genetic code. J Mol Evol 1989;29:288–93.
[3] Haig D, Hurst LD. A quantitative measure of error minimization in the genetic-code. J Mol Evol 1991;33:412–7.
[4] Ardell DH. On error minimization in a sequential origin of the standard genetic code. J Mol Evol 1998;47:1–13.
[5] Freeland SJ, Hurst LD. Load minimization of the genetic code: history does not explain the pattern. Proc R Soc B Biol Sci 1998;265:2111–9.
[6] Freeland SJ, Hurst LD. The genetic code is one in a million. J Mol Evol 1998;47: 238–48.
[7] Ardell DH, Sella G. On the evolution of redundancy in genetic codes. J Mol Evol 2001;53:269–81.
[8] Freeland SJ, Wu T, Keulmann N. The case for an error minimizing standard genetic code. Origins Life Evol B 2003;33:457–77.
[9] Błażej P, Miasojedow B, Grabińska M, Mackiewicz P. Optimization of mutation pressure in relation to properties of protein-coding sequences in bacterial genomes. PLoS One 2015;10(6):e0130411.
[10] Błażej P, Mackiewicz D, Grabińska M, Wnętrzak M, Mackiewicz P. Optimization of amino acid replacement costs by mutational pressure in bacterial genomes. Sci Rep 2017;7(1):1061.
[11] Blazej P, Wnetrzak M, Mackiewicz P. The role of crossover operator in evolutionary-based approach to the problem of genetic code optimization. Biosystems 2016;150: 61–72.
[12] de Oliveira LL, de Oliveira PS, Tinos R. A multiobjective approach to the genetic code adaptability problem. BMC Bioinformatics 2015;16:52.
[13] Seligmann H. Do anticodons of misacylated tRNAs preferentially mismatch codons coding for the misloaded amino acid? BMC Mol Biol 2010;11:41.
[14] Seligmann H. Error compensation of tRNA misacylation by codon-anticodon mismatch prevents translational amino acid misinsertion. Comput Biol Chem 2011;35(2):82–95.
[15] Seligmann H. Coding constraints modulate chemically spontaneous mutational replication gradients in mitochondrial genomes. Curr Genomics 2012;13(1): 38–52.
[16] Barthélémy RM, Seligmann H. Cryptic tRNAs in chaetognath mitochondrial genomes. Comput Biol Chem 2016;62:119–32.
[17] Seligmann H, Pollock DD. The ambush hypothesis: hidden stop codons prevent off-frame gene reading. DNA Cell Biol 2004;23(10):701–5.
[18] Itzkovitz S, Alon U. The genetic code is nearly optimal for allowing additional information within protein-coding sequences. Genome Res 2007;17(4):405–12.
[19] Seligmann H. Cost minimization of ribosomal frameshifts. J Theor Biol 2007; 249(1):162–7.
[20] Seligmann H. The ambush hypothesis at the whole-organism level: off frame, 'hidden' stops in vertebrate mitochondrial genes increase developmental stability. Comput Biol Chem 2010;34(2):80–5.
[21] Singh TR, Pardasani KR. Ambush hypothesis revisited: evidences for phylogenetic trands. Comput Biol Chem 2009;33(3):239–44.
[22] Tse H, Cai JJ, Tsoi HW, Lam EP, Yuen KY. Natural selection retains overrepresented out-of-frame stop codons against frameshift peptides in prokaryotes. BMC Genomics 2010;11:491.
[23] Křížek M, Křížek P. Why has nature invented three stop codons of DNA and only one start codon? J Theor Biol 2012;304:183–7.
[24] Gilis D, Massar S, Cerf NJ, Rooman M. Optimality of the genetic code with respect to protein stability and amino-acid frequencies. Genome Biol 2001;2 [RESEARCH0049].
[25] Guilloux A, Jestin JL. The genetic code and its optimization for kinetic energy conservation in polypeptide chains. Biosystems 2012;109(2):141–4.
[26] Guilloux A, Caudron B, Jestin JL. A method to predict edge strands in beta-sheets from protein sequences. Comput Struct Biotechnol J 2013;7:e201305001.
[27] Wong JT. A co-evolution theory of the genetic code. Proc Natl Acad Sci U S A 1975; 72:1909–12.
[28] Di Giulio M. On the origin of the genetic code. J Theor Biol 1997;187:573–81.
[29] Di Giulio M. The coevolution theory of the origin of the genetic code. J Mol Evol 1999;48:253–5.
[30] Di Giulio M. An extension of the coevolution theory of the origin of the genetic code. Biol Direct 2008;3:37.
[31] Wong JT. The coevolution theory at age thirty. Bioessays 2005;27(4):416–25.
[32] Guimarães RC. Metabolic basis for the self-referential genetic code. Orig Life Evol Biosph 2011;41(4):357–71.
[33] Morgens DW, Cavalcanti ARO. An alternative look at code evolution: using non-canonical codes to evaluate adaptive and historic models for the origin of the genetic code. J Mol Evol 2013;76:71–80.
[34] Guimarães RC. The self-referential genetic code is biologic and includes the error minimization property. Orig Life Evol Biosph 2015;45:69–75.
[35] Di Giulio M. The lack of foundation in the mechanism on which are based the physico-chemical theories for the origin of the genetic code is counterposed to the credible and natural mechanism suggested by the coevolution theory. J Theor Biol 2016;399:134–40.
[36] Di Giulio M. Some pungent arguments against the physico-chemical theories of the origin of the genetic code and corroborating the coevolution theory. J Theor Biol 2017;414:1–4.
[37] Higgs PG, Pudritz RE. A thermodynamic basis for prebiotic amino acid synthesis and the nature of the first genetic code. Astrobiology 2009;9(5):483–90.
[38] Novozhilov AS, Koonin EV. Exceptional error minimization in putative primordial genetic codes. Biol Direct 2009;4:44.
[39] Santos J, Monteagudo A. Genetic code optimality studied by means of simulated evolution and within the coevolution theory of the canonical code organization. Nat Comput 2009;8:719.
[40] Tlusty T. A colorful origin for the genetic code: information theory, statistical mechanics and the emergence of molecular codes. Phys Life Rev 2010;7(3):362–76.
[41] Di Giulio M. The origin of the genetic code: matter of metabolism or physicochemical determinism? J Mol Evol 2013;77:131–3.
[42] Banhu AV, Aggarwal N, Sengupta S. Revisiting the physico-chemical hypothesis of code origin: an analysis based on code-sequence coevolution in a finite population. Orig Life Evol Biosph 2013;43:465–89.
[43] Seligmann H, Amzallag GN. Chemical interactions between amino acid and RNA: multiplicity of the levels of specificity explains origin of the genetic code. Naturwissenschaften 2002;89(12):542–51.
[44] Woese CR, Dugre, Saxinger WC, Dugre SA. The molecular basis for the genetic cocde. Proc Natl Acad Sci U S A 1978;55:966–74.
[45] Weber AL, Lacey JC. Genetic code correlations: amino acids and their anticodon nucleotides. J Mol Evol 1966;11:199–210.
[46] Shu JJ. A new integrated symmetrical table for genetic codes. Biosystems 2017;151: 21–6.
[47] Nemzer LR. A binary representation of the genetic code. Biosystems 2017;155:10–9.
[48] Gonzalez DL, Giannerini S, Rosa R. Strong short-range correlations and dichotomic codon classes in coding DNA sequences. Phys Rev E Stat Nonlin Soft Matter Phys 2008;78(5 Pt 1):051918.

[49] Castro-Chavez F. A tetrahedral representation of the genetic code emphasizing aspects of symmetry. BIOcomplexity 2012;2012(2):1–6.

[50] Fujimoto M. Tetrahedral codon stereo-table[4,702,704. U.S. Patent]; 1987. [http://www.google.com/patents/US4702704].

[51] Arquès DG, Michel CJ. A complementary circular code in the protein coding genes. J Theor Biol 1996;182(1):45–58.

[52] Michel CJ. The maximal C³ self-complementary trinucleotide circular code X in genes of bacteria, eukaryotes, plasmids and viruses. J Theor Biol 2015;380: 156–77.

[53] Michel CJ. The maximal C(3) self-complementary trinucleotide circular code X in genes of bacteria, eukaryotes, plasmids and viruses. Life 2017;7(2):e20.

[54] Ahmed A, Frey G, Michel CJ. Frameshift signals in genes associated with the circular code. In Silico Biol 2007;7(2):155–68.

[55] Ahmed A, Frey G, Michel CJ. Essential molecular functions associated with the circular code evolution. J Theor Biol 2010;264(2):613–22.

[56] Michel CJ. Circular code motifs in transfer and 16S ribosomal RNAs: a possible translation code in genes. Comput Biol Chem 2012;37:24–37.

[57] Michel CJ. Circular code motifs in transfer RNAs. Comput Biol Chem 2013;45: 17–29.

[58] El Soufi K, Michel CJ. Circular code motifs in the ribosome decoding center. Comput Biol Chem 2014;52:9–17.

[59] El Soufi K, Michel CJ. Circular code motifs near the ribosome decoding center. Comput Biol Chem 2015;59(Pt A):158–76.

[60] Michel CJ, Seligmann H. Bijective transformation circular codes and nucleotide exchanging RNA transcription. Biosystems 2014;118:39–50.

[61] El Houmami N, Seligmann H. Evolution of nucleotide punctuation marks: from structural to linear signals. Front Genet 2017;8:36.

[62] Fimmel E, Strüngmann L. Codon distribution in error-detecting circular codes. Life 2016;6(1):e14.

[63] Rumer YB. About the codon systematization in the genetic code. Proc Acad Sci USSR 1966;167:1393–4.

[64] Shsherbak VI. Rumer's rule and transformation in the context of the co-operative symmetry of the genetic code. J Theor Biol 1989;139(2):271–6.

[65] Gumbel M, Fimmel E, Danielli A, Strüngmann L. On models of the genetic code generated by binary dichotomic algorithms. Biosystems 2015;128:9–18.

[66] Seligmann H. Polymerization of non-complementary RNA: systematic symmetric nucleotide exchanges mainly involving uracil produce mitochondrial RNA transcripts coding for cryptic overlapping genes. Biosystems 2013;111(3): 156–74.

[67] Seligmann H. Systematic asymmetric nucleotide exchanges produce human mitochondrial RNAs cryptically encoding for overlapping protein coding genes. J Theor Biol 2013;324:1–20.

[68] Seligmann H. Species radiation by DNA replication that systematically exchanges nucleotides? J Theor Biol 2014;363:216–22.

[69] Seligmann H. Mitochondrial swinger replication: DNA replication systematically exchanging nucleotides and short 16S ribosomal DNA swinger inserts. Biosystems 2014;125:22–31.

[70] Seligmann H. Sharp switches between regular and swinger mitochondrial replication: 16S rDNA systematically exchanging nucleotides A↔T + C↔G in the mitogenome of Kamimuria wangi. Mitochondrial DNA A DNA Mapp Seq Anal 2016;27(4):2440–6.

[71] Seligmann H. Translation of mitochondrial swinger RNAs according to tri-, tetra- and pentacodons. Biosystems 2016;140:36–48.

[72] Delarue M. An asymmetric underlying rule in the assignment of codons. RNA 2007; 13:161–9.

[73] Eriani G, Delarue M, Poch O, Gangloff J, Moras D. Partition of tRNA synthetases into two classes based on mutually exclusive sets of sequence motifs. Nature 1990;347: 203–6.

[74] Cusack S. Aminoacyl-tRNA synthetases. Curr Opin Struct Biol 1997;7:881–9.

[75] Sprinzl M, Cramer F. Site of aminoacylation of tRNAs from Escherichia coli with respect to the 2′- or 3′3′-hydroxyl group of the terminal adenosine. Proc Natl Acad Sci U S A 1975;72:3049–53.

[76] Arnez JG, Moras D. Aminoacyl-tRNA synthetase tRNA recognition. Oxford: IRL Press; 1994 61–81.

[77] Jestin JL, Soulé C. Symmetries by base substitutions in the genetic code predict 2′2′ or 3′3′ aminoacylation of tRNAs. J Theor Biol 2007;247(2):391–4.

[78] Seligmann H. Overlapping genes coded in the 3′-to-5′-direction in mitochondrial genes and 3′-to-5′ polymerization of non-complementary RNA by an 'invertase'. J Theor Biol 2012;315:38–52.

[79] Seligmann H. Triplex DNA:RNA, 3′-to-5′ inverted RNA and protein coding in mitochondrial genomes. J Comput Biol 2013;20(9):660–71.

[80] Seligmann H. Systematic exchanges between nucleotides: genomic swinger repeats and swinger transcription in human mitochondria. J Theor Biol 2015;384:70–7.

[81] Seligmann H. Swinger RNAs with sharp switches between regular transcription and transcription systematically exchanging ribonucleotides: case studies. Biosystems 2015;135:1–8.

[82] Seligmann H. Swinger RNA self-hybridization and mitochondrial non-canonical swinger transcription, transcription systematically exchanging nucleotides. J Theor Biol 2016;399:84–91.

[83] Nozawa K, O'donoghue P, Gundllapalli S, Araiso Y, Ishitani R, Umehara T, et al. Pyrrolysyl-tRNA synthetase-tRNA^Pyl structure reveals the molecular basis of orthogonality. Nature 2009;457:1163–7.

[84] Ashraf SS, Guenther R, Agris PF. Orientation of the tRNA anticodon in the ribosomal P-site: quantitative footprinting with U33-modified, anticodon stem and loop domains. RNA 1999;5(9):1191–9.

[85] Dale T, Fahlman RP, Olejniczak M, Uhlenbeck OC. Specificity of the ribosomal A site for aminoacyl-tRNAs. Nucleic Acids Res 2009;37(4):1202–10.

[86] Krasheninnikov IA, Komar AA, Adzhubei IA. Nonuniform size distribution of nascent globin peptides, evidence for pause localization sites, and a contranslational protein-folding model. J Protein Chem 1991;10(5):445–53.

[87] Fedorov AN, Baldwin TO. Contribution of cotranslational folding to the rate of formation of native protein structure. Proc Natl Acad Sci U S A 1995;92(4):1227–31.

[88] Kolb VA, Makeyev EV, Kommer A, Spirin AS. Cotranslational folding of proteins. Biochem Cell Biol 1995;73(11 – 12):1217–20.

[89] Gross M. Linguistic analysis of protein folding. FEBS Lett 1996;390(3):249–52.

[90] Fedorov AN, Baldwin TO. Cotranslational protein folding. J Biol Chem 1997; 272(52):32715–8.

[91] Kolb VA. Cotranslational protein folding. Mol Biol 2001;35:584–90.

[92] Dana A, Tuller T. Determinants of translation elongation speed and ribosomal profiling biases in mouse embryonic stem cells. PLoS Comput Biol 2012;8(11):e1002755-5.

[93] O'Brien EP1, Vendruscolo M, Dobson CM. Prediction of variable translation rate effects on cotranslational protein folding. Nat Commun 2012;3:868.

[94] Nissley DA1, O'Brien EP. Timing is everything: unifying codon translation rates and nascent proteome behavior. J Am Chem Soc 2014;136(52):17892–8.

[95] O'Brien EP1, Ciryam P, Vendruscolo M, Dobson CM. Understanding the influence of codon translation rates on cotranslational protein folding. Acc Chem Res 2014; 47(5):1536–44.

[96] Ray SK, Baruah VJ, Satapathy SS, Banerjee R. Cotranslational protein folding reveals the selective use of synonymous codons along the coding sequence of a low expression gene. J Genet 2014;93(3):613–7.

[97] Trovato F, O'Brien EP. Insights into cotranslational nascent protein behavior from computer simulations. Annu Rev Biophys 2016;45:345–69.

[98] Lu HM, Liang J. A model study of protein nascent chain and cotranslational folding using hydrophobic-polar residues. Proteins 2008;70(2):442–9.

[99] Ugrinov KG1, Clark PL. Cotranslational folding increases GFP folding yield. Biophys J 2010;98(7):1312–20.

[100] Ciryam P, Morimoto RI, Vendruscolo M, Dobson CM, O'Brien EP. In vivo translation rates can substantially delay the cotranslational folding of the Escherichia coli cytosolic proteome. Proc Natl Acad Sci U S A 2013;110(2):E132-0.

[101] Sander IM1, Chaney JL, Clark PL. Expanding Anfinsen's principle: contributions of synonymous codon selection to rational protein design. J Am Chem Soc 2014; 136(3):858–61.

[102] Holtkamp W, Kokic G, Jäger M, Mittelstaet J, Komar AA, Rodnina MV. Cotranslational protein folding on the ribosome monitored in real time. Science 2015;350(6264):1104–7.

[103] Nilsson OB, Nickson AA, Hollins JJ, Wickles S, Steward A, Beckmann R, et al. Cotranslational folding of spectrin domains via partially structured states. Nat Struct Mol Biol 2017;24(3):221–5.

[104] O'Brien EP, Vendruscolo M, Dobson CM. Kinetic modelling indicates that fast-translating codons can coordinate cotranslational protein folding by avoiding misfolded intermediates. Nat Commun 2014;5:2988.

[105] Cabrita LD, Cassaignau AM, Launay HM, Waudby CA, Wlodarski T, Camilloni C, et al. A structural ensemble of a ribosome-nascent chain complex during cotranslational protein folding. Nat Struct Mol Biol 2016;23(4):278–85.

[106] Trovato F, O'Brien EP. Fast protein translation can promote co- and posttranslational folding of misfolding-prone proteins. Biophys J 2017;112(9):1807–19.

[107] Eichmann C1, Preissler S, Riek R, Deuerling E. Cotranslational structure acquisition of nascent polypeptides monitored by NMR spectroscopy. Proc Natl Acad Sci U S A 2010;107(20):9111–6.

[108] Han Y, David A, Liu B, Magadán JG, Bennink JR, Yewdell JW, et al. Monitoring cotranslational protein folding in mammalian cells at codon resolution. Proc Natl Acad Sci U S A 2012;109(31):12467–72.

[109] Ellis JJ1, Huard FP, Deane CM, Srivastava S, Wood GR. Directionality in protein fold prediction. BMC Bioinformatics 2010;11:172.

[110] Srivastava S, Patton Y, Fisher DW, Wood GR. Cotranslational protein folding and terminus hydrophobicity. Adv Bioinformatics 2011;2011:176813.

[111] Focke PJ, Hein C, Hoffmann B2, Matulef K, Bernhard F, Dötsch V, et al. Combining in vitro folding with cell free protein synthesis for membrane protein expression. Biochemistry 2016;55(30):4212–9.

[112] Li G-W, Oh E, Weissman JS. The anti-Shine-Dalgarno sequence drives translational pausing and codon choice in bacteria. Nature 2012;484:538–41.

[113] Ta T, Argos P. Protein secondary structural types are differentially coded on messenger RNA. Protein Sci 1996;5(10):1973–83.

[114] Brule CE, Grayhack EJ. Synonymous codons: choose wisely for expression. Trends Genet 2017;33(4):283–97.

[115] Oresic M, Shalloway D. Specific correlations between relative synonymous codon usage and protein secondary structure. J Mol Biol 1998;281(1):31–48.

[116] Saunders R, Deane CM. Synonymous codon usage influences the local protein structure observed. Nucleic Acids Res 2010;38(19):6719–28.

[117] Phoenix DA, Korotkov E. Evidence of rare codon clusters within Escherichia coli coding regions. FEMS Microbiol Lett 1997;155(1):63–6.

[118] Clarke TF, Clark PL. Rare codons cluster. PLoS One 2008;3(10):e3412-.

[119] Chartier M, Gaudreault F, Najmanovich R. Large scale analysis of conserved rare codon clusters suggests an involvement in co-translational molecular recognition events. Bioinformatics 2012;28(11):1438–45.

[120] Kimchi-Sarfaty C, Oh JM, Kim I-W, Sauna ZE, Calcagno AM, Ambudkar SV, et al. A "silent" polymorphism in the MDR1 gene changes substrate specificity. Science 2007;315(5811):525–8.

[121] Komar AA. Silent SNPs: impact on protein function and phenotype. Pharmacogenomics 2007;8(8):1075–80.

[122] Agashe D, Martinez-Gomez NC, Drummond DA, Marx CJ. Good codons, bad transcript: large reductions in gene expression and fitness arising from synonymous mutations in a key enzyme. J Mol Evol 2012;30(3):549–60.

[123] Fu J, Murphy KA, Zhou M, Li YH, Lam VH, Tabuloc CA, et al. Codon usage affects the structure and function of the *Drosophila* circadian clock protein period. Genes Dev 2016;30(15):1761–75.

[124] Morton BR. Selection at the amino acid level can influence synonymous codon usage: implications for the study of codon adaptation in plastid genes. Genetics 2001;159:347–58.

[125] Błażej P, Mackiewicz D, Wnętrzak M, Mackiewicz P. The impact of selection at the amino acid level on the usage of synonymous codons. G3 (Bethesda) 2017;7(3): 967–81.

[126] Ikemura T. Correlation between the abundance of *Escherichia coli* transfer RNAs and the occurrence of the respective codons in its protein genes: a proposal for a synonymous codon choice that is optimal for the *E. coli* translational system. J Mol Biol 1981;151:389–409.

[127] Bennetzen JL, Hall BD. Codon selection in yeast. J Biol Chem 1982;257:3026–31.

[128] Gouy M, Gautier C. Codon-usage in bacteria: correlation with gene expressivity. Nucleic Acids Res 1982;10:7055–74.

[129] Ikemura T. Codon usage and tRNA content in unicellular and multicellular organisms. Mol Biol Evol 1985;2:13–34.

[130] Kanaya S, Yamada Y, Kudo Y, Ikemura T. Studies of codon usage and tRNA genes of 18 unicellular organisms and quantification of *Bacillus subtilis* tRNAs: gene expression level and species-specific diversity of codon usage based on multivariate analysis. Gene 1999;238:143–55.

[131] Akashi H. Translational selection and yeast proteome evolution. Genetics 2003; 164:1291–303.

[132] Ghaemmaghami S, Huh WK, Bower K, Howson RW, Belle A, Dephoure N, et al. Global analysis of protein expression in yeast. Nature 2003;425:737–41.

[133] Goetz RM, Fuglsang A. Correlation of codon bias measures with mRNA levels: analysis of transcriptome data from *Escherichia coli*. Biochem Biophys Res Commun 2005;327:4–7.

[134] Pechmann S1, Frydman J. Evolutionary conservation of codon optimality reveals hidden signatures of cotranslational folding. Nat Struct Mol Biol 2013;20(2):237–43.

[135] Lopez D, Pazos F. Protein functional features are reflected in the patterns of mRNA translation speed. BMC Genomics 2015;16:513.

[136] Deane CM, Dong M, Huard FP, Lance BK, Wood GR. Cotranslational protein folding—fact or fiction? Bioinformatics 2007;23(13):i142–8.

[137] Evans MS1, Sander IM, Clark PL. Cotranslational folding promotes beta-helix formation and avoids aggregation in vivo. J Mol Biol 2008;383(3):683–92.

[138] Kelkar DA, Khushoo A, Yang Z, Skach WR. Kinetic analysis of ribosome-bound fluorescent proteins reveals an early, stable, cotranslational folding intermediate. J Biol Chem 2012;287(4):2568–78.

[139] Kim SJ, Yoon JS, Shishido H, Yang Z, Rooney LA, Barral JM, et al. Protein folding. Translational tuning optimizes nascent protein folding in cells. Science 2015; 348(6233):444–8.

[140] Paslawski W, Lillelund OK, Kristensen JV, Schafer NP, Baker RP, Urban S, et al. Cooperative folding of a polytopic α-helical membrane protein involves a compact N-terminal nucleus and nonnative loops. Proc Natl Acad Sci U S A 2015;112(26): 7978–83.

[141] Siemion IZ, Stefanowicz P. Periodical changes of amino acid reactivity within the genetic code. Biosystems 1992;18:297–303.

[142] Chou PY, Fasman GD. Prediction of the secondary structure of proteins from their amino acid sequence. Adv Enzymol Relat Areas Mol Biol 1978;47:45–147.

[143] Levitt M. Conformational preferences of amino acids in globular proteins. Biochemistry 1978;17(20):4277–85.

[144] Deléage G, Roux B. An algorithm for protein secondary structure prediction based on class prediction. Protein Eng 1987;1(4):289–94.

[145] Chen H, Gu F, Huang Z. Improved Chou-Fasman method for protein secondary structure prediction. BMC Bioinformatics 2006;7(Suppl. 4):S14.

[146] Lifson S, Sander C. Antiparallel and parallel β-strands differ in amino acid residue preferences. Nature 1979;282:109–11.

[147] Caudron B, Jestin JL. Sequence criteria for the anti-parallel character of protein beta-strands. J Theor Biol 2012;315:146–9.

[148] Seligmann H. Positive and negative cognate amino acid bias affects compositions of aminoacyl-tRNA synthetases and reflects functional constraints on protein structure. BIO 2012;2:11–26.

[149] Zhu CT, Zeng XB, Huang WD. Codon usage decreases the error minimization within the genetic code. J Mol Evol 2003;57(5):533–7.

[150] Archetti M. Codon usage bias and mutation constraints reduce the level of error minimization of the genetic code. J Mol Evol 2004;59(2):258–66.

[151] Marquez R, Smit S, Knight R. Do universal codon-usage patterns minimize the effects of mutation and translation error? Genome Biol 2005;6(11):R91.

[152] Mackiewicz P, Biecek P, Mackiewicz D, Kiraga J, Baczkowski K, Sobczynski M, et al. Optimisation of asymmetric mutational pressure and selection pressure around the universal genetic code. Comput Sci – ICCS 2008;Pt 3(5103):100–9.

[153] Seligmann H. Phylogeny of genetic codes and punctuation codes within genetic codes. Biosystems 2015;129:36–43.

[154] Seligmann H. Avoidance of antisense, antiterminator tRNA anticodons in vertebrate mitochondria. Biosystems 2010;101:42–50.

[155] Seligmann H. Undetected antisense tRNAs in mitochondrial genomes? Biol Direct 2010;5:39.

[156] Seligmann H. Pathogenic mutations in antisense mitochondrial tRNAs. J Theor Biol 2011;269(1):287–96.

[157] Seligmann H. Two genetic codes, one genome: frameshifted primate mitochondrial genes code for additional proteins in presence of antisense antitermination tRNAs. Biosystems 2011;105(3):271–85.

[158] Seligmann H. Putative protein-encoding genes within mitochondrial rDNA and the D-loop region. In: Lin Z, Liu W, editors. Ribosomes: molecular structure, role in

[159] biological functions and implications for genetic diseases. Nova Science Publishers; 2013. p. 67–86.

[159] Faure E, Delaye L, Tribolo S, Levasseur A, Seligmann H, Barthélémy RM. Probable presence of an ubiquitous cryptic mitochondrial gene on the antisense strand of the cytochrome oxidase I gene. Biol Direct 2011;6:56.

[160] Seligmann H. An overlapping genetic code for frameshifted overlapping genes in *Drosophila* mitochondria: antisense antitermination tRNAs UAR insert serine. J Theor Biol 2012;298:51–76.

[161] Seligmann H. Overlapping genetic codes for overlapping frameshifted genes in Testudines, and *Lepidochelys olivacea* as special case. Comput Biol Chem 2012;41: 18–34.

[162] Seligmann H. Codon expansion and systematic transcriptional deletions produce tetra-, pentacoded mitochondrial peptides. J Theor Biol 2015;387:154–65.

[163] Seligmann H. Chimeric mitochondrial peptides from contiguous regular and swinger RNA. Comput Struct Biotechnol J 2016;14:283–97.

[164] Seligmann H. Natural chymotrypsin-like-cleaved human mitochondrial peptides confirm tetra-, pentacodon, non-canonical RNA translations. Biosystems 2016; 147:78–93.

[165] Seligmann H. Unbiased mitoproteome analyses confirm non-canonical RNA, expanded codon translations. Comput Struct Biotechnol J 2016;14:391–403.

[166] Seligmann H. Natural mitochondrial proteolysis confirms transcription systematically exchanging/deleting nucleotides, peptides coded by expanded codons. J Theor Biol 2017;414:76–90.

[167] Srivastava S, Lal SB, Mishra DC, Angadi UB, Chaturvedi KK, Rai SN, et al. An efficient algorithm for protein structure comparison using elastic shape analysis. Algorithms Mol Biol 2016;11:27.

[168] Knecht C, Mort M, Junge O, Cooper DN, Krawczak M, Caliebe A. IMHOTEP—a composite score integrating popular tools for predicting the functional consequences of non-synonymous sequence variants. Nucleic Acids Res 2017;45(3):e13.

[169] Elzanowski A, Ostell J. The genetic codes. November 2016 update https://www. ncbi.nlm.nih.gov/Taxonomy/Utils/wprintgc.cgi.

[170] Moraes EM, Spressola VL, Prado PRRR, Costa LF, Sene FM. Divergence in wing morphology among sibling species of the *Drosophila buzzatii* cluster. J Zool Syst Evol Res 2004;42(2):154–8.

[171] Renaud S, Chevret P, Michaux J. Morphological vs. molecular evolution: exology and phylogeny both shape the mandible of rodents. Zool Scr 2007;36(5):525–35.

[172] Davies TJ, Savolainen V. Neutral theory, phylogenies, and the relationship between phenotypic change and evolutionary rates. Evolution 2006;60(3):476–83.

[173] Seligmann H. Positive correlations between molecular and morphological rates of evolution. J Theor Biol 2010;264(3):799–807.

[174] Graham BE. Animal evolution: trilobites on speed. Curr Biol 2013;23(19):R878-0.

[175] Sessions SK, Larson A. Developmental correlates of genome size in plethodontid salamanders and their implications for genome evolution. Evolution 1987;41(6): 1239–51.

[176] Licht LE, Lowcock LA. Genome size and metabolic-rate in salamanders. Comp Biochem Physiol B Biochem Mol Biol 1991;100(1):83–92.

[177] Roth G, Blanke J, Wake DB. Cell-size predicts morphological complexity in the brains of frogs and salamanders. Proc Natl Acad Sci U S A 1994;91(11):4796–800.

[178] Roth G, Walkowiak W. The influence of genome and cell size on brain morphology in amphibians. Cold Spring Harb Perspect Biol 2015;7(9):a019075.

[179] Seligmann H. Cost minimization of amino acid usage. J Mol Evol 2003;56(2): 151–61.

[180] Seligmann H. Putative mitochondrial polypeptides coded by expanded quadruplet codons, decoded by antisense tRNAs with unusual anticodons. Biosystems 2012; 110(2):84–106.

[181] Seligmann H. Pocketknife tRNA hypothesis: anticodons in mammal mitochondrial tRNA side-arm loops translate proteins? Biosystems 2013;113(3):165–75.

[182] Seligmann H. Putative anticodons in mitochondrial tRNA sidearm loops: pocketknife tRNAs? J Theor Biol 2014;340:155–63.

[183] Seligmann H, Labra A. Tetracoding increases with body temperature in Lepidosauria. Biosystems 2013;114(3):155–63.

[184] Seligmann H, Krishnan NM. Mitochondrial replication origin stability and propensity of adjacent tRNA genes to form putative replication origins increase developmental stability in lizards. J Exp Zool B Mol Dev Evol 2006;306:433–49.

[185] Seligmann H. Error propagation across levels of organization: from chemical stability of ribosomal RNA to developmental stability. J Theor Biol 2006;242(1): 69–80.

[186] Chwastyk M, Cieplak M. Cotranslational folding of deeply knotted proteins. J Phys Condens Matter 2015;27(35):354105.

[187] Felsenstein J. Phylogenies and the comparative method. Am Nat 1985;125:1–15.

[188] Sato Y, Fujiwara T, Kimura H. Expression and function of different guanine-plus-cytosine content 16S rRNA genes in *Haloarcula hispanica* at different temperatures. Front Microbiol 2017;8:482.

[189] Musto H, Naya H, Zaval A, Romero H, Alvarez-Valin F, Bernardi G. Genomic GC level, optimal growth temperature, and genome size in prokaryotes. Biochem Biophys Res Commun 2006;347(1):1–3.

[190] Zheng H, Wu HW. Gene-centric association analysis for the correlation between the guanine-cytosine content levels and temperature range conditions of prokaryotic species. BMC Bioinformatics 2010;11:S7.

[191] Ream RA, Johns GC, Somero GN. Base compositions of genes encoding alpha-actin and lactate dehydrogenase-A from differently adapted vertebrates show no temperature-adaptive variation in G + C content. Mol Biol Evol 2003;20(1): 105–10.

[192] Marashi SA, Ghalanbor Z. Correlations between genomic GC levels and optimal growth temperatures are not 'robust'. Biochem Biophys Res Commun 2004; 20(1):381–3.

[193] Wang HC, Susko E, Roger AJ. On the correlation between genomic G + C content and optimal growth temperature in prokaryotes: data quality and confounding factors. Biochem Biophys Res Commun 2006;342:681–4.

[194] Seligmann H. Evidence that minor directional asymmetry is functional in lizard hindlimbs. J Zool 1998;248:205–8.

[195] Seligmann H. Evolution and ecology of developmental processes and of the resulting morphology: directional asymmetry in hindlimbs of Agamidae and Lacertidae (Reptilia: Lacertilia). Biol J Linn Soc 2000;69:461–81.

[196] Seligmann H, Beiles A, Werner YL. Avoiding injury or adapting to survive injury? Two coexisting strategies in lizards. Biol J Linn Soc 2003;78:307–24.

[197] Seligmann H, Beiles A, Werner YL. More injuries in left-footed lizards. J Zool 2003; 260:129–44.

[198] Seligmann H, Moravec J, Werner YL. Morphological, functional and evolutionary aspects of tail autotomy and regeneration in the 'living fossil' Sphenodon (Reptilia: Rhynchocephalia). Biol J Linn Soc 2008;93(4):721–43.

[199] Bilal E, Rabadan R, Alexe G, Fuku N, Ueno H, Nishigaki Y, et al. Mitochondrial DNA haplogroup D4a is a marker for extreme longevity in Japan. PLoS One 2008; 3(6):e2421.

[200] Tanak M, Cabrera M, Gonzalez M, Larruga M, Takeyasu T, Fuku N, et al. Mitochondrial genome variation in eastern Asia and the peopling of Japan. Genome Res 2004;14:1832–50.

[201] Alexe G, Fuku N, Bilal E, Ueno H, Nishigaki Y, Fujita Y, et al. Enrichment of longevity phenotype in mtDNA haplogroups D4b2b, D4a, and D5 in the Japanese population. Hum Genet 2007;121:347–56.

[202] Hansen TF, Pienaar J, Orzack SH. A comparative method for studying adaptation to a randomly evolving environment. Evolution 2007;62:1965–77.

[203] Bailey LJ, Doherty AJ. Mitochondrial DNA replication: a PrimPol perspective. Biochem Soc Trans 2017;45(2):513–29.

[204] Reyes A, Gissi C, Pesole G, Saccone C. Asymmetrical directional mutation pressure in the mitochondrial genome of mammals. Mol Biol Evol 1998;15(8): 957–66.

[205] Krishnan NM, Seligmann H, Raina SZ, Pollock DD. Detecting gradients of asymmetry in site-specific substitutions in mitochondrial genomes. DNA Cell Biol 2004; 23(10):707–14.

[206] Seligmann H, Krishnan NM, Rao BJ. Possible multiple origins of replication in primate mitochondria: alternative role of tRNA sequences. J Theor Biol 2006;241(2): 321–32.

[207] Seligmann H, Krishnan NM, Rao BJ. Mitochondrial tRNA sequences as unusual replication origins: pathogenic implications for Homo sapiens. J Theor Biol 2006; 243(3):375–85.

[208] Seligmann H. Hybridization between mitochondrial heavy strand tDNA and expressed light strand tRNA modulates the function of heavy strand tDNA as light strand replication origin. J Mol Biol 2008;379(1):188–99.

[209] Seligmann H. Coding constraints modulate chemically spontaneous mutational replication gradients in mitochondrial genomes. Curr Genomics 2012;13(1): 37–54.

[210] Helfenbein KG, Brown WM, Boore JL. The complete mitochondrial genome of the articulate brachiopod Terebratalia transversa. Mol Biol Evol 2001;18:1734–44.

[211] Hassanin A, Leger N, Deutsch J. Evidence for multiple reversals of asymmetric mutational constraints during the evolution of the mitochondrial genome of Metazoa, and consequences for phylogenetic inferences. Syst Biol 2005;54(2):277–98.

[212] Fonseca MM, Froufe E, Harris DJ. Mitochondrial gene rearrangements and partial genome duplications detected by multigene asymmetric compositional bias analysis. J Mol Evol 2006;63(5):654–61.

[213] Min XJ, Hickey DA. DNA asymmetric strand bias affects the amino acid composition of mitochondrial proteins. DNA Res 2007;14:201–6.

[214] Fonseca MM, Harris DJ. Relationship between mitochondrial gene rearrangements and stability of the origin of light strand replication. Genet Mol Biol 2008;31(2): 566–74.

[215] Fonseca MM, Posada D, Harris DJ. Inverted replication of vertebrate mitochondria. Mol Biol Evol 2008;25:805–8.

[216] Masta SE, Longhorn SJ, Boore JL. Arachnid relationships based on mitochondrial genomes: asymmetric nucleotide and amino acid bias affects phylogenetic analyses. Mol Phylogenet Evol 2009;50:117–28.

[217] Wei SJ, Shi M, Chen XX, Sharkey MJ, van Achterberg C, Ye GY, et al. New views on strand asymmetry in insect mitochondrial genomes. PLoS One 2010;5(9):e12708.

[218] Fonseca MM, Harris DJ, Posada D. The inversion of the control region in three mitogenomes provides further evidence for an asymmetric model of vertebrate mtDNA replication. PLoS One 2014;9(9):e106654.

[219] Seligmann H. Mitochondrial tRNAs as light strand replication origins: similarity between anticodon loops and the loop of the light strand replication origin predicts initiation of DNA replication. Biosystems 2010;99(2):85–93.

[220] Seligmann H, Labra A. The relation between hairpin formation by mitochondrial WANCY tRNAs and the occurrence of the light strand replication origin in Lepidosauria. Gene 2014;542(2):248–57.

[221] Seligmann H. Mutation patterns due to converging mitochondrial replication and transcription increase lifespan and cause growth rate-longevity tradeoffs. Chapter 4 , In: Seligmann H, editor. DNA replication—current advances. InTech; 2011. http://dx.doi.org/10.5772/24319.

[222] Perneger TV. What's wrong with Bonferroni adjustments. Br Med J 1998; 316(7139):1236–8.

[223] Bender R, Lange S. Multiple test procedures other than Bonferroni's deserve wider use. Br Med J 1999;318(7183):600.

[224] Benjamini Y, Hochberg Y. Controlling the false discovery rate: a practical and powerful approach to multiple testing. J R Stat Soc B 1995;57(1):289–300. http://dx.doi.org/10.2307/2346101.

[225] Käll L, Storey JD, MacCoss MJ, Noble WS. Posterior error probabilities and false discovery rates: two sides of the same coin. J Proteome Res 2008;7(01):40–4.

[226] Petoukhov SV. Genetic coding and united-hypercomplex systems in the models of algebraic biology. Biosystems 2017;158:31–46.

[227] Cairns-Smith AG, Hartman H. Clay minerals and the origin of life. Cambridge University Press; 1986 197.

[228] Gonzalez DL, Giannerini S, Rosa R. On the origin of the mitochondrial genetic code: Towards a unified mathematical framework for the management of genetic information. Available from nature Precedings; 2012. http://dx.doi.org/10.1038/npre. 2012.7136.1.

6

LRSim: A Linked-Reads Simulator Generating Insights for Better Genome Partitioning

Ruibang Luo [a,b,*], Fritz J. Sedlazeck [a], Charlotte A. Darby [a], Stephen M. Kelly [c], Michael C. Schatz [a,b,d]

[a] *Department of Computer Science, Johns Hopkins University, United States*
[b] *Center for Computational Biology, McKusick-Nathans Institute of Genetic Medicine, Johns Hopkins University School of Medicine, United States*
[c] *Center for Health Informatics and Bioinformatics, New York University School of Medicine, United States*
[d] *Simons Center for Quantitative Biology, Cold Spring Harbor Laboratory, United States*

ARTICLE INFO

Keywords:
Linked-read
Molecular barcoding
Reads partitioning
Phasing
Reads simulation
Genome assembly
10X Genomics

ABSTRACT

Linked-read sequencing, using highly-multiplexed genome partitioning and barcoding, can span hundreds of kilobases to improve *de novo* assembly, haplotype phasing, and other applications. Based on our analysis of 14 datasets, we introduce LRSim that simulates linked-reads by emulating the library preparation and sequencing process with fine control over variants, linked-read characteristics, and the short-read profile. We conclude from the phasing and assembly of multiple datasets, recommendations on coverage, fragment length, and partitioning when sequencing genomes of different sizes and complexities. These optimizations improve results by orders of magnitude, and enable the development of novel methods. LRSim is available at https://github.com/aquaskyline/LRSIM.

1. Background

Haplotype-resolved or phased genomes are desirable for obtaining insight into diploid or polyploid genomes and studying allele specific expression, allele-specific regulation, and many other important genomic features [1]. However, most of the genomes assembled to date are only a single haploid 'mosaic' consensus sequence with parental alleles merged arbitrarily [2]. Only a few studies have reported true diploid *de novo* assemblies so far. One of the first such studies successfully combined Illumina short-read sequencing, PacBio sequencing, and BioNano Genomics, generated a phased assembly of the widely studied human sample NA12878 [3]. The second introduced the FALCON-Unzip algorithm and applied it to *de novo* assemble three phased diploid genomes of *Arabidopsis thaliana*, *Vitis vinifera* (grape), and the coral fungus *Clavicorona pyxidata*, relying exclusively on PacBio sequencing [4]. A third approach generated 605,566 fosmid clones on the YH1 human genome and mixed them into 30 clones per pool, each pool containing 0.04% of the diploid genome [5]. However, the sample requirements and the associated high cost of the three studies preclude their widespread use.

More recently, Zheng et al. demonstrated a high-throughput low-cost method marketed as GemCode and its successor Chromium by 10X Genomics for creating human reference-based phased genomes [6]. It uses an automated microfluidic system to isolate high molecular weight DNA molecules inside millions of partitions containing sequencing primers and a unique barcode to prepare a library that can be used for Illumina paired-end read sequencing. Each partition contains several DNA molecules spanning up to 100 kbp or more, and will share the same barcode. The library can be generated with as little as 1 ng of high molecular weight DNA, far less than alternative approaches. And as the sequencing is performed using inexpensive Illumina short-read sequencing, the overall cost is at least two orders of magnitude lower than pooled fosmid sequencing for a diploid genome and significantly less expensive than current long-read sequencing alternatives [7]. The method has been widely used to study novel species such as the Hawaiian Monk Seal [8] and the Pepper [9], as well as additional resequencing analysis in human samples [10].

Simulated sequencing data has proved indispensable for guiding tool development and evaluating tool performance [11]. Especially for the complex and unique workflow involved in constructing linked-reads, it is essential to develop simulation software that can produce linked-reads that capture the most essential characteristics of genome partitioning. To our knowledge, no alternative read simulator for linked-

* Corresponding author at: Department of Computer Science, Johns Hopkins University, United States.

E-mail addresses: rluo5@jhu.edu (R. Luo), fritz.sedlazeck@gmail.com (F.J. Sedlazeck), cdarby@jhu.edu (C.A. Darby), stephen.kelly@nyumc.org (S.M. Kelly), mschatz@jhu.edu (M.C. Schatz).

reads is available. Thus, we developed LRSim, a Linked-Reads Simulator, which simulates whole genome sequencing modeled on Chromium Linked-Read technology but is flexible enough to represent alternative technologies. We modeled the characteristics of all the relevant steps of the Chromium protocol so that it can be used to study linked-read sequencing of different genomes, mutation rates, input libraries, and short-read sequencing conditions *in silico*. We tested LRSim with the 10X Genomics LongRanger variant identification and phasing application and the 10X Genomics SuperNova genome assembler [7] as well as the independent HapCUT2 phasing algorithm [12] to confirm that alignment, variant identification, phasing, and *de novo* assembly are supported and deliver results similar to those of real data. After studying simulated datasets with multiple parameter combinations, we concluded that 1) the best phase block size of human genome with 50x linked-reads sequencing coverage and 1.5M partitions (barcodes) can be achieved with a molecule size between 150 kbp and 200 kbp; and 2) the standard library preparation protocol tailor-made for the mammalian-sized genomes needs to be adjusted regarding the number of partitions (barcodes) before it can be efficiently used for other genomes of significantly different size, such as *A. thaliana*. We also recognize that linked reads can produce less contiguous genome assemblies than long reads in highly repetitive genomes like maize.

2. Results

2.1. Characteristics of Real 10× Genomic Sequences

We analyzed 13 publicly available real datasets processed by Chromium's LongRanger analysis pipeline to derive models and characteristics (Additional file 1: Supplementary Note). We identified ~1.5 (1.42) million partitions with >100 reads each in NA12878 (Fig. 1). The Chromium uses 16 bp barcode sequence. The barcodes are located at the beginning of reads, thus having a higher error rate compared with the non-barcode bases. We observed 2.09% total base errors, which is slightly higher than the 1.78% estimated by base quality (Additional file 1: Supplementary Table 1). Interestingly, two of the barcodes were found to be consistently over-represented in all 13 samples ("GTATCTTCAGATCTGT", "GTGCCTTCAGATCTGT", Additional file 1: Supplementary Table 2).

One of the advantages of linked-reads is that they can be mapped into repetitive sequences not accessible by standard short read sequencing. Considering all regions ≥50 base-pairs that had zero coverage

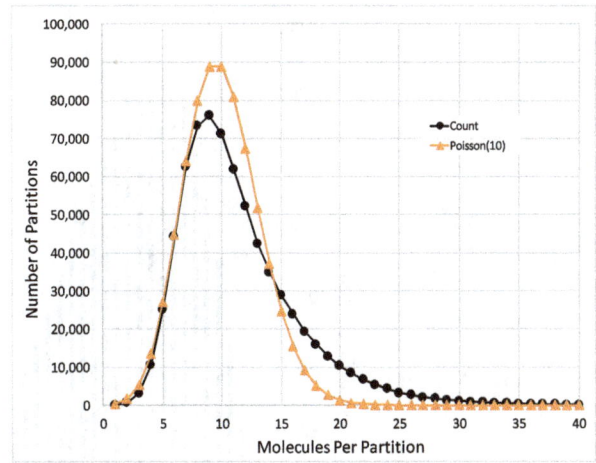

Fig. 2. Distribution of the number of molecules per partition for NA12878.

(excluding ambiguous genomics regions denoted by 'N'), we found that the Chromium 80x NA12878 dataset left 164 Mb (5.47%) uncovered by any read alignment, versus 209 Mb (6.97%) that had zero coverage in the NIST 300x NA12878 Illumina only paired-end read dataset (Additional file 1: Supplementary Table 3, considering all alignments including MQ0). In the NA12878 sample specifically, a median 10 DNA molecules were allocated to each partition (Fig. 2) and the weighted molecule length peaked at around 40-50 kbp (Fig. 3). The distribution of molecule coverage peaked at 0.2x coverage (Fig. 4). Reads were generally uniformly distributed along the genome, as well as along the molecule, although surprisingly, we observed the chromosome 21 with unexpectedly high coverage in all samples possibly due to the existence of abundant Ribosomal DNA repeats around 9Mbp in the chromosome 21 according to the Gencode v26 gene annotations [13] and LongRanger analysis pipeline's way to deal with repetitive segments (Fig. 5; detailed views in Additional file 1: Supplementary Fig. 1a, b).

2.2. Parameters of Linked-Read

The overview of the technology for generating linked-reads and the available parameters of linked-reads were introduced in Zheng et al. [6]. By checking the variability of the available parameters in the 13 real

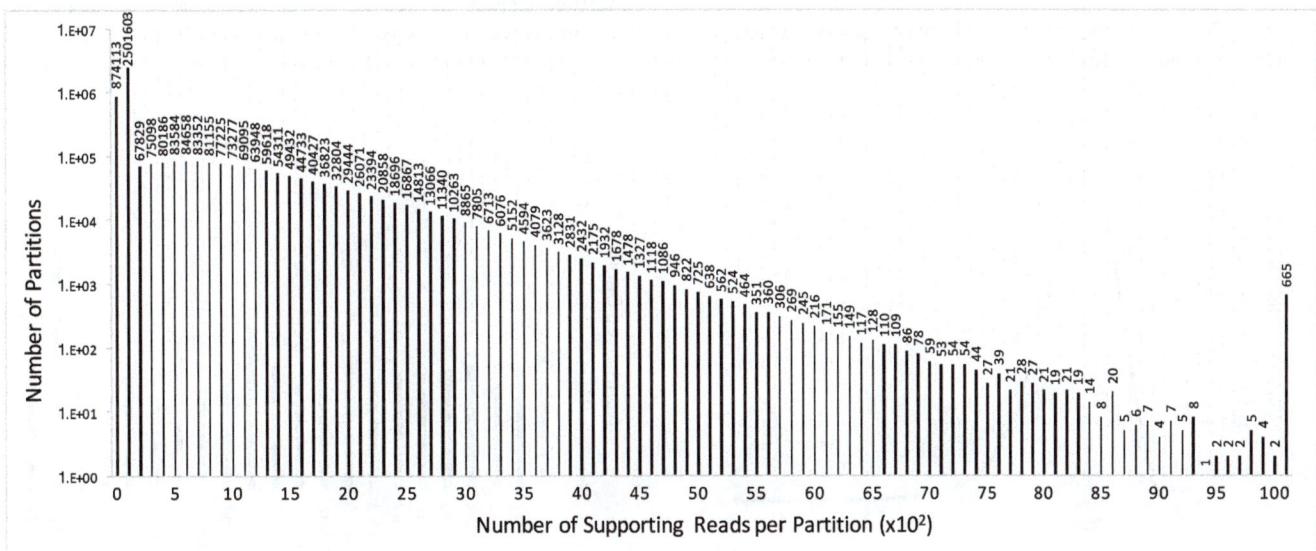

Fig. 1. The distribution of number of supporting reads per partition. About 1.5 million partitions are supported by more than 100 reads. The rightmost column shows the number of partitions with the number of supporting reads per partition >100.

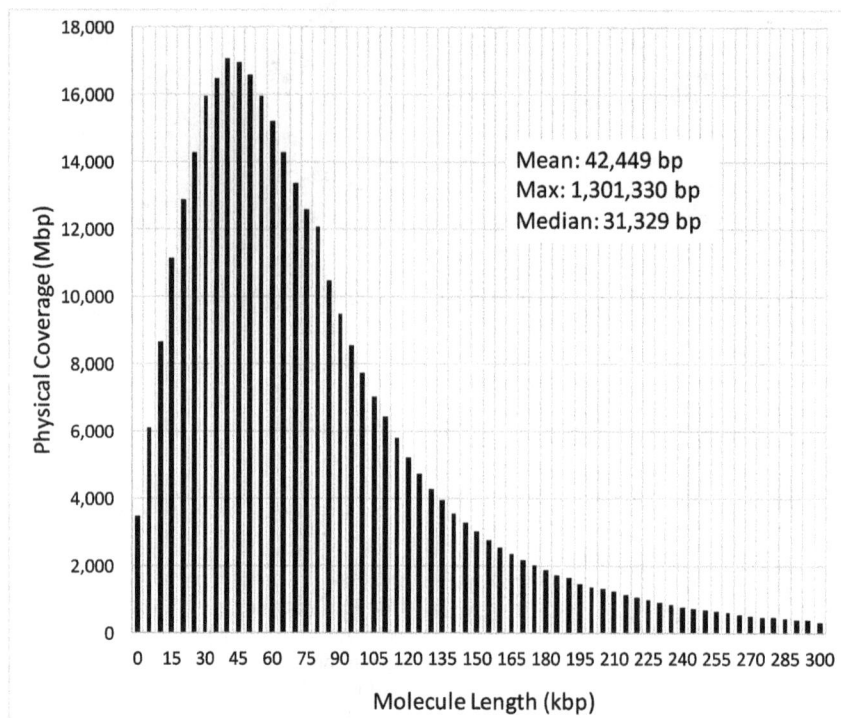

Fig. 3. Weighted molecule length distribution for NA12878. Physical Coverage equals $\sum_{i=1}^{n} l_i$, where l_i is the length of a molecule and n is the number of molecules in that size range.

datasets, we identified four important parameters and a dependent parameter of linked-read. The important parameters are: number of read pairs (x); number of partitions (t); mean molecule length (f); and mean number of molecules per partition (m). The dependent parameter is sequencing coverage per molecule (c). Given a genome size and a fixed x, the other four parameters t, f, m and c correlate with each other inversely. For example, with x, f and m remain constants, a higher number of partitions (t) will lead to lower sequencing coverage per molecule (c). The relationships between the parameters are introduced starting from the next subsection.

2.3. Effect of Molecule Size (f)

One of the critical requirements of linked-read construction is extracting high-quality, high-molecular weight DNA from the sample.

To study how the molecule size changes the performance of linked-read sequencing, using human reference genome GRCh38, we simulated six datasets of different mean molecule sizes (f: 20, 50, 100, 150, 200 and 250 kbp), with 600 million read pairs (x), 1.5 million partitions (t) and 10 molecules per partition (m). Here we are using a normal distribution for the molecule lengths, although LRSim allows for arbitrarily complicated distributions as well. Instead of simulating random variants using SURVIVOR [14], which may not mimic the characteristics of real variants, we used 3.2M phased SNPs and indels identified from NA12878 (Additional file 1: Supplementary Note). The datasets were processed by LongRanger and phased by HapCUT2 [12] using a 48-core Intel E7–8857 v2 @3GHz machine with 1TB memory, running on average 1.5 days each. The sum of bases of different phase block sizes for six simulated datasets and NA12878 (Additional file 1: Supplementary Note) are shown in Fig. 6. The results show that the NA12878's performance lies between molecule size 50 kbp and 100 kbp, corroborating

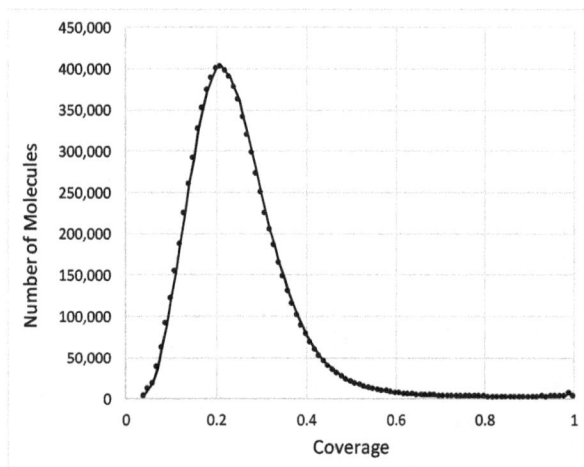

Fig. 4. Distribution of molecule coverage for NA12878.

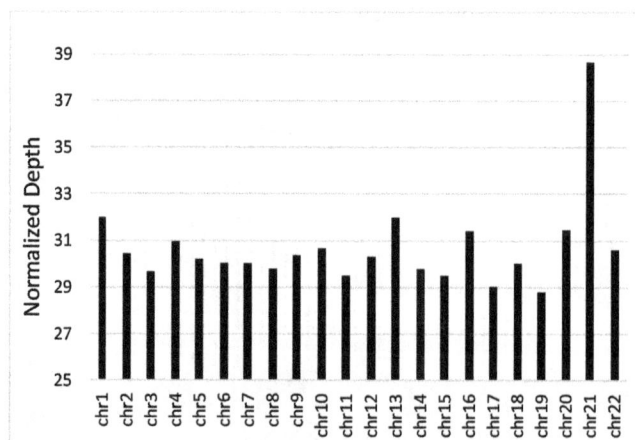

Fig. 5. Average sequencing coverage of 13 samples per chromosome. The coverages were normalized to the sample with the lowest average coverage (NA24149, 30.36x).

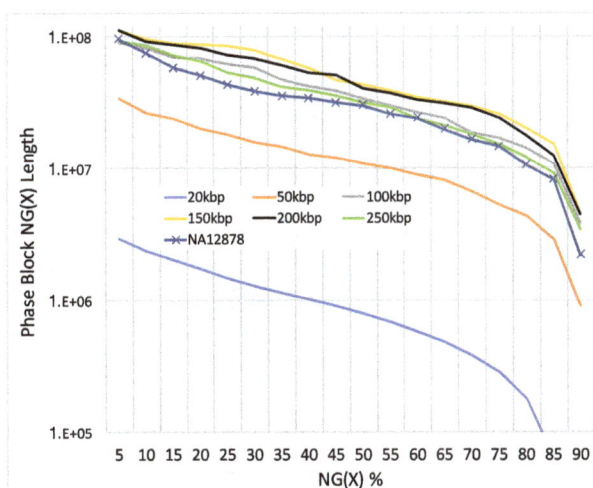

Fig. 6. NG graph showing an overview of phased block sizes of 7 datasets. NG(X) is defined as X% of the genome is in phased blocks equals to or larger than the NG(X) length.

our observation in the weighted molecule length distribution of NA12878 (Fig. 4). The divergence between 50 kbp and NA12878 can be explained by the fact that the real data is platykurtic and has a longer tail on long molecules, which is an outcome that highly depends on the quality and length of DNA input.

We further noted that the Phase Block N50 sizes did not monotonically increase with longer molecules and instead plateaued at 200 kbp molecules. On investigation, we determined the cause to be that given a constant number of total reads, the coverage per molecule decreases proportionally to the molecule size. For example, if we increase the molecule size from 50 kbp to 250 kbp, the coverage decreases from ~0.2x to ~0.04x, and the average distance and standard deviation of the distances between reads rises, thus leading to shorter phase blocks.

2.4. Effect of the Number of Partitions (t)

We simulated six datasets with varying numbers of partitions (t), including 15k, 20k, 30k, 50k, 100k and the standard 1.5 million, to study the impact of the number of partitions on the assembly using the A. thaliana genome (TAIR10).

We kept the other parameters constant at 18 million read pairs (x), 50 kbp mean molecule size (f) and 10 molecules per partition (m). Using the same computer, SuperNova finished each assembly within 2 hours. The best assembly, as measured by contig N50, scaffold N50 or Phase Block N50, was from the dataset with $t = 20,000$ partitions (Table 1). Intriguingly, using 1.5 million partitions, which is the default for the 10X Chromium platform, the results are orders of magnitude worse (e.g. 14.6 kb vs. ~2 Mbp scaffold N50) for this reduced genome size.

One of the major reasons for the deficiency in the assembly performance is attributed to insufficient coverage per molecule. For a 3 Gbp genome with the default parameters, the coverage per molecule is 0.2x on average. For A. thaliana, where the genome size is 20 times

smaller (150 Mbp vs. 3 Gbp), if 3 parameters including f, t and m remain the default, the number of reads allotted to each molecule will be 20 times smaller, i.e. 0.01x coverage. The excessively low coverage increases the mean distance and its standard deviation between reads, which confounds the genome assembler; it also largely removes the chance for reads belonging to the same molecule to cover multiple heterozygous variants, which is essential for phasing. Therefore, we suggest adjusting the number of partitions according to the genome size if possible. On the current 10X Chromium instrument, it is not possible for the operator to directly control the number of partitions used, so alternatively we and 10× Genomics recommend increasing the overall sequencing coverage and subsample the partitions proportionally. This approach also results in sufficient coverage per partition, which greatly improves the assembly results using linked-reads for smaller genomes.

2.5. Effect of Sequencing Coverage (x)

Genome assembly using Illumina short reads requires careful control of the sequencing coverage. Shallow coverage decreases the maximum usable kmer-size (to achieve the minimum requirement for kmer depth), thus limiting the ability to disentangle repetitive sequences. Excessively deep coverage leads to a lower signal-to-noise ratio because of the saturation of authentic sequences and the accumulation of more random errors in the 'assembly graph', thus decreasing the performance of the assembly outcome [15,16]. The best practice for sequencing coverage ranges from around 30x to 100x coverage, depending on affordability (a longer kmer-size can be used with greater sequencing coverage, but this is limited by the length of read input and read errors) and the genomic nature of different species, including their heterozygosity, heterogeneity and repetitiveness. It is less clear how the sequencing coverage of linked-reads changes the performance of genome assembly.

Using the A. thaliana genome, we simulated four datasets with three different numbers of read pairs (x), 9, 18 and 27 million, which equates to 17-, 34- and 51-fold of the genome, respectively, and two different numbers of partitions, 20,000 and 30,000 for $x = 27$. The molecule length (f) was held constant at 50 kbp, as was the number of molecules per partition at $m = 10$. We used SuperNova to assemble the four datasets. The results are shown in Table 2. We found that 18 million read pairs (34-fold) with 20,000 partitions achieved the best Contig N50, Phase Block N50 and Scaffold N50. Interestingly, the assembly result of 27 million read pairs (51-fold) was worse than 18 million on all three metrics, and only improved slightly on Contig N50 and Phase Block N50 after increasing the number of partitions to 30,000 (to keep the molecule coverage the same as 18 million read pairs). This indicates that the sequencing coverage itself rather than the molecule coverage makes a difference in linked-reads genome assembly using the SuperNova assembler.

2.6. Effect of the Number of Molecules per Partition (m)

The number of molecules per partition is usually determined by sample preparation technologies and cannot be easily modified except by carefully controlling the total amount of input DNA. Thus, the number needs to be carefully selected and verified before production. A

Table 1
Assembly statistics of different number of partitions.

No. of partitions (×1,000)	Contig N50	Phase Block N50	Scaffold N50
15	198,485	1,146,590	1,016,017
20	265,543	2,881,040	2,796,090
30	230,711	1,945,051	1,880,870
50	215,743	1,472,710	1,459,816
100	177,635	1,471,806	1,271,685
1500	14,597	1588	14,685

The Contig N50, Phase Block N50 and Scaffold N50 of the A. thaliana genome with 6 different partition numbers.

Table 2
Assembly statistics of different sequencing coverage.

No. of read pairs (M)	No. of partitions (×1,000)	Contig N50	Phase Block N50	Scaffold N50
9 (17-fold)	20	233,233	1,027,768	899,826
18 (34-fold)	20	265,543	2,881,040	2,796,090
27 (51-fold)	20	221,680	1,971,701	1,896,517
27 (51-fold)	30	241,319	1,979,723	1,688,453

Contig N50, Phase Block N50 and Scaffold N50 of the A. thaliana genome with 4 different combinations of number of read pairs and number of partitions.

lower number of molecules per partition requires a larger number of barcodes to arrive at the same number of molecules. A higher number of molecules per partition requires fewer barcodes, but increases the chance of two molecules coming from the two haplotypes in the same genome position forming a "collision" that can lead to phase errors or reduced phase block sizes. Given a certain number of barcodes, the number of molecules per partition will increase or decrease the coverage per molecule, which changes the performance of genome assembly and phasing.

Using the *A. thaliana* genome, we simulated 6 datasets with 1, 4, 7, 10, 15 and 20 molecules per partition, with the number of read pairs ($x = 18$ million), molecule length ($f = 50,000$) and number of partitions ($t = 20,000$). The assembly results are shown in Table 3. The Phase Block N50 and Scaffold N50 peaked with 10 molecules per partition, while Contig N50 peaked with 7. The metrics change insignificantly in the range 4 to 20 molecules per partition and went down significantly with only 1 molecule per partition; the reason remains unclear. We speculate that it is related to the average molecule coverage increasing to 2x with only 1 molecule per partition; this could confound phasing algorithms oblivious to conflicting alleles caused by sequencing error within a molecule. Also, the total span of the genomic regions being covered decreases with the same number of partitions but less molecules per partition.

2.7. Validation of LRSim in a Repetitive Plant Genome

To illustrate the versatility of LRSim, we also analyzed linked-read sequencing of the highly repetitive maize genome line NC350 (SRA accession number PRJNA380806). NC350 is a temperate-adapted all-tropical inbred with broad adaptation and has a genome size of about 2.3 Gbp. Assembling 1036.45 million reads from this sample with SuperNova results in an assembly with a 25.68 kbp contig N50, 0.23 Mbp Phase Block N50 and 0.35 Mbp scaffold N50 (Fig. 7). Assembling the same number of simulated reads from the high-quality reference, established from a related line B73, results in a 41.03 kb Contig N50, 0.21 Mbp Phase Block N50 and 0.16 Mbp scaffold N50. While B73 is the closest species with an available reference genome, the modestly larger Contig N50 and modestly smaller scaffold N50 of the simulated data reflect the differences between the two genomes (NC350 vs. B73). Simulating reads from the incomplete reference genome also introduces some bases since linked reads will never span beyond the available contigs even though they would in the real sample. Nevertheless, a detailed comparison in Fig. 7 on both the contig and scaffold length distributions shows that the simulated data perform on par with the real data across the entire size range. In the figure, we also illustrate a Pacbio long read based assembly of the B73 genome [17] that demonstrates substantially higher contiguity than the linked-reads based B73 and NC350 scaffolds. These results indicate that a deficiency in assembly performance is expected when using the 10x Genomics' Chromium technology or other similar linked-reads technologies on genomes as repetitive and complicated as Maize.

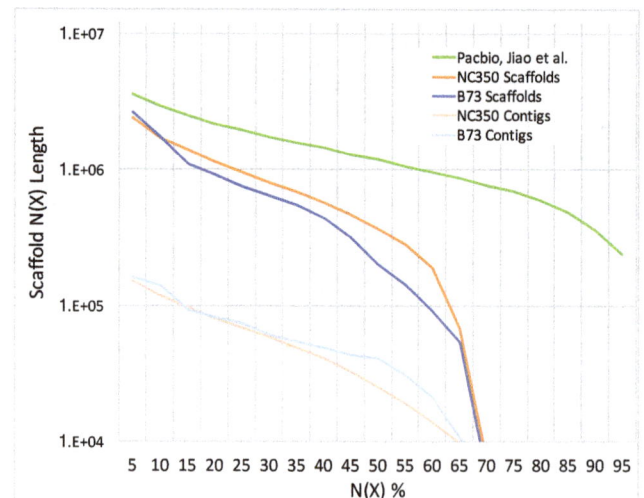

Fig. 7. NG graph showing an overview of scaffold sizes of three datasets, including a simulated linked-reads for B73, real linked-reads for NC350 and a B73 assembly using PacBio long reads by Jiao et al. NG(X) is defined as X% of the genome is in phased blocks equals to or larger than the NG(X) length.

3. Discussion and Conclusions

In this paper, we presented an analysis of 14 real datasets of linked-reads generated by 10x Genomics technology. Based on this analysis, we implemented a linked-read simulator named LRSim to allow fine tuning of both the type and number of variants and Illumina read specifications, and full control of important parameters for linked-reads we identified in real datasets, including 1) the number of read pairs; 2) the number of partitions; 3) the mean molecule length; and 4) the mean number of molecules per partition. We validated the performance of the simulator by the high concordance to the phasing and assembly results in the real NA12878 human dataset as well as in the highly repetitive maize genome. We concluded that from the phasing results of 6 simulated datasets with different mean molecule lengths and a real dataset of NA12878 that if constrained at a certain sequencing coverage, the best molecule size to achieve the best phase block size needs to be meticulously chosen. This can be done by wet-lab experiments, but would be more efficient with a simulator *in silico*. We also performed experiments on 6 simulated *A. thaliana* datasets with a different number of partitions and demonstrated a substantial degradation in assembly performance with an improper number of partitions, which leads to insufficient coverage per molecule. We concluded an appropriate sequencing coverage needs to be chosen for different applications and species before sequencing to achieve the best performance out of linked-reads.

In our study, linked-reads enabled much longer contigs, scaffolds and phase blocks on both the human genome and *A. thaliana* than using Illumina short-reads for genome assembly. The better outcomes, in turn, broaden the horizons for studies of allele specific expression, allele-specific regulation, and many other important genomic features critical to precision medicine. Furthermore, numerous other sequencing applications, such as improved structural variation analysis, epigenetics, metagenomics and RNA-seq, could potentially benefit from linked-read data. Linked-read technology is promising, and we believe that more complex genomics workflows will include and benefit from it. In 10x Genomics technology, "the number of partitions" of a certain product (such as GemCode and Chromium) is predetermined. Users can change two variables including "the molecule length" and "the number of read pairs", and a dependent variable "the number of molecules per partition" on the total amount of DNA loaded for the experiment. We therefore encourage users to use LRSim to aid in the development of these new workflows and fully utilize the potential of the new linked-

Table 3
Assembly statistics of different number of molecules per partition.

No. of molecules per partition	Contig N50	Phase Block N50	Scaffold N50
1	46,993	72,400	54,769
4	249,371	2,105,517	2,020,687
7	274,133	1,869,856	2,074,127
10	265,543	2,881,040	2,796,090
15	232,708	2,860,223	2,416,675
20	245,894	1,920,175	1,878,113

Contig N50, Phase Block N50 and Scaffold N50 of the *A. thaliana* genome with 6 different molecule numbers per partition.

read technologies such as 10X Genomics and IGenomX [18], which follow similar molecular preparations but having very different properties with respect to the number of possible partitions and barcodes possible.

4. Methods

4.1. Identify Mismatches in Barcodes, the Number of Partitions, Molecule Lengths, Number of Molecules per Partition and Molecule Coverages in 13 Real Datasets

The BAM (Binary sequence Alignment/Map format) file of 13 samples generated by LongRanger was downloaded from the links in Additional file 1: Supplementary Note. For each pair of a paired-end read in a BAM file, the RX tag gives the raw barcode sequence, which is subject to sequencing errors. The BX tag gives the barcode sequence that is error-corrected and confirmed against a list of known-good barcode sequences. Differences between the two barcode sequences provided by RX and BX tag respectively are identified as mismatches in barcodes. The position, allele and base quality are extracted from the mismatches to study the error profile of barcodes.

Using the barcode sequences provided by the BX tags, reads were grouped according to their corresponding barcode sequence. Each group of reads were assembled using the 'targetcut' command of SAMtools [19]. Notice that the default parameters of 'targetcut' were optimized for Fosmid Pooling. Since a typical molecule coverage of linked-read sequencing is below 1, which is much lower than Fosmid Pooling, the parameters need to be fine-tuned to allow gaps in the assembled molecules. The detailed commands and a set of parameters empirically optimized for linked-read sequencing are presented in Additional file 1: Supplementary Note. The total number of partitions is determined as the number of unique barcodes in BX tags. Molecule lengths are the length of assembled sequences. The number of molecules per partition are the number of assembled sequences per partition. The molecule coverages are calculated as the percentage of non-gap regions of the assembled sequences.

4.2. Simulator Design and Performance

We included the following parameters in our simulator: x: number of read pairs; t: number of partitions; f: mean molecule length; and m: mean number of molecules per partition. The Poisson distribution is used to sample data from f and m by default; it can easily be switched to other functions. LRSim also allows for fine control of the short-read sequencing error profile and for biological variants to be introduced into the reference genome.

Fig. 8 shows the overview of the LRSim workflow. Briefly, LRSim first uses SURVIVOR [14] to simulate homozygous and heterozygous SNPs,

indels and structural variants within the user-specified genome sequence. Second, LRSim uses DWGSIM (GitHub: nh13/DWGSIM) to mimic the error profile of Illumina paired-end reads and generates 50% more reads than the user requested. Default parameters for running DWGSIM are shown in the Additional file 1: Supplementary Note. PCR duplicates are not simulated. Third, considering the Illumina reads as a pool, LRSim emulates the process of linked-read sequencing by attaching barcodes to reads randomly selected from the pool. It is worth mentioning that LRSim allows both user-specified genome variants, or rely on SURVIVOR to randomly simulate SNPs, indels and structural variants at a user specified rate. The DWGSIM we used in LRSim downstream to SURVIVOR is also capable in simulating SNPs and indels, but not structural variants. Users can also disable SURVIVOR and enable DWGSIM for variant simulation by change just a few lines of code.

LRSim is available open-source in Github under an MIT License. The LRSim pipeline is written in perl, and uses a few genomics tools as components for extracting sequences and simulating reads including SAMtools, SURVIVIOR, and DWGSIM. For a human genome using default parameters, the memory consumption peaks at 48 GB, and starting from scratch takes about five hours to finish using 8 threads, or 1.5 hours if only the linked-reads related parameters x, f, t or m are altered, thus avoiding rerunning the expensive DWGSIM stage. In exchange for lower memory consumption, longer running time is required. The peak memory can be further decreased by decreasing the copies of DWGSIM run in parallel.

Declarations

Acknowledgements

We thank Steven Salzberg for his valuable comments and editing the manuscript. We also thank the team of 2016 CSHL-NCBI Hackathon for helpful discussions. We would also like to thank Deanna Church, David Jaffe, and Patrick Marks from 10X Genomics for their helpful discussions during the development of LRSim. We would also like to thank the MaizeCode Project for providing access to the maize sequencing data.

Funding

This work has been supported by the NSF [DBI-1350041 and IOS-1445025] and the NIH [R01-HG006677] to Michael C. Schatz.

Availability of Data and Materials

Project name: LRSim.
Project homepage: https://github.com/aquaskyline/LRSIM
Archived version: https://github.com/aquaskyline/LRSIM/releases/tag/1.0
Example scripts: https://github.com/aquaskyline/LRSIM/tree/master/test
Operating system: Platform independent.
Programming language: Perl and C++.
Other requirements: See GitHub page.
License: MIT.
DOI: https://doi.org/10.5281/zenodo.808913, 10.24433/CO.8e25703d-92df-4eb5-8683-d1108100b39c.
Any restrictions to use by non-academics: None.

Authors' Contribution

RL, FJS and MCS conceived the study. RL developed and implemented the LRSim algorithm. RL, SMK and CAD performed the evaluation on different parameters. CAD modified the SURVIVOR software package to allow for SNP simulation. RL, FJS and MCS wrote the paper. All authors have read and approved the final version of the manuscript.

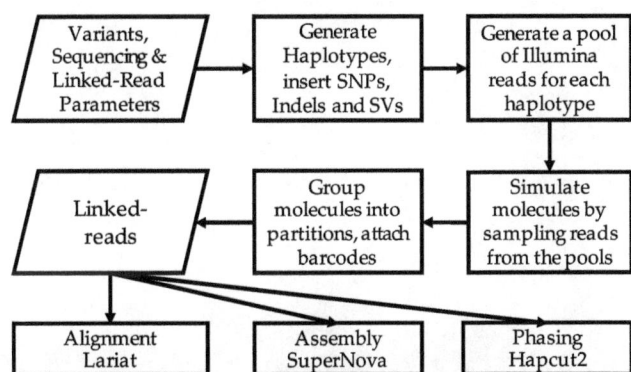

Fig. 8. LRSim workflow. Lariat, SuperNova and HapCUT2 are three tools downstream to LRSim. Lariat is an aligner module of LongRanger specified for linked-read alignment. SuperNova is a genome assembler specified for lined-read. HapCUT2 is a phasing algorithm that works with linked-read. LRSim provides an option to skip variant simulation with SURVIVOR and take a user-provided variant file.

Competing Interests

The authors declare that they have no competing interests.

References

[1] Snyder MW, Adey A, Kitzman JO, Shendure J. Haplotype-resolved genome sequencing: experimental methods and applications. Nat Rev Genet 2015;16:344–58.

[2] Luo R, Liu B, Xie Y, Li Z, Huang W, Yuan J, et al. SOAPdenovo2: an empirically improved memory-efficient short-read de novo assembler. Gigascience 2012;1:18.

[3] Pendleton M, Sebra R, Pang AW, Ummat A, Franzen O, Rausch T, et al. Assembly and diploid architecture of an individual human genome via single-molecule technologies. Nat Methods 2015;12:780–6.

[4] Chin CS, Peluso P, Sedlazeck FJ, Nattestad M, Concepcion GT, Clum A, et al. Phased diploid genome assembly with single-molecule real-time sequencing. Nat Methods 2016;13:1050–4.

[5] Cao H, Wu H, Luo R, Huang S, Sun Y, Tong X, et al. De novo assembly of a haplotype-resolved human genome. Nat Biotechnol 2015;33:617–22.

[6] Zheng GX, Lau BT, Schnall-Levin M, Jarosz M, Bell JM, Hindson CM, et al. Haplotyping germline and cancer genomes with high-throughput linked-read sequencing. Nat Biotechnol 2016;34:303–11.

[7] Weisenfeld NI, Kumar V, Shah P, Church DM, Jaffe DB. Direct determination of diploid genome sequences. Genome Res 2017;27:757–67.

[8] Mohr DW, Naguib A, Weisenfeld N, Kumar V, Shah P, Church DM, et al. Improved de novo genome assembly: linked-read sequencing combined with optical mapping produce a high quality mammalian genome at relatively low cost. bioRxiv; 2017.

[9] Hulse-Kemp AM, Maheshwari S, Stoffel K, Hill TA, Jaffe D, Williams S, et al. Reference quality assembly of the 3.5 Gb genome of *Capsicum annuum* from a single linked-read library. bioRxiv; 2017.

[10] Spies N, Weng Z, Bishara A, McDaniel J, Catoe D, Zook JM, et al. Genome-wide reconstruction of complex structural variants using read clouds. bioRxiv; 2016.

[11] Earl D, Bradnam K, St John J, Darling A, Lin D, Fass J, et al. Assemblathon 1: a competitive assessment of de novo short read assembly methods. Genome Res 2011;21:2224–41.

[12] Edge P, Bafna V, Bansal V. HapCUT2: robust and accurate haplotype assembly for diverse sequencing technologies. Genome Res 2016;27:801–12.

[13] Harrow J, Frankish A, Gonzalez JM, Tapanari E, Diekhans M, Kokocinski F, et al. GENCODE: the reference human genome annotation for the ENCODE project. Genome Res 2012;22:1760–74.

[14] Jeffares DC, Jolly C, Hoti M, Speed D, Shaw L, Rallis C, et al. Transient structural variations have strong effects on quantitative traits and reproductive isolation in fission yeast. Nat Commun 2017;8:14061.

[15] Li Z, Chen Y, Mu D, Yuan J, Shi Y, Zhang H, et al. Comparison of the two major classes of assembly algorithms: overlap–layout–consensus and de-Bruijn-graph. Brief Funct Genomics 2012;11:25–37.

[16] Zerbino DR, Birney E. Velvet: algorithms for de novo short read assembly using de Bruijn graphs. Genome Res 2008;18:821–9.

[17] Jiao Y, Peluso P, Shi J, Liang T, Stitzer MC, Wang B, et al. Improved maize reference genome with single-molecule technologies. Nature 2017;546:524–7.

[18] iGenomX. http://www.igenomx.com/.

[19] Li H, Handsaker B, Wysoker A, Fennell T, Ruan J, Homer N, et al. Genome project data processing S: the sequence alignment/map format and SAMtools. Bioinformatics 2009;25:2078–9.

7

Fourier Analysis of Conservation Patterns in Protein Secondary Structure

Ashok Palaniappan[a],*, Eric Jakobsson[b],*

[a]Dept of Biotechnology, Sri Venkateswara College of Engineering, Post Bag No. 3, Pennalur, Sriperumbudur 602117, India
[b]University of Illinois at Urbana–Champaign, IL 61820, USA

ARTICLE INFO

Keywords:
Periodicity
Secondary structure
Evolution
Moment of conservation
Fourier transform
Potassium channel

ABSTRACT

Residue conservation is a common observation in alignments of protein families, underscoring positions important in protein structure and function. Though many methods measure the level of conservation of particular residue positions, currently we do not have a way to study spatial oscillations occurring in protein conservation patterns. It is known that hydrophobicity shows spatial oscillations in proteins, which is characterized by computing the hydrophobic moment of the protein domains. Here, we advance the study of moments of conservation of protein families to know whether there might exist spatial asymmetry in the conservation patterns of regular secondary structures. Analogous to the hydrophobic moment, the conservation moment is defined as the modulus of the Fourier transform of the conservation function of an alignment of related protein, where the conservation function is the vector of conservation values at each column of the alignment. The profile of the conservation moment is useful in ascertaining any periodicity of conservation, which might correlate with the period of the secondary structure. To demonstrate the concept, conservation in the family of potassium ion channel proteins was analyzed using moments. It was shown that the pore helix of the potassium channel showed oscillations in the moment of conservation matching the period of the α-helix. This implied that one side of the pore helix was evolutionarily conserved in contrast to its opposite side. In addition, the method of conservation moments correctly identified the disposition of the voltage sensor of voltage-gated potassium channels to form a 3_{10} helix in the membrane.

1. Introduction

Amino-acid conservation is an evolutionary property. Physical properties of amino acid side-chains exhibit a higher-order moment (also known as periodicity) in the context of repetitive secondary structures, such as the α-helix and β-sheet. A notable physical property whose moments turned out to be significant is the hydrophobicity [1]. The disposition of structured domains in the protein is strongly correlated with the overall hydrophobicity and the amphiphilicity of the domains. These properties stabilize the structure of the protein and, for membrane proteins, the protein's association with the membrane. For α-helical membrane proteins, the strength of the hydrophobic moment is maximal at the period of the helix (i.e, 100°). Similarly, for beta-barrel membrane proteins, the hydrophobic moment is maximal at the period of the beta sheet (i.e, 160°–180°). In both cases, the surface of the secondary structure element which is in contact with lipid exhibits a strong hydrophobicity to allow for partitioning into the membrane. Domains exposed to the electrolyte on either side of the membrane exhibit strong amphiphilicity, or a high hydrophobic moment. The periodicity of residue properties of α-helical proteins can be visualized using the helical wheel representation. Spatial asymmetry in the distribution of hydrophobicity, say, on the helical wheel would imply the fine-tuning of protein function via the achievement of amphiphilicity. Amphiphilicity has turned out to be key to the activity of antimicrobial peptides. Most native and engineered antimicrobial peptides face the amphiphilicity requirement to successfully insert into and permeabilize the bacterial membrane [2].

The use of sequence profiles improved the ability of hydrophobicity to predict the formation of α-helices [3]. Analogous to hydrophobicity, we consider that the residue conservation in a protein alignment displays a first-order moment. Residue conservation is directly correlated with general functional importance. The moment of residue conservation would likely contain information not captured by a linear residue-by-residue conservation. The evolutionary basis of the moment of conservation is as follows: one face of an α-helix involved in critical interatomic interactions must be

* Corresponding authors.
  E-mail addresses: aplnppn@gmail.com, ashok@svce.ac.in (A. Palaniappan),
jake@illinois.edu (E. Jakobsson).

conserved, while the diametric face might not be equally constrained and evolve with neutral drift. In order to detect and quantify this spatially oscillatory constraint in the protein secondary structure, we introduce a measure called 'conservation moment' and illustrate its applications.

2. Material and Methods

2.1. Calculation of the Zeroth Moment of Conservation

The zeroth moment of conservation is the sum of the conservation values of the residues based on a profile of homologous sequences. The profile is built using homology detection methods and multiple sequence alignment. The conservation c_n of each column n of the alignment could then be computed using, for e.g., Shannon entropy:

$$c_n = -\sum_i p_i \ln p_i \tag{1}$$

where the p_i's are the probabilities of finding residue i in column n and the summation is over all the 20 amino acids. The c_n's are scaled from 0 to 1, 0 denoting a column of all different residues and 1 denoting a column of all identical residues. The resulting one-dimensional function of conservation values over the length of the alignment is called the conservation vector. The zeroth conservation moment C_0 of an alignment segment of length N is equal to the sum of the c_n's of the columns of the alignment segment.

$$C_0 = \sum_{n=1}^{N} c_n \tag{2}$$

C_0 is a measure of the net conservation of an alignment segment. A contiguous sequence of conserved residues in a protein family would give rise to a high C_0.

2.2. Calculation of First-order Conservation Moment

To detect an asymmetry in the conservation pattern of an alignment segment, we search for periodicities in the corresponding conservation vector. The moment of the conservation vector at a given periodicity is a measure of the signal strength at that periodicity, and

is known as the first-order conservation moment, $C_1(\theta)$. For a given period θ,

$$C_1(\theta) = \left\{ \left[\sum_{n=1}^{N} C_n \sin(\theta n) \right]^2 + \left[\sum_{n=1}^{N} C_n \cos(\theta n) \right]^2 \right\}^{\frac{1}{2}} \tag{3}$$

where N is the length of the alignment segment, and the period θ is measured in radian. An evolutionary asymmetry in the α-helix structure would be manifested as a strong conservation moment at the period of the α-helix. This corresponds to $\theta = 2\pi/100° = 3.6$ rad. Similarly, an evolutionary moment in the β-sheet structure would give rise to a maximal signal at the period of the β-sheet ($=160°$–$180°$). Eq. (3) could be rewritten as the modulus of the fourier transform of the conservation vector.

$$C_1(\theta) = \left| \sum_{n=1}^{N} C_n e^{i\theta n} \right| \tag{4}$$

3. Results and Discussion

When the protein secondary structure is known, from a crystal structure or otherwise, $C_1(\theta)$ could be calculated for each secondary structure element at its respective period to detect any spatial asymmetry in evolutionary pressure. Periodicity in evolutionary pressure is valuable for transmembrane structures which accommodate hydrophobic constraints to be stable in the lipid bilayer. This might enable the transmembrane structure to achieve a higher-order functional specificity. An illustrative secondary structure element is the pore helix of the potassium ion (K^+) channel.

Potassium channels are tetrameric transmembrane (TM) structures with two TM helices per subunit [6]. In addition, each subunit has a pore helix that spans half the membrane before looping back. These pore-helices are under an interesting evolutionary constraint. By virtue of scaffolding the 'selectivity filter' of potassium channels, their packing interfaces are evolutionarily constrained. This sidedness of conservation could be detected using the first-order conservation moment. A profile of all the human potassium channel sequences was constructed in the following manner. A representative sequence of each potassium channel subfamily was chosen, and used as a query in PSI-BLAST with an E-value of 0.001 until convergence [7]. After eliminating duplicates, each hit was screened for the presence of selectivity filter characteristic of a potassium-selective channel, to obtain 123 channels (available as a supporting information).

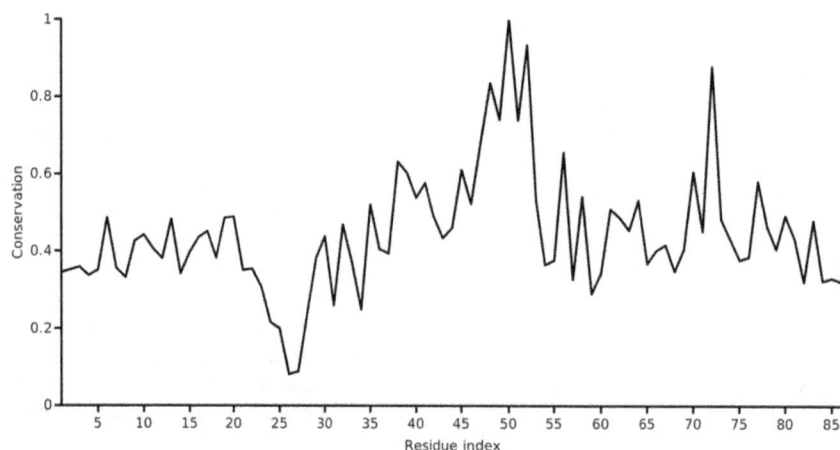

Fig. 1. Profile of the conservation of each position in the KcsA potassium channel sequence, as calculated using Scorecons. A peak (conservation $= 1.0$) corresponding to the selectivity filter of the K^+-channel could be observed at position 50.

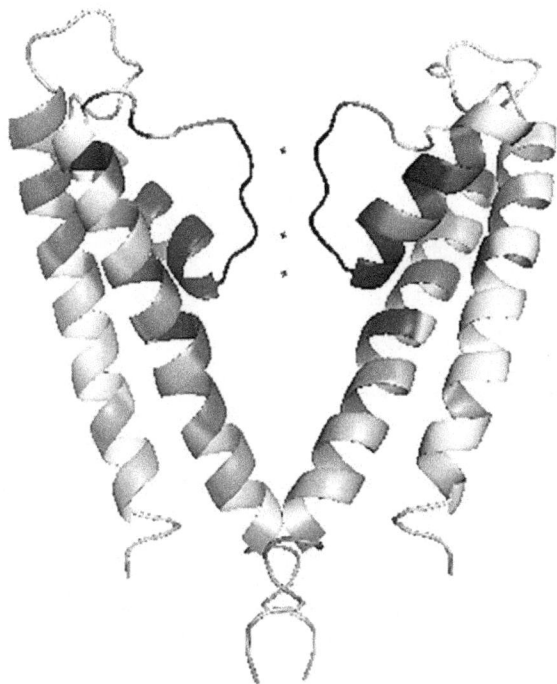

Fig. 2. Structural mapping of conservation of human K^+-channel sequences, using the Consurf algorithm [4]. Imbalance in conservation could be observed. Only two diametric KcsA channel monomers are shown.
Source: Data for this figure obtained from ref. [5].

Since potassium channels are highly heterogeneous in their domain composition, the permeation pathways of the channels were extracted for further analysis by pivoting about the selectivity filter. It must be noted that the two-pore channels contain two distinct permeation pathways to form a 'tetramer' via a hetero-dimer of homo-dimers. After sorting by sequence lengths, multiple rounds of profile–profile alignments were needed to gradually build the global alignment of all human potassium channels [8]. Owing to the variable extracellular turret region, the alignment was manually edited to register the pore-helix and the surrounding TM helices. Finally the KcsA (PDB 1bl8) K^+-channel sequence was aligned with the rest of the sequences. This alignment was used to calculate the one-dimensional conservation function for each position in the KcsA sequence.

The conservation of each column of the final alignment was calculated using Scorecons [9], which uses a residue substitution matrix and sequence-weighting to arrive at its final score. These biological refinements to calculating c_n enhance the entropic formulation in Eq. (1). The score for a particular column of the alignment is normalized to the range [0,1] in the order of increasing conservation. Scorecons assessed the informativeness of the alignment and estimated a diversity of 94.6% (higher the diversity, the more informative the alignment). Fig. 1 shows the one-dimensional conservation vector as a function of the residue index of the KcsA channel. The location of the selectivity filter which is essential in establishing the precise selectivity of the channel is evident. A glimpse of the asymmetry in the conservation patterns for the potassium channel could be revealed by mapping the conservation metric of each KcsA residue over its 3D structure. A plot which maps the 'emphasis' of conservation on the structure is shown in Fig. 2.

The first-order conservation moment, $C_1(\theta)$, was computed using Eq. 4. The block length was set to the length of the pore helix (=11 residues) and periodicity of interest was the helical periodicity ($\theta = 3.6$). Fig. 3 shows the computed conservation moment for a sliding window of 11 residues over the full length of the protein at a periodicity of 100°. This produced a profile of the conservation moment as a function of position (i.e, the center of the sliding window). It was observed that the moment oscillated periodically between low and high values about the pore helix region. These initial observations were in accord with the case for a conservation moment of the pore helix. To investigate whether the period of oscillation coincided with the period of the α-helix, we determined the $C_1(\theta)$ of the pore helix region alone (an ungapped column subset of the alignment of 11 residues) at various periodicities ranging from 2.0 to 5.0 radian in steps of 0.1 radian (shown in Fig. 4). Close to 3.6 radian, it was found that the $C_1(\theta)$ reached a maximum, validating the helical periodicity of the conservation. To ascertain the variation of $C_1(\theta)$ with the window size, the window size was varied from 5 to 31 residues around the central residue of the pore helix (default = 11 residues) and the $C_1(\theta)$'s at $\theta = 3.6$ was computed (shown in Fig. 5). Close to the true length of 11 residues, the maximum moment was observed.

Fig. 6 shows a plot of the conservation moment ($C_1(\theta)$) against the normalized net conservation using a sliding window of 11 residues. This is a plot of the first-order moment against the normalized zeroth moment, both defined for a window size = 11. Normalization of the net conservation was achieved by the mean of C_0 for the given length (which is identical to $C_1(\theta)$ for an infinite period). In the region of the pore helix, a moderate residue conservation coupled with significant oscillations in the moment of conservation could be observed. It appeared that the sidedness property of conservation was more important than the residue conservation *per se*. This would

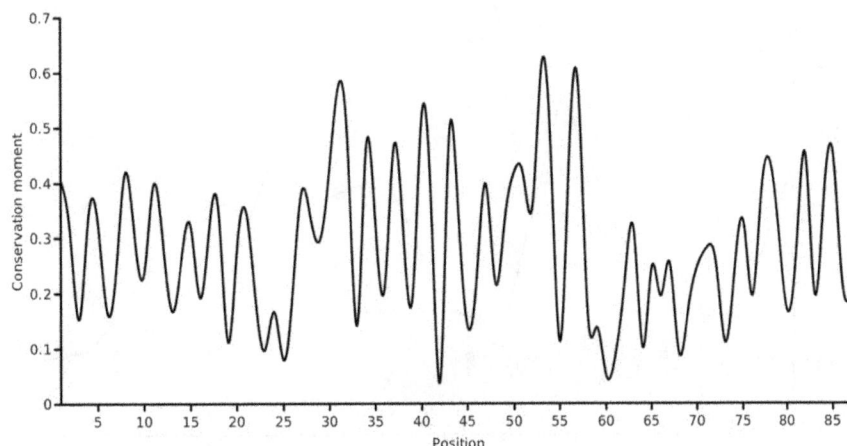

Fig. 3. Conservation moment $C_1(\theta = 3.6)$ of the KcsA potassium channel sequence, computed using Eq. (4) and a sliding window of 11 residues. The pore helix spans positions 35 to 45.

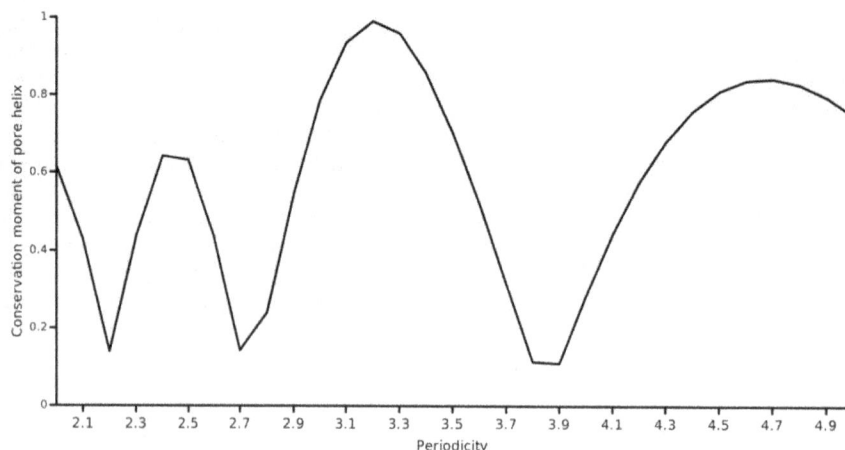

Fig. 4. Profile of the conservation moment $C_1(\theta)$ over the pore helix as a function of periodicity. The conservation moment was estimated about the central residue position of the pore helix using a block size = 11.

imply that the moment of conservation was more important than the conservation of the identity of the residue.

To further analyze the utility of the conservation moment, the voltage sensing module of voltage-gated potassium channels was investigated. The voltage sensor is the gating module in ion channels that underlies their steep dependence on membrane voltage for channel opening and closing. The precise mechanism of voltage-gating has remained uncertain [10]. All voltage-gated potassium channels (Kv channels) contain the voltage sensor. A database of voltage-gated potassium channels was constructed using a representative from each of the Kv subfamilies as the query of a PSI-BLAST search with an E-value of 0.001 until convergence. To create a non-redundant dataset, the Kv channels were clustered at 90% sequence identity [11]. A total of 147 Kv channels were obtained, and their voltage sensors were extracted based on the known motif [12]. Since these were 18 residues in length, an ungapped alignment was obtained (available as a supporting information). The conservation moments for this functional region were calculated at periodicities corresponding to three different secondary structures: the regular α helix, the 3_{10} helix and the β sheet. Fig. 7 shows the profile of these three conservation moments over the voltage sensor using a block of 10 residues. Two observations emerged from our analysis. First, the conservation moments of the α helix and 3_{10} helix rise much above the average conservation (which could assume a value of 1.0 at the maximum) whereas the β strand conformation is disfavored. Conservation moments that exceed the maximum possible

conservation would reflect a selection for the moment, indicating possible functional significance. Second, the conservation moment of the 3_{10} helix exceeds that of the α helix at position 3 which contains the conserved arginine residue of the gating pore. The segment is uncertain in its preference for the α helix, especially at the positions containing the conserved positively charged residues. This suggested a dominant preference for the 3_{10} helical conformation over the alpha helix over the length of canonical motif. Surprisingly, crystallographic studies of the voltage-gated potassium channel have determined this region to adopt an unusual transmembrane 3_{10} helix, stretching out inside an α helical conformation at the ends of the voltage sensor [13]. This local 3_{10} helical conformation accounted for the energetics of the voltage sensor movement in a hydrophobic lipid membrane environment. The S4 helix (i.e, the voltage sensor) maintained an entire face of spatially oriented positively charged residues, which could interact with conserved acidic residues from other TM helices, forming stabilizing ion pairs. The opposite face of the S4 helix was variable, maintaining a hydrophobic character that would have preferred the α helical conformation, if it were not for the voltage sensing motif. Our analysis was able to detect this asymmetry crucial to the gating pore and seemed to provide support for mechanisms of gating that involve the formation of ion pairs with 3_{10} helices in the S4 voltage sensor. It is clear from this example that the conservation moment captured a feature of evolution that would not have been apparent from an examination of residue-by-residue conservation.

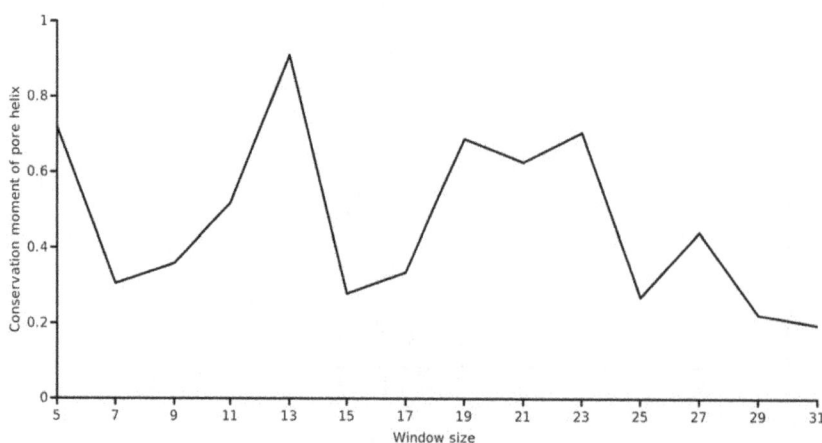

Fig. 5. Profile of the conservation moment at $C_1(\theta = 3.6)$ over the pore helix, as a function of the sliding window size.

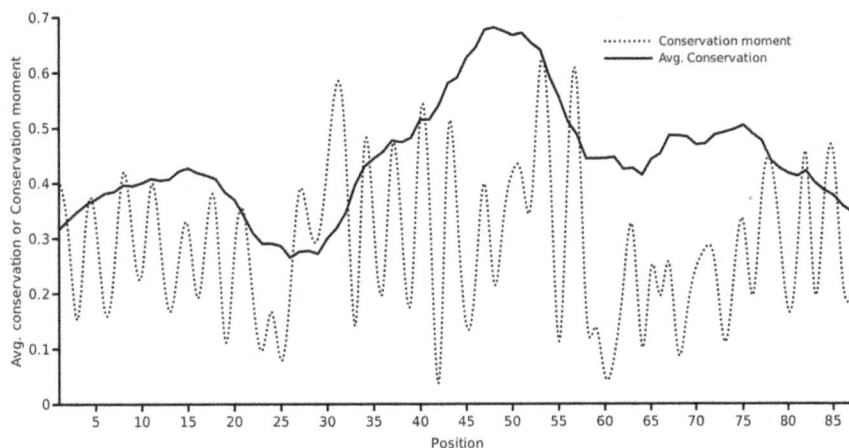

Fig. 6. A plot of $C_1(\theta = 3.6)$ versus the mean conservation (using a window size = 11 residues). In the region from positions 32 to 45, moderate residue conservation is coupled with oscillations in the moment of conservation.

4. Conclusion

The proposed conservation moment demonstrated its effectiveness in the analysis of the pore helix and the voltage sensor of potassium channels. It was observed that oscillations in conservation moments matched the period of the α-helix enabling differential conservation of packing interfaces of the pore helix. In the case of the voltage sensor, the method of conservation moments detected the preference for the rare 3_{10}-helix over the α-helix. Two conclusions could be made from the above. Differential moments for the periodicities corresponding to different secondary structures would be predictive of the 'momentous' secondary structure. Second, facially differential conservation within secondary structures (i.e, the existence of a significant conservation moment in the secondary structure) would be diagnostic of regions of functional activity. The profile of conservation moments of a protein sequence calculated using an appropriate profile would be useful in detecting both the spatially asymmetric conservation and the secondary structure preference. It would be a valuable tool in interrogating structure–function relationships in proteins and its potential for the automated detection of functionally important regions in proteins could be explored in the future. The conservation moment embodies an enrichment of the information contained in residue conservation. The implemented algorithm could be applied with little modification to calculate the strength of Fourier components and detect periodicity in the one-dimensional function of any residue property including hydrophobicity and packing. By combining the information and moments of both physical and evolutionary properties, higher-order trends could be found.

References

[1] Eisenberg D, Weiss R, Terwilliger T. The hydrophobic moment detects periodicity in protein hydrophobicity. Proc Natl Acad Sci USA 1984;81:140–4.

[2] Manzo G, Scorciapino M, Wadhwani P, Bürck J, Montaldo N, Pintus M. et al. Enhanced amphiphilic profile of a short β-stranded peptide improves its antimicrobial activity. PLoS One 2015;10:e0116379. http://dx.doi.org/10.1371/journal.pone.0116379.

[3] Cheng W, Yan C. Detecting periodicity associated with the alpha-helix structure using Fourier transform. Computational Molecular Bioscience 2012;2:109–14. http://dx.doi.org/10.4236/cmb.2012.24011.

[4] Ashkenazy H, Abadi S, Martz E, Chay O, Mayrose I, Pupko T. et al. Consurf 2016: an improved methodology to estimate and visualize evolutionary conservation in macromolecules. Nucleic Acids Res 2016;44:W344–W350. http://dx.doi.org/10.1093/nar/gkw408.

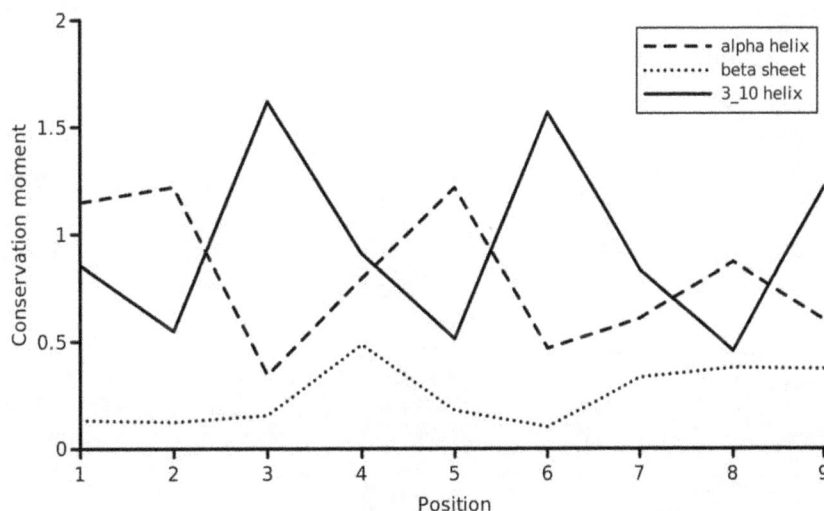

Fig. 7. Profile of the conservation moment of the 18-residue voltage sensing module in the S4 transmembrane region of voltage-gated potassium channels, using a window size = 10. Three different secondary structures are probed at their respective periodicities. For the 3_{10} helix, $\theta = 3.0$.

[5] Palaniappan A, Jakobsson E. Evolutionary analysis of biological excitabil-ity.2009, arXiv q-bio/0609023 URL https://arxiv.org/abs/q-bio/0609023

[6] Doyle D, Morais Cabral J, Pfuetzner R, Kuo A, Gulbis J, Cohen S. et al. The structure of the potassium channel: molecular basis of K^+ conduction and selectivity. Science 1998;280:69–77.

[7] Altschul S, Madden T, Schäffer A, Zhang J, Zhang Z, Miller W. et al. Gapped BLAST and PSI-BLAST: a new generation of protein database search programs. Nucleic Acids Res 1997;25(17):3389–402.

[8] Jeanmougin F, Thompson J, Gouy M, Higgins D, Gibson T. Multiple sequence alignment with ClustalX. Trends Biochem Sci 1998;23:403–5.

[9] Valdar W. Scoring residue conservation. Proteins 2002;48:227–41.

[10] Catterall W. Ion channel voltage sensors: structure, function, and pathophysiol-ogy. Neuron 2010;67(6):915. http://dx.doi.org/10.1016/j.neuron.2010.08.021.

[11] Li W, Godzik A. Cd-hit: a fast program for clustering and comparing large sets of protein or nucleotide sequences. Bioinformatics 2006;22(13):1658–9.

[12] Noda M, Shimizu S, Tanabe T, Takai T, Kayano T, Ikeda T. et al. Primary struc-ture of Electrophorus electricus sodium channel deduced from cDNA sequence. Nature 1984;312:121.

[13] Long S, Tao X, Campbell E, MacKinnon R. Atomic structure of a voltage-dependent K^+ channel in a lipid membrane-like environment. Nature 2007;450:376–82.

8

Evaluation of multiple approaches to identify genome-wide polymorphisms in closely related genotypes of sweet cherry (*Prunus avium* L.)

Seanna Hewitt [a,b,1], Benjamin Kilian [a,b,1], Ramyya Hari [b,1], Tyson Koepke [a,b,2], Richard Sharpe [a,b], Amit Dhingra [a,b,*]

[a] *Molecular Plant Sciences Graduate Program, Washington State University, Pullman, WA 99164, United States*
[b] *Department of Horticulture, Washington State University, Pullman, WA 99164-6414, United States*

ARTICLE INFO

Keywords:
Polymorphisms
Prunus avium
Next-generation sequencing
Target region amplification polymorphism (TRAP)
Genetic diversity
SNParray
Reduced representation sequencing
Whole genome sequencing (WGS)

ABSTRACT

Identification of genetic polymorphisms and subsequent development of molecular markers is important for marker assisted breeding of superior cultivars of economically important species. Sweet cherry (*Prunus avium* L.) is an economically important non-climacteric tree fruit crop in the Rosaceae family and has undergone a genetic bottleneck due to breeding, resulting in limited genetic diversity in the germplasm that is utilized for breeding new cultivars. Therefore, it is critical to recognize the best platforms for identifying genome-wide polymorphisms that can help identify, and consequently preserve, the diversity in a genetically constrained species. For the identification of polymorphisms in five closely related genotypes of sweet cherry, a gel-based approach (TRAP), reduced representation sequencing (TRAPseq), a 6k cherry SNParray, and whole genome sequencing (WGS) approaches were evaluated in the identification of genome-wide polymorphisms in sweet cherry cultivars. All platforms facilitated detection of polymorphisms among the genotypes with variable efficiency. In assessing multiple SNP detection platforms, this study has demonstrated that a combination of appropriate approaches is necessary for efficient polymorphism identification, especially between closely related cultivars of a species. The information generated in this study provides a valuable resource for future genetic and genomic studies in sweet cherry, and the insights gained from the evaluation of multiple approaches can be utilized for other closely related species with limited genetic diversity in the breeding germplasm.

1. Introduction

Plants are fundamental to continued life on this planet as they are the basis of food production and an essential part of the global ecosystem. Application of different molecular tools and access to plant genomes has facilitated identification of genome-wide polymorphisms and thus, development of molecular markers that can be utilized in breeding programs [1,2]. Next-generation sequencing now allows genomic information to be obtained, even for non-model plant systems, further accelerating the development of molecular markers and genetic research [3,4]. Efforts to efficiently develop desirable genotypes by establishing an association of important agronomic traits, such as yield, nutritional content, and timing of flowering and fruit ripening with

specific polymorphic regions of the genome, are ongoing in various plant species [5,6].

Sweet cherry (*Prunus avium* L.) is a member of the Rosaceae family, which represents many other important crop species, including apple, peach, plum, almond, strawberry, raspberry and rose [7]. Despite an estimated genome size of 225–330 Mb [8,9], sweet cherry is lacking in genomic information in comparison with other prominent Rosaceae members, including peach and apple [10,11]. Linkage maps and molecular markers have been developed for sweet cherry [12] as well as peach and almond, two other members of the sub-family Prunoideae [13–15], and a comprehensive and advanced draft of the peach genome serves as the foundation for several comparative studies [10]. Recently, a draft genome of sweet cherry cultivar 'Stella' was released [16]. To advance diversity and genetics-related studies, efforts were made to evaluate the transferability of the molecular markers from one member of Rosaceae family to other members with mixed success [17–19].

In addition to lack of comprehensive genetic information, domesticated sweet cherry cultivars exhibit a genetic bottleneck as a result of breeding. Despite the prevalence of several wild landraces [20], there are only three chloroplast haplotypes represented in the commercial

* Corresponding author at: Department of Horticulture, Washington State University, Pullman, WA 99164-6414, United States.
E-mail address: adhingra@wsu.edu (A. Dhingra).
[1] These authors contributed equally to this manuscript.
[2] Present address: Phytelligence Inc., 1615 NE Eastgate Blvd #3, Pullman, WA 99163.

cultivars indicating a very narrow maternal parental lineage in sweet cherry [21,22]. Given the genetic closeness, it can be difficult to identify genetic diversity unless comprehensive approaches are utilized. A recent study in tree genus *Milica*, where population structure was analyzed using nuclear SNPs, SSRs and DNA sequences, revealed hidden species diversity in closely related species [23]. In sweet cherry, a previous study compared and evaluated the utility of 7 simple sequence repeat (SSR) molecular markers versus 40 single nucleotide polymorphism (SNP) molecular markers to determine the genetic diversity and relatedness in 99 cultivated genotypes of sweet cherry [24]. SSRs were found to generate a higher average number of alleles per locus, mean observed heterozygosity, expected heterozygosity, and polymorphic information content values; however, the SNPs allowed for finer resolution of a closely related genotype, which was indistinguishable with SSRs. Despite the higher resolution of SNPs, both sets of markers produced a similar genetic relatedness for all the accessions tested [24].

In this study, the efficiency of different genotyping approaches was evaluated to differentiate between five sweet cherry cultivars. The cultivars selected for diversity analysis are suspected to be very closely related, and their interrelatedness was not tested in the previous study that included 99 cultivars [24]. The genotypes included a newly identified cultivar named 'Glory,' which was proposed to be an open-pollinated seedling of 'Sonata'. However, it has also been proposed that it is the same cultivar as '13S2009' 'Staccato', owned by Summerland Variety Corporation, Canada [25–27]. Similarly, 'Kimberly' and 'Bing' were selected since it has been proposed that the former may have been derived from the latter as a random mutation or sport [28]. 'Sweetheart' was selected as it is the parent of 'Staccato' [29]. The newly released cultivars 'Glory' and 'Kimberly' represent late maturing cultivars, like 'Staccato' and 'Sweetheart', making them highly desirable cultivars. The similarity in late maturing phenotype across the four cultivars has led to the notion that the new cultivars may share a close genetic relationship, or that they may in fact be the same as previously released cultivars. In order to resolve the identity conundrum and understand the genetic relationship between these cultivars and genetically distinguish them from each other, a gel-based, Targeted Region Amplified Polymorphism (TRAP) approach [30], a reduced representation or genotype by sequencing (GBS) approach called TRAPseq, a *Prunus* SNParray [31], and a whole genome sequencing (WGS) approach were evaluated for their relative effectiveness.

2. Methods

2.1. Plant Material Source and Preparation

Five sweet cherry genotypes used in this study were obtained from VanWell Nursery, East Wenatchee, WA. Emerging leaf samples were collected for each genotype following fruit harvest and flash frozen in liquid nitrogen. All samples were pulverized under liquid nitrogen using SPEX SamplePrep® FreezerMill 6870 (Metuchen, NJ, USA) and kept frozen at −80 °C prior to processing.

2.2. Genomic DNA Extraction

Total genomic DNA was extracted from young leaf tissue using cetyltrimethylammonium bromide (CTAB) phenol chloroform extraction method [32]. Extracted DNA pellets were air dried and suspended in 50 μl of nuclease-free water and incubated at 37 °C with DNase free RNAse for 30 min. RNAse was inactivated by incubating the tubes at 65 °C for 10 min. DNA was quantified using Nanodrop 8000 spectrophotometer (Thermo Scientific, Waltham, MA, USA) and 50 ng of extracted genomic DNA was electrophoresed on a 1% agarose gel and compared to Lambda DNA dilution series (100, 80, 60, 40, 20, 10 ng) to confirm quality and quantity.

2.3. TRAP – Target Region Amplification Polymorphism

PCR was conducted with a final reaction volume of 10 μl in a BioRad ICycler (Bio-Rad Laboratories, Hercules, CA) with components in the following final concentrations: 10 ng DNA, 1.5 mM $MgCl_2$, 0.2 mM dNTPs, 0.02 mM 700- and 800-IR dye-labeled arbitrary primers, 0.2 mM fixed primer (BRK 393 or BRK 394, Table 1), and 1 U Taq DNA polymerase and 1× corresponding polymerase buffer (Biolase). PCR was carried out by initially denaturing the template DNA at 94 °C for 2 min. The thermocycle profile consisted of five cycles of 94 °C for 45 s, 35 °C for 45 s, and 72 °C for 1 min, followed by 35 cycles at 94 °C for 45 s, 50 °C for 45 s, and 72 °C for 1 min. The final extension step was at 72 °C for 7 min. Thereafter, 5 μl of IR stop dye was added and the product was denatured at 4 °C for 4 min. A 6.5% polyacrylamide gel (KB-PLUS, LI-COR) was cast, the reactions loaded, and the PCR product electrophoresed at 1500 V for 2.5 h in a Li-COR 4300 DNA Analyzer (LI-COR Biosciences, Lincoln, NE). Images were captured by the Li-COR instrument and analyzed using LI-COR 4300 DNA Analyzer image software to identify polymorphisms.

2.4. TRAPseq and Read Processing Using Stacks and BLAST2GO Analysis

Genomic DNA (~1 μg) was isolated from 'Glory' and 'Staccato' young leaf tissue. The reduced representation of the genome was achieved by performing TRAP PCR with fixed primers targeting MADS-box, PPR1, and PPR2 gene families (Table 1). Amplification was followed by generation of NGS sequence data from the products (Ion Torrent PGM, Thermo Fisher Scientific, Inc., Waltham, MA). The short read sequence data generated from TRAPseq was submitted to NCBI under the following accession numbers: SRS1706064 - Glory_Trapseq and SRS1706056 - Staccato_Trapseq. The fixed MADS primer was selected because the MADS-box gene family is predicted to contain polymorphic regions even in closely-related plant cultivars [33,34]. The TRAP PCR parameters used were identical to the TRAP protocol described above, except for the 5-min denaturing step. Following TRAP amplification and PCR cleanup, the reduced representation sample library was prepared using the NEBNext® Fast DNA Library Prep Set as per the manufacturer's instructions with the following modifications. TRAP PCR products from each reaction were sheared with NEB Next Fragmentase. After heat disabling the fragmentase, each sample was processed for A-tailing by adding 0.2 mM dATP (1 mM stock), 1 U of Taq polymerase (5 U/μl), 1.6 mM of $MgCl_2$ (50 mM), and 1× Taq polymerase buffer (10× stock). Complementary, custom adaptors were then annealed to the sheared DNA, the annealed product was purified and extracted according the NEBNext FastDNA Library Prep protocol. The libraries were quantified, pooled, and sequenced using the Ion Torrent PGM (Life Technologies, Inc.). The sequencing run included 850 flows on a 318C chip producing single reads of various lengths.

The sequenced libraries (Ion Torrent PGM, Thermo Fisher Scientific, Inc., Waltham, MA) generated ~230 Mb combined data for 'Glory' and 'Staccato' genotypes, comprised of 795k reads with an average read length of 145 nucleotides. The sequencing data was processed through the Stacks program to identify loci containing polymorphisms [35]. This allowed for the generation of an output file containing the Stacks catalog ID and corresponding genotype for 'Glory' and 'Staccato' at each locus (Supplementary File 1). Each combination of nucleotides at the polymorphic loci was assigned a numeric code of 1–16. All loci originally identified in Stacks were run through the Blast2GO sequence alignment, gene ontology (GO) mapping, and functional annotation pipeline [36,37]. The output file is available as Supplementary File 2. Sequences were processed through BLAST against the Viridiplantae database using an e-value cutoff of 1.0e − 3 [38].

Table 1
TRAP and TRAPseq primers. Information regarding method, genomic target, primer type, and nucleotide sequence are provided.

Name	Method	Target	Type	Sequence
BKP-383	TRAP	VRN2	Fixed	GCGCCAATTCCAAATACAGT
BKP-384	TRAP	VRN2	Fixed	TTTTGTGACCCAATTCGACA
SA12	TRAP	–	Arbitrary	AminoC6 + DY78...TAATCCAACAACA
GA5	TRAP/TRAPseq	–	Arbitrary	AminoC6 + DY68...AAACACACATGAAGA
MADS-box	TRAPseq	MADS-box gene family	Fixed	TGGCCTCTTCAAGAAGGC
PPR1	TRAPseq	Pentatricopeptide repeat 1 gene family	Fixed	ATGGTTGATCTTCTTGGC
PPR2	TRAPseq	Pentatricopeptide repeat 2 gene family	Fixed	AATGATTGGGCGAAGGC
ODD15	TRAPseq	–	Arbitrary	AminoC6 + DY...GGATGCTACTGGTT

2.5. SNParray

For this experiment, 'Bing', 'Sweetheart', 'Glory', 'Kimberly' and 'Staccato' sweet cherry cultivars were analyzed using the sweet cherry 6k Infinium II SNParray [31]. The output data were analyzed with GenomeStudio v. 1.0, Genotyping module (Illumina, Inc., San Diego, CA), which determines cluster positions of the AA/AB/BB genotypes for each putative SNP. Default quality metrics for GenomeStudio were used in the assay: GenTrain score ≥ 0.5, minor allelic frequency (MAF) ≥ 0.15 and call rate of >80%. The resulting data show pair-wise comparisons between each cultivar for each specific SNP. A subset of the predicted SNPs was evaluated in silico by using BLAST to compare twenty SNPs from NCBI with the de novo assembly from each genotype. All twenty SNPs tested were confirmed using this method (Supplementary File 3).

The identified SNPs were filtered to remove missing data, assigned numeric codes corresponding to respective AA/AB/BB genotype, and categorized for downstream population structure analysis.

2.6. WGS and Genetic Diversity Analysis Using Stacks

For all the genotypes, approximately 25× coverage sequence data represented by 2 × 100 paired end reads, were generated with the Illumina HiSeq 2000 sequencing platform. All short read sequenced data was submitted to NCBI under the following accession numbers: SRS1706059 - Bing_Illumina; SRS1706061 - Sweetheart_Illumina; SRS1706060 - Staccato_Illumina; SRS1706062 - Glory_Illumina; SRS1706063 - Kimbery_Illumina. Stacks [35] was used to identify SNPs from the short-read sequence genomic data. This was accomplished through building artificial loci from the raw data ('stacks' of reads). An internal module (Process_shortreads) was used which filters reads with uncalled bases, discards reads with low quality scores and removes any traces of remaining inline barcodes. Thereafter, the dataset was processed by running the de novo map wrapper, which includes ustacks, cstacks, sstacks, populations (map). Ustacks builds stacks, forms loci, and looks for SNPs. Cstacks merges identified loci together across a population based on the consensus sequence from each locus. Then, sstacks creates a map between the loci in the population that match the catalog and assigns respective catalog IDs to these loci [35]. SNPs were detected at each locus using a maximum likelihood framework by iteratively comparing loci for each sweet cherry genotype in a pairwise comparison against other genotypes.

2.7. Population Structure Analysis Using STRUCTURE and NTSys

A SNP-based population structure analysis was conducted for both the SNParray and the Stacks data using STRUCTURE [39] and NTSys [40]. Loci with missing data were omitted from the final analyses, as were loci with the same score for each of the 5 genotypes. For the SNParray data, the cherry genotypes were assigned a numeric code of 1–6, corresponding to the respective AA/AB/BB genotype at each polymorphic locus. This was the input file for the subsequent STRUCTURE analysis (Supplementary File 4). For the WGS data, a structure.tsv file

from the Stacks 'populations' output was modified. Numbers 1, 2, 3, and 4 were used to code for A, C, G, and T, respectively, and '0' was used to indicate missing data. The Stacks output file contained information regarding the replicates and separate paired end reads for each allele, therefore, to consolidate data, the most frequent non-zero nucleotide code was identified for each genotype (Supplementary Files 5 and 6). The modified SNParray and WGS Stacks files were saved as *.csv files for input into STRUCTURE (Supplementary File 7). The parameters for the preparation of data upload to STRUCTURE were as follows: row of marker names = TRUE, individuals = 5, ploidy = 2, loci = 9029. Additional parameters for running the population structure algorithm were specified as follows: Length of Burnin Period = 20,000, Number of MCMC Reps after Burnin = 20,000, Use Admixture Model = TRUE, Allele Frequencies Correlated = TRUE, Compute probability of the data (for estimating K) = TRUE, Print Q-hat = TRUE.

Analysis of K values from 1 to 5 was specified, along with 5 iterations of the defined STRUCTURE analysis. Upon completion of the Structure run, Structure Harvester was used for identification of most likely K-value based on the data [41].

The NTSys software [40] was used to produce a tree dendrogram and to determine sample order for the population structure output. The latter is used for running of CLUMPP [42] and DISTRUCT [43] clustering and visualization programs. The SNParray and WGS data files for input into NTSys were prepared by modifying the STRUCTURE files (Supplementary Files 8 and 9). In the case of the Stacks data, the alleles were assigned an ID of 'a' or 'b' and were listed under their respective genotypes to be treated as separate markers in the NTSys analysis. This was not necessary for the SNParray data, as the allele combinations were assigned numeric codes, as previously stated.

To run NTSys, the input files were uploaded, and the following functions run: 1.) Qualitative data Dis/Similarity method, 2.) SAHN UPGMA clustering method 3.) Tree plot graphic generation function. The result is a tree dendrogram representing WGS SNP-based genetic relationships (Fig. 6). The K2 and K4 indfiles from the Structure Harvester output were then run through CLUMPP and DISTRUCT [42,43] according to an in-house workflow to produce a graphic representing population structure.

2.8. Validation of NTSys and STRUCTURE Results

To validate the NTSys and STRUCTURE outputs, Excel was used to calculate the number of SNPs in pairwise comparisons between each genotype, with Bing as the reference genotype. The resulting data was prepared as a distance matrix— genetic distance (or genotypic variation) increases as the number of SNPs increases.

The data was saved as a *.txt file and imported into R studio as a "dist" object for further analysis. A dendrogram similar to the one generated by NTSys was produced using the R "plot" and "hclust" functions. As with NTSys, the UPGMA ("average") method of hierarchical clustering was employed to generate a Euclidian distance-based tree dendrogram which could be compared to the results of the NTSys output (Supplementary Files 10 and 11).

3. Results and Discussion

3.1. Pedigree Information of the Five Genotypes and Genomics Approaches Evaluated

Given the documented lack of genetic diversity within the cultivars of sweet cherry, it is important to understand the pedigree information regarding the five genotypes used in this study namely, 'Bing', 'Sweetheart', 'Staccato', 'Glory' and 'Kimberly'. 'Sweetheart' is known to be the maternal parent of 'Staccato' while the paternal parent is unknown as it was developed via open pollination. 'Van' and 'Newstar' (pollinator) are the parents of 'Sweetheart', but 'Sweetheart' and 'Staccato' have no known familial relationship to the other three genotypes used in this study. Previously published SNP marker analysis has shown the paternal parent of 'Bing' to likely be 'Napoleon' [44]. 'Napoleon' is also the paternal grandparent of 'Stella' (Fig. 1). Therefore, 'Bing' and 'Stella', for which the reference genome is available, share Napoleon in their pedigree as a paternal parent and grandparents respectively. 'Kimberly' and 'Glory' were serendipitous discoveries in orchards based on their delayed fruit maturation phenotype and therefore have unknown lineage. Three of the known sweet cherry cultivars used for analysis in this study belong to different self-incompatibility S-allele genotypes [45].

The first approach, TRAP assay, is a PCR-based technique that uses one fixed primer targeting a conserved DNA sequence usually representing a gene family across the genome and one or two arbitrary primers with either an AT- or GC-rich core that anneal to an intron or an exon, respectively [30]. The 5′ end of the arbitrary primers is fluorescently labeled to enable laser-mediated detection of DNA fragments during electrophoresis and subsequent polymorphism identification. Since it has been proposed that 'Glory' and 'Staccato' are the same genotypes, this approach was first employed to evaluate if there are any differences between the two genotypes using fixed primers targeting the flowering-related genes as based on shared ontogeny with the process of fruit development such genes may influence time of fruit maturation. The second approach, TRAPseq was developed as part of this study and is a modified reduced representation sequencing method derived from the TRAP assay. This method was also tested for its capacity to identify any differences between 'Glory' and 'Staccato'. In the third approach, all five genotypes were analyzed using a sweet cherry SNParray. This 6K Infinium II array contains 5696 predicted genome-wide SNPs, 4214 from diploid sweet cherry (*P. avium*) and 1482 from allotetraploid sour cherry (*P. cerasus*) accessions [31]. For the final, and the highest-resolution approach, WGS was performed on the five genotypes followed by processing of short reads and identification of polymorphisms using Stacks [35]. Subsequent population structure analyses were performed using the SNParray data and Stacks output from the WGS data to determine the genetic relatedness of the genotypes based on the identified SNPs.

3.2. Evaluation of Gel-based Approach, TRAP

By specifically targeting a flowering-related gene family, we were able to identify polymorphisms between 'Glory' and 'Staccato' using the TRAP approach [30]. The fixed primer targeted the VRN2 gene, which has been implicated in temperature-induced induction of flowering [46,47]. Two polymorphic regions were identified out of a total of 45 amplified loci (Table 1, Fig. 2). This corresponds to a 4.4% rate of polymorphism detection (Table 3). It is important to consider, however, that selection of fixed primer targets is particularly important when analyzing highly similar genotypes. As delayed maturation of the fruit is the only observable phenotypic difference between 'Glory' and 'Staccato,' TRAP primers were designed to target flowering related genes with the presumption that during the ontogenic progression, these genes may influence fruit maturation. Relationship between VRN2 and Polycomb-group Proteins, which work in concert to regulate fruit maturation in tomato has been reported recently [48]. It is premature to comment on the direct role of VRN2 in regulating fruit maturation in non-climacteric sweet cherry based on this result. However, when non-flowering gene-targeted primers were used no polymorphisms were detected (data not shown). This speaks to the utility of TRAP as a cost-effective and preliminary method for identification of genome wide polymorphisms only when fixed primers are specifically targeted to putative genes underlying an observable phenotype. While this method is the easiest to implement, it is a low-throughput approach that requires prior information about the trait and putative genes that may underlie the observable phenotype. TRAP is an empirical approach that may have limited success in identifying polymorphic loci since it each primer set provides access to a very small fraction of the genome.

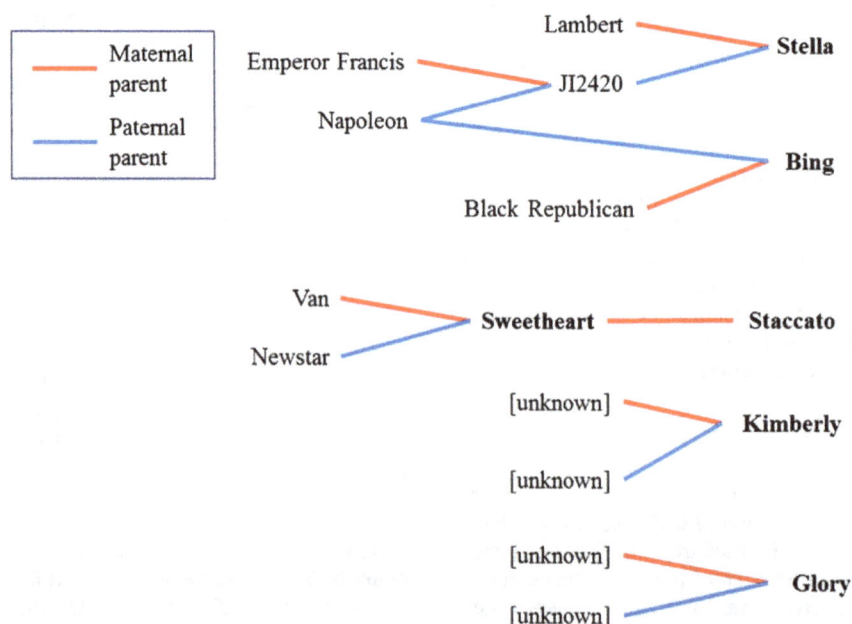

Fig. 1. Pedigree relationships of five of the sweet cherry cultivars analyzed in this study Pedigree of the sweet cherry cultivars used for SNP development. The maternal parent is marked by a red line and the parental parent by a blue line.

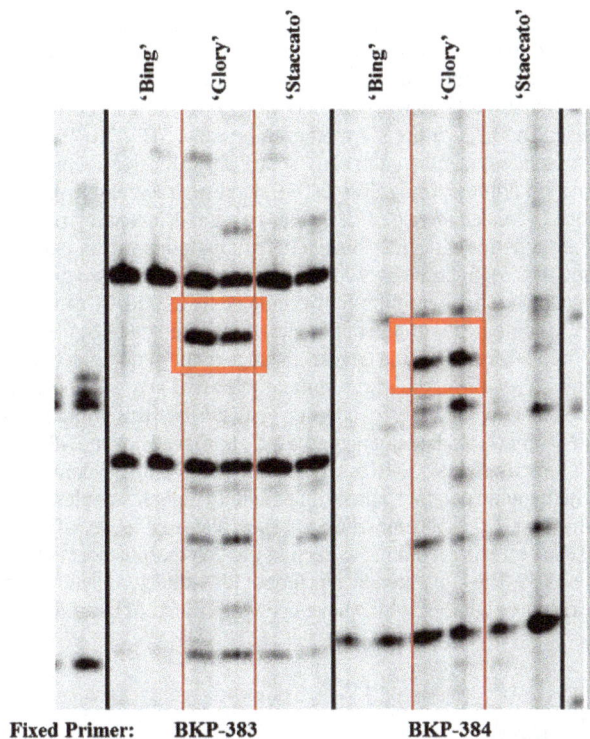

Fig. 2. TRAP analysis of Bing, Glory and Staccato sweet cherry cultivars. Experiment was performed in duplicate. Primer screen was performed using fixed primers BKP-383, 384 and arbitrary primers SA12, GA5. Primer sequences are provided in Table 1. Red boxes are indicative of polymorphic loci. The size of the unique BKP-383 and BKP-384 'Glory' amplicons is approximately 336 and 330 bp, respectively.

3.3. Evaluation of TRAPseq — Modified Reduced Representation Sequencing to Identify Polymorphisms

The reduced representation of the genome was achieved by performing TRAP PCR, followed by generating NGS sequence data from the amplified products. By applying the Stacks pipeline and populations map to the TRAPseq data, 942 polymorphic loci corresponding to SNPs between 'Glory' and 'Staccato' out of 24,984 total loci were identified (Supplementary File 1). This corresponds to a 3.8% rate of polymorphism detection, slightly less than the polymorphism detection rate of the gel-based TRAP analysis, but more representative of genome-wide polymorphisms (Table 3). In terms of genome representation, TRAPseq accessed 0.01% of the genome whereas TRAP only accessed 0.0002% of the genome and that too without any sequence information. These results indicate the importance of identifying appropriate target genes for the fixed primer. While somewhat of a high-throughput approach, it provides a limited coverage of the genome. To enhance coverage, multiple primer sets may need to be utilized. One could utilize the TRAP gel approach to first assess the primer sets that provide the most polymorphic loci and then utilize the same primer sets for TRAPseq to enhance the identification of the number of polymorphic loci.

The Blast2GO gene annotation suite was used to identify the top NCBI Blast hit corresponding to each of the polymorphic loci identified via the TRAPseq analysis. Among the annotated loci were: G-type lectin S-receptor-like serine threonine- kinases, which have been implicated in drought, salinity and cold tolerance [49], ATPase WRNIP1 (ATXAB2), which may play a role in DNA UV damage repair [50,51], HIPP proteins, which are responsive to cold and drought conditions [52], SKP1 proteins, previously implicated in cell cycle progression and floral organ development [53,54], DES1 protein homologues, which may interact with FLC in *Arabidopsis* to regulate flowering time [55], and succinate dehydrogenase complex subunit coding genes. As these sequences were identified via processing of short reads using Stacks, and were not

extensive in length, increased stringency parameters ensured that only sequences of highest similarity to their top blast hit (e-value cutoff of $1.0e-3$) were annotated. In the case of 'Glory' v. 'Staccato', where delayed fruit maturity is the only observable difference at the phenotypic level, it is promising that several polymorphic sequences were identified in genes associated with flowering time, cold induction of developmental processes, and floral organ development. While further investigation is necessary to correlate the annotated gene fragments with the delayed fruit maturity phenotype between 'Glory' and 'Staccato', this analysis has demonstrated that functional annotation of polymorphic sequences can be of use in further understanding the genetic basis for phenotypic differences.

3.4. Evaluation of Cherry SNParray

SNParray analysis enabled the identification of 1385 polymorphic loci out of the 5696 representative loci in the five cultivars namely 'Bing', 'Sweetheart', 'Glory', 'Kimberly', and 'Staccato'. This corresponds to a 24.3% SNP detection rate. The SNParray has been used previously to genotype sweet cherry cultivars and determine their genetic relatedness [24]. The putative polymorphisms represented on the array are spread relatively evenly across each chromosome, but their finite number derived from a pre-selected set of genotype indicates that only a representative subset of potential SNPs can be examined from the sweet cherry genome. Since the SNParray represents a limited number of SNPs derived from the originally represented genotypes, the efficacy of polymorphism detection is far greater for the represented genotype 'Bing'. Approximately 600 SNPs were identified when 'Glory' and 'Staccato', were compared to 'Bing' however, only 66 SNPs were identified when the two genotypes were compared to 'Sweetheart' and 'Kimberly'. The SNParray failed to detect any SNPs between 'Glory'/'Staccato' and 'Sweetheart'/'Kimberly' (Table 2). Furthermore, 174 unique SNPs (3.1%) were detected for 'Bing', whereas no unique SNPs were detected for 'Glory', 'Staccato', 'Sweetheart', or 'Kimberly' (Table 3). While a SNParray is a great analysis tool for repeat polymorphism detection in reference genotypes or samples that were originally represented on the array, it does have some major limitations when the target sample is different from the references sample set. The latter situation leads to the introduction of ascertainment bias [56] a statistical term that describes the deviation observed between real results versus expected results due to the use of non-reference samples. While there are approaches to overcome ascertainment bias, they may not be applicable in non-model plant systems as they lack vast amount of genomic data across the genera as in case of model systems.

Table 2
Shared and unique SNPs identified using SNParray and WGS methods. Pairwise SNP comparison (top left) and number of unique SNPs (top right) for five sweet cherry genotypes analyzed using WGS approaches. Pairwise SNP comparison (bottom left) and number of unique SNPs (bottom right) for the five sweet cherry genotypes analyzed in the SNParray. Using the latter method, no SNPs were found between 'Glory' and 'Staccato' or 'Kimberly' and 'Sweetheart'.

WGS pairwise SNP comparison						WGS unique SNPs	
	Bing	Glory	Staccato	Sweetheart	Kimberly		
Bing	0	2251	2150	2142	2217	Bing	956
Glory	2251	0	1569	1665	1771	Kimberly	496
Staccato	2150	1569	0	1620	1704	Glory	450
Sweetheart	2142	1665	1620	0	1701	Staccato	436
Kimberly	2217	1771	1704	1701	0	Sweetheart	390

SNParray pairwise SNP comparison						SNParray unique SNPs	
	Bing	Glory	Staccato	Sweetheart	Kimberly		
Bing	0	600	600	559	559	Bing	174
Glory	600	0	0	66	66	Glory	0
Staccato	600	0	0	66	66	Staccato	0
Sweetheart	559	66	66	0	0	Sweetheart	0
Kimberly	559	66	66	0	0	Kimberly	0

Table 3

Summary of methods employed in genome-wide polymorphism detection. Total number of loci, number of identified polymorphisms, detection efficiency and percentage genome coverage for each method (total loci sampled per 250 MB estimated genome size) were calculated for each method.

	TRAP	TRAPseq	SNParray	WGS
Samples analyzed	Glory Staccato	Glory Staccato	Glory Staccato Sweetheart Bing Kimberly	Glory Staccato Sweetheart Bing Kimberly
Total loci sampled	45	24984	5696	1239693
Polymorphic loci identified	2	942	1385	2071
Detection efficiency	4.44%	3.77%	24.32%	0.17%
% genome coverage	0.00000018	0.009994	0.00002278	0.00495877
Total loci sampled/250 MB (estimated genome size)				

3.5. Evaluation of WGS to Identify Polymorphisms

For each of the five genotypes analyzed using SNParray, 22.2× average coverage of Illumina HiSeq paired end read data, or 4.6–5.5 Gb of sequence data were generated. SNPs were identified using the Stacks workflow [35,57]. Stacks generated loci from short read Illumina data and identified polymorphisms within the genotype-specific loci. Overall, 2071 polymorphic loci were identified among the compared genotypes out of 1,239,693 catalog loci matching the generated stacks representing 0.5% of the sweet cherry genome. STRUCTURE analysis and subsequent identification of most probable ΔK values, representing population number, using STRUCTURE Harvester's Evanno method calculations revealed increased ΔK values at 2 and 4, indicating that there are four genetically distinct sweet cherry subgroups within two larger groups (Fig. 5). In both cases, 'Bing' segregated into its own group and subgroup. The final graphics files produced by DISTRUCT can be seen in Fig. 3, combined with the dendrogram produced by NTSys (Fig. 6).

While WGS enables the largest coverage of the genome, sequencing of random regions reduces the comparable areas across samples. Perhaps enhancing the depth of coverage can alleviate this limitation. The major strength of all sequencing based approaches over SNParray is that it directly couples SNP discovery with genotyping by identification of genome wide polymorphisms directly in the target samples.

The Blast2GO gene annotation suite was also used to identify the top NCBI Blast hit corresponding to each of the polymorphic loci identified by the WGS. The annotated loci included: RNA-directed DNA polymerases, receptor kinases, which have been implicated in brassinosteroid

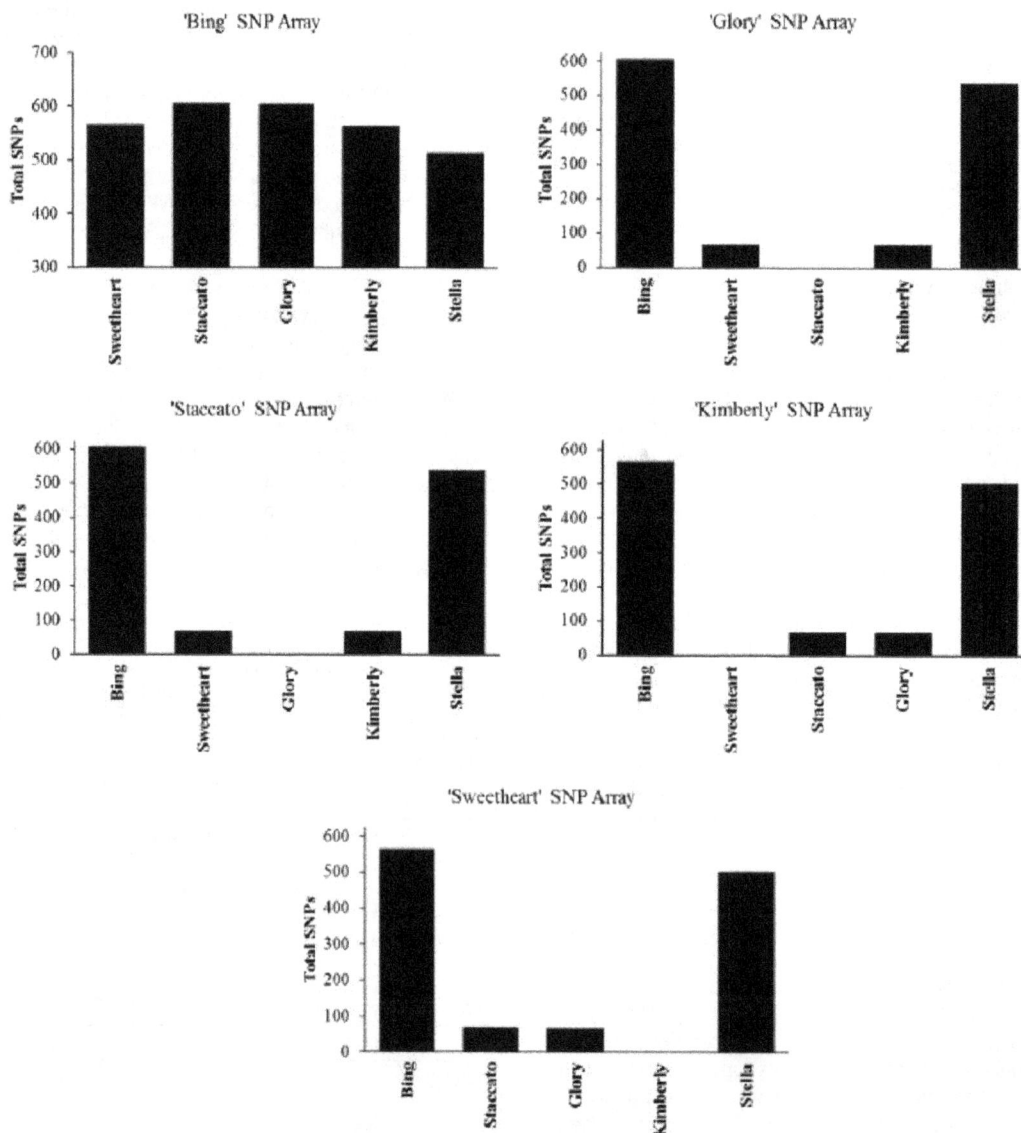

Fig. 3. SNParray, individual genotype comparisons of total SNPs. The title of each subfigure indicates the reference by which the other genotypes were compared.

signaling [58], and numerous genes encoding plastid targeted proteins–NADH dehydrogenase subunits, NAD(P)H quinone oxidoreductase subunits, Rubisco subunits, and cytochrome b6 f complex precursors (Supplementary File 12). A large portion of the identified genes corresponding to polymorphic sequences are both plastid-targeted and plastid-encoded in nature. This is intriguing considering there are only three maternal haplotypes reported for all sweet cherry cultivars [21].

3.6. Comparison of Population Structures Derived from WGS and SNParray Data

STRUCTURE and NTSys were used to analyze and produce graphical representations of population structure respectively (Figs. 4 and 6). In the case of both SNParray and WGS, 'Bing' forms an outgroup relative to the other four genotypes, which display much higher genetic similarity. This is consistent with the results of shared and unique SNP counts (Table 2) where 'Bing' displayed the greatest number of unique SNPs, whereas 'Glory', 'Staccato', 'Sweetheart', and 'Kimberly' possessed far fewer (0 in the case of the SNParray). While both approaches produced similar results, the greater efficiency of polymorphism detection of the WGS approach is evident. Using this method, combined with the STRUCTURE and Structure Harvester analyses, we identified 4 distinct subgroups ('Bing', 'Glory'/'Staccato', 'Sweetheart', 'Kimberly') within two larger groups; group 1 represented by 'Bing' and group 2 represented by 'Glory', 'Staccato', 'Sweetheart' and 'Kimberly', as shown in the ΔK graph (Fig. 5). The data from SNParray produced a similar cluster dendrogram as did the WGS approach; however STRUCTURE did not resolve differences between 'Glory'/'Staccato' and 'Sweetheart'/'Kimberly' in case of SNParray.

4. Conclusion

Multiple methods of polymorphism detection were evaluated across five closely related genotypes of sweet cherry. Each of the described approaches resulted in detection of polymorphisms, although certain ones provided higher resolution of detection between closely related genotypes.

The TRAP method allowed for identification of polymorphic regions between 'Glory' and 'Staccato'. This represents the first gel-based evidence of genetic differences between these two genotypes, which were previously only distinguished by delayed fruit maturity phenotype. The observed 4.4% rate of polymorphism detection, however, is not necessarily representative of the detection rate for the TRAP approach in general. The efficiency of polymorphism identification for this method is largely dependent upon both the genetic similarity of cultivars tested as well as the specificity of the fixed primer target. While it has been demonstrated that polymorphic regions can be detected even

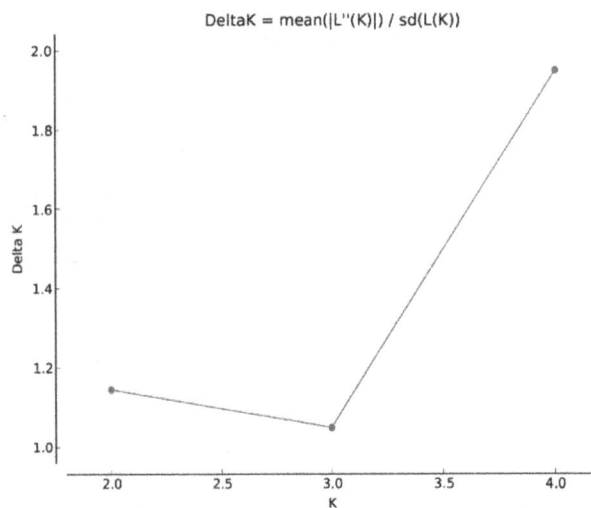

Fig. 5. Evanno method based calculations for population number ΔK. ΔK values were highest for K = 2 and K = 4 indicating greatest likelihood of two larger groups comprised of 4 distinct subgroups.

among highly genetically similar cultivars, this success was largely dependent upon the design of primers targeting the flowering-related VRN2 gene. We recommend a primer screen of various putative gene targets in order to identify the most promising fixed primer candidates for this analysis.

The TRAPseq approach allowed for identification of 942 polymorphisms between 'Glory' and 'Staccato' using Stacks [35,57]. As with the gel-based TRAP approach, fixed primer design is an important factor for consideration; however, TRAPseq is expected to have a broader genomic range of SNP detection when the fixed primers are designed to target diverse and rapidly evolving gene families, such as the MADS-box and PPR1 and PPR2 gene families. These genes are known to be widely distributed across the genome and represent a large family across the plant kingdom [33,34] which are likely to contain polymorphisms when comparing closely related species. Many MADS-box genes have arisen via duplication events and have since acquired new functions [59]. Among the acquired functions is regulation of endodormancy release [60] which makes the MADS-box genes particularly useful in comparing the selected cultivars as the genotypes exhibit a late fruit maturation phenotype. Because this is a sequence based method, single nucleotide polymorphisms, which may not be visible using the gel-based approach, can be easily detected. The application of the Stacks program following sequencing of the TRAPseq PCR product allowed us to consider only those fragments that contained putative

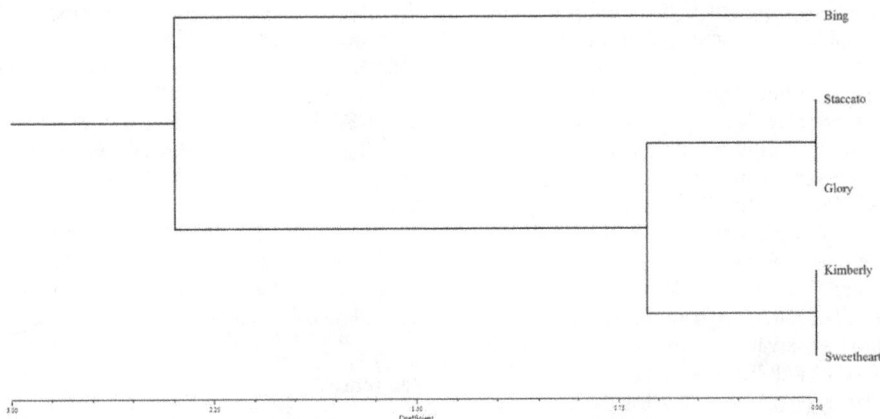

Fig. 4. Tree dendrogram generated from SNParray data. 1385 polymorphic loci in an array of with 5696 loci were not able to distinguish between 'Staccato' vs. 'Glory' and 'Kimberly' vs. 'Sweetheart' most likely due to limited genome coverage and use of non-referenced samples resulting in ascertainment bias.

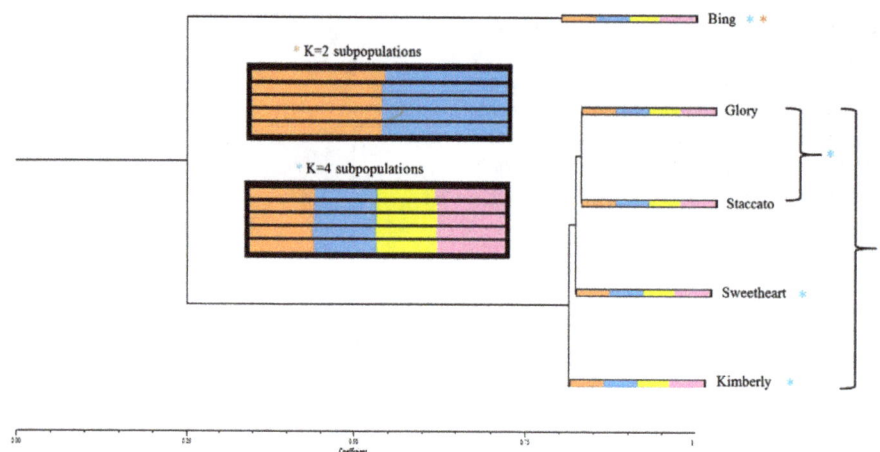

Fig. 6. Dendrogram depicting genetic relatedness of Bing, Glory, Staccato, Sweetheart, and Kimberly based on 18,058 unique SNPs (dendrogram generated from NTSys). Colored bars represent proportion of an individual belonging to a distinct group or subgroup, based on shared and unique SNPs (generated using STRUCTURE, CLUMPP, DISTRUCT, and NTSys). Orange asterisks denote the two larger groups, and blue asterisks denote the four distinct subgroups.

SNPs. Even though TRAPseq analysis only allows for a representation of specific primer targets throughout the genome, our evaluation demonstrates that it is able to generate quality data to identify polymorphisms between highly similar genotypes, with an observed detection rate of 3.8%.

The cherry SNParray represented 5696 SNPs derived from sweet and sour cherry accessions [31]. This method facilitated detection of 1385 SNPs when 'Bing', 'Glory', 'Staccato', 'Sweetheart' and 'Kimberly' genotypes were considered, an overall SNP detection rate of 24.32%. This appears far more efficient than either the gel-based TRAP approach or the TRAPseq approach. However, due to the inherent limitations of only detecting fixed, representative polymorphisms and ascertainment bias introduced due to analysis of non-referenced samples [56], the SNParray failed to identify SNPs present in the closely related genotypes. This is evident by the lack of SNP detection when 'Glory'/'Staccato' and 'Kimberly'/'Sweetheart' were compared. In such cases, a gel-based, reduced representation, and/or WGS based approach were more informative.

The WGS approach, not surprisingly, provided the highest resolution of polymorphism detection among the five genotypes analyzed. This method is advantageous in that it provides genome wide coverage and can be easily implemented in species with little or no genomic information. WGS can be limited by the depth of coverage and assembly methodology. This is especially true around polymorphic repeat regions of the genome. However, when combined with the Stacks short-read approach, the effectiveness of polymorphism detection of the WGS approach greatly increases. Processing of short reads in Stacks allowed the consideration of only regions with putative polymorphisms, which could then be used in population structure analysis of the five genotypes.

'Bing' is the most genetically distinct from the other genotypes analyzed, as supported by the results of NTSys, STRUCTURE, and the R clustering algorithms (Figs. 4 and 6). This was expected, as more unique SNPs (almost twice as many) were identified for Bing than for any of the other cultivars analyzed (Table 2). The STRUCTURE and NTSys analyses of WGS data suggest that 'Glory' and 'Staccato' segregate together into their own subgroup, despite displaying high degree of genetic similarity to both 'Sweetheart' and 'Kimberly' (Fig. 6).

The only previously described difference between 'Glory' and 'Staccato' is based on phenotypic observation of delayed fruit maturity. Using three different methods, TRAP, TRAPseq, and WGS, it has been demonstrated that these two genotypes are subtly distinct from one another and 'Glory' is most likely a spontaneous mutation or 'sport' derived from Staccato. Thus, it seems that 'Glory' and 'Staccato', despite their high genetic similarity, are indeed distinct genotypes. Further

analysis will allow us to determine whether polymorphisms between 'Glory' and 'Staccato' arose from a mutation(s) in a flowering related gene(s), as is suggested by the TRAP assay.

In summary, the sequencing based approaches evaluated in this study have generated a robust dataset of predicted polymorphisms in sweet cherry. We expect that the described methods, used in conjunction with one another, will be highly useful in genetics and genomics –based research in other closely related species of agronomic importance.

Competing Interests

The authors declare no competing interests.

Authors' Contributions

Conceived and designed the experiments: AD, BK, SH, TK, RS.
Performed the experiments: SH, BK, TK.
Analyzed the data: SH, RH, BK, TK, AD.
Contributed reagents/materials/analysis tools: AD, RS, TK.
Wrote the paper: AD, SH, BK, RH.

Acknowledgements

We are grateful to Audrey Sebolt in the Iezzoni lab at Michigan State University for facilitating the SNParray experiment and Dr. Yunyang Zhao in the Oraguzie lab at Washington State University for her help with interpretation of the SNParray output data. The authors are grateful to the Van Well Nursery, Wenatchee, USA (http://www.vanwell.net/) for making the TRAPseq and whole genome sequence data, generated at Phytelligence Inc. (www.Phytelligence.com), publicly available for this study. SLH and BRK acknowledge the support of NIH/NIGMS through institutional training grant award T32-GM008336. The contents of the publication are solely the responsibility of the authors and do not necessarily represent the official views of the NIGMS or NIH. This research was funded in part by Washington State University Agriculture Research Center Hatch funds to AD.

References

[1] Yang H, Tao Y, Zheng Z, Li C, Sweetingham MW, Howieson JG. Application of next-generation sequencing for rapid marker development in molecular plant breeding: a case study on anthracnose disease resistance in *Lupinus angustifolius* L. BMC Genomics 2012;1:318.

[2] Yumurtaci A. Utilization of diverse sequencing panels for future plant breeding. In: Al-Khayri JM, Jain SM, Johnson DV, editors. Advances in Plant Breeding Strategies: Breeding, Biotechnology and Molecular Tools. Cham: Springer International Publishing; 2015. p. 539–61.

[3] Unamba CI, Nag A, Sharma RK. Next generation sequencing technologies: the doorway to the unexplored genomics of non-model plants. Front Plant Sci 2015;1074.

[4] Koepke T, Schaeffer S, Krishnan V, Jiwan D, Harper A, Whiting M, et al. Rapid gene-based SNP and haplotype marker development in non-model eukaryotes using 3′ UTR sequencing. BMC Genomics 2012;18.

[5] Zhang P, Liu X, Tong H, Lu Y, Li J. Association mapping for important agronomic traits in core collection of rice (Oryza sativa L.) with SSR markers. PLoS One 2014;10: e111508.

[6] Mora F, Castillo D, Lado B, Matus I, Poland J, Belzile F, et al. Genome-wide association mapping of agronomic traits and carbon isotope discrimination in a worldwide germplasm collection of spring wheat using SNP markers. Mol Breed 2015;2:69.

[7] Hummer KE, Janick J. Rosaceae: taxonomy, economic importance, genomics. In: Folta KM, Gardiner SE, editors. Genetics and Genomics of Rosaceae. New York, NY: Springer New York; 2009. p. 1–17.

[8] Arumuganathan K, Earle ED. Nuclear DNA content of some important plant species. Plant Mol Biol Report 1991:208–18.

[9] Carrasco B, Meisel L, Gebauer M, Garcia-Gonzales R, Silva H. Breeding in peach, cherry and plum: from a tissue culture, genetic, transcriptomic and genomic perspective. Biol Res 2013:219–30.

[10] International Peach Genome I, Verde I, Abbott AG, Scalabrin S, Jung S, Shu S, et al. The high-quality draft genome of peach (Prunus persica) identifies unique patterns of genetic diversity, domestication and genome evolution. Nat Genet 2013;5:487–94.

[11] Velasco R, Zharkikh A, Affourtit J, Dhingra A, Cestaro A, Kalyanaraman A, et al. The genome of the domesticated apple (Malus × domestica Borkh.). Nat Genet 2010; 10:833–9.

[12] Guajardo V, Solis S, Sagredo B, Gainza F, Munoz C, Gasic K, et al. Construction of high density sweet cherry (Prunus avium L.) linkage maps using microsatellite markers and snps detected by genotyping-by-sequencing (GBS). PLoS One 2015;5:e0127750.

[13] Tavassolian I, Rabiei G, Gregory D, Mnejja M, Wirthensohn MG, Hunt PW, et al. Construction of an almond linkage map in an Australian population nonpareil × lauranne. BMC Genomics 2010;1:551.

[14] Bielenberg DG, Rauh B, Fan S, Gasic K, Abbott AG, Reighard GL, et al. Genotyping by sequencing for SNP-based linkage map construction and QTL analysis of chilling requirement and bloom date in peach [Prunus persica (L.) Batsch]. PLoS One 2015;10: e0139406.

[15] Fan S, Bielenberg DG, Zhebentyayeva TN, Reighard GL, Okie WR, Holland D, et al. Mapping quantitative trait loci associated with chilling requirement, heat requirement and bloom date in peach (Prunus persica). New Phytol 2010.

[16] Dhingra A. Pre-publication Release of Rosaceae Genome Information. https://genomics.wsu.edu/research/; 2013.

[17] M-y Zhang, Fan L, Q-z Liu, Song Y, S-w Wei, Zhang S-I Wu J. A novel set of EST-derived SSR markers for pear and cross-species transferability in Rosaceae. Plant Mol Biol Report 2014;1:290–302.

[18] Park YH, Ahn SG, Choi YM, Oh HJ, Ahn DC, Kim JG, et al. Rose (Rosa hybrida L.) EST-derived microsatellite markers and their transferability to strawberry (Fragaria spp.). Sci Hortic 2010;4:733–9.

[19] Zhou Y, Li J, Korban SS, Han Y. Apple SSRs present in coding and noncoding regions of expressed sequence tags show differences in transferability to other fruit species in Rosaceae. Can J Plant Sci 2013;2:183–90.

[20] Ganopoulos I, Moysiadis T, Xanthopoulou A, Ganopoulou M, Avramidou E, Aravanopoulos FA, et al. Diversity of morpho-physiological traits in worldwide sweet cherry cultivars of GeneBank collection using multivariate analysis. Sci Hortic 2015:381–91.

[21] Mariette S, Tavaud M, Arunyawat U, Capdeville G, Millan M, Salin F. Population structure and genetic bottleneck in sweet cherry estimated with ssrs and the gametophytic self-incompatibility locus. BMC Genet 2010;77.

[22] Campoy JA, Lerigoleur-Balsemin E, Christmann H, Beauvieux R, Girollet N, Quero-Garcia J, et al. Genetic diversity, linkage disequilibrium, population structure and construction of a core collection of Prunus avium L. landraces and bred cultivars. BMC Plant Biol 2016;49.

[23] Daïnou K, Blanc-Jolivet C, Degen B, Kimani P, Ndiade-Bourobou D, Donkpegan ASL, et al. Revealing hidden species diversity in closely related species using nuclear SNPs, SSRs and DNA sequences — a case study in the tree genus Milicia. BMC Evol Biol 2016;1:259.

[24] Fernandez i Marti A, Athanson B, Koepke T, Font i Forcada C, Dhingra A, Oraguzie N. Genetic diversity and relatedness of sweet cherry (Prunus avium L.) cultivars based on single nucleotide polymorphic markers. Front Plant Sci 2012:1–13.

[25] Warner G. Glory Be. Good Fruit Grower; 2011.

[26] Well PV. 'Glory' and 'Staccato' Cherries Are Claimed to be the Same Sweet Cherry Cultivars; 2011.

[27] Kappel F, MacDonald RA, Brownlee R. 13s2009 (staccato™) sweet cherry. Can J Plant Sci 2006;4:1239–41.

[28] Well PV. 'Kimberly' is a Random Limb Mutation of 'Bing' Sweet Cherry Variety; 2014.

[29] Lane WD, MacDonald RA. Sweetheart sweet cherry. Can J Plant Sci 1996;1:161–3.

[30] Hu J, Vick BA. Target region amplification polymorphism: a novel marker technique for plant genotyping. Plant Mol Biol Report 2003:289–94.

[31] Peace C, Bassil N, Main D, Ficklin S, Rosyara UR, Stegmeir T, et al. Development and evaluation of a genome-wide 6k SNP array for diploid sweet cherry and tetraploid sour cherry. PLoS One 2012;7(12):e48305.

[32] Healey A, Furtado A, Cooper T, Henry RJ. Protocol: a simple method for extracting next-generation sequencing quality genomic DNA from recalcitrant plant species. Plant Methods 2014;21.

[33] Becker A, Theissen G. The major clades of MADS-box genes and their role in the development and evolution of flowering plants. Mol Phylogenet Evol 2003;3:464–89.

[34] O'Toole N, Hattori M, Andres C, Iida K, Lurin C, Schmitz-Linneweber C, et al. On the expansion of the pentatricopeptide repeat gene family in plants. Mol Biol Evol 2008;6:1120–8.

[35] Catchen J, Hohenlohe PA, Bassham S, Amores A, Cresko WA. Stacks: an analysis tool set for population genomics. Mol Ecol 2013:3124–40.

[36] Conesa A, Gotz S. Blast2go: a comprehensive suite for functional analysis in plant genomics. Int J Plant Genomics 2008;619832.

[37] Gotz S, Garcia-Gomez JM, Terol J, Williams TD, Nagaraj SH, Nueda MJ, et al. High-throughput functional annotation and data mining with the blast2go suite. Nucleic Acids Res 2008;10:3420–35.

[38] Altschul S, Gish W, Miller W, Myers EW, Lipman DJ. Basic local alignment search tool. J Mol Biol 1990;215:403–10.

[39] Pritchard JK, Stephens M, Donnelly P. Inference of population structure using multilocus genotype data. Genetics 2000;2:945–59.

[40] Rohlf FJ. NTSYS-pc: microcomputer programs for numerical taxonomy and multivariate analysis. Am Stat 1987;4:330.

[41] Earl DA, vonHoldt BM. STRUCTURE HARVESTER: a website and program for visualizing STRUCTURE output and implementing the Evanno method. Conserv Genet Resour 2012;2:359–61.

[42] Jakobsson M, Rosenberg NA. Clumpp: a cluster matching and permutation program for dealing with label switching and multimodality in analysis of population structure. Bioinformatics 2007;14:1801–6.

[43] Rosenberg NA. Distruct: a program for the graphical display of population structure. Mol Ecol Notes 2004;1:137–8.

[44] Rosyara UR, Sebolt AM, Peace C, Iezzoni AF. Identification of the paternal parent of 'bing' sweet cherry and confirmation of descendants using single nucleotide polymorphism markers. J Am Soc Hortic Sci 2014:148–56.

[45] Schuster M. Incompatible (S-) genotypes of sweet cherry cultivars (Prunus avium L.). Sci Hortic 2012:59–73.

[46] Castède S, Campoy JA, Le Dantec L, Quero-García J, Barreneche T, Wenden B, et al. Mapping of candidate genes involved in bud dormancy and flowering time in sweet cherry (Prunus avium). PLoS One 2015:e0143250.

[47] Castede S, Campoy JA, Garcia JQ, Le Dantec L, Lafargue M, Barreneche T, et al. Genetic determinism of phenological traits highly affected by climate change in Prunus avium: flowering date dissected into chilling and heat requirements. New Phytol 2014:703–15.

[48] Liu D-D, Zhou L-J, Fang M-J, Dong Q-L, An X-H, You C-X, et al. Polycomb-group protein slmsi1 represses the expression of fruit-ripening genes to prolong shelf life in tomato. Sci Rep 2016;31806.

[49] Vaid N, Macovei A, Tuteja N. Knights in action: lectin receptor-like kinases in plant development and stress responses. Mol Plant 2013;5:1405–18.

[50] Kunz BA, Anderson HJ, Osmond MJ, Vonarx EJ. Components of nucleotide excision repair and DNA damage tolerance in Arabidopsis thaliana. Environ Mol Mutagen 2005;2-3:115–27.

[51] Ganpudi AL, Schroeder DF. UV Damaged DNA Repair & Tolerance in Plants, Selected Topics in DNA Repair. In: Chen Clark, editor. InTech; 2011. http://dx.doi.org/10.5772/22138. Available from: https://www.intechopen.com/books/selected-topics-in-dna-repair/uv-damaged-dna-repair-tolerance-in-plants.

[52] de Abreu-Neto JB, Turchetto-Zolet AC, de Oliveira LF, Zanettini MH, Margis-Pinheiro M. Heavy metal-associated isoprenylated plant protein (HIPP): characterization of a family of proteins exclusive to plants. FEBS J 2013;7:1604–16.

[53] Zhao D, Ni W, Feng B, Han T, Petrasek MG, Ma H. Members of the Arabidopsis-SKP1-like gene family exhibit a variety of expression patterns and may play diverse roles in Arabidopsis. Plant Physiol 2003;1:203–17.

[54] Soltis DE, Soltis PS, Albert VA, Oppenheimer DG, dePamphilis CW, Ma H, et al. Missing links: the genetic architecture of flowers [correction of flower] and floral diversification. Trends Plant Sci 2002;1:22–31 [dicussion 31–24].

[55] Shibuya K, Nagata M, Tanikawa N, Yoshioka T, Hashiba T, Satoh S. Comparison of mRNA levels of three ethylene receptors in senescing flowers of carnation (Dianthus caryophyllus L.). J Exp Bot 2002;368:399–406.

[56] Lachance J, Tishkoff SA. SNP ascertainment bias in population genetic analyses: why it is important, and how to correct it. Bioessays 2013;9:780–6.

[57] Catchen JM, Amores A, Hohenlohe P, Cresko W, Postlethwait JH, De Koning D-J. Stacks: building and genotyping loci de novo from short-read sequences. Genes Genomes Genet 2011:171–82.

[58] Li J, Wen J, Lease KA, Doke JT, Tax FE, Walker JC. BAK1, an Arabidopsis LRR receptor-like protein kinase, interacts with bri1 and modulates brassinosteroid signaling. Cell 2002;2:213–22.

[59] Smaczniak C, Immink RGH, Angenent GC, Kaufmann K. Developmental and evolutionary diversity of plant MADS-domain factors: insights from recent studies. Development 2012;17:3081–98.

[60] Wells CE, Vendramin E, Jimenez Tarodo S, Verde I, Bielenberg DG. A genome-wide analysis of MADS-box genes in peach [Prunus persica (L.) Batsch]. BMC Plant Biol 2015;1:41.

9

Evaluation of the novel algorithm of flexible ligand docking with moveable target-protein atoms

Alexey V. Sulimov [a,b], Dmitry A. Zheltkov [c], Igor V. Oferkin [a], Danil C. Kutov [a,b], Ekaterina V. Katkova [a,b], Eugene E. Tyrtyshnikov [c,d], Vladimir B. Sulimov [a,b,*]

[a] *Dimonta, Ltd, Nagornaya Street 15, Bldg. 8, Moscow 117186, Russia*
[b] *Research Computer Center, Moscow State University, Leninskie Gory 1, Bldg. 4, Moscow 119992, Russia*
[c] *Faculty of Computational Mathematics and Cybernetics of Lomonosov Moscow State University, Leninskie Gory 1, Bldg. 52, Moscow 119992, Russia*
[d] *Institute of Numerical Mathematics of Russian Academy of Sciences, Gubkin Street 8, Moscow, 119333, Russia*

A R T I C L E   I N F O

ABSTRACT

Keywords:
Docking
Tensor train
Protein-ligand complex
Protein moveable atoms
Flexible ligand
Drug design

We present the novel docking algorithm based on the Tensor Train decomposition and the TT-Cross global optimization. The algorithm is applied to the docking problem with flexible ligand and moveable protein atoms. The energy of the protein-ligand complex is calculated in the frame of the MMFF94 force field in vacuum. The grid of precalculated energy potentials of probe ligand atoms in the field of the target protein atoms is not used. The energy of the protein-ligand complex for any given configuration is computed directly with the MMFF94 force field without any fitting parameters. The conformation space of the system coordinates is formed by translations and rotations of the ligand as a whole, by the ligand torsions and also by Cartesian coordinates of the selected target protein atoms. Mobility of protein and ligand atoms is taken into account in the docking process simultaneously and equally. The algorithm is realized in the novel parallel docking SOL-P program and results of its performance for a set of 30 protein-ligand complexes are presented. Dependence of the docking positioning accuracy is investigated as a function of parameters of the docking algorithm and the number of protein moveable atoms. It is shown that mobility of the protein atoms improves docking positioning accuracy. The SOL-P program is able to perform docking of a flexible ligand into the active site of the target protein with several dozens of protein moveable atoms: the native crystallized ligand pose is correctly found as the global energy minimum in the search space with 157 dimensions using 4700 CPU ∗ h at the Lomonosov supercomputer.

1. Introduction

The initial stage of new drug development is a search of the molecules which are inhibitors of a given target protein. Inhibitors block the active site of the protein associated with a disease and the disease is cured. A quick and effective solution of this problem decreases considerably material costs and duration of the whole drug development process. Nowadays, this problem can be addressed effectively with the help of computer simulations [1,2]. Reliable predictions of the target protein inhibition by a low molecular weight ligand are defined by the accuracy of the docking programs. Docking programs carry out positioning of the ligand in the active site of the protein and calculate the protein-ligand binding free energy. The accuracies of positioning and the binding energy calculation are closely linked: faulty positioning cannot result in the high accuracy of the binding energy calculation based on the found ligand poses. The positioning accuracy of many existing docking

programs is satisfactory and unpredictable positioning failures take place rather rarely. However, the accuracy of binding energy calculations for a randomly selected target protein is too bad: for the effective development of new inhibitors this accuracy should be better than 1 kcal/mol [3]. High accuracy of the protein-ligand binding energy calculations with docking programs is the key problem that should be solved in order to increase considerably effectiveness of the use of molecular modeling for the new inhibitors' development. This accuracy depends on many factors: the force field choice for modeling intra- and inter-molecular interactions, the solvent model, target protein and ligand models, the docking algorithm, the free energy calculation method, respective approximations and computer resources required for docking of one ligand.

In the frame of the docking procedure the protein-ligand binding energy ΔG_{bind} should be calculated as the difference between the free energy of the protein-ligand complex G_{PL} and the sum of free energies of the unbound protein G_P and the unbound ligand G_L:

$$\Delta G_{bind} = G_{PL} - G_P - G_L.$$

* Corresponding author at: Dimonta, Ltd, Nagornaya Street 15, Bldg. 8, Moscow 117186, Russia.
E-mail address: vladimir.sulimov@gmail.com (V.B. Sulimov).

Free energies of the protein, the ligand and their complex are described by respective energy landscapes and they can be calculated through the configuration integrals over the respective phase space. In the thermodynamic equilibrium the molecular system occupies its low energy minima. The configuration integral will come to the sum of configuration integrals over the separate low energy minima if these minima are separated by sufficiently high energy barriers [4,5]. Thus, the docking accuracy is defined by the completeness of finding the low energy minima and by the accuracy of the configuration integral calculation in each of these minima.

The target protein model defines complexity and the volume of calculations and in many docking programs the rigid protein approximation is adopted. Moreover, in some docking programs, e.g. AutoDock [6,7], ICM [8], DOCK [9], SOL [10], the grid of preliminary calculated potentials of the ligand probe atoms Coulomb and van der Waals interactions with the target protein is used in the main docking procedure. This results in the increase of computing speed but at the expense of restrictions on the docking performance and of worsening of the accuracy of binding energy calculations. The protein model with the preliminary calculated grid of potentials has a number of limitations. Firstly, this approach obviously cannot take into account mobility of the protein atoms. Secondly, such approach makes impossible carrying out the local optimization of the protein-ligand energy with the variation of coordinates of ligand and protein atoms. Thirdly, the local potentials in the grid nodes cannot represent the non-locality of the interaction of solute atom charges with polarized charges induced on the solvent excluded surface in implicit solvent models; as a result the interaction of the protein and the ligand with water cannot be treated accurately. Finally, the ligand poses found in this docking approach do not correspond to any energy minima because the local optimization of the energy is not performed.

Some time ago we decided to reject the docking procedure with preliminary calculated energy grid in the attempt to increase the accuracy of the protein-ligand binding energy calculations. Docking without the preliminary calculated energy grid requires much more computational resources even for vacuum calculations since one has to find low energy minima on the complicated multi-dimensional energy surface computing the energy in the frame of the whole given force field for each system conformation appearing in the minima search algorithm. Such docking programs, FLM [5] and SOL-T [11], have been developed for the rigid target protein and the flexible ligand. The parallel FLM program can perform the comprehensive minima search either in vacuum or with the rigorous implicit solvent model [12,13]. However FLM requires too large supercomputer resources and it can be used mainly for finding low energy reference minima of protein-ligand complexes for the validation of docking algorithms [11] and force fields [5,14]. The parallel SOL-T program employs the novel tensor train global optimization algorithm and it requires much less supercomputer resources than FLM. The docking positioning accuracy of FLM and SOL-T in vacuum for the rigid protein is comparable with one another at least for some test complexes [11]. The TT-docking algorithm was compared [15] with the genetic algorithm realized in the SOL program [10] with one and the same energy function on the preliminary calculated energy grid for rigid proteins and flexible ligands. In this case the ability to find the global energy minimum and the native (crystallized) position is close but the TT-docking algorithms perform about 10 times faster [15]. Further, it is demonstrated [5] that the ligand positioning accuracy is much better when the force field is used with a continuum solvent model. The ligand positioning accuracy is much better when the recent quantum chemical semiempirical methods, PM7 [14] and PM6 [16], are used instead of classical force fields.

However, proteins are flexible and dynamic molecular systems. A noticeable difference between protein's unbound (*apo*) and bound (*holo*) structures is sometimes observed. Ligand binding may cause a small side-chain rearrangement or individual atom's motions as well as significant conformational changes connected with domain motions.

Thus, the protein flexibility can have a major impact on the molecular modeling results. It is reasonable to assume that the protein flexibility can significantly improve the docking positioning accuracy as well as the accuracy of the protein-ligand binding energy calculation on the base of docking results.

There are several methods to take protein flexibility into account [17–20].

Soft docking [21] is the simplest method of protein flexibility accounting. It simulates the mobility of protein atoms by reducing the steric components of the scoring function ("softening" of van der Waals potentials). However, this approach can increase the number of false positives [22].

The ensemble docking approach is the docking into the ensemble of receptor conformations instead of docking into a single one. This method is popular because there is no need to change the existing docking algorithms in order to take protein flexibility into consideration. Multiple conformations are generated before docking and can be obtained from X-ray crystallography, of NMR spectroscopy or can be produced by molecular modeling, e.g. molecular dynamics. Moreover, ensemble docking can be carried out sequentially into each protein structure ("multiple-run" docking) [18,23] or into one averaged structure [24] or into the dynamic pharmacophore model [25] ("single-run" docking). The composite structure also can be created on the basis of the ensemble of conformations and it consists of different parts of the original ensemble. Such composite structures are generated directly during the docking process [26,27].

In the case of selective methods a few "critical" atoms or amino acid residues can move explicitly to explore the protein flexibility. Certain side chains of the active site are often chosen as the protein's degrees of freedom. Variation of their positions can be performed due to rotations around torsional degrees of freedom. Such rotation can be either continuous [28,29] or discrete when the angles of rotation are determined on the basis of well-known libraries of rotamers [30,31]. Selective methods also include the approach when only hydrogen atoms' reorientation is performed to optimize hydrogen bonds between protein and ligand [32,33]. Some implementations of selective methods vary the protein conformations after the ligand optimization in the rigid protein [34]. There is also an approach that allows optimization of both the ligand and the side chains of protein simultaneously. However, this can be done only for a strongly restricted number (no more than 22) of protein and ligand degrees of freedom [35].

Protein flexibility can also be investigated in the context of post-docking ("induced-fit" methods): first, the ligand position is found using rigid docking or soft docking, and then the additional optimization of the protein conformation is performed using a selective approach [36]. Sometimes this post-optimization can be performed by the Monte Carlo method or molecular dynamics [37,38] to take into account flexibility of the whole protein. A more refined docking approach combines multiple local optimizations with the subsequent global optimization in vicinities of picked out local minima [39]. Initially, 1000 local minima were found with the help of the energy gradient optimization with variations of coordinates of ligand and protein atoms (more than 1000 atoms) using randomly selected initial poses of the ligand in the active site of the target proteins. Then the global optimization by the Monte Carlo method was performed in the close vicinity of most perspective local minima [39].

There is also the Monte Carlo docking procedure [40] where random target protein side-chain perturbations are followed by the local energy optimization with variations of coordinates of ligand and protein atoms and this procedure is repeated iteratively. The docking method of "molecular relaxations" [41] employs the molecular dynamics (MD) approach, but this method is supplanted now by more accurate and more computationally expensive MD methods of the protein-ligand binding energy calculation, e.g. the free energy perturbation procedure [42].

Algorithms of most modern docking programs are based on the docking paradigm [5,11,14]. This paradigm assumes that the ligand binding pose in the active site of the target protein corresponds to the global minimum of the protein-ligand energy function or is near it. Due to this paradigm the docking problem is reduced to the search of the global minimum on the multi-dimensional protein-ligand energy surface. The dimensionality of this surface (d) is defined by the number of protein-ligand system degrees of freedom. Docking of small molecules into the rigid target protein is reliable when the number of ligand degrees of freedom (translations and rotations as a whole and torsions) is not more than 20–25 [10]. For larger dimensionality of the search space, i.e. for larger number of protein-ligand system degrees of freedom, commonly used docking algorithms, e.g. the genetic algorithm, are not able to perform docking successfully. Therefore inclusion of coordinates of moveable protein atoms into the docking procedure increases significantly the dimensionality of the global minimum search space and the solution of the docking problem requires more effective global optimization algorithms.

Is it possible to perform the global optimization of the protein-ligand energy considering the ligand flexibility and the mobility of protein atoms simultaneously and equally at least for several dozens of protein moveable atoms? The present study demonstrates that the answer is positive: yes, it is possible to perform successfully such docking employing the novel tensor train global optimization algorithm [11]. In this study we describe the main features of this novel algorithm, the respective program SOL-P for docking flexible ligands into target proteins with moveable atoms [43] and the results of validation of the ligand positioning accuracy for a test set of 30 protein-ligand complexes [11]. However, the protein-ligand binding energy calculation is out of the scope of this work. It is demonstrated here that even limited mobility of protein atoms results in considerable improvement of the docking positioning accuracy. Whilst the present results were obtained for the MMFF94 force field [44] in vacuum, the performance of SOL-P allows including in the docking procedure one of either rigorous (PCM or COSMO) or heuristic (Generalized Born) solvent models [45]. The ability to perform docking with the PCM solvent model has been already demonstrated by the FLM program which demands more computing resources [5]. Although the SOL-P program does not outperform existing docking programs either in terms of positioning accuracy or speed of calculation, it opens the way for the accurate calculation of the protein-ligand binding free energy by employing the sets of low-energy minima of the molecular systems (the target protein, the ligand and their complex) which are carefully found for a given force field with a continuum solvent model. If low energy minima are found, the whole configuration integral defining the free energy of the respective molecular system can be accurately calculated as a sum of configuration integrals over these separated minima [4,5]. Such an accurate approach cannot be handled by commercial, superfast software that runs on laptops in seconds.

2. Materials and methods

For the realization of the novel docking algorithm we use the MMFF94 force field [44] in vacuum. While looking for low-energy minima, ligands are considered to be fully flexible and some of protein atoms are moveable. The force field determines the energy of the protein-ligand complex for its every conformation. The MMFF94 force field combines sufficiently good parameterization based on ab initio quantum-chemical calculations of a broad spectrum of organic molecules and the well-defined procedure of atom typification applicable to an arbitrary organic compound. This force field is not worse than many other popular force fields such as: AMBER [46,47], OPLS-AA [48], CHARMM [49] etc. MMFF94 is implemented in the SOL docking program [10] used successfully for new inhibitors' development [50–52]. Moreover, it has been recently shown that the docking paradigm is true for some protein-ligand complexes, if the energy of the

complex is calculated in the frame of the MMFF94 force field in vacuum [5]. The docking paradigm is not satisfied for many complexes, if the energy is calculated with MMFF94 in vacuum [5], but accounting for solvent in the frame of an implicit solvent model improves the situation significantly [5]. However, it is found in [5] and later is supported in the quasi-docking procedure [14] that the recent quantum-chemical semi-empirical PM7 method with solvent is much better than MMFF94 with solvent. The same finding is presented independently in [16] comparing the PM6-D3H4X semiempirical method with eight different force fields including AMBER [46,47] and several empirical and knowledge-based force fields. Unfortunately, these quantum-chemical methods are much slower than force fields. Keeping all this in mind we investigate here the influence of protein atoms' mobility in the docking procedure on the quality of ligand positioning using only the MMFF94 field in vacuum. The results will be much better, if either MMFF94 is used with the solvent model or PM7 is used with the solvent model.

2.1. TT-docking

The novel docking algorithm (TT-docking) utilizes the TT global optimization method. It is based on the novel methods of tensor analysis. The detailed description of this algorithm is presented elsewhere [11,15] and here we describe only its main features.

The Tensor Train decomposition for d-dimensional tensors was introduced to numerical analysis in 2009 [53] as a means to fight against the so-called *curse of dimensionality*, given by the fact that the number of entries of a d-dimensional tensor grows exponentially in d and can easily exceed the number of atoms in the universe even for a kind of "small sizes", i.e. for $d = 300$ and 2 points at each dimension. Consequently, the list of entries cannot be used for practical computations. The Tensor Train (TT) is a decomposition in which the number of the tensor representation parameters grows in d just linearly. Moreover, despite some other classical decompositions (such as CPD — the Canonical Polyadic Decomposition [54]), the TT algorithms reduce all computations to structured low-rank matrices associated with the given tensor. In our optimization procedure this structure is used to navigate in the space for where to search for better minima. This procedure is essentially based on the TT Cross algorithm [55] that constructs a TT decomposition using only a small portion of the entries of the given tensor. Eventually the number of those entries used during the optimization depends on d only polynomially, and the *curse of dimensionality* mentioned above is no longer an obstacle.

The continuous protein-ligand energy function is transformed into the multi-dimensional array (tensor) and the novel tensor analysis methods are applied for the search of the tensor element with the maximal absolute value: obviously, the docking problem is the global minimization problem but it can be easily transformed to an equivalent problem of the magnitude maximization. If d is the number of degrees of freedom of the protein-ligand complex, then we can introduce the grid in the configuration space with n_i nodes in each direction $i = 1, 2 \ldots d$. If the grid is fine enough, then the solutions of continuous and discrete problems are expected to be close.

The basis of this consideration is the Tensor Train (TT) decomposition [53,56] of a tensor $A \in \mathbb{R}^{n_1 \times \ldots \times n_d}$ in the form:

$$A(i_1, \ldots, i_d) = \sum_{\alpha_1 = 1, \ldots, \alpha_{d-1} = 1}^{r_1, \ldots, r_d} G_1(i_1, \alpha_1) G_2(\alpha_1, i_2, \alpha_2) \ldots G_{d-1}(\alpha_{d-2}, i_{d-1}, \alpha_{d-1}) G_d(\alpha_{d-1}, i_d)$$

The numbers $r_1, \ldots, r_{d-1}$ are called TT-ranks of the tensor; for convenience, dummy ranks $r_0 \equiv r_d \equiv 1$ are also introduced. The 3-dimensional tensors $G_i \in \mathbb{R}^{r_{i-1} \times n_i \times r_i}$ are called cores or carriages of the tensor train. If TT-ranks are reasonably small, then the TT decomposition possesses several very useful properties [53,56]. However, we cannot afford computing or storing all the elements for large tensors. Therefore, it becomes crucial to have for tensors a fast approximation method utilizing only a small number of their elements. Such a method was proposed and called the TT-Cross method [55]. It heavily exploits the matrix cross interpolation [57–61] algorithm applied cleverly, although heuristically, to selected submatrices in the unfolding matrices of the given tensor. The matrix $A_k \in \mathbb{R}^{n^k \times n^{d-k}}$, $A_k(i_1 \ldots i_k, i_{k+1} \ldots i_d) = A(i_1, i_2, \ldots, i_d)$ is called the k-th unfolding matrix of the tensor A. Such matrices are highly connected with TT-decomposition, TT-rank r_k is just the rank of the matrix A_k.

The TT-Cross method iteratively improves the sets of interpolation points searching for submatrices of larger volume (determinant in modulus) and consequently the elements of larger magnitude. This property allows one to take it as a base for the global optimization method [11].

The TT-docking iteratively performs the following steps:

1. Generation of submatrices of unfolding matrices using sets of tensor elements.
2. Interpolation of submatrices using TT-Cross method with rank $\leq r_{max}$.
3. A set of interpolation points for each submatrix contains elements with large values in modulus.
4. Rough local optimization of interpolation points (protein-ligand poses) by the simplex method, addition of optimized point projections to the tensor and to the interpolation point sets.
5. Updating of each set of interpolation points of the unfolding matrix by merging the interpolation points of the previous unfolding matrix and ones of the subsequent unfolding matrix.
6. Addition of the best points (ligand poses) to the interpolation point set of the unfolding matrix, and transition to step 1 using the obtained point set as the tensor elements.

The complexity of the TT global optimization method is $O(dnr_{max}^2)$ functional evaluations, $O(dr_{max})$ local optimizations and $O(dnr_{max}^3)$ arithmetic operations, where r_{max} is the maximal rank of the Tensor Train decomposition, n is the initial grid size along one dimension and d is the number of dimensions. It is easy to see that operations for different unfolding matrices could be performed independently, and we need synchronization only when constructing the new points at the end of each iteration. Moreover, a parallel implementation of the matrix cross method is also available [62]. In the result, we have a parallel version of the TT global optimization algorithm with parallel complexity $O(r_{max})$ functional evaluations, $O(1)$ local optimizations and $O(d + r_{max}^2)$ arithmetic operations.

2.2. SOL-P docking program

The parallel SOL-P docking program is constructed on the base of the TT-docking algorithm (see above). The SOL-P program is developed for finding the low energy local minima spectrum of protein-ligand complexes, proteins or ligands including the respective global energy minimum. The energy of each molecule conformation is calculated directly in the frame of the MMFF94 force field [44] in vacuum without any simplification or fitting parameters. The conformation space of the system coordinates is formed by translations and rotations of the ligand as a whole, by the ligand torsions and also by Cartesian coordinates of the selected target protein atoms. The description of the ligand flexibility with torsions is used as a basic approach in many docking programs (AutoDock [6,7], ICM [8], DOCK [9], SOL [10] and GOLD [63]) to decrease the dimensionality of the search space. Certainly, in this approach some features of the ligand flexibility, e.g. the macrocyclic system flexibility, are not taken into account. The flexibility of the target protein is described here by the variations of Cartesian coordinates of the protein atoms located near the ligand atoms for certain ligand poses. This is the first step to the approach of the protein flexibility and it is chosen here only for the uniformity of consideration of different proteins and ligands and to keep restricted the change of the initial protein configuration taken from Protein Data Bank (PDB) [64]. While solving a particular

docking problem for a given target protein it is better to choose moveable protein atoms more cleverly, by sampling configurations of whole groups of the covalently bound protein atoms, such as side chains or loops, selected on the base of a priori knowledge. But such detailed investigation is out of the scope of the present work. The parallel MPI (message passing interface) based SOL-P program is written on C++ with usage of BLAS and LAPACK libraries. Main SOL-P parameters are: the maximal rank r_{max} of the TT-Cross approximation method, the power m of the discretization degree of the search space (the initial grid size is equal to $n = 2^m$ along one dimension) and the number of iterations of the TT global optimization algorithm. The initial grid is introduced in the d-dimensional search space to transform the continuous global optimization problem to the discrete one: finding the maximal in magnitude element of the d-dimensional tensor. Each point in the search space corresponds to a certain pose of the ligand in a certain configuration of the active site of the target protein and each element of the d-dimensional tensor corresponds to the MMFF94 energy of the protein-ligand complex in a given node of the grid. The total number of nodes in the grid (2^{md}) is made large enough (see Section 2.6) to keep smoothness of the continuous MMFF94 energy function in the discrete problem: energy values in neighboring nodes are close to one another. Moreover, it is convenient to apply the TT magnitude maximization to the functional $f(x, E_*) = \exp\{100 \operatorname{arccot}[E(x) - E_*]\}$, where $E(x)$ is the dimensionless MMFF94 energy for the given configuration x of the protein-ligand complex, E_* is the currently found global minimum. This function transforms the minimization problem to the maximization one. This function also zeroes large positive MMFF94 energy values arising due to the van der Waals repulsion of closely located atoms and it better separates low energy minima. As it is mentioned in the previous section there is a rough local energy optimization in the TT-docking algorithm by the Nelder-Mead simplex method [65] within the Subplex algorithm [66] implemented as Sbplx program in NLOpt library [67].

2.3. Moveable atoms

The ligand is considered as flexible with variations of its torsions, and also some protein atoms are moveable. In the present consideration a protein atom is moveable when it is close to at least one of reference ligand poses. The protein atom is close to a ligand pose when the distance between this protein atom and at least one ligand atom is less than a given threshold. In one extreme case, only the nonoptimized native (crystallized) ligand pose can be included into the set of reference ligand poses. In another extreme case, the reference poses of the ligand can be taken from the set of ligand poses corresponding to low-energy minima of the protein-ligand system which were found by SOL-P for the flexible ligand and the rigid protein. In this case the maximal number of protein atoms will be moveable. In the present work we took three ligand poses as reference ones: the ligand pose corresponding to the global protein-ligand energy minimum found by the FLM program [11] for the rigid protein, the locally optimized native ligand pose and the nonoptimized native ligand pose. None of movements of whole side chains is considered in this study. Such choice of the reference ligand poses is taken here only for the uniformity of consideration of all different proteins and ligands of the test set. Determination of moveable protein atoms is carried out by our original specially written program Mark-PMA (Mark Protein Moveable Atoms) with the MLT (Moveable Layer Thickness) parameter defining the threshold distance. The MLT parameter is taken up to 3 Å in the present investigation.

2.4. Docking procedure

The molecular data of the ligand and the protein with the marked moveable atoms are the input of the SOL-P program (shown in I stage in Fig. 1). The SOL-P program uses a cube centered in the geometrical

I stage

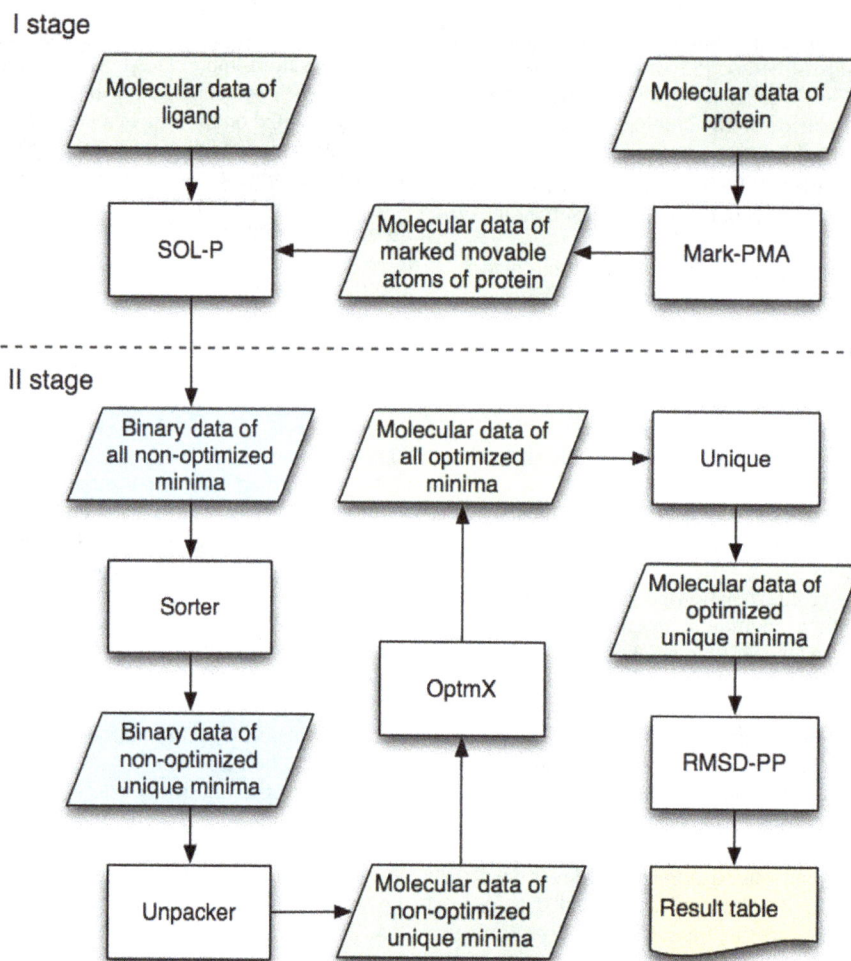

Fig. 1. Flowgraph of the program complex for low energy local minima search with flexible ligand and moveable target protein atoms. Stage I: the data preparation and TT global energy minima search with the SOL-P program. Stage II: the analysis of binary data with the "non-optimized minima" obtained from the SOL-P program and preparation of the table with the results and the final minima set.

center of the native ligand position in the crystallized protein-ligand complex as the spatial region for the low-energy minima search: all found ligand positions have their geometrical centers inside this cube (the docking cube). The cube is aligned along the Cartesian axes of the protein-ligand system. Each of the moveable protein atoms can move inside its own small cube centered in the initial atom position taken from the crystallized protein-ligand complex. Geometrical characteristics of the big docking cube and small cubes of moveable protein atoms are specified in the parameter file of the SOL-P program. In this work we set the docking cube edge equal to 10 Å and the small cube edge equal to 1 Å. We restrict motions of the moveable protein atoms in such a way that their Cartesian coordinates can change in the range of ± 0.5 Å from their positions in the crystallized structure because even small changes can make big differences in the protein-ligand energetics [3]. The position and the size of the docking cube in the active site of the target protein are usually defined by binding sites which can be interesting from a pharmacological point of view. For the positioning accuracy validation in this study we choose the center of the docking cube in the geometrical center of the native (crystallized) ligand position of the respective complex. Such choice of the docking cube implies that all low energy minima, which were found in the docking procedure, correspond to a single locus of the ligand binding. The SOL-P program performs MPI-parallelized search for the low-energy minima of protein-ligand complexes by the TT-docking algorithm containing the rough local optimization by the simplex method. The ligand has six rotational-translational degrees of freedom as a whole rigid body

plus torsional degrees of freedom for each single non-cyclic bond; each of the protein moveable atoms has three degrees of freedom — its Cartesian coordinates. The optimized target function is the protein-ligand complex total energy calculated by the MMFF94 force field in vacuum without any simplification or fitting parameters. Data about all found low-energy minima including protein-ligand configurations are too large to be saved in the molecular data format. These configurations are saved as the binary data (shown in Fig. 1 as "Binary data of all non-optimized minima").

2.5. Analysis of local minima

At stage II in Fig. 1 the post-processing of low energy configurations stored in the binary data is performed with the Sorter program. The Sorter program sorts the "nonoptimized minima" by their MMFF94 energies in vacuum and excludes minima with similar ligand positions — only one minimum with the lowest energy is being kept. Two ligand positions are considered similar if RMSD between them is less than a given threshold (0.1 Å), where RMSD is calculated atom-to-atom without chemical symmetry accounting. Thus, all the remaining low-energy configurations ("unique non-optimized minima" in Fig. 1) have different ligand positions. Then, the Unpacker program performs exporting all unique low-energy configurations from the binary file to the file with molecular format MOL2. The post-processing of low energy protein-ligand configurations consists of the performance of two programs: OptmX and Unique (Fig. 1). The OptmX program locally optimizes all of the "unique non-optimized minima". For these purposes, the OptmX program uses the L-BFGS algorithm [68,69] applied to the local optimization of the MMFF94 energy function in vacuum with variations of Cartesian coordinates of all ligand atoms and moveable protein atoms. Optimization of different minima is MPI-parallelized. After this optimization, the "all optimized minima" (Fig. 1) set is obtained. However, many of these minima may become similar again. Therefore, we need to re-exclude similar minima. The Unique program excludes similar minima from the "all optimized minima" set as follows. Among several close configurations only the minimum with the lowest energy is being kept as it is made in the binary data file post-processing by the Sorter program. However, in contrast to the Sorter program the protein moveable atoms are also taken into account in RMSD calculation, and the RMSD is calculated with chemical symmetry analysis.

Analysis of the local minima remaining after post-processing is carried out by the RMSD-PP program which calculates RMSD (with respect to all ligand atoms) between the ligand pose in a certain energy minimum of the protein-ligand complex and the ligand pose in the energy minimum corresponding to the native ligand position obtained after the local optimization from its configuration in the crystallized complex. The RMSD here is calculated taking into account the approximate chemical symmetry analysis as follows. A special attribute (so called "chemical digest"; in the present implementation it is the 32-bit integer number) is assigned to each atom, depending only on the MMFF94 type of this atom and the MMFF94 types of the adjacent atoms bound with this atom by chemical bonds, as follows. The selected atoms, including the analyzed atom, are ordered to a sequence, where atom "A" precedes atom "B" if "A" is closer to the analyzed atom (i.e. number of separating bonds from the analyzed atom is less for "A") or, in case of equally distanced "A" and "B", if "A" has a lower MMFF94-type (an integer from 1 to 99). Then, this sequence of MMFF94-types is processed by a hash function; in the present implementation, we used the CRC32 (32-bit Cyclic Redundancy Check) algorithm [70]. The obtained hash function value is the "chemical digest". The neighbors are analyzed by the breadth-first search [71] until the given depth (we set this parameter equal to 13) will be reached. So, chemically symmetric atoms have the same "chemical digest". Unfortunately, not every one-to-one atom mapping, keeping the "chemical digest" invariant, can preserve the whole chemical structure. Nevertheless, the "chemical digest" heuristic can filter off many of the wrong atom-to-atom mapping during the RMSD calculation. After the "chemical digest" calculation, all atoms with the same "chemical digest" are grouped. Within the group, all possible squared distances are calculated, where the first atom position belongs to the first configuration and the second atom position belongs to the second configuration. Then, the atom-to-atom assignment is searched by the Hungarian method [72]. So, the calculated RMSD doesn't exceed (and in many cases equal to) the lowest possible RMSD with keeping the chemical structure atom-to-atom mapping. This RMSD with approximate chemical symmetry accounting is a good metric to estimate the geometrical difference between two configurations of a protein-ligand complex; it can correctly discard geometrical pseudo-differences such as phenyl residue flip, comparing to the native atom-to-atom RMSD calculation.

As a result the RMSD-PP program creates in its output (Fig. 1) the resulting table containing: the minimum index, the minimum energy, RMSD from the optimized native configuration and the distance from the ligand geometric center in the given minimum to the ligand geometric center in the optimized native configuration. The energy minima are sorted by their energy in the ascending order; that is, every minimum gets its own index equal to its number in this sorted list of minima. The lowest energy minimum has the index equal to 1.

Some minima from the list might be close in space to the optimized native ligand position. We designate the index of the minimum having RMSD from the optimized native ligand position less than 2 Å as "Index of the minimum Near Optimized Native" or "INON." If there are several

such minima which are close to the optimized native ligand position, we will choose the minimum with the lowest energy (with the lowest index) as "INON". When INON = 1 the docking paradigm is satisfied: the global minimum of the protein-ligand energy is near the native configuration. If there are no minima with the ligand pose near the optimized native configuration among all minima found by the SOL-P program, we use notation INON = inf.

It is useful to enhance the requirement on the minimum situated near the optimized native ligand position including the restriction on its energy and to introduce another index (EN) as the energy index of the minimum being near the optimized native ligand in space (RMSD <2 Å as it is used in the definition of INON) and in energy (in the energy interval ± 1 kcal/mol from the energy of the optimized native ligand). If there are several such minima, we will choose the minimum with the lowest energy (with the lowest index) as EN. Index EN demonstrates how far from the global minimum is the energy of the minimum found near the optimized native ligand pose. If EN is equal to a small positive integer, it means that the docking program finds a minimum near the optimized native ligand position and its energy is one of the lowest among the whole found minima spectrum. Index EN is useful when the energy of the optimized native ligand pose differs strongly from the energy of the global minimum.

In the present consideration we compare the energy minima found by the SOL-P program with ones obtained by the FLM program [11] with the same target function — energy in the frame of the MMFF94 force field in vacuum. FLM performs exhaustive search of low energy local minima of protein-ligand complexes in the rigid protein and flexible ligand approximation performing massive parallel energy minima search and employing large computing resources (about 20,000 CPU * h per one complex) available at supercomputer Lomonosov of Moscow State University [73].

2.6. Optimal SOL-P parameters

To choose optimal parameters of the SOL-P program we execute two sets of test calculations. First, calculations for the selection of the optimal parameters of the TT global optimization method (TT-docking) are performed. Second, calculations for selection of the optimal number of the moveable protein atoms are carried out. The first set of test calculations are carried out for 7 different protein-ligand complexes with rigid proteins (they are shown in Table 1). TT-docking performance is investigated with different values of two parameters: the maximal rank $r_{max} = \{4, 8, 16\}$ and the initial grid size $n = \{2^8, 2^{16}\}$.

Results of this testing demonstrate that for the higher initial grid size even the lowest tested maximal rank $r_{max} = 4$ is enough to find the optimum reliably and precisely. However, the increase of the initial grid size leads to slower convergence of the method and the iteration number must be larger (for $n = 2^{16}$ from 10 to 15 iterations need to be performed). The high grid size for ranks 8 and 16 makes computations significantly slower, thus the initial grid size of 2^{12} is used for such ranks. For such initial grid size the computation time is reduced

Table 1

Complexes for testing parameters of the TT-docking algorithm. PDB ID is the ID of the respective protein-ligand complex taken from Protein Data Bank [64].

Protein name	PDB ID	Number of ligand atoms including hydrogen ones	Number of ligand torsions
Urokinase	1C5Y	20	2
	1F5L	24	6
	1VJA	61	17
	1VJ9	74	19
CHK1 (checkpoint kinase 1)	4FTA	35	6
Thrombin	1TOM	64	10
ERK2 (extracellular signal-regulated kinase 2)	4FV6	57	12

Table 2

Values of INON index (Index of the minimum Near Optimized Native) for three protein-ligand complexes with different numbers of protein moveable atoms. PDB ID is the ID of the respective protein-ligand complex taken from Protein Data Bank [64].

PDB ID (number of ligand torsions)	Number of protein moveable atoms	INON		
		$r_{max} = 4$ $n = 2^{16}$	$r_{max} = 8$ $n = 2^{12}$	$r_{max} = 16$ $n = 2^{12}$
1SQO (3)	0, 6, 15, 27, 35	1	1	1
3CEN (7)	0, 6	inf	inf	inf
	13	1	inf	17
	26	1	1	2
	48	2	2	1
4FT9 (5)	0	16	24	29
	6	17	21	21
	13	17	18	19
	25	15	15	15
	42	inf	inf	inf

by 1.5 times and the number of iteration decreases. Finally, three sets of optimal parameters are chosen: the first set with $r_{max} = 4$ and $n = 2^{16}$, the second set with $r_{max} = 8$ and $n = 2^{12}$, and the third set with $r_{max} = 16$ and $n = 2^{12}$. For all sets the same number of iterations equal to 15 is used.

Second testing calculations are carried out for 3 different complexes (Table 2) with different numbers of moveable protein atoms. The MARK-PMA program defines different numbers of moveable protein atoms for these complexes. Numbers of moveable protein atoms for respective complexes and the calculated values of INON index are presented in Table 2.

It can be seen that INON = 1 for all sets of protein moveable atoms for 1SQO complex. It means that the ligand pose corresponding to the global energy minimum is situated near the optimized native configuration and the docking paradigm is satisfied for the case of the rigid protein as well as for all selected cases of moveable protein atoms. For the 3CEN complex SOL-P does not find the energy minimum near the optimized native configuration for the rigid protein as well as for 6 protein moveable atoms. However, for 13, 26 and 48 protein moveable atoms INON is equal to 1 or 2 corresponding to cases when SOL-P finds the minimum near the optimized native configuration and its energy is lowest (its index is 1or 2) among energies of all other minima found by SOL-P. For 4FT9 complex SOL-P finds the minimum close to the native

configuration ($INON \neq inf$) but there are many minima with energies lower than energy of this close to the native configuration minimum. This means that the target energy function defined by the MMFF94 force field in vacuum is not adequate for this complex. Moreover, for 42 protein moveable atoms SOL-P cannot find the minimum close to the native configuration (INON = inf). Probably, in this case the TT-docking algorithm cannot find respective minima due to the high number of degrees of freedom for the given system: 137 = 126 (protein) + 11 (ligand). Strictly speaking the docking paradigm is not satisfied for the 4FT9 complex. So, we see that for some complexes (e.g. 1SQO) the docking paradigm is satisfied for the rigid protein as well as for the protein up to 35 moveable atoms. For some complexes (e.g. 3CEN) the docking paradigm is satisfied only for a sufficiently large number (13, 26, 48) of protein moveable atoms and SOL-P is able to find the global energy minimum in the search configuration space of 157 = 144 (protein) + 13 (ligand) degrees of freedom. For other complexes (e.g. 4FT9) the MMFF94 force field energy in vacuum is not adequate and the energy surface is so complicated that for the too large number of protein moveable atoms (42) SOL-P is not able to find minima near the native configuration. Computing resources needed for the native ligand docking using the SOL-P program with different TT-docking parameters and different numbers of protein moveable atoms are presented in Fig. 2.

Comparing computing resources in Fig. 2 and results of INON calculations in Table 2 two cases of optimal numbers of protein moveable atoms are chosen (13–18 and 25–35 atoms depending on the complex) in the present study for more broad validation.

2.7. Validation set of protein-ligand complexes

For low-energy local minima search we use 30 protein-ligand complexes with experimentally known 3D structures [11] (see Table 3). All protein-ligand complexes are chosen with good resolution from PDB [64]. The ligand variety covers a wide range from small and rigid ligands (e.g. the ligand of the 1C5Y complex) to big and flexible ones (e.g. the ligand of the 1VJ9 complex). For all these complexes the locally optimized ligand native position has RMSD from the original (crystallized) native pose less than 1.5 Å. Thus the locally optimized ligand native position still can represent the native ligand pose.

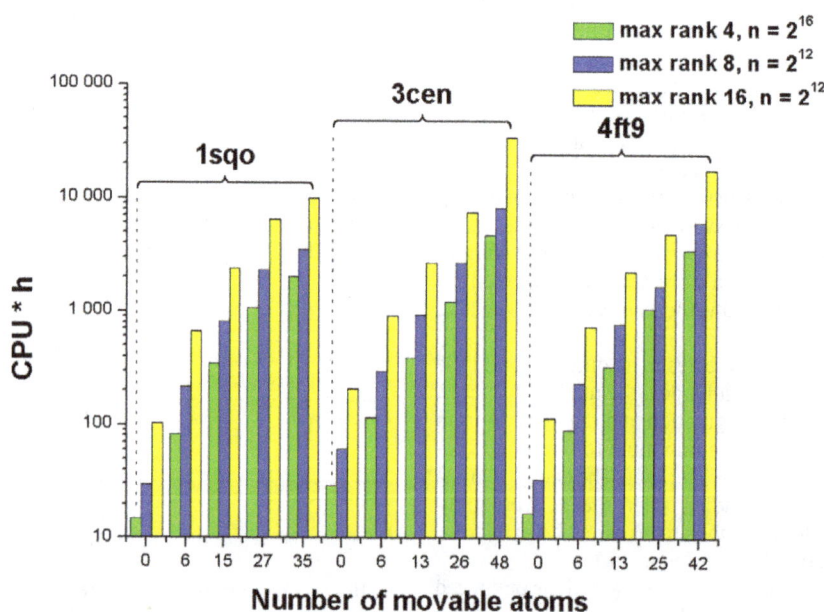

Fig. 2. Dependence of computing resources on the number of protein moveable atoms for the native ligand docking by the SOL-P program with different sets of TT-docking parameters. Integer n is the initial grid size.

Table 3
Validation set of protein-ligand complexes. Numbers of atoms includes hydrogen ones. N_P is the total number of the protein moveable atoms. N_H is the number of the protein moveable hydrogen atoms.

Protein name	PDB ID	Num. of ligand torsions	Numbers of ligand atoms	Numbers of moveable protein atoms 13–18 N_P/N_H	Numbers of moveable protein atoms 25–35 N_P/N_H
Urokinase	1C5Y	2	20	14/8	26/17
	1SQO	4	34	15/8	27/17
	1F5L	6	24	16/8	27/14
	1O3P	6	46	17/11	28/20
	1VJA	17	61	16/8	28/15
	1VJ9	19	74	16/9	30/18
CHK1 (checkpoint kinase 1)	4FSW	0	26	15/11	29/22
	4FT0	3	42	15/11	26/20
	4FT9	5	32	13/10	25/20
	4FTA	6	35	15/11	31/22
Factor Xa	1MQ6	7	54	14/10	30/18
	2P94	7	60	13/10	29/21
	3CEN	7	50	13/9	26/17
	1LQD	8	61	17/10	31/22
Poly(ADP-ribose) polymerase	2PAX	1	24	14/7	20/10
	1EFY	3	33	14/12	27/18
	3PAX	3	20	14/9	25/15
ERK2 (extracellular signal-regulated kinase 2)	4FV5	8	52	16/13	33/25
	4FV6	12	57	18/11	26/19
Thrombin	1TOM	10	64	14/6	25/15
	1DWC	12	71	16/12	26/19
Trypsin	1PPC	6	69	15/8	25/16
	1K1J	10	68	13/8	27/19
GNC92H2 antibody	1I7Z	5	44	17/16	29/26
Apolipoprotein	3KIV	6	22	13/5	30/10
Beta-1,4-xylanase	1J01	6	35	15/8	29/18
Ricin	1BR5	7	29	15/9	27/18
Neuraminidase	1B9V	10	50	16/12	33/23
Hen egg-white lysozyme	1LZG	11	56	15/10	26/19
HIV-1 protease	1HPV	14	70	14/9	31/21

Protein structures are prepared as follows. All the records corresponding to atoms, ions and molecules which are not a part of the protein structure are eliminated from the PDB files of the complexes. Hydrogen atoms are added to this structure by the APLITE program [10]. The APLITE program adds hydrogen atoms according to the standard amino acid protonation states at pH = 7 and performs the protein energy optimization with variations of positions of all hydrogen atoms in the frame of the MMFF94 force field keeping fixed all protein heavy atoms. Ligands are also taken from PDB files. Hydrogen atoms are added to ligands by the Avogadro program [74].

As can be seen from Table 3 the largest part of all protein moveable atoms are hydrogen ones almost for all test complexes. Movements of hydrogen atoms during the docking process can be favorable for the hydrogen bond formation. However, we do not consider that properties of the MMFF94 force field [44] enable SOL-P to reproduce hydrogen bonds with high precision and there is no sense to analyze their formation in the present study.

3. Results

The total number of low energy minima found by the SOL-P program, N_{tot}, for each complex varies considerably for different complexes depending on the complexity of the protein-ligand energy surface. This number can be as small as $N_{tot} = 25$ for the rigid protein of the 1C5Y complex and it can be as large as $N_{tot} = 7149$ for 1VJ9 with 30 protein moveable atoms. The N_{tot} number expands with the increase of the number of protein moveable atoms for any tested complex when the dimensionality of minima search space increases, e.g. for the 1I7Z

complex $N_{tot} = 362$ for the rigid protein and $N_{tot} = 1437$ for 29 moveable protein atoms. Values of N_{tot} found by SOL-P and FLM are comparable for many complexes.

Computing resources for all 30 test protein-ligand complexes are 5–120, 110–1300 and 600–3200 CPU $*$ h for the rigid proteins, for the cases of 13–18 and 25–35 protein moveable atoms, respectively. So, docking with 13–18 protein moveable atoms needs dozens of times more computing resources as compared with the rigid protein case and docking with 25–35 protein moveable atoms needs several times more resources as against docking with 13–18 protein moveable atoms.

A priori there is one special local energy minimum in the protein-ligand energy minima spectrum for any energy function calculated in the frame of any force field either in vacuum or in solvent. It is the minimum obtained by the local optimization of the protein-ligand energy beginning from the ligand pose in the crystallized protein-ligand complex. The ligand pose in this local minimum we call optimized native ligand pose. The local energy optimization is performed with variations of either only ligand atoms or ligand and moveable protein atoms. Due to the docking paradigm this local minimum must be in the low energy part of the whole energy minima spectrum and the docking program must find it. The ability to find this energy minimum is one of indicators of the high quality of the low energy minima search algorithm: finding this minimum is the necessary condition of the thoroughness of the docking program performance. The SOL-P program finds such minimum for 10, 14 and 13 complexes (out of 30 complexes) for docking into the rigid protein, into the protein with 13–18 moveable atoms and 25–35 moveable atoms, respectively. We see that moveable protein atoms improve the ability of the SOL-P program to find the optimized native ligand pose. However, this feature of the SOL-P program is worse than one of the FLM programs which finds the optimized native ligand pose for 17 complexes of the same test set performing the exhaustive low energy minima search [11]. For 7 complexes (1C5Y, 1I7Z, 1O3P, 2PAX, 3PAX, 4FSW and 4FT0) both SOL-P for rigid proteins and for proteins with moveable atoms and FLM (for rigid proteins) find the optimized native ligand minimum. For 3 complexes (4FTA, 1SQO and 1EFY) SOL-P can find and FLM cannot find the optimized native ligand minimum. For 10 complexes neither SOL-P (with and without protein moveable atoms) nor FLM (for the rigid proteins) can find the optimized native ligand minimum. This result shows that the low energy minima search by either SOL-P or FLM docking programs is not perfect for some complexes. These failures could be partly due to the non-adequate target energy function: search algorithms look for low energy minima and can miss the optimized native ligand pose if its energy is too high.

The validation shows that SOL-P finds either the global minimum or one of low energy minima corresponding to the ligand pose being near the optimized native ligand pose for the rigid protein and/or for the protein with moveable atoms for more than two thirds of the whole test set of protein-ligand complexes (for 22 out 30) (see Table 4): for these 22 complexes INON = 1 or INON ≤ 25 and the docking paradigm is fulfilled for them in the frame of the MMFF94 force field in vacuum. The test complexes are collected in groups in respect with values of their INON index in Fig. 3. This assertion is true also for FLM performance for the rigid proteins practically for the same complexes (see Table 4).

Taking into account protein atoms' mobility is crucial for 4 complexes (1J01, 1K1J, 1MQ6 and 3CEN) out of 30. SOL-P does not find any minima near the optimized native ligand pose for docking into the rigid protein (INON = inf). However, when mobility of protein atoms is taken into account, the docking procedure finds near the optimized native ligand pose either the global minimum (INON = 1) or one of the lowest energy minima (INON ≤ 25). Moreover, SOL-P with 25–35 protein moveable atoms always finds energy minima corresponding to the ligand pose near the optimized native ligand pose.

Table 4

Indexes EN and INON for the SOL-P program with different numbers of protein moveable atoms (0, 13–18 and 25–35) and for the FLM program with rigid proteins.

Complex id	EN/INON, SOL-P 0	EN/INON, SOL-P 13–18	EN/INON, SOL-P 25–35	EN/INON, FLM
1B9V	inf/344	513/353	inf/333	inf/inf
1BR5	144/45	241/23	inf/29	inf/309
1C5Y	1/1	1/1	1/1	1/1
1DWC	inf/20	inf/289	inf/98	inf/377
1EFY	72/46	38/20	40/16	158/81
1F5L	1/1	2/1	2/1	1/1
1HPV	inf/1	6/1	2/1	98/1
1I7Z	1/1	1/1	1/1	1/1
1J01	inf/inf	1/1	1/1	1/1
1K1J	inf/inf	inf/inf	inf/19	1/4
1LQD	inf/5	1/1	1/1	1/1
1LZG	inf/inf	inf/1270	inf/771	inf/inf
1MQ6	inf/inf	2/2	inf/3	7/4
1O3P	13/11	13/2	2/1	16/14
1PPC	inf/inf	inf/26	inf/51	1/1
1SQO	1/1	1/1	1/1	1/1
1TOM	inf/181	inf/465	inf/570	inf/inf
1VJ9	inf/26	inf/29	inf/23	48/1
1VJA	inf/50	inf/inf	inf/127	41/4
2P94	inf/2	inf/2	27/2	36/2
2PAX	1/1	1/1	1/1	1/1
3CEN	inf/inf	inf/1	inf/1	94/1
3KIV	9/1	5/1	4/1	12/1
3PAX	2/1	2/1	2/1	2/1
4FSW	6/5	6/5	6/5	8/7
4FT0	21/20	20/15	15/9	32/30
4FT9	inf/23	35/21	40/22	46/29
4FTA	176/176	370/370	415/415	inf/inf
4FV5	inf/231	87/87	122/84	189/122
4FV6	inf/337	inf/213	inf/325	inf/inf

On the other hand, for rigid proteins SOL-P and FLM cannot find such minima (INON = inf) for 6 and 5 complexes, respectively. It is worth to note that SOL-P is able to find the minimum near the optimized native ligand pose for all 5 complexes where FLM is not able to do this.

The FLM program (performing the exhaustive minima search) finds the global energy minimum near the optimized native ligand pose (INON = 1) for 13 complexes: 1C5Y, 1F5L, 1HPV, 1I7Z, 1J01, 1LQD, 1PPC, 1SQO, 1VJ9, 2PAX, 3CEN, 3KIV and 3PAX. The SOL-P program (with and without protein moveable atoms) finds also the global energy minimum near the optimized native ligand pose (INON = 1) for almost all these complexes except only two complexes: 1VJ9 and 1PPC.

Further, SOL-P finds not more than 10 minima near the optimized native ligand pose for most of the test complexes and only for few complexes the number of such minima is 11–36. Moreover, some of such minima are global energy minima (INON = 1) and their energies are close to the energies of respective optimized native ligand minima (EN = 1) for 5 or 6 complexes depending on mobility of protein atoms, e.g. for complexes 1C5Y, 1I7Z, 1J01, 1LQD, 1SQO and 2PAX with moveable protein atoms (see Table 4). There are 8 such minima found by the FLM program.

Therefore, we can say that in tote the SOL-P program (with and without protein moveable atoms) works not worse than the FLM program and much faster than the latter.

Our observation that neither SOL-P nor FLM can find any minimum near the optimized native ligand pose for 11 complexes (out of 30) is connected with inadequacy of the energy target function calculated in the frame of the MMFF94 force field in vacuum. It has been previously demonstrated [5] that protein-ligand energy calculation in the frame of the MMFF94 force field in solvent (with an implicit model) improves docking performance of the FLM program for the rigid proteins and with such target energy function SOL-P should also work better.

4. Conclusions

The validation results of the novel supercomputer SOL-P docking program are presented. This program performs docking of a flexible ligand into the protein with moveable atoms on the base of the search of the low-energy minima spectrum of a protein-ligand complex. Protein and ligand atoms' mobility is taken into account simultaneously and equally in the docking procedure. During this search the energy of

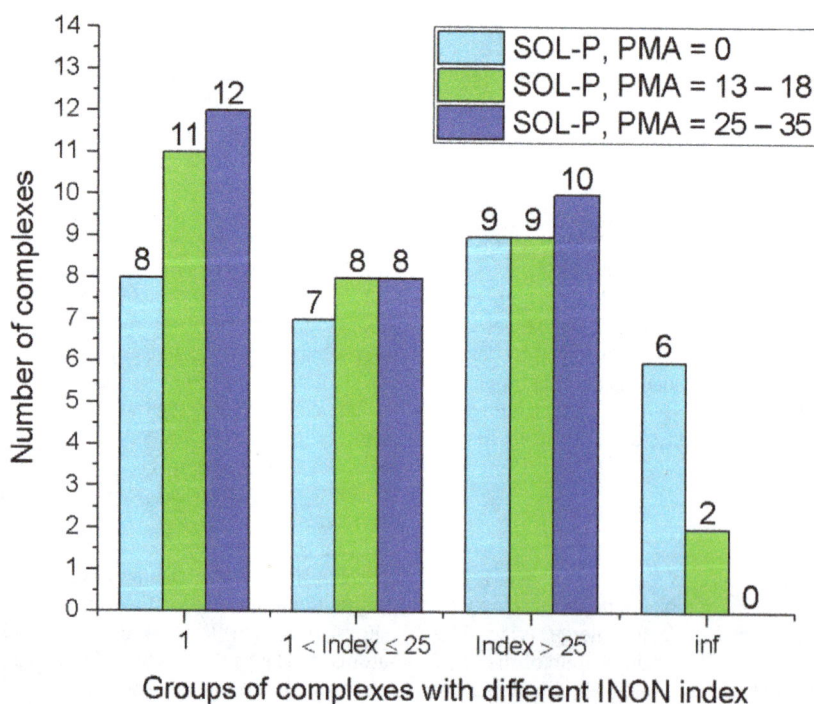

Fig. 3. Numbers of complexes with different values of INON index. PMA indicates the range of protein moveable atoms for the SOL-P program. INON is the index of the minimum having RMSD from the optimized native ligand position less than 2 Å; if there are several such minima, the minimum with the lowest energy (with the lowest index) should be taken.

each configuration of the protein-ligand complex is computed directly with the MMFF94 force field without simplifications and any fitting parameters. The grid of precalculated energy potentials of probe ligand atoms in the field of target protein atoms is not used. For the docking positioning validation energies of low-energy minima and their spatial locations corresponding to the ligand poses are carefully analyzed. Low-energy minima spectra of 30 protein-ligand complexes are investigated in the frame of the MMFF94 force field in vacuum.

It is shown that the program is able to perform docking of a flexible ligand into the active site of the target protein taking mobility of assigned protein atoms into account: up to 157 degrees of freedom in the conformation space using about 9 h at 512 core of the Lomonosov supercomputer [73]. As far as we know this is the first time when the docking program is able to perform successfully the global energy minimum search in the conformational space with such a large dimensionality. This result is achieved due to the usage of the novel docking algorithms (TT-docking) which are based on the so-called Tensor Train decomposition of multi-dimensional arrays (tensors) and the TT global optimization method [11,15]. TT-docking does not suffer from the curse of dimensionality. In principle the SOL-P program has no restrictions (except the availability of supercomputer resources) to perform docking with a larger number of moveable protein atoms including side-chain mobility and/or docking the very flexible ligands such as oligopeptides.

It is found that docking performance of the SOL-P program is comparable with one of the FLM program, which executes the exhaustive energy minima search for rigid target proteins due to employment of much larger computing resources. It is demonstrated that in some cases docking results are being improved even when small movements of protein atoms are taken into account in the docking procedure.

It is demonstrated that the docking paradigm is fulfilled for the target energy function calculated in the frame of the MMFF94 force field in vacuum for a flexible ligand and for target proteins with 25–35 moveable atoms for two thirds of the whole test set of protein-ligand complexes. Taking into account an implicit solvent model in the calculation of the energy of the protein-ligand complexes should improve the positioning performance of the SOL-P docking program as it is observed for the FLM program [5].

The SOL-P docking program can be used for finding spectra of low-energy minima of the protein, the ligand and their complex in the frame of a given force field, and these spectra can be used for the binding free energy calculation through the configuration integrals over separated minima of the respective systems. This approach should improve accuracy of the protein-ligand binding energy calculations and it is similar to the "mining minima" method [4]. However our approach differs from the "mining minima" method mainly by more uniform and exhaustive low-energy local minima search instead of the exploration of the configuration space along a combination of low-frequency modes as it is made by the "mining minima" method [4].

The present investigations became possible due to the computing resources of M.V. Lomonosov Moscow State University supercomputer Lomonosov [73].

Abbreviations

TT	tensor train
PDB	protein data bank
NMR	nuclear magnetic resonance
	MMFF94 Merck molecular force field
MPI	message passing interface
BLAS	basic linear algebra subprograms
LAPACK	linear algebra package
RMSD	root-mean-square deviation
INON	index of the minimum near the optimized native ligand position

EN   index of the minimum being near the optimized native ligand in space and in energy

Acknowledgment

The work was financially supported by the Russian Science Foundation, Agreement no. 15-11-00025.

References

[1] Sliwoski G, Kothiwale S, Meiler J, Lowe EW. Computational methods in drug discovery. Pharmacol Rev 2014;66:334–95. http://dx.doi.org/10.1124/pr.112.007336.

[2] Sadovnichii VA, Sulimov VB. Supercomputing technologies in medicine. In: Voevodin VV, Sadovnichii VA, Savin GI, editors. Supercomputing technologies in science, education, and industry. Moscow University Publishing; 2009. p. 16–23 [in Russian].

[3] Mobley DL, Dill KA. Binding of small-molecule ligands to proteins: "what you see" is not always "what you get". Structure 2009;17(4):489–98. http://dx.doi.org/10.1016/j.str.2009.02.010.

[4] Chen W, Gilson MK, Webb SP, Potter MJ. Modeling protein-ligand binding by mining minima. J Chem Theory Comput 2010;6(11):3540–57.

[5] Oferkin IV, Katkova EV, Sulimov AV, Kutov DC, Sobolev SI, Voevodin VV, et al. Evaluation of docking target functions by the comprehensive investigation of protein-ligand energy minima. Adv Bioinforma 2015;2015:12. http://dx.doi.org/10.1155/2015/126858, 126858.

[6] Morris GM, Goodsell DS, Halliday RS, Huey R, Hart WE, Belew RK, et al. Automated docking using a Lamarckian genetic algorithm and an empirical binding free energy function. J Comput Chem 1998;19:1639–62. http://dx.doi.org/10.1002/(SICI)1096-987X(19981115)19:14<1639::AID-JCC10>3.0.CO;2-B.

[7] Huey R, Morris GM, Olson AJ, Goodsell DS. A semiempirical free energy force field with charge-based desolvation. J Comput Chem 2007;28:1145–52.

[8] Neves MAC, Totrov M, Abagyan R. Docking and scoring with ICM: the benchmarking results and strategies for improvement. J Comput Aided Mol Des 2012;26:675–86. http://dx.doi.org/10.1007/s10822-012-9547-0.

[9] Allen WJ, Balius TE, Mukherjee S, Brozell SR, Moustakas DT, Lang PT, et al. DOCK 6: impact of new features and current docking performance. J Comput Chem 2014;36(15):1132–56.

[10] Sulimov AV, Kutov DC, Oferkin IV, Katkova EV, Sulimov VB. Application of the docking program SOL for CSAR benchmark. J Chem Inf Model 2013;53(8):1946–56. http://dx.doi.org/10.1021/ci400094h.

[11] Oferkin IV, Zheltkov DA, Tyrtyshnikov EE, Sulimov AV, Kutov DC, Sulimov VB, et al. Evaluation of the docking algorithm based on tensor train global optimization. Bull South Ural State Univ Ser Math Model Program Comput Softw 2015;8(4):83–99. http://dx.doi.org/10.14529/mmp150407.

[12] Romanov AN, Jabin SN, Martynov YB, Sulimov AV, Grigoriev FV, Sulimov VB. Surface generalized born method: a simple, fast and precise implicit solvent model beyond the Coulomb approximation. J Phys Chem A 2004;108:9323–7.

[13] Sulimov VB, Mikhalev AYu, Oferkin IV, Oseledets IV, Sulimov AV, Kutov DC, et al. Polarized continuum solvent model: considerable acceleration with the multicharge matrix approximation. Int J Appl Eng Res 2015;10(24):44815–30.

[14] Sulimov AV, Kutov DC, Katkova EV, Sulimov VB. Combined docking with classical force field and quantum chemical semiempirical method PM7. Adv Bioinforma. Accepted 22 December 2016. Volume 2017, Article ID 7167691, 6 pp., http://dx.doi.org/10.1155/2017/7167691.

[15] Zheltkov DA, Oferkin IV, Katkova EV, Sulimov AV, Sulimov VB, Tyrtyshnikov EE, et al. TTDock: docking method based on tensor train. Vychislitelnie Metody Programmirovanie (Numer Methods Program) 2013;14:279–91 [(in Russian) Available: http://num-meth.srcc.msu.ru/english/zhurnal/tom_2013/v14r131.html. Accessed 2017 January 23].

[16] Pecina A, Meier R, Fanfrlík J, Lepšík M, Řezáč J, Hobza P, et al. The SQM/COSMO filter: reliable native pose identification based on the quantum-mechanical description of protein–ligand interactions and implicit COSMO solvation. Chem Commun 2016;52:3312–5.

[17] Antunes DA, Devaurs D, Kavraki LE. Understanding the challenges of protein flexibility in drug design. Expert Opin Drug Discovery Dec 2015;10(12):1301–13. http://dx.doi.org/10.1517/17460441.2015.1094458.

[18] Fischer M, Coleman RG, Fraser JS, Shoichet BK. Incorporation of protein flexibility and conformational energy penalties in docking screens to improve ligand discovery. Available: http://www.nature.com/nchem/journal/v6/n7/full/nchem.1954.html; 2014. [Accessed 2017 January 23].

[19] B-Rao C, Subramanian J, Sharma SD. Managing protein flexibility in docking and its applications. Available: http://www.sciencedirect.com/science/article/pii/S1359644609000063; 2009. [Accessed 2017 January 23].

[20] Sousa SF, Ribeiro AJ, Coimbra JT, Neves RP, Martins SA, Moorthy NSHN, et al. Protein-ligand docking in the new millennium – a retrospective of 10 years in the field. Curr Med Chem 2013;20(18):2296–314. http://dx.doi.org/10.2174/0929867311320180002.

[21] Jiang F, Kim SH. "Soft docking": matching of molecular surface cubes. J Mol Biol 1991;219:79–102. http://dx.doi.org/10.1016/0022-2836(91)90859-5.

[22] Buonfiglio R, Recanatini M, Masetti M. Protein flexibility in drug discovery: from theory to computation. ChemMedChem 2015;10:1141–8. http://dx.doi.org/10.1002/cmdc.201500086.

[23] Martinez-Ramos F, Fonseca-Sabater Y, Soriano-Ursua MA, Torres E, Rosales-Hernandez MC, Trujillo-Ferrara JG, et al. o-Alkylselenenylated benzoic acid accesses several sites in serum albumin according to fluorescence studies, Raman spectroscopy and theoretical simulations. Protein Pept Lett 2013;20:705–14. http://dx.doi.org/10.2174/0929866511320060009.

[24] Osterberg F, Morris GM, Sanner MF, Olson AJ, Goodsell DS. Automated docking to multiple target structures: incorporation of protein mobility and structural water heterogeneity in AutoDock. Proteins 2002;46:34–40. http://dx.doi.org/10.1002/prot.10028.

[25] Carlson HA, Masukawa KM, Rubins K, Bushman FD, Jorgensen WL, Lins RD, et al. Developing a dynamic pharmacophore model for HIV-1 integrase. J Med Chem 2000; 43:2100–14. http://dx.doi.org/10.1021/jm990322h.

[26] Claussen H, Buning C, Rarey M, Lengauer T. FlexE: efficient molecular docking considering protein structure variations. J Mol Biol 2001;308:377–95. http://dx.doi.org/10.1006/jmbi.2001.4551.

[27] Corbeil CR, Englebienne P, Moitessier N. Docking ligands into 11 flexible and solvated macromolecules. 1. Development and validation of FITTED 1.0. J Chem Inf Model 2007;47:435–49. http://dx.doi.org/10.1021/ci6002637.

[28] Abagyan R, Totrov M, Kuznetsov D. ICM—a new method for protein modeling and design: applications to docking and structure prediction from the distorted native conformation. J Comput Chem 1994;15:488–506. http://dx.doi.org/10.1002/jcc.540150503.

[29] Schnecke V, Kuhn LA. Virtual screening with solvation and ligand induced complementarity. Perspect Drug Discovery Des 2000;20:171–90. http://dx.doi.org/10.1023/A:1008737207775.

[30] Leach AR. Ligand docking to proteins with discrete side-chain flexibility. J Mol Biol 1994;235:345–56. http://dx.doi.org/10.1016/S0022-2836(05)80038-5.

[31] Ding F, Yin S, Dokholyan NV. Rapid flexible docking using a stochastic rotamer library of ligands. J Chem Inf Model 2010;50:1623–32. http://dx.doi.org/10.1021/ci100218t.

[32] Jones G, Willett P, Glen RC. Molecular recognition of receptor sites using a genetic algorithm with a description of desolvation. J Mol Biol 1995;245:43–53. http://dx.doi.org/10.1016/S0022-2836(95)80037-9.

[33] Smiesko M. DOLINA — docking based on a local induced-fit algorithm: application toward small-molecule binding to nuclear receptors. J Chem Inf Model 2013;53: 1415–23. http://dx.doi.org/10.1021/ci400098y.

[34] Schnecke V, Swanson CA, Getzoff ED, Tainer JA, Kuhn LA. Screening a peptidyl database for potential ligands to proteins with side-chain flexibility. Proteins 1998;33: 74–87.

[35] Morris GM, Huey R, Lindstrom W, Sanner MF, Belew RK, Goodsell DS, et al. AutoDock4 and AutoDockTools4: automated docking with selective receptor flexibility. J Comput Chem 2009;30:2785–91. http://dx.doi.org/10.1002/jcc.21256.

[36] Sherman W, Day T, Jacobson MP, Friesner RA, Farid R. Novel procedure for modeling ligand/receptor induced fit effects. J Med Chem 2006;49:534–53. http://dx.doi.org/10.1021/jm050540u.

[37] Sokkar P, Sathis V, Ramachandran M. Computational modeling on the recognition of the HRE motif by HIF-1: molecular docking and molecular dynamics studies. J Mol Model 2012;18:1691–700. http://dx.doi.org/10.1007/s00894-011-1150-0.

[38] Schaffer L, Verkhivker GM. Predicting structural effects in HIV-1 protease mutant complexes with flexible ligand docking and protein side-chain optimization. Proteins 1998;33:295–310.

[39] Apostolakis J, Pluckthun A, Caflisch A. Docking small ligands in flexible binding sites. J Comput Chem 1998;19:21–37. http://dx.doi.org/10.1002/(SICI)1096-987X(19980115)19:1<21::AID-JCC2>3.0.CO;2-0.

[40] Borrelli KW, Cossins B, Guallar V. Exploring hierarchical refinement techniques for induced fit docking with protein and ligand flexibility. J Comput Chem 2010;31: 1224–35. http://dx.doi.org/10.1002/jcc.21409.

[41] Tsfadia Y, Friedman R, Kadmon J, Selzer A, Nachliel E, Gutman M, et al. Molecular dynamics simulations of palmitate entry into the hydrophobic pocket of the fatty acid binding protein. FEBS Lett 2007;581:1243–7. http://dx.doi.org/10.1016/j.febslet.2007.02.033.

[42] Klimovich PV, Shirts MR, Mobley DL. Guidelines for the analysis of free energy calculations. J Comput Aided Mol Des 2015;29(5):397–411. http://dx.doi.org/10.1007/s10822-015-9840-9.

[43] Sulimov A, Zheltkov D, Oferkin I, Kutov D, Tyrtyshnikov E, Sulimov V, et al. Novel gridless program SOL-P for flexible ligand docking with moveable protein atoms. 21st EuroQSAR where molecular simulations meet drug discovery, September 4–8, 2016. Aptuit Conference Center, Verona Italy, abstract book, OC15; 2016. p. 52 [www.euroqsar2016.org].

[44] Halgren TA. Merck molecular force field. I. Basis, form, scope, parameterization and performance of MMFF94. J Comput Chem 1996;17:490–519. http://dx.doi.org/10.1002/(SICI)1096-987X(199604)17:5/6<490::AID-JCC1>3.0.CO;2-P.

[45] Katkova EV, Onufriev AV, Aguilar B, Sulimov VB. Accuracy comparison of several common implicit solvent models and their implementations in the context of protein-ligand binding. J Mol Graph Model March 2017;72:70–80. http://dx.doi.org/10.1016/j.jmgm.2016.12.011.

[46] Cornell WD, Cieplak P, Bayly CI, Gould IR, Merz Jr KM, Ferguson DM, et al. A second generation force field for the simulation of proteins, nucleic acids, and organic molecules. J Am Chem Soc 1995;117:5179–97. http://dx.doi.org/10.1021/ja00124a002.

[47] Wang J, Wolf RM, Caldwell JW, Kollman PA, Case DA. Development and testing of a general amber force field. J Comput Chem 2004;25:1157.

[48] Jorgensen WL, Maxwell DS, Tirado-Rives J. Development and testing of the OPLS all-atom force field on conformational energetics and properties of organic liquids. J Am Chem Soc 1996;118(45):11225–36. http://dx.doi.org/10.1021/ja9621760.

[49] Vanommeslaeghe K, Hatcher E, Acharya C, Kundu S, Zhong S, Shim J, et al. CHARMM general force field: a force field for drug-like molecules compatible with the CHARMM all-atom additive biological force fields. J Comput Chem 2010;31(4): 671–90. http://dx.doi.org/10.1002/jcc.21367.

[50] Sinauridze EI, Romanov AN, Gribkova IV, Kondakova OA, Surov SS, Gorbatenko AS, et al. New synthetic thrombin inhibitors: molecular design and experimental verification. PLoS One 2011;6(5):e19969. http://dx.doi.org/10.1371/journal.pone.0019969.

[51] Sulimov VB, Katkova EV, Oferkin IV, Sulimov AV, Romanov AN, Roschin AI, et al. Application of molecular modeling to Urokinase inhibitors development. Biomed Res Int 2014;2014. http://dx.doi.org/10.1155/2014/625176, 625176 [15 pp.].

[52] Sulimov VB, Gribkova IV, Kochugaeva MP, Katkova EV, Sulimov AV, Kutov DC, et al. Application of molecular modeling to development of new factor Xa inhibitors. Available: https://www.hindawi.com/journals/bmri/2015/120802/; 2015. [Accessed 2017 January 23].

[53] Oseledets IV, Tyrtyshnikov EE. Breaking the curse of dimensionality, or how to use SVD in many dimensions. SIAM J Sci Comput 2009;31(5):3744–59. http://dx.doi.org/10.1137/090748330.

[54] Bader B, Kolda T. Tensor decompositions and applications. SIAM Rev 2009;51: 455–500.

[55] Oseledets IV, Tyrtyshnikov EE. TT-Cross approximation for multidimensional arrays. Linear Algebra Appl 2010;432(1):70–88. http://dx.doi.org/10.1016/j.laa.2009.07.024.

[56] Oseledets IV. Tensor-train decomposition. SIAM J Sci Comput 2011;33(5):2295–317. http://dx.doi.org/10.1137/090752286.

[57] Goreinov SA, Tyrtyshnikov EE, Zamarashkin NL. Pseudo-skeleton approximations of matrices. Reports of Russian Academy of Sciences, Vol. 342(2); 1995. p. 151–2. http://dx.doi.org/10.1016/S0024-3795(96)00301-1.

[58] Goreinov SA, Tyrtyshnikov EE, Zamarashkin NL. A theory of pseudo-skeleton approximations. Linear Algebra Appl 1997;261:1–21. http://dx.doi.org/10.1016/S0024-3795(96)00301-1.

[59] Tyrtyshnikov EE. Incomplete cross approximation in the mosaic-skeleton method. Comput Secur 2000;64(4):367–80. http://dx.doi.org/10.1007/s006070070031.

[60] Goreinov SA, Tyrtyshnikov EE. The maximal-volume concept in approximation by low-rank matrices. Contemp Math 2001;208:47–51.

[61] Goreinov SA, Oseledets IV, Savostyanov DV, Tyrtyshnikov EE, Zamarashkin NL. How to find a good submatrix. Research report 8–10. Kowloon Tong, Hong Kong: ICM HKBU; 2008. http://dx.doi.org/10.1142/9789812836021_0015.

[62] Zheltkov DA, Tyrtyshnikov EE. Parallel implementation of matrix cross method. Vychislitelnye Metody Programmirovanie (Numer Methods Program) 2015;16: 369–75 [in Russian].

[63] Cole JC, Nissink JWM, Taylor R. Protein-ligand docking and virtual screening with GOLD. In: Alvarez J, Shoichet B, editors. Virtual screening in drug discovery. Taylor & Francis Group, LLC; 2005. p. 379–415.

[64] Berman HM, Westbrook J, Feng Z, Gilliland G, Bhat TN, Weissig H, et al. The protein data bank. Nucleic Acids Res 2000;28(1):235–42 [http://www.rcsb.org/pdb/home/home.do].

[65] Nelder JA, Mead R. A simplex method for function minimization. Comput J 1965;7: 308–13.

[66] Rowan T. Functional stability analysis of numerical algorithms. [Ph.D. thesis] Austin: Department of Computer Sciences, University of Texas; 1990.

[67] Johnson Steven G. The NLopt nonlinear-optimization package. http://ab-initio.mit.edu/nlopt.

[68] Byrd RH, Lu P, Nocedal J, Zhu C. A limited memory algorithm for bound constrained optimization. SIAM J Sci Comput 1995;16(5):1190–208. http://dx.doi.org/10.1137/0916069.

[69] Zhu C, Byrd RH, Lu P, Nocedal J. Algorithm 778: L-BFGS-B: Fortran subroutines for large-scale bound-constrained optimization. ACM Trans Math Softw 1997;23(4): 550–60. http://dx.doi.org/10.1145/279232.279236.

[70] Press WH, Teukolsky SA, Vetterling WT, Flannery BP. Section 22.4 Cyclic redundancy and other checksums. Numerical recipes: the art of scientific computing. 3rd ed. New York: Cambridge University Press978-0-521-88068-8; 2007.

[71] Donald EK. The art of computer programming vol 1. 3rd ed. Boston: Addison-Wesley 0-201-89683-4; 1997.

[72] Kuhn HW. The Hungarian method for the assignment problem. Nav Res Logist Q 1955;2:83–97. http://dx.doi.org/10.1002/nav.3800020109 [Kuhn's original publication].

[73] Sadovnichy VA, Tikhonravov AV, Voevodin VV, Opanasenko V. "Lomonosov": supercomputing at Moscow State University. Contemporary high performance computing: from petascale toward exascale. CRC Press; 2013. p. 283–307.

[74] Avogadro: an open-source molecular builder and visualization tool. Version 1. XX. Available: http://avogadro.cc/wiki/Main_Page. [Accessed 2017 January 23].

10

Exploring Components of the CO_2-Concentrating Mechanism in Alkaliphilic Cyanobacteria Through Genome-Based Analysis

Amornpan Klanchui [a], Supapon Cheevadhanarak [b], Peerada Prommeenate [c,*], Asawin Meechai [d,*]

[a] Biological Engineering Program, Faculty of Engineering, King Mongkut's University of Technology Thonburi, Bangkok 10140, Thailand
[b] Division of Biotechnology, School of Bioresources and Technology, King Mongkut's University of Technology Thonburi, Bangkok 10150, Thailand
[c] Biochemical Engineering and Pilot Plant Research and Development (BEC) Unit, National Center for Genetic Engineering and Biotechnology, National Science and Technology Development Agency at King Mongkut's University of Technology Thonburi, Bangkok 10150, Thailand
[d] Department of Chemical Engineering, Faculty of Engineering, King Mongkut's University of Technology Thonburi, Bangkok 10140, Thailand

ARTICLE INFO

Keywords:
Inorganic carbon uptake
CO_2-concentrating mechanism
Carbonic anhydrase
Carboxysomes
Alkaliphilic cyanobacteria
Genomic data

ABSTRACT

In cyanobacteria, the CO_2-concentrating mechanism (CCM) is a vital biological process that provides effective photosynthetic CO_2 fixation by elevating the CO_2 level near the active site of Rubisco. This process enables the adaptation of cyanobacteria to various habitats, particularly in CO_2-limited environments. Although CCM of freshwater and marine cyanobacteria are well studied, there is limited information on the CCM of cyanobacteria living under alkaline environments. Here, we aimed to explore the molecular components of CCM in 12 alkaliphilic cyanobacteria through genome-based analysis. These cyanobacteria included 6 moderate alkaliphiles; *Pleurocapsa* sp. PCC 7327, *Synechococcus* spp., *Cyanobacterium* spp., *Spirulina subsalsa* PCC 9445, and 6 strong alkaliphiles (i.e. *Arthrospira* spp.). The results showed that both groups belong to β-cyanobacteria based on β-carboxysome shell proteins with form 1B of Rubisco. They also contained standard genes, *ccmKLMNO* cluster, which is essential for β-carboxysome formation. Most strains did not have the high-affinity Na^+/HCO_3^- symporter SbtA and the medium-affinity ATP-dependent HCO_3^- transporter BCT1. Specifically, all strong alkaliphiles appeared to lack BCT1. Beside the transport systems, carboxysomal β-CA, CcaA, was absent in all alkaliphiles, except for three moderate alkaliphiles: *Pleurocapsa* sp. PCC 7327, *Cyanobacterium stranieri* PCC 7202, and *Spirulina subsalsa* PCC 9445. Furthermore, comparative analysis of the CCM components among freshwater, marine, and alkaliphilic β-cyanobacteria revealed that the basic molecular components of the CCM in the alkaliphilic cyanobacteria seemed to share more degrees of similarity with freshwater than marine cyanobacteria. These findings provide a relationship between the CCM components of cyanobacteria and their habitats.

1. Introduction

CO_2-concentrating mechanism (CCM) is an important process that maximizes the efficiency of inorganic carbon (C_i; CO_2 and HCO_3^-) uptake and CO_2 fixation in cyanobacteria and eukaryotic algae [1]. It elevates CO_2 level near the active site of Ribulose-1,5-bisphosphate carboxylase/oxygenase (Rubisco) enclosed in a polyhedral microcompartment called carboxysomes, thus enhancing photosynthetic performance [2]. In cyanobacteria, CCM is the key process that enables them to adapt to their diverse ranges of CO_2-limited aquatic environments such as freshwater, marine, and alkaline lakes [3–5]. Insights into the basic molecular components of cyanobacterial CCM in relation to their

habitats may provide us with an efficient strategy for improvement of photosynthetic CO_2 fixation and biomass yield in these organisms [6,7] and crop plants [8,9].

In general, the cyanobacterial CCM consists of two primary components –C_i uptake systems and carboxysomes– as described below.

1.1. C_i Uptake Systems

The C_i uptake systems are comprised of two uptake systems of CO_2 [10,11] and three transport systems of HCO_3^- [12,13]. The CO_2 uptake systems, located at the thylakoid membrane, convert cytosolic CO_2 into HCO_3^- [14]. These systems are based on NAD(P)H dehydrogenase type 1 (NDH-1) complexes comprising of NDH-1_3 and NDH-1_4 protein complexes. NDH-1_3 is the low-CO_2 inducible high-affinity CO_2 uptake system, encoded by *ndhD3*, *ndhF3*, and *cupA*(*chpY*). On the other hand, NDH-1_4 protein complex is the constitutive low-affinity CO_2 uptake

* Corresponding authors.
E-mail addresses: peerada.pro@biotec.or.th (P. Prommeenate), asawin.mee@kmutt.ac.th (A. Meechai).

system encoded by *ndhD4*, *ndhF4*, and *cupB*(*chpX*) genes [15]. While protein subunits NdhD and NdhF are responsible for CO_2 uptake [10], CupA and CupB catalyze the hydration reaction of CO_2 into HCO_3^- [16]. For the transport of HCO_3^- it is facilitated by three transporters, located at the plasma membrane, including BicA (a SulP-type sodium dependent HCO_3^- transporter), SbtA (a sodium-dependent HCO_3^- symporter), and BCT1 (an ATP-binding cassette (ABC)-type HCO_3^- transporter). These three transporters have different properties. BicA has low affinity for bicarbonate ($K_m = 70–350\ \mu M$), but high flux of HCO_3^- uptake, while SbtA has high affinity for bicarbonate ($K_m < 5\ \mu M$), but low flux of HCO_3^- uptake [12,17]. BCT1 has medium substrate affinity for bicarbonate ($K_m = 10–15\ \mu M$) and low flux of HCO_3^- uptake [18]. The operation of the C_i uptake systems ends up with a cytosolic C_i pool in the form of HCO_3^-, which is subsequently diffused into carboxysomes.

1.2. Carboxysomes

Carboxysomes are specialized sub-cellular compartments composing of protein shells and two encapsulated enzymes, Rubisco and carbonic anhydrase (CA) [19,20]. In carboxysomes, CA catalyzes HCO_3^- into CO_2, which is a substrate for Rubisco [21]. There are two types of carboxysomes, α- and β-. The *cso*-type of shell proteins, encoded by *cso* operon, is termed α-carboxysomes, while the *ccm*-type of shell polypeptides, encoded by *ccmKLMNO* operon, is termed β-carboxysomes. Based on this criterion, the cyanobacterial species carrying form 1A of Rubisco within α-carboxysomes are classified as α-cyanobacteria while the species containing form 1B of Rubisco within β-carboxysomes are classified as β-cyanobacteria [22,23]. Although the two carboxysome types are different in gene organization, formation, and species distribution, they have similar functions which are to limit CO_2 leaking, reduce the risk of photorespiration, and enhance the carboxylase activity of Rubisco [20,24]. Among the β-carboxysome proteins, which have been extensively studied, CcmK, CcmL, and CcmO were proposed to be in the outer shell layer [25,26], while CcmM and CcmN were proposed to localize in the inner shell [27]. Concerning on CA, various carboxysomal CA have been reported. They are named β-CA (CcaA) and γ-CA (CcmM) in β-cyanobacteria [28] and named β-CA (CsoSCA) in α-cyanobacteria [29]. β-cyanobacterial species also contain two types of non-carboxysomal CAs, β-CA (EcaB) and α-CA (EcaA), localized in the cell membrane or in the periplasmic space [30]. However, the specific function of EcaA/B has not yet been confirmed. For Rubisco, it catalyzes CO_2 fixation reaction to generate 3-phosphoglycerate as a precursor for the Calvin-Benson-Bassham cycle. This enzyme consists of eight small (RbcS; 12-18 kDa) and eight large (RbcL, 50–55 kDa) subunits [31]. Assembly of Rubisco requires chaperone proteins [32]. RbcX encoded by *rbcX* is a Rubisco assembly chaperone, which interacts with RbcL to facilitate the assembly of RbcL and RbcS to form Rubisco holoenzyme [33,34]. It has been reported that RbcX is highly conserved in organisms having form 1B Rubisco [35].

Cyanobacteria tend to have different sets of CCM components depending on their habitats [1]. Studies have shown that α- and β-cyanobacteria occupy different environments [36]. Most of the α-cyanobacteria such as *Prochlorococcus* and *Synechococcus* strains inhabit marine while β-cyanobacteria such as *Synechocystis* sp. PCC 6803 [37–39], *Anabaena variabilis* [40], and *Synechococcus elongatus* PCC 7942 [41] live mainly in freshwater. The two distinct environments differ mainly in their conditions such as pH, C_i content, and salinity. The factor that affects CCM the most is pH because it is strongly linked to the equilibrium of C_i species (H_2CO_3, CO_2, HCO_3^-, and CO_3^{2-}) in a system [42]. At high pH (>9), C_i content is usually high with dominant CO_3^{2-} and HCO_3^- ions, while pH 6–8, HCO_3^- is mostly present. At low pH (<6), C_i content is low with dominant CO_2 and H_2CO_3 ions. It is reported that the C_i concentration in marine environment (pH $\approx$ 8.2) is fairly constant around 2 mM [3]. The C_i availability in freshwater (pH $\approx$ 7) is however lower and fluctuates [43]. Based on C_i content, freshwater cyanobacteria tend to have complete C_i uptake systems which allow

them to cope with the C_i fluctuation, whereas marine strains appear to lack some C_i uptake systems because they mainly experience with stable environment [3]. In addition, some cyanobacteria can also survive in alkaline environments (pH = 8.5–11) [44]. An example of alkaline environments is soda lake which is characterized by the strong alkaline (pH $\geq$ 9.5) and high C_i concentration dominated with HCO_3^- and CO_3^{2-} ions [43]. Although CCM of freshwater and marine cyanobacteria are well studied, only a few observations of CCM in alkaliphilic cyanobacteria have been reported [43,45]. Some researchers have hypothesized that CCM might not be necessary in the alkaliphilic cyanobacteria because of unlimited supply of inorganic carbon in the form of HCO_3^- and CO_3^{2-} in the alkaline environments. Nevertheless, CCM components have recently been identified in some alkaliphilic cyanobacteria. In 2007, Dudoladova et al. [46] discovered α- and β-classes of CA and their sub-cellular localization in *Rhabdoderma lineare*. Later, Mikhodyuk et al. [47] studied the transport systems for carbonate of the natronophilic cyanobacterium *Euhalothece* sp. Z-M001. Recently, with the availability of a complete genome sequence of *Microcoleus* sp. IPPAS B-353, Kupriyanova et al. [48] identified a whole set of putative CCM components in this alkaliphilic organism living in soda lakes. They found that composition of the CCM components of the *Microcoleus* strain is similar to that of *Synechocystis* sp. PCC 6803 and *Synechococcus* PCC 7002 which are freshwater and marine β-cyanobacteria, respectively.

To further explore alkaliphilic cyanobacterial CCM, we aimed to probe unique features of molecular components of CCM in 12 alkaliphilic strains and relationship with their habitat. All the candidate genes/proteins involved in C_i uptake systems and carboxysomes of 12 alkaliphilic strains, including those inhabiting moderate (pH 8.5–9.4) and strong alkaliphilic (pH $\geq$ 9.5) environments, were identified. Computational identification of orthologous proteins was performed between the selected alkaliphiles and the 'model' β-cyanobacterium, *Synechocystis* sp. PCC 6803, whose CCM has been well studied. By sequence-based analysis, the variation of CCM components and potential orthologous sequences associated with such components was proposed. Comparative analyses within alkaliphilic, freshwater, and marine β-cyanobacteria were investigated, and the relationship between CCM components and ecological adaptation of alkaliphilic cyanobacteria were also emphasized. Since CCM is the crucial mechanism for CO_2 fixation and photosynthesis in cyanobacteria, we believe that a better understanding of the CCM components could pave the way for future research towards cellular improvement of economically important cyanobacteria such as *Arthrospira* spp.

2. Materials and Methods

2.1. Protein Sequence Retrieval

Amino acid sequences of 27 proteins involved in CCM of β-cyanobacterium, *Synechocystis* sp. PCC 6803 (Assembly ID GCA_000009725.1), were retrieved from the CyanoBase (http://genome.kazusa.or.jp/cyanobase) as a reference target protein set. These proteins were NdhD4 (gene ID *sll0027*), NdhF4 (gene ID *sll0026*), CupB (gene ID *slr1302*), NdhD3 (gene ID *sll1733*), NdhF3 (gene ID *sll1732*), CupA (gene ID *sll1734*), BicA1 (gene ID *sll0834*), BicA2 (gene ID *slr0096*), SbtA (gene ID *slr1512*), SbtB (gene ID *slr1513*), CmpA (gene ID *slr0040*), CmpB (gene ID *slr0041*), CmpC (gene ID *slr0043*), CmpD (gene ID *slr0044*), CcmK1 (gene ID *sll1029*), CcmK2 (gene ID *sll1028*), CcmK3 (gene ID *slr1838*), CcmK4 (gene ID *slr1839*), CcmL (gene ID *sll1030*), CcmM (gene ID *sll1031*), CcmN (gene ID *sll1032*), CcmO (gene ID *slr0436*), RbcL (gene ID *slr0009*), RbcS (gene ID *slr0012*), RbcX (gene ID *slr0011*), CcaA (gene ID *slr1347*), and EcaB (gene ID *slr0051*).

The genome, protein sequences, and annotation data of the 12 selected alkaliphiles were retrieved from the database of the National Center for Biotechnology Information (NCBI) in October, 2016. The analyzed alkaliphilic cyanobacteria included *Pleurocapsa* sp. PCC

7327 (P7; GenBank CP003590) [49], *Synechococcus* sp. JA-2-3B′a(2–13) (S2; GenBank NC_007776) [50], *Synechococcus* sp. JA-3-3Ab (S3; GenBank NC_007775) [50], *Cyanobacterium* PCC 7702 (CP; GenBank NZ_KB235926) [49], *Cyanobacterium stranieri* PCC 7202 (CS; GenBank CP003940) [49], *Spirulina subsalsa* PCC 9445 (SS; GenBank NZ_JH980292) [49], *Arthrospira platensis* C1 (AC; GenBank NZ_CM001632) [51], *Arthrospira platensis* NIES-39 (AN; GenBank NC_016640) [52], *Arthrospira platensis* str. Paraca (AP; GenBank ACSK00000000) [53], *Arthrospira maxima* CS-328 (AM; GenBank ABYK00000000) [54], *Arthrospira* sp. PCC 8005 (A8; GenBank NZ_FO818640) [55], and *Arthrospira* sp. TJSD091 (AT; GenBank LAYT00000000) [56].

2.2. Identification of Orthologous Proteins

The bidirectional sequence alignment approach, namely reciprocal BLASTP [57], was employed to identify proteins of 12 studied species, which are homologous to the reference proteins of *Synechocystis* sp. PCC 6803. To avoid under- and over-estimation of sequence similarity of these related species, the candidate orthologous proteins were determined based on BLAST statistics with the E-value threshold ($\leq 10^{-6}$) [58], the identity (≥ 30) [58], and coverage percentage (≥ 60) [58]. Only protein sequences with the BLASTP scores above the set critical values were further analyzed for the conserved domain using the Pfam database 27.0, provided by the Sanger Centre, UK (http://pfam.xfam.org/search) [59]. The default E-value cut-off of 1.0 was applied for this study [60]. The GUIDANCE web-server tool (http://guidance.tau.ac.il/) [61] was used to evaluate a confidence score of multiple sequence alignments. Additionally, the genomic features were visualized by GView [62].

2.3. Phylogenetic Analysis

A phylogenetic tree of the 12 selected strains and reference cyanobacteria was constructed based on Rubisco large subunit (RbcL) amino acid sequences, which were used to infer the protein function and classification among the strains. Other phylogenetic trees based on protein sequences of CmpABCD of the HCO_3^- transporter BCT1 and

sequences of NrtABCD of the nitrite/nitrate transporter were constructed to confirm the identity between the proteins. The reference species were selected according to types of carboxysomes (α- and β-classes), the existence of both CmpABCD and NrtABCD transporters in genomes, or their habitats. These strains included freshwater (*Anabaena* sp. PCC 7120, *Anabaena variabilis* ATCC 29413, *Cyanothece* sp. PCC 8801, *Cyanothece* sp. PCC 8802, *Nostoc punctiforme* ATCC 29133, *Synechococcus* sp. PCC 7942, and *Synechocystis* sp. PCC 6803) and marine (*Lyngbya* sp. PCC 8106, *Trichodesmium erythraeum* IMS101, *Synechococcus* sp. PCC 7002, *Synechococcus* sp. CC9605, *Synechococcus* sp. CC9902, *Prochlorococcus marinus* AS9601, NATL1A, and NATL2A, and *Prochlorococcus marinus* MIT 9211, 9215, 9301, 9303, 9312, 9313, and 9515) cyanobacteria. Their corresponding amino acid sequences were retrieved from the public databases, including the CyanoBase (http://genome.kazusa.or.jp/cyanobase) and the GenBank (http://www.ncbi.nlm.nih.gov/genbank) databases. A phylogenetic tree was created by performing multiple sequence alignment with MUSCLE [63,64], and then constructed based on the Maximum Likelihood [65] through the MEGA 6.0 software [66]. The reliability of the trees/branches was estimated via the bootstrap method [67], with 3000 replications.

3. Results

3.1. Strains and Classification of Alkaliphilic Cyanobacterial CCM

In this study, 12 selected alkaliphilic cyanobacterial strains were defined based on their ability to grow in an alkaline environment (pH roughly 8.5–11). The chosen strains included both unicellular and filamentous blue-green algae, which have different original habitats. According to the habitat pH values, we classified the selected strains into two main groups: moderately alkaliphilic cyanobacteria (pH 8.5–9.4) and strongly alkaline cyanobacteria (pH ≥ 9.5) (Table 1). The first group, moderately alkaline cyanobacteria, was comprised of two subgroups: alkali-thermophile and alkali-mesophile. The subgroup alkali-thermophile consisted of four species isolated from alkaline hot spring environments (pH 8.5–8.8, 50–70 °C), P7, S2, S3, and CP. The alkali-mesophile group of cyanobacteria was composed of two euryhaline cyanobacteria, CS and SS, living under a saline and alkaline

Table 1
Ecological niches of selected alkaliphilic cyanobacteria whose genome sequences are available (October 2016).

Species	Isolation site	Characteristics and habitats	Classification	Genome status	Reference
Pleurocapsa sp. PCC 7327 (P7)	Hunters hot spring, Oregon, USA	A unicellular nitrogen-fixing cyanobacterium. It is a stenohaline strain that can only survive within a narrow range of salinities. It has been found in alkaline water, hot spring (53 °C pH 8.5).	Moderate alkali-thermophile	Finished	[49]
Synechococcus sp. JA-2-3B′a(2–13) (S2) *Synechococcus* sp. JA-3-3Ab (S3)	Octopus Spring, Yellowstone National Park	A group of small (2–5 μm) unicellular cyanobacteria. Non-nitrogen-fixing cyanobacteria. The strain is dominant in alkaline siliceous hot springs (50–70 °C pH 8.5).	Moderate alkali-thermophile	Finished	[50]
Cyanobacterium PCC 7702 (CP)	Alkaline hot spring, near Reykjavik, Iceland	A practically unicellular cyanobacterium. The strain is isolated from alkaline siliceous hot springs (50–70 °C pH 8.8). Nitrogen-fixing and non-motile.	Moderate alkali-thermophile	Permanent Draft	[49]
Cyanobacterium stranieri PCC 7202 (CS)	Alkaline pond, Chad	A unicellular non-nitrogen-fixing cyanobacterium capable of growth in both freshwater and seawater media. Thus, it is able to adapt to a wide range of salinities (euryhaline). Non-motile.	Moderate alkali-mesophile	Finished	[49]
Spirulina subsalsa PCC 9445 (SS)	Alkaline-saline volcanic lake, Pantelleria, Italy	A motile filamentous cyanobacterium. It is capable of growth under a saline and alkaline environment.	Moderate alkali-mesophile	Permanent draft	[49]
Arthrospira platensis C1 (AC) *Arthrospira platensis* NIES-39 (AN) *Arthrospira platensis* str. Paraca (AP) *Arthrospira maxima* CS-328 (AM) *Arthrospira* sp. PCC 8005 (A8) *Arthrospira* sp. TJSD091 (AT)	Unknown Lake Chad, Chad, East Africa Africa Unknown Lake Chad, Chad, East Africa Seaside wetland, China, Bohai Unknown	A group of filamentous cyanobacteria that have an important role in industrial applications. Non-heterocyst-forming and non-nitrogen-fixing cyanobacteria; hydrogen-producing strains. They grow naturally in a high-salt alkaline (carbonate/bicarbonate) open pond system. The optimum pH for growth of ordinary *Arthrospira* strains is in the range of 9.0–9.5. AC is a laboratory strain.	Strong alkaliphile	Permanent draft Permanent draft Permanent draft Permanent draft Permanent draft Permanent draft	[51–56]

environment (pH 8.5–8.8, 30–45 °C). The strongly alkaline cyanobacteria group was comprised of six *Arthrospira* strains (AC, AN, AP, AM, A8, and AT), a dominant genus found in natural soda lakes, with growth optimum at pH range of 9.5–10.5 [68,69]. More details about the ecological niches of all studied alkaliphilic strains are given in Table 1.

To examine the carboxysome type operating in the 12 investigated cyanobacteria, a phylogenetic tree was constructed based on RbcL amino acid sequences. RbcL was chosen because it is a well-conserved enzyme for CO_2 fixation and has been used for the classification of cyanobacteria groups before [70]. A total of 36 protein sequences from 36 cyanobacteria were analyzed; 12 chosen alkaliphilic cyanobacteria

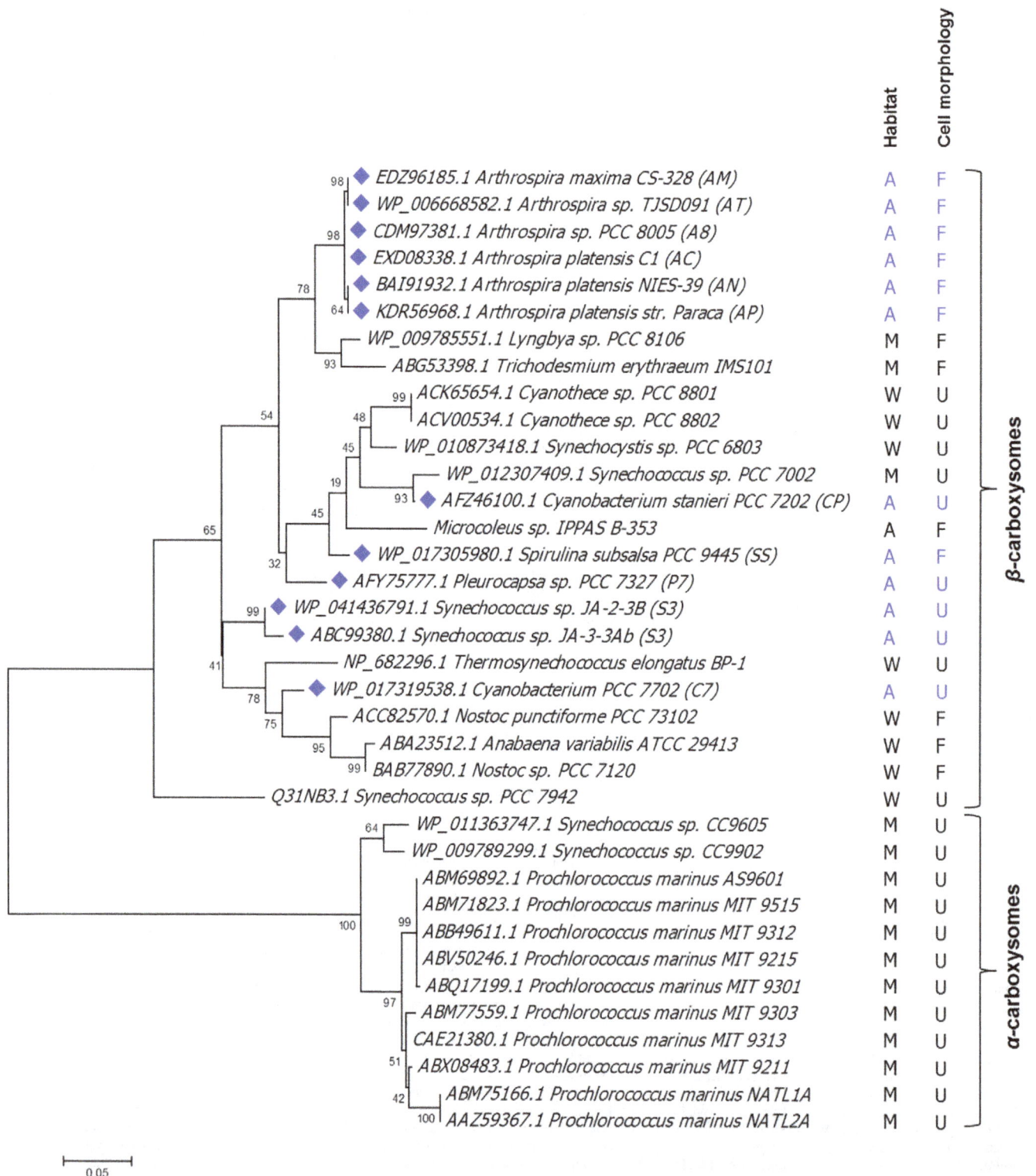

Fig. 1. Phylogenetic tree based on Rubisco large subunit protein sequences. The 12 alkaliphilic cyanobacterial strains examined in this study are identified by the blue diamond, while other reference species are represented without diamond. Cyanobacterial habitat, cell arrangement, and carboxysome type (α- or β-) are displayed. Within the column for habitat, freshwater strains are denoted by W, marine by M, and alkaline niche by A. Unicellular and filamentous cell arrangement is represented by U and F, respectively.

and 24 reference species consisting of 8 freshwater, 15 marine, and a haloalkaliphilic species. The results showed that all 36 cyanobacterial strains were divided into two main groups, β- and α-cyanobacteria, according to their Rubisco forms (Fig. 1). All 12 studied alkaliphilic cyanobacteria were clustered together in the β-cyanobacteria branch, reflecting the presence of the Rubisco 1B form. However, they were not completely grouped in the same cluster, following neither to their habitat nor morphology. For instance, four moderately alkali-thermophilic cyanobacteria, P7, S2, S3, and CP, were located in different clusters. In addition, SS which is a filamentous cyanobacterium appeared to be in the same cluster with the unicellular species.

3.2. Identification of Orthologous Proteins and Genes in Alkaliphilic Strains

Proteins corresponding to the CCM components of the 12 chosen alkaliphilic strains were identified as described in the Materials and Methods. All identified proteins had different degree of identity (40–80%, E-value threshold of $\leq 10^{-20}$) with the reference sequences. They also contained conserved domain regions which were similar to the reference proteins. Annotation details, including gene and protein accession number, annotation scores, protein domain analysis, are available in Supplementary File. The presence and absence of genes encoding the CCM components of alkaliphilic strains are shown in Table 2. The comparison of molecular CCM components from the *Synechocystis* sp. PCC 6803 and from the analyzed species revealed that approximately 20 orthologous genes are present in the investigated alkaliphiles. Furthermore, the results showed that the moderately alkaline group possesses more CCM components than the strongly alkaline ones.

Focusing on the C_i transport systems, there were up to five systems identified in the 12 alkaliphilic cyanobacteria: i) a low-affinity NDH-1_4 complex (NdhD4/NdhF4/CupB); ii) a high-affinity NDH-1_3 complex (NdhD3/NdhF3/CupA); iii) a SulP-type low-affinity Na$^+$-dependent HCO$_3^-$ BicA transporter; iv) a high-affinity Na$^+$/HCO$_3^-$ symporter SbtA;

and v) a high-affinity ATP-binding cassette BCT1(CmpABCD). All protein sequences of NDH-1_4 and NDH-1_3 showed a high sequence similarity with the reference sequences. Genes encoding each NDH-1 complex were localized together (Fig. 2). All studied strains showed high degree of homology with the BicA of *Synechocystis* sp. PCC 6803 ($\geq 60\%$ of amino acid identity). However, the orthologs of the SbtA transporter were found only in SS, AN, and AP. *SbtB* gene encoding SbtB protein that possibly functions as SbtA regulator [71] was also found nearby *sbtA* in the opposite direction in these three strains (Fig. 2). For the third HCO$_3^-$ transporter, BCT1, the orthologs of CmpABCD and NrtABCD (nitrate/nitrite transport system) cluster were observed in all studied alkaliphiles. They exhibited a moderate sequence similarity (55–71%) with the reference proteins. Both CmpABCD and NrtABCD protein sequences contained a similar protein domain, PBP2_NrtA_CpmA. In addition, a confidence score of multiple sequence alignments from the GUIDANCE web-server tool (http://guidance.tau.ac.il/) [61] showed highly conserved regions among these two protein clusters. Since the CmpABCD and NrtABCD protein sequences have been previously reported to share high similarity in sequences belonging to the same ABC transporter family [72], the experimentally confirmed proteins CmpABCD of *Synechococcus* sp. PCC 7942, were included in the subsequent analysis to verify the previous annotation. The BLAST's results showed a high homology between CmpABCD of *Synechococcus* sp. PCC 7942 and the sequences retrieved from four species, P7, S2, S3, and CP, with ~65–75% identity and an E-value of ~10^{-100}. Moreover, the sequences were further identified using the phylogenetic analysis (Fig. 3). The trees showed that the candidate sequences of CmpABCD of four species, P7, S2, S3, and CP, were clustered into the CmpABCD of *Synechococcus* sp. PCC 7942 and the other reference cyanobacteria, while the putative NrtABCD sequences of eight cyanobacteria, CS, SS, and six *Arthrospira* spp., were clustered into the NrtABCD of the reference cyanobacterium. As such, it is likely that BCT1 is present only in the 4 out of the 12 studied alkaliphilic cyanobacteria. Nevertheless, an

Table 2
Variation of the genes involved in CO$_2$-concentrating mechanism among alkaliphilic cyanobacterial strains. *Pleurocapsa* sp. PCC 7327 (P7), *Synechococcus* sp. JA-2-3B′a(2–13) (S2), *Synechococcus* sp. JA-3-3Ab (S3), *Cyanobacterium* PCC 7702 (CP), *Cyanobacterium stranieri* PCC 7202 (CS), *Spirulina subsalsa* PCC 9445 (SS), *A. platensis* C1 (AC), *A. platensis* NIES-39 (AN), *A. platensis* Paraca (AP), *A. maxima* CS-328 (AM), *Arthrospira* sp. PCC 8005 (A8), and *Arthrospira* sp. TJSD091 (AT) Numbers represent copy number of genes; nf is referred to not found; genes coding for putative NrtABCD are denoted by the symbol "?".

Component		Gene		Alkaliphilic strains											
				Moderate alkaliphilic cyanobacteria						Strong alkaliphilic cyanobacteria					
				Thermophiles				Mesophiles							
				P7	S2	S3	CP	CS	SS	AC	AN	AP	AM	A8	AT
C_i uptake systems	CO$_2$ uptake	NDH-1_4 complex	*ndhF4*	1	1	1	1	1	1	1	1	1	1	1	1
			ndhD4	1	1	1	1	1	1	1	1	1	1	1	1
			cupB	1	1	1	1	1	1	1	1	1	1	1	
		NDH-1_3 complex	*ndhF3*	1	1	1	1	1	1	1	1	1	1	1	1
			ndhD3	1	1	1	1	1	1	1	1	1	1	1	1
			cupA	1	1	1	1	1	1	1	1	1	1	1	1
	HCO$_3^-$ transport	BicA	*bicA1*	1	1	1	1	1	1	1	1	1	1	1	1
			bicA2	nf	nf	nf	nf	nf	1	1	1	1	1	1	nf
		SbtA SbtA regulator	*sbtA*	nf	nf	nf	nf	nf	1	nf	1	1	nf	nf	nf
			sbtB	nf	nf	nf	nf	nf	1	nf	1	1	nf	nf	nf
		BCT1	*cmpA*	1	1	1	1	?	?	?	?	?	?	?	?
			cmpB	1	1	1	1	?	?	?	?	?	?	?	?
			cmpC	1	1	1	1	?	?	?	?	?	?	?	?
			cmpD	1	1	1	1	?	?	?	?	?	?	?	?
Carboxysomes	Shell proteins	β-Carboxysomal shell proteins	*ccmK1*	1	1	1	1	1	1	1	1	1	1	1	1
			ccmK2	1	1	1	1	1	1	1	1	1	1	1	1
			ccmK3	1	nf	nf	1	1	1	1	1	1	1	1	1
			ccmK4	1	nf	nf	1	1	1	1	1	1	1	1	1
			ccmL	1	1	1	1	1	1	1	1	1	1	1	1
			ccmM (containing γ-CA domain)	1	1	1	1	1	1	1	1	1	1	1	1
			ccmN	1	1	1	1	1	1	1	1	1	1	1	1
			ccmO	1	1	1	1	1	1	1	1	1	1	1	1
	Encapsulated enzymes	Rubisco	*rbcL*	1	1	1	1	1	1	1	1	1	1	1	1
			rbcS	1	1	1	1	1	1	1	1	1	1	1	1
		β-CA	*ccaA*	1	nf	nf	nf	1	1	nf	nf	nf	nf	nf	nf

experimental study of the specificity of BCT1 to a certain substrate should be further performed to clarify the function of putative protein subunits (CmpABCD) in these four alkaliphilic cyanobacteria.

According to the observed C_i uptake systems (Table 2), the 12 analyzed alkaliphilic strains could be divided into three genotypes: I) strains containing NDH-1_3, NDH-1_4, BicA and BCT1, II) strains containing NDH-1_3, NDH-1_4, and BicA, and III) strains containing NDH-1_3, NDH-1_4, BicA, and SbtA. While the moderate alkali-thermophiles possessed genotype I, the moderate alkali-mesophiles (euryhaline) and strong alkaliphiles seemed to possess either genotype II or III. These results revealed that all alkaliphilic strains shared the same CO_2 uptake systems. However, their distinctions were observed by the presence of BCT1, a HCO_3^- transporter. This transporter appeared to exist only in the moderate alkali-thermophiles, but absent in all moderate alkali-mesophiles and strong alkaliphiles.

From the organization of the CCM genes, genes encoding carboxysome shell proteins in all alkaliphiles, except P7 and SS, were found to arrange in a cluster, ccmKLMNO, consisting of ccmK1, ccmK2, ccmL, ccmM, ccmN and ccmO (Fig. 2). In addition, ccmK3 and ccmK4 were also found to be present in the 10 strains, but S2 and S3. The protein sequences shared significant moderate similarity with the reference sequences retrieved from the model organism. The observed maximal homology was around ~60–70% identity, with an E-value of ~10^{-50} (see Supplementary File). Of these, weak homologs (~40% similarity) were found only for CcmN sequences. Protein domain analysis showed that CcmK1-K4 and CcmO contained bacterial microcompartment (BMC) domain (Pfam00936), whereas CcmL (BMC-P) contained ethanolamine utilization (EutN) domain (Pfam03319). Regarding to CcmN, although low similarity was observed, multiple protein sequence alignment among all examined cyanobacteria with Synechococcus sp. PCC 7942 revealed two functionally conserved distinct regions at N- and C-terminals, which were separated by a poorly conserved linker. These results supported the functions of CcmKLMNO as carboxysome shell proteins in alkaliphilic cyanobacteria. Beside the shell proteins, the amino acid sequences of Rubisco subunits, RbcL and RbcS, were moderately conserved (60% identity with the reference sequences) in their sequences; this was

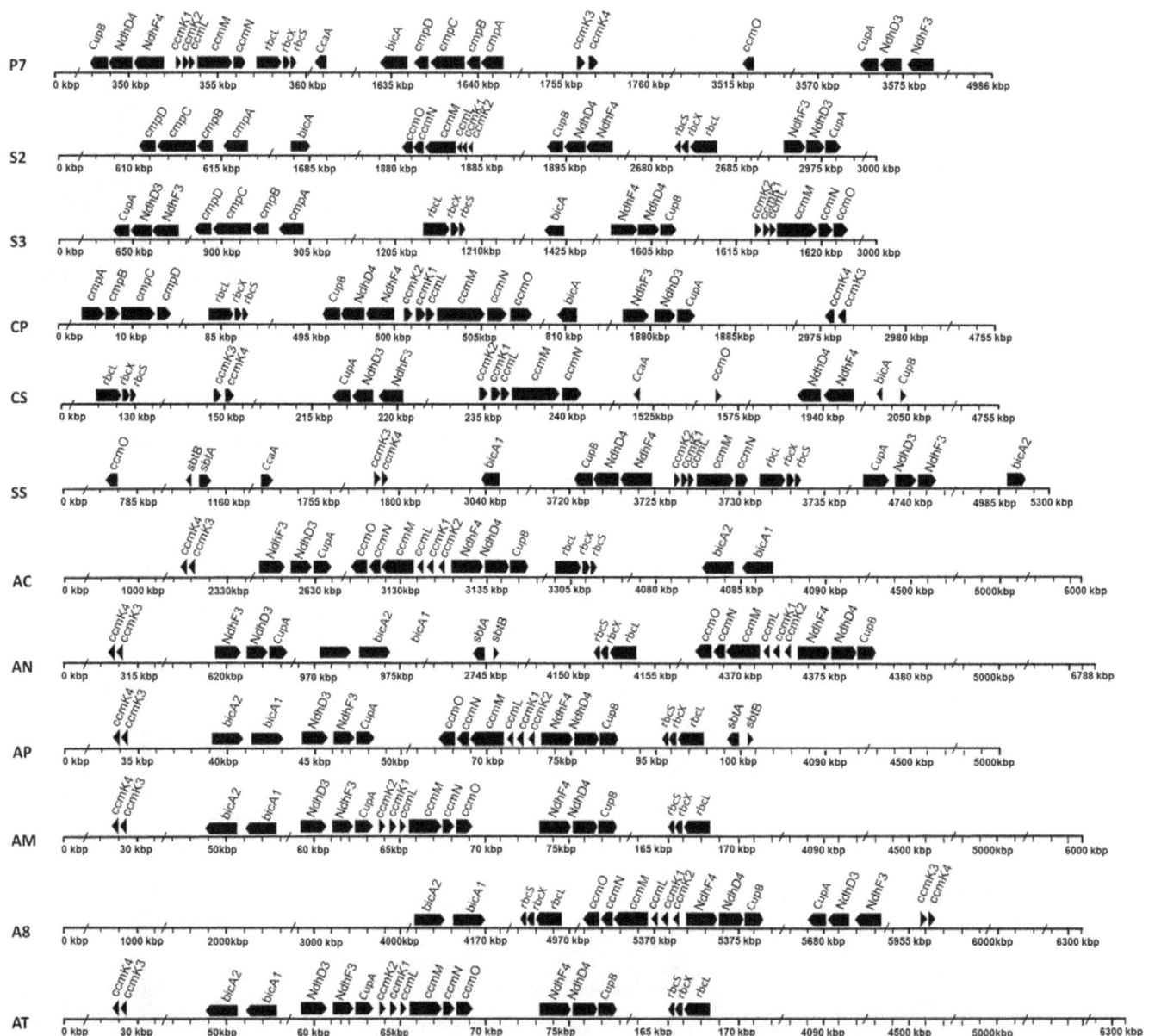

Fig. 2. Comparative genomic structure and gene organization of the CCM in all 12 alkaliphilic cyanobacteria. Solid arrow boxes indicate genes and the direction of transcription. The completed genome sequence was available for P7, S2, S3, and CS, while for other strains there is only permanent draft genome information.

compared with the assembly chaperone RbcX protein (~45% identity with the reference sequences). The *rbcLSX* gene clusters were found to appear in all investigated genomes, located up- and down-streams of the *ccmKLMNO* cluster as shown in Fig. 2.

Focusing on β-CAs, which are enclosed in the carboxysome (CcaA) or localized in periplasmic space of β-cyanobacteria (EcaB), moderate similarity of amino acid sequence (~60% identity with reference protein) was found for CcaA proteins in only three studied cyanobacteria, P7, CS, and SS (Table 2). EcaB orthologs were not detected in any of the studied organisms. However, CcmM proteins of all alkaliphilic strains were found to have the γ-CA-like domain at N-terminal region. We further searched for other recognized CA classes in all 12 alkaliphilic cyanobacteria by using the protein sequences of α-CA, EcaA (*all2929*) of

Anabaena sp. PCC 7120. The results showed no homologs sequences of EcaA in all 12 studied cyanobacteria. As a result, we concluded that all alkaliphilic species possessed γ-CA (CcmM), of which three moderate alkaliphiles contained additional β-CA (CcaA). To further evaluate a potential function of CcmM as an active CA, the comparative analysis of the γ-CA-like domain in N-terminal protein sequence was performed. The CcmM sequences from *Thermosynechococcus elongatus* BP-1 and *Nostoc* sp. PCC 7120 were included as functional γ-CA [73,74]. The CcmM from *S. elongatus* PCC 7942 and *Synechocystis* sp. PCC 6803 were also included as a non-functional γ-CA [28,75]. Fig. 4 shows the important amino acid residues in γ-CA-like domain of the 11 alkaliphilic strains, except SS, which were structurally similar to those of active CcmM in *T. elongatus* BP-1 [73] and *Nostoc* sp. PCC 7120 [74]. This result

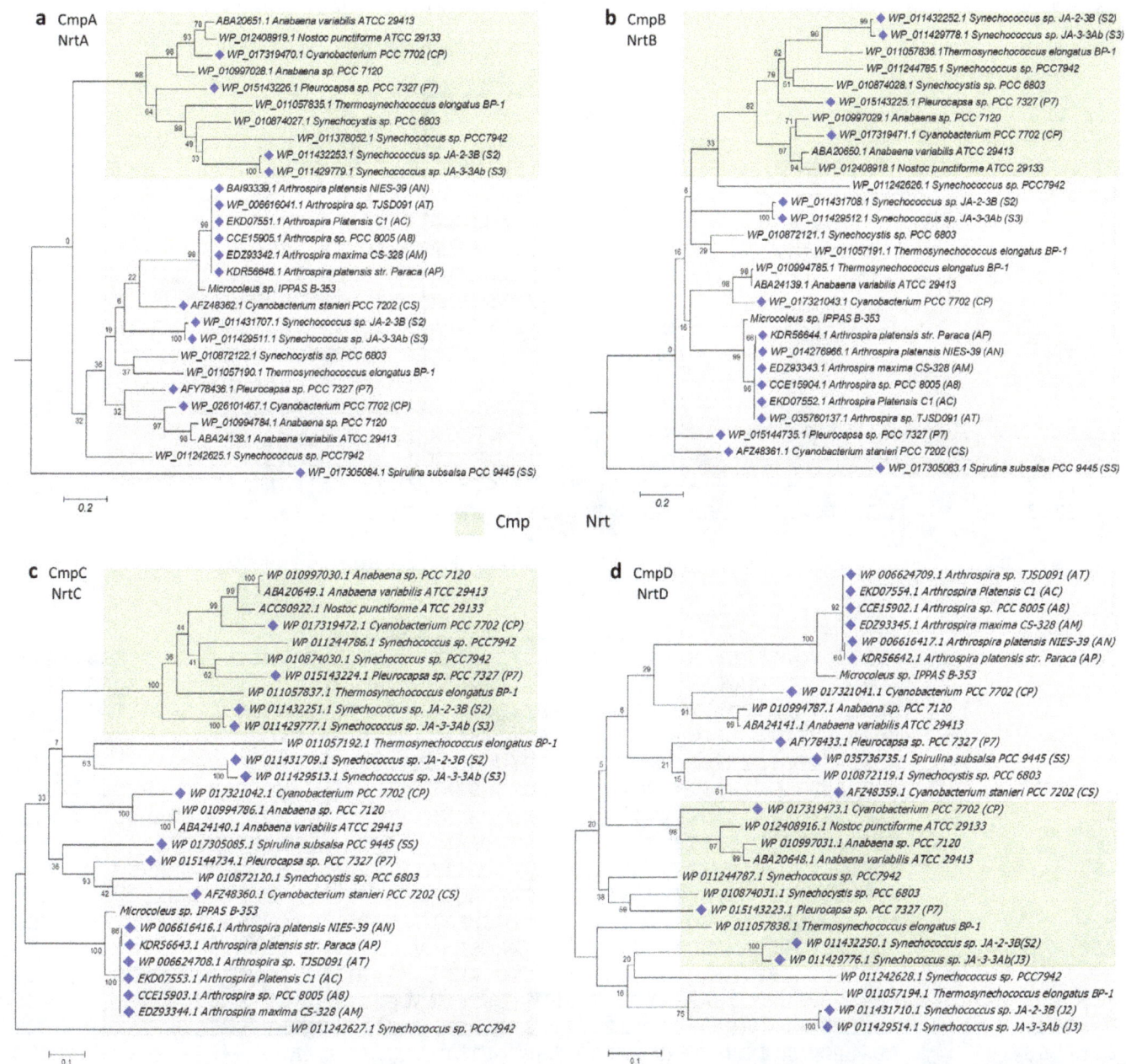

Fig. 3. Phylogenetic trees of the CmpABCD proteins and NrtABCD proteins of the selected alkaliphilic cyanobacteria. The outgroup cyanobacteria includes *Synechococcus* sp. PCC 7942, which has an experimental study of HCO₃⁻ transporter BCT1 and nitrite/nitrate transport system, NRT [12]. (a) Phylogenetic tree based on the CmpA and NrtA protein sequences. (b) Phylogenetic tree based on the CmpB and NrtB protein sequences. (c) Phylogenetic tree based on the CmpC and NrtC protein sequences. (d) Phylogenetic tree based on the CmpD and NrtD protein sequences. The alkaliphilic cyanobacteria are identified by the blue diamond, respectively. Cmp and Nrt families are highlighted in orange and green, respectively. Bootstrap values with 3000 replicates are shown at the nodes of the tree. The scale bars indicate the number of nucleotide substitutions per site.

implied that the CcmM proteins of such 11 species might potentially have CA activity when the carboxysomal β-CA, CcaA, was missing.

3.3. Comparative Analysis of CCM Components Among β-Cyanobacteria

Comparative analysis of CCM components among β-cyanobacteria, living in freshwater (pH ~7), marine (pH ~8.2), and alkaliphilic (pH 8.5–11) strains were performed. Fig. 5 shows different CCM components among the three groups. The overall compositions of CCM components in alkaliphilic cyanobacteria were more similar to the freshwater than the marine groups. The cyanobacteria inhabiting freshwater and alkaline ecological niches possessed both CO_2 uptake systems, NDH-1_3 and NDH-1_4, while most strains inhabiting marine habitats seemed to lack the NDH-1_3. Focusing on the HCO_3^- transport system, the results showed that marine and some alkaliphilic cyanobacteria consistently lacked the BCT1 type of the HCO_3^- transporter. In addition, the

freshwater β-cyanobacteria possessed the highest abundance of CAs, β-CA (CcaA and EcaB), α-CA (EcaA), and γ-CA (CcmM), while the alkaliphilic cyanobacteria were likely to possess only two conventional CAs, carboxysomal β-CA (CcaA) and γ-CA (CcmM). However, it should be noted that nine out of the twelve investigated alkaliphiles appeared to have only γ-CA (CcmM).

4. Discussion

Based on Rubisco phylogeny, all studied alkaliphilic cyanobacteria fall into β-cyanobacteria group (Fig. 1). It is obviously that the phylogeny of the studied alkaliphiles based on RbcL sequences can classify the cyanobacterial types; however, the tree is insufficient to elucidate the evolutionary relationship within the group based on cell morphology and habitats. Komarek et al. [76] previously performed a phylogenetic tree of 146 cyanobacterial OTUs using 31 conserved protein sequences

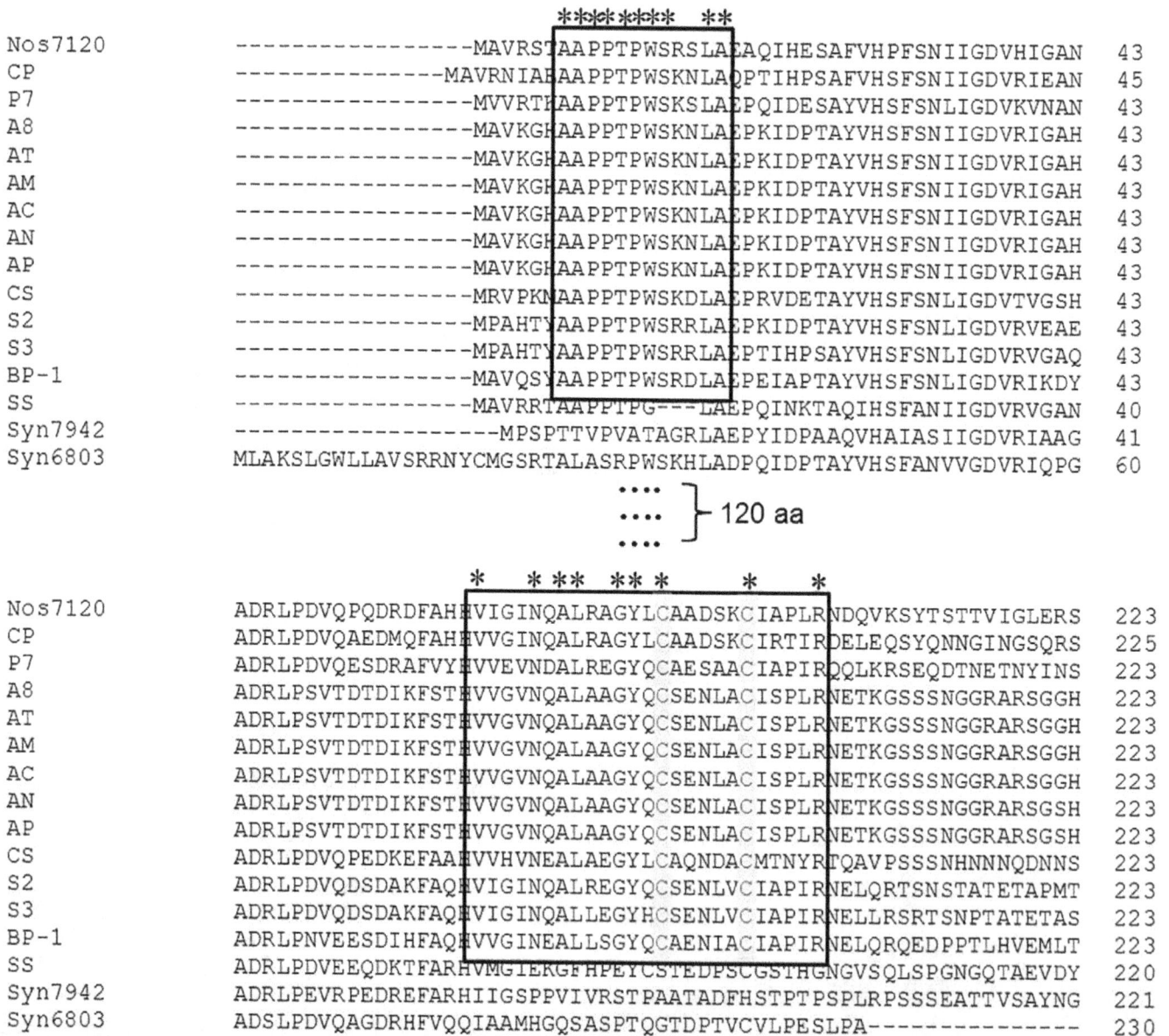

Fig. 4. Partial alignment of CcmM amino acid sequences of 12 studied alkaliphiles with those of *Synechococcus elongatus* PCC 7942 (Syn7942), *Synechocystis* sp. PCC 6803 (Syn6803), *Thermosynechococcus elongatus* BP-1 (BP-1), and *Nostoc* sp. PCC 7120 (Noc7120) (GenBank accession no. BAA16773.1, Q03513.1, NP_681734, and BAB72822.1, respectively). The sequence order is based on the alignment. Boxes represent conserved regions of the N-terminal domain of CcmM that are assumed to be necessary for CA activity, according to Pena et al. [73]. Shaded cysteine amino acids showed essential residues participating in the disulfide bond in the C-termini of active CcmM protein. *Asterisks* indicate conserved amino acids inside such regions.

and reported that the tree could not be clustered based on their morphology. Thus, phylogenetic analysis may not be an appropriate technique to unveil evolutionary relationship of cyanobacterial morphology and environments.

The presence of the *ccmKLMNO* cluster (Fig. 2) in all the 12 studied strains indicates the genes conserved in the *ccm* cluster of β-carboxysomes. Our finding suggests that all the investigated alkaliphilic cyanobacteria possess complete standard genes, which are essential for carboxysome formation. In addition, since CcmK3 and CcmK4 were considered as an accessory protein improving the functionality of the shell [26], the 10 studied strains found to possess CcmK1–4 would have a better shell protein function than the others. However, there is no obvious correlation between numbers of *ccmK* genes and environment niche (moderately to strongly alkaline habitat) of the examined strains.

Two systems of CO_2 uptake, NDH-1_3 and NDH-1_4 complexes, were identified in all 12 analyzed strains (Fig. 5). Recently, Kupriyanova et al. [48] confirmed the presence of NDH-1_3 and NDH-1_4 in an alkaliphilic cyanobacterium *Microcoleus* sp. IPPAS B-353 and showed that genes corresponding to the NDH-1_3 were transcribed and probably constitutively expressed. Both CO_2 uptake systems were also observed in freshwater β-cyanobacteria and 20 strains of *M. aeruginosa* living in brackish waters and eutrophic lakes [77]. In contrast, the absence of NDH-1_3 and/or NDH-1_4 was reported in the oceanic α-cyanobacteria, *Prochlorococcus* species [36], and the marine β-cyanobacteria, *Trichodesmium erythraeum* species [3]. Thus, existence of NDH-1_3 and NDH-1_4 is apparently related to environments with varying CO_2 availability. This is due to the distinct property of each complex in that NDH-1_3 has a higher substrate affinity for CO_2 than NDH-1_4 [16]. The presence of both complexes in all 12 strains of alkaliphilic cyanobacterial seems to have an essential role in the survival and maintenance under CO_2 fluctuation, particularly in alkaline environments (i.e. hot spring and soda lake).

Our finding for HCO_3^- transport systems, BicA and SbtA, revealed that several alkaliphilic cyanobacteria strains have only one of the two transporters, preferably BicA. This may be due to the difference in their affinity for bicarbonate. BicA is a low-affinity and high flux rate bicarbonate transport system ($K_m = 70–350\ \mu M$), while SbtA is high-affinity ($K_m < 5\ \mu M$) HCO_3^- transporter [17]. Possibly, the high-affinity SbtA is not necessary in most alkaliphilic strains typically inhabiting HCO_3^- rich environment. If so, the organisms would most likely possess BicA rather than SbtA. Meanwhile, the presence of high-affinity transporter SbtA in SS, AN, and AP may, though indirectly, indicates that these organisms are able to face low concentrations of exogenous C_i. Therefore, the presence of both BicA and SbtA in such strains might give a selective advantage in HCO_3^- uptake and allow cell growth, enabling them to adapt in response to different C_i concentrations. These cyanobacteria might have a greater ability to maintain their higher growth rate than the other alkaliphiles, particularly when they face a wide dynamic range in HCO_3^- availability, i.e. during cyanobacterial bloom, by utilizing BicA at high HCO_3^- and SbtA at low HCO_3^- conditions.

In regards to BCT1, it has been reported as an inducible transporter under C_i limitation [12,78] and high-light stress [79]. This transporter

β-Cyanobacteria

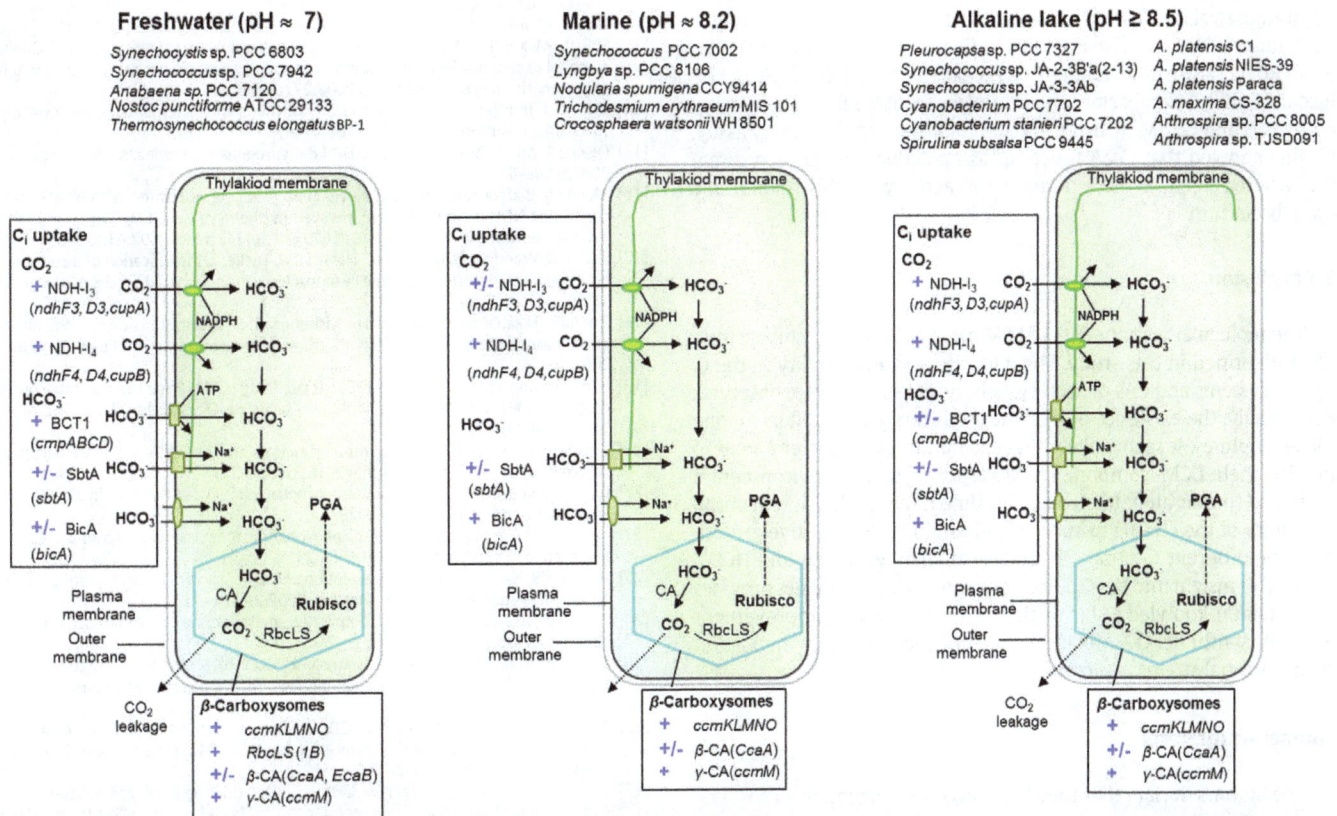

Freshwater (pH ≈ 7)

Synechocystis sp. PCC 6803
Synechococcus sp. PCC 7942
Anabaena sp. PCC 7120
Nostoc punctiforme ATCC 29133
Thermosynechococcus elongatus BP-1

Marine (pH ≈ 8.2)

Synechococcus PCC 7002
Lyngbya sp. PCC 8106
Nodularia spumigena CCY9414
Trichodesmium erythraeum MIS 101
Crocosphaera watsonii WH 8501

Alkaline lake (pH ≥ 8.5)

Pleurocapsa sp. PCC 7327   *A. platensis* C1
Synechococcus sp. JA-2-3B'a(2-13)   *A. platensis* NIES-39
Synechococcus sp. JA-3-3Ab   *A. platensis* Paraca
Cyanobacterium PCC 7702   *A. maxima* CS-328
Cyanobacterium stanieri PCC 7202   *Arthrospira* sp. PCC 8005
Spirulina subsalsa PCC 9445   *Arthrospira* sp. TJSD091

Fig. 5. Diversity in characteristic components of the cyanobacterial CCM living in three different pH environments; freshwater (pH ~7), marine (pH ~8.2), and alkaline (pH > 8.5). The scheme is based on the literature data and is depicted for β-cyanobacteria. The species that were used to derive the groups are shown on the figure. CCM components of freshwater and marine cyanobacteria are adapted from [3]. CCM components of high alkaliphilic cyanobacterial type were identified in this study. + and ± indicate that the particular component is 'always present' and 'sometimes present', respectively. Designation: NDH-1_4, low-affinity CO_2 uptake system NDH-1_4 complex; NDH-1_3, low CO_2-inducible high-affinity CO_2 uptake system NDH-1_3 complex; BCT1, ATP-binding cassette (ABC)-type high-affinity HCO_3^- transporter; SbtA, high-affinity sodium-dependent HCO_3^- symporter; BicA, SulP-type low-affinity sodium dependent HCO_3^- transporter; CA, carbonic anhydrase; Rubisco, ribulose-1,5-bisphosphate carboxylase/oxygenase.

is the medium substrate affinity class ($K_m = 10–15\ \mu M$) [18]. Comparative genome analysis showed that this transporter is found only in freshwater β-cyanobacteria and four strains of the moderate alkali-thermophiles, but not in other moderate alkali-mesophiles, strong alkaliphiles, and the marine cyanobacteria (Table 2 and Fig. 5). Since the ATP binding cassette transporter BCT1 helps facilitating bicarbonate transportation, its presence is crucial for cyanobacteria living in freshwater where the contents of inorganic carbon and ions are extremely low. However, the reason why BCT1 is required in all the moderate alkali-thermophiles is not obvious. It is speculated that high temperature might limit the solubility of inorganic carbon and ions in such environment. This is supported by Kamennaya et al. [80] who reported that some inorganic carbons can form insoluble carbonate, of which its solubility is decreased with increasing temperature. The lack of BCT1 in all of the alkali-mesophiles and highly-alkaliphiles indicates the unnecessity of this transporter and the adaptation of such cyanobacterial groups residing in the saline alkaline environments with enriched carbonate ion.

Finally, diversity of CAs was observed among freshwater, marine, and alkaliphilic β-cyanobacteria. While carboxysomal γ-CA, CcmM, was observed in all cyanobacteria, β-CA, CcaA, was only found in some cyanobacteria (Fig. 5). Nine out of the twelve investigated alkaliphilic strains were found with the absence of CcaA. The reason why the strains living in the high pH conditions tend to lose CcaA is still unclear. On the contrary, the existence of CcmM in all studied strains is not surprising given the role it plays in β-carboxysomes. The evolution of CcaA and CcmM within the carboxysomes of these alkaliphiles remains to be evaluated. Thus far, it has been believed that CcmM not only functions as a shell protein for carboxysome but also as CA activity for the strains lacking CcaA. Peña et al. in 2010 [73] reported the CA activity in *Thermosynechococcus elongatus* BP-1 possessing only CcmM, and later de Araujo et al. in 2014 [74] suggested that activity of γ-CA might be regulated by RbcS-like domains in CcmM. Recently, Kupriyanova et al. 2016 [48] has attempted to reveal the function of CA in the haloalkaliphilic cyanobacterium *Microcoleus* sp. IPPAS B-353 possessing both CcaA and CcmM by using Western blotting and CA activity assay. Results showed that CcaA functions as an active non-carboxysomal CA, whereas CcmM did not have CA activity in this alkaliphilic cyanobacterium.

5. Conclusion

The molecular components of CCM in 12 alkaliphilic cyanobacteria were identified in this study. The diversity and adaptability in the C_i uptake systems and CAs of such cyanobacterial species were observed. Remarkably, the existence of HCO_3^- transporters greatly differs among the alkaliphiles. It seems likely that alkaliphilic cyanobacteria tend to modify their CCM components in response to the environmental influence (moderately to strongly alkaline habitat). These reflect the capability of the strains to survive and establish competitive growth by using different C_i uptake strategies at changes of CO_2 and HCO_3^- levels. This insight into the CCM components of the alkaliphiles provides fundamental knowledge for further research towards improvement of photosynthetic CO_2 fixation in some economically important cyanobacterial strains and crops.

Competing Interests

The authors declare that they have no competing interests.

Acknowledgements

The authors would like to thank Tayvich Vorapreeda for his suggestion in sequence analysis and all members of the Systems Biology and Bioinformatics (SBI) research group at King Mongkut's University of Technology Thonburi for their assistance. AK was supported by grant P-11-01089, which was provided by the National Center for Genetic Engineering and Biotechnology (BIOTEC), NSTDA, Thailand.

References

[1] Badger MR, Price GD, Long BM, Woodger FJ. The environmental plasticity and ecological genomics of the cyanobacterial CO_2 concentrating mechanism. J Exp Bot 2006;57:249–65.

[2] Badger MR, Andrews TJ, Whitney SM, Ludwig M, Yellowlees DC, Leggat W, et al. The diversity and coevolution of Rubisco, plastids, pyrenoids, and chloroplast-based CO_2-concentrating mechanisms in algae. Can J Bot 1998;76:1052–71.

[3] Price GD, Badger MR, Woodger FJ, Long BM. Advances in understanding the cyanobacterial CO_2-concentrating-mechanism (CCM): functional components, C_i transporters, diversity, genetic regulation and prospects for engineering into plants. J Exp Bot 2008;59:1441–61.

[4] Burnap RL, Hagemann M, Kaplan A. Regulation of CO_2 concentrating mechanism in cyanobacteria. Life (Basel) 2015;5:348–71.

[5] Hagemann M, Kern R, Maurino VG, Hanson DT, Weber AP, Sage RF, et al. Evolution of photorespiration from cyanobacteria to land plants, considering protein phylogenies and acquisition of carbon concentrating mechanisms. J Exp Bot 2016;67:2963–76.

[6] Mangan N, Brenner M. Systems analysis of the CO_2 concentrating mechanism in cyanobacteria. Elife 2014;10(7554).

[7] Gaudana SB, Zarzycki J, Moparthi VK, Kerfeld CA. Bioinformatic analysis of the distribution of inorganic carbon transporters and prospective targets for bioengineering to increase C_i uptake by cyanobacteria. Photosynth Res 2015;126:99–109.

[8] Hanson MR, Lin MT, Carmo-Silva AE, Parry MA. Towards engineering carboxysomes into C_3 plants. Plant J 2016;87:38–50.

[9] Long BM, Rae BD, Rolland V, Forster B, Price GD. Cyanobacterial CO_2-concentrating mechanism components: function and prospects for plant metabolic engineering. Curr Opin Plant Biol 2016;31:1–8.

[10] Shibata M, Ohkawa H, Kaneko T, Fukuzawa H, Tabata S, Kaplan A, et al. Distinct constitutive and low-CO_2-induced CO_2 uptake systems in cyanobacteria: genes involved and their phylogenetic relationship with homologous genes in other organisms. Proc Natl Acad Sci 2001;98:11789–94.

[11] Ma W, Ogawa T. Oxygenic photosynthesis-specific subunits of cyanobacterial NADPH dehydrogenases. IUBMB Life 2015;67:3–8.

[12] Omata T, Price GD, Badger MR, Okamura M, Gohta S, Ogawa T. Identification of an ATP-binding cassette transporter involved in bicarbonate uptake in the cyanobacterium *Synechococcus* sp. strain PCC 7942. Proc Natl Acad Sci U S A 1999;96:13571–6.

[13] Shibata M, Katoh H, Sonoda M, Ohkawa H, Shimoyama M, Fukuzawa H, et al. Genes essential to sodium-dependent bicarbonate transport in cyanobacteria: function and phylogenetic analysis. J Biol Chem 2002;277:18658–64.

[14] Price GD. Inorganic carbon transporters of the cyanobacterial CO_2 concentrating mechanism. Photosynth Res 2011;109:47–57.

[15] Ogawa T, Mi H. Cyanobacterial NADPH dehydrogenase complexes. Photosynth Res 2007;93:69–77.

[16] Maeda S, Badger MR, Price GD. Novel gene products associated with NdhD3/D4-containing NDH-1 complexes are involved in photosynthetic CO_2 hydration in the cyanobacterium, *Synechococcus* sp. PCC7942. Mol Microbiol 2002;43:425–35.

[17] Price GD, Woodger FJ, Badger MR, Howitt SM, Tucker L. Identification of a SulP-type bicarbonate transporter in marine cyanobacteria. Proc Natl Acad Sci U S A 2004;101:18228–33.

[18] Omata T, Takahashi Y, Yamaguchi O, Nishimura T. Structure, function and regulation of the cyanobacterial high-affinity bicarbonate transporter, BCT1. Funct Plant Biol 2002;29.

[19] Rae BD, Long BM, Whitehead LF, Forster B, Badger MR, Price GD. Cyanobacterial carboxysomes: microcompartments that facilitate CO_2 fixation. J Mol Microbiol Biotechnol 2013;23:300–7.

[20] Kerfeld CA, Melnicki MR. Assembly, function and evolution of cyanobacterial carboxysomes. Curr Opin Plant Biol 2016;31:66–75.

[21] Espie GS, Kimber MS. Carboxysomes: cyanobacterial RubisCO comes in small packages. Photosynth Res 2011;109:7–20.

[22] Tabita FR. Microbial ribulose 1,5-bisphosphate carboxylase/oxygenase: a different perspective. Photosynth Res 1999;60:1–28.

[23] Badger MR, Hanson D, Price GD. Evolution and diversity of CO_2 concentrating mechanisms in cyanobacteria. Funct Plant Biol 2002;29:161–73.

[24] Kinney JN, Axen SD, Kerfeld CA. Comparative analysis of carboxysome shell proteins. Photosynth Res 2011;109:21–32.

[25] Samborska B, Kimber MS. A dodecameric CcmK2 structure suggests beta-carboxysomal shell facets have a double-layered organization. Structure 2012;20:1353–62.

[26] Rae BD, Long BM, Badger MR, Price GD. Structural determinants of the outer shell of beta-carboxysomes in *Synechococcus elongatus* PCC 7942: roles for CcmK2, K3-K4, CcmO, and CcmL. PLoS One 2012;7:e43871.

[27] Long BM, Tucker L, Badger MR, Price GD. Functional cyanobacterial beta-carboxysomes have an absolute requirement for both long and short forms of the CcmM protein. Plant Physiol 2010;153:285–93.

[28] Cot SS, So AK, Espie GS. A multiprotein bicarbonate dehydration complex essential to carboxysome function in cyanobacteria. J Bacteriol 2008;190:936–45.

[29] Cannon GC, Heinhorst S, Kerfeld CA. Carboxysomal carbonic anhydrases: structure and role in microbial CO_2 fixation. Biochim Biophys Acta 1804;2010:382–92.

[30] So AK, Espie GS. Cloning, characterization and expression of carbonic anhydrase from the cyanobacterium *Synechocystis* PCC6803. Plant Mol Biol 1998;37:205–15.

[31] Andersson I, Taylor TC. Structural framework for catalysis and regulation in ribulose-1,5-bisphosphate carboxylase/oxygenase. Arch Biochem Biophys 2003;414:130–40.

[32] Liu C, Young AL, Starling-Windhof A, Bracher A, Saschenbrecker S, Rao BV, et al. Coupled chaperone action in folding and assembly of hexadecameric Rubisco. Nature 2010;463:197–202.

[33] Saschenbrecker S, Bracher A, Rao KV, Rao BV, Hartl FU, Hayer-Hartl M. Structure and function of RbcX, an assembly chaperone for hexadecameric Rubisco. Cell 2007;129:1189–200.

[34] Bracher A, Starling-Windhof A, Hartl FU, Hayer-Hartl M. Crystal structure of a chaperone-bound assembly intermediate of form I Rubisco. Nat Struct Mol Biol 2011;18:875–80.

[35] Tabita FR. Rubisco: the enzyme that keeps on giving. Cell 2007;129:1039–40.

[36] Badger MR, Price GD. CO2 concentrating mechanisms in cyanobacteria: molecular components, their diversity and evolution. J Exp Bot 2003;54:609–22.

[37] Xu M, Bernát G, Singh A, Mi H, Rögner M, Pakrasi HB, et al. Properties of mutants of Synechocystis sp. strain PCC 6803 lacking inorganic carbon sequestration systems. Plant Cell Physiol 2008;49:1672–7.

[38] Zhang S, Spann KW, Frankel LK, Moroney JV, Bricke TM. Identification of two genes, sll0804 and slr1306, as putative components of the CO2-concentrating mechanism in the cyanobacterium Synechocystis sp. strain PCC 6803. J Bacteriol 2008;190:8234–7.

[39] Burnap RL, Nambudiri R, Holland S. Regulation of the carbon-concentrating mechanism in the cyanobacterium Synechocystis sp. PCC6803 in response to changing light intensity and inorganic carbon availability. Photosynth Res 2013;118:115–24.

[40] Kaplan A, Badger MR, Berry JA. Photosynthesis and the intracellular inorganic carbon pool in the bluegreen alga Anabaena variabilis: response to external CO2 concentration. Planta 1980;149:219–26.

[41] Long BM, Badger MR, Whitney SM, Price GD. Analysis of carboxysomes from Synechococcus PCC7942 reveals multiple Rubisco complexes with carboxysomal proteins CcmM and CcaA. J Biol Chem 2007;282:29323–35.

[42] Kalff J. Limnology: inland water ecosystems. San Francisco, United States: Pearson education (US); 2002 218–22.

[43] Kupriyanova EV, Samylina OS. CO2-concentrating mechanism and its traits in haloalkaliphilic cyanobacteria. Microbiol Mol Biol Rev 2015;84:112–24.

[44] Horikoshi K. Alkaliphiles: some applications of their products for biotechnology. Microbiol Mol Biol Rev 1999;63:735–50 [table of contents].

[45] Diaz MM, MaBerly SC. Carbon-concentrating mechanisms in acidophilic algae. Phycologia 2009;48:77–85.

[46] Dudoladova MV, Kupriyanova EV, Markelova AG, Sinetova MP, Allakhverdiev SI, Pronina NA. The thylakoid carbonic anhydrase associated with photosystem II is the component of inorganic carbon accumulating system in cells of halo- and alkaliphilic cyanobacterium Rhabdoderma lineare. Biochim Biophys Acta 1767; 2007:616–23.

[47] Mikhodyuk OS, Zavarzin GA, Ivanovsky RN. Transport systems for carbonate in the extremely natronophilic cyanobacterium Euhalothece sp. Microbiology 2008;77:412–8.

[48] Kupriyanova EV, Cho SM, Park YI, Pronina NA, Los DA. The complete genome of a cyanobacterium from a soda lake reveals the presence of the components of CO2-concentrating mechanism. Photosynth Res 2016.

[49] Shih PM, Wu D, Latifi A, Axen SD, Fewer DP, Talla E, et al. Improving the coverage of the cyanobacterial phylum using diversity-driven genome sequencing. Proc Natl Acad Sci U S A 2013;110:1053–8.

[50] Bhaya D, Grossman AR, Steunou AS, Khuri N, Cohan FM, Hamamura N, et al. Population level functional diversity in a microbial community revealed by comparative genomic and metagenomic analyses. ISME J 2007;1:703–13.

[51] Cheevadhanarak S, Paithoonrangsarid K, Prommeenate P, Kaewngam W, Musigkain A, Tragoonrung S, et al. Draft genome sequence of Arthrospira platensis C1 (PCC9438). Stand Genomic Sci 2012;6:43–53.

[52] Fujisawa T, Narikawa R, Okamoto S, Ehira S, Yoshimura H, Suzuki I, et al. Genomic structure of an economically important cyanobacterium, Arthrospira (Spirulina) platensis NIES-39. DNA Res 2010;17:85–103.

[53] Lefort F, Calmin G, Crovadore J, Falquet J, Hurni JP, Osteras M, et al. Whole-genome shotgun sequence of Arthrospira platensis strain Paraca, a cultivated and edible cyanobacterium. Genome Announc 2014;2.

[54] Carrieri D, Ananyev G, Lenz O, Bryant DA, Dismukes GC. Contribution of a sodium ion gradient to energy conservation during fermentation in the cyanobacterium Arthrospira (Spirulina) maxima CS-328. Appl Environ Microbiol 2011;77:7185 94.

[55] Janssen PJ, Morin N, Mergeay M, Leroy B, Wattiez R, Vallaeys T, et al. Genome sequence of the edible cyanobacterium Arthrospira sp. PCC 8005. J Bacteriol 2010; 192:2465–6.

[56] Dong S, Chen J, Wang S, Wu Y, Hou H, Li M, et al. Draft genome sequence of cyanobacteria Arthrospira sp. TJSD091 isolated from seaside wetland. Mar Genomics 2015;24:197–8.

[57] Altschul SF, Gish W, Miller W, Myers EW, Lipman DJ. Basic local alignment search tool. J Mol Biol 1990;215:403–10.

[58] Pearson WR. An introduction to sequence similarity ("homology") searching. Current protocols in bioinformatics/editoral board, Andreas D. Baxevanis... [et al.]; 2013[Chapter 3:Unit 3 1].

[59] UniProt C. UniProt: a hub for protein information. Nucleic Acids Res 2015;43:D204–12.

[60] Finn RD, Bateman A, Clements J, Coggill P, Eberhardt RY, Eddy SR, et al. Pfam: the protein families database. Nucleic Acids Res 2014;42:D222–30.

[61] Penn O, Privman E, Ashkenazy H, Landan G, Graur D, Pupko T. GUIDANCE: a web server for assessing alignment confidence scores. Nucleic Acids Res 2010;38:W23–8.

[62] Petkau A, Stuart-Edwards M, Stothard P, Van Domselaar G. Interactive microbial genome visualization with GView. Bioinformatics 2010;26:3125–6.

[63] Edgar RC. MUSCLE: a multiple sequence alignment method with reduced time and space complexity. BMC Bioinf 2004;5:113.

[64] Edgar RC. MUSCLE: multiple sequence alignment with high accuracy and high throughput. Nucleic Acids Res 2004;32:1792–7.

[65] Felsenstein J. Evolutionary trees from DNA sequences: a maximum likelihood approach. J Mol Evol 1981;17:368–76.

[66] Hall BG. Building phylogenetic trees from molecular data with MEGA. Mol Biol Evol 2013;30:1229–35.

[67] Felsenstein J. Confidence limits on phylogenies: an approach using the bootstrap. Evolution 1985;39:783–91.

[68] Vonshak A. Spirulina platensis (Arthrospira): physiology, cell-biology and biotechnology. London: Taylor & Francis; 1997.

[69] Dadheech PK, Glockner G, Casper P, Kotut K, Mazzoni CJ, Mbedi S, et al. Cyanobacterial diversity in the hot spring, pelagic and benthic habitats of a tropical soda lake. FEMS Microbiol Ecol 2013;85:389–401.

[70] Tomitani A, Knoll AH, Cavanaugh CM, Ohno T. The evolutionary diversification of cyanobacteria: molecular-phylogenetic and paleontological perspectives. Proc Natl Acad Sci U S A 2006;103:5442–7.

[71] Du J, Forster B, Rourke L, Howitt SM, Price GD. Characterisation of cyanobacterial bicarbonate transporters in E. coli shows that SbtA homologs are functional in this heterologous expression system. PLoS One 2014;9:e115905.

[72] Omata T. Structure, function and regulation of the nitrate transport system of the cyanobacterium Synechococcus sp. PCC7942. Plant Cell Physiol 1995;36:207–13.

[73] Pena KL, Castel SE, de Araujo C, Espie GS, Kimber MS. Structural basis of the oxidative activation of the carboxysomal gamma-carbonic anhydrase, CcmM. Proc Natl Acad Sci U S A 2010;107:2455–60.

[74] de Araujo C, Arefeen D, Tadesse Y, Long BM, Price GD, Rowlett RS, et al. Identification and characterization of a carboxysomal gamma-carbonic anhydrase from the cyanobacterium Nostoc sp. PCC 7120. Photosynth Res 2014;121:135–50.

[75] So AKC, Espie GS. Cyanobacterial carbonic anhydrases. Can J Bot 2005;83:721–34.

[76] Komarek J, Kastovsky J, Mares J, Johansen JR. Taxonomic classification of cyanoprokaryotes (cyanobacterial genera) 2014, using a polyphasic approach. Preslia 2014;86:295–335.

[77] Sandrini G, Matthijs HCP, Verspagen JMH, Muyzer G, Huisman J. Genetic diversity of inorganic carbon uptake systems causes variation in CO2 response of the cyanobacterium Microcystis. ISME J 2014;8:589–600.

[78] Omata T, Ogawa T. Biosynthesis of a 42 kDa polypeptide in the cytoplasmic membrane of the cyanobacterium Anacystis nidulans strain-R2 during adaptation to low CO2 concentration. Plant Physiol 1986;80:525–30.

[79] Reddy KJ, Masamoto K, Sherman DM, Sherman LA. DNA sequence and regulation of the gene (cbpA) encoding the 42 kilodalton cytoplasmic membrane carotenoprotein of the cyanobacterium Synechococcus sp. strain PCC7942. J Bacteriol 1989;171:3486–93.

[80] Kamennaya NA, Ajo-Franklin Caroline M, Northen T, Jansson C. Cyanobacteria as biocatalysts for carbonate mineralization. Minerals 2012;2:338–64.

11

Molecular Mechanism of Binding between 17β-Estradiol and DNA

Tamsyn A. Hilder [a,b,*], Justin M. Hodgkiss [a,c,**]

[a] School of Chemical and Physical Sciences, Victoria University of Wellington, Wellington 6040, New Zealand
[b] Computational Biophysics Group, Research School of Biology, Canberra, ACT 0200, Australia
[c] The MacDiarmid Institute of Advanced Materials and Nanotechnology, New Zealand

ARTICLE INFO

Keywords:
17β-estradiol
Estrogen response element
Molecular dynamics
Intercalation

ABSTRACT

Although 17β-estradiol (E2) is a natural molecule involved in the endocrine system, its widespread use in various applications has resulted in its accumulation in the environment and its classification as an endocrine-disrupting molecule. These molecules can interfere with the hormonal system, and have been linked to various adverse effects such as the proliferation of breast cancer. It has been proposed that E2 could contribute to breast cancer by the induction of DNA damage. Mass spectrometry has demonstrated that E2 can bind to DNA but the mechanism by which E2 interacts with DNA has yet to be elucidated. Using all-atom molecular dynamics simulations, we demonstrate that E2 intercalates (inserts between two successive DNA base pairs) in DNA at the location specific to estrogen receptor binding, known as the estrogen response element (ERE), and to other random sequences of DNA. Our results suggest that excess E2 has the potential to disrupt processes in the body which rely on binding to DNA, such as the binding of the estrogen receptor to the ERE and the activity of enzymes that bind DNA, and could lead to DNA damage.

1. Introduction

17β-estradiol (or E2) is a natural steroidal hormone and is also commonly used in therapeutics such as postmenopausal estrogen replacement therapy and the treatment of Alzheimer's disease. Unfortunately, it has also become one of the most widely encountered endocrine-disrupting molecules in the environment [1]. Endocrine-disrupting molecules, either natural or synthetic, can interfere with the hormonal system. For example, they have the potential to cause cancerous tumours, birth defects, and developmental disorders in humans, while low concentrations (ng/L) of natural and synthetic estrogen hormones have been shown to have a harmful effect on fish [2]. Unfortunately, their widespread use in applications such as therapeutics and plastics has resulted in the accumulation of endocrine disrupting molecules in groundwater, rivers and lakes. Environmental exposure to endocrine-disrupting molecules such as E2 has the potential to cause adverse effects to the ecosystem and to humans [2–5], particularly through their presence in drinking water.

In the presence of normal levels of E2, the consensus response mechanism involves E2 first binding to an estrogen receptor (ER) protein. Then, this E2-ER complex forms a dimer and binds to the estrogen response element (ERE) on the DNA strand to initiate a hormone response. This ERE is a palindromic DNA sequence, and similar sequences have been identified in numerous sequences involving estrogen action such as oxytocin [6]. The E2-ER complex measures both the spacing and helical repeat of it's ERE, thus greatly increasing the specificity of the interaction [7]. A disruption to this spacing and the helical repeat may be sufficient to disturb this very delicate conformational equilibrium and cause unwanted side effects.

In the presence of elevated levels of E2, it has been suggested that E2 can directly interact with the ERE and interfere with normal signalling. Estrogen signalling drives cell proliferation in 60–70% of breast cancers that express the estrogen receptor [8], and anti-estrogen therapy is prescribed to the majority of these patients to prevent breast cancer recurrence. Estrogen exposure is now widely accepted as a risk factor in breast cancer development, but the mechanisms through which estrogens induce breast carcinogenesis have not been fully elucidated [9]. Typically, research has demonstrated that endocrine disrupting molecules, such as bisphenol-A inhibit the hormone binding pocket on the estrogen receptor [10–12]. However, this may not be the only mechanism by which endocrine disruptors cause harm. Caldon [8] state that

---

* Correspondence to: T.A. Hilder, School of Chemical and Physical Sciences, Victoria University of Wellington, Wellington 6040, New Zealand.
** Correspondence to: J.M. Hodgkiss, The MacDiarmid Institute of Advanced Materials and Nanotechnology, New Zealand.
E-mail addresses: Tamsyn.Hilder@vuw.ac.nz (T.A. Hilder), Justin.Hodgkiss@vuw.ac.nz (J.M. Hodgkiss).

high levels of estrogen are a major risk for breast cancer, and that one mechanism by which estrogen could contribute to breast cancer is via the induction of DNA damage. Using mass spectrometry, Heger et al. [9] found that E2 binds to DNA and leads to destabilization of hydrogen bonds between nitrogenous bases of DNA strands resulting in a decrease of their melting temperature. Their results revealed that E2 forms non-covalent physical complexes with DNA, and they suggest that these interactions could trigger mutations leading to unwanted side effects [9]. DNA damage as a result of exposure to E2 has also been demonstrated in barnacle larvae [13] and rodents [14], for example. In rodents, this DNA damage ultimately led to tumours in estrogen-responsive tissues [14]. Zhang et al. [15] demonstrated experimentally that bisphenol-A intercalates between adjacent base pairs of DNA.

Despite the risks associated with estrogen exposure the exact mechanisms by which estrogen contributes to the initiation and progression of breast cancer remains elusive [8]. However, a major mechanism is potentially the induction of DNA damage as estrogen treatment leads to double stranded DNA breaks and genomic instability [8]. We use all-atom molecular dynamics (MD) simulations to elucidate how E2 binds to DNA, and to obtain critical understanding of the effect of this binding on the DNA structure. We demonstrate that excess E2 could disrupt estrogenic processes in our bodies by binding directly to the ERE.

2. Material and Methods

2.1. System Setup

The initial coordinates for the ER DNA-binding domain (DBD) were taken from the Protein Data Bank (PDB) with the entry 1HCQ, determined to 2.4 Å resolution [7]. The crystal structure contains the ER symmetric dimer (ER-α, and ER-β) bound to DNA at its palindromic binding site, as shown in Fig. 1. The palindromic binding site consists of two 6 base pair (bp) consensus half sites with 3 intervening bps, illustrated in Table 1. This sequence is referred to as erDNA.

2.2. Molecular Docking

We used the rigid body docking program ZDOCK 3.0.1 [16] to generate a set of conformations of E2 bound to erDNA in the absence of the ER protein. The erDNA sequence in Table 1 was isolated from the ER-DNA complex and the double stranded erDNA was used as input into the

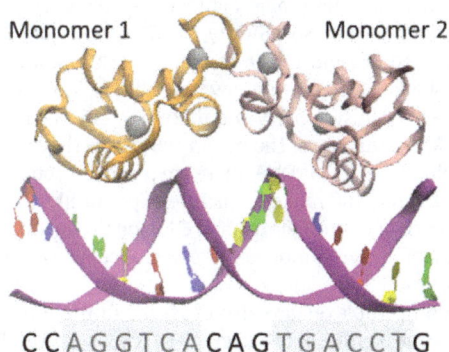

Fig. 1. ER dimer-DNA complex. Atomic coordinates are taken from Protein Data Bank (PDB) entry 1HCQ (12) and image is created using VMD (18). Each monomer is represented as a ribbon (orange and pink, respectively). Grey spheres represent the two zinc ions associated with each monomer. The DNA sequence of one strand is given below the ER-DNA complex. The two half-sites are highlighted in grey. Colours are used to help identify DNA backbone (purple), and bases adenosine (blue), guanine (green), thymine (yellow) and cytosine (red). Note that for clarity bases are only shown on the strand of DNA given in the sequence shown.

Table 1

The DNA sequence taken from PDB ID: 1HCQ (12). The two 6 base pair consensus half sites are highlighted in grey. Arrows indicate bases involved in binding to E2.

1	2	3	4	5	6	7	8	9	10	11	12	13	14	15	16	17	18
5'- C	C	A	G	G	T	C	A	C	A	G	T	G	A	C	C	T	G -3'
				↑	↑								↑	↑			

rigid docking program ZDOCK [16] to generate a set of likely bound complexes with E2. The atomic coordinates of E2 were obtained from the Protein Data Bank (PDB) entry 1FDS [17], and the chemical structure of E2 is shown in Fig. 2A [18]. Chemicalize.org was used to obtain the atomic partial charges, January 2015, chemicalize.org and ChemAxon (http://www.chemaxon.com). Although ZDOCK is typically used for protein-ligand docking, Fanelli and Ferrari [19] have proven the effectiveness of ZDOCK for DNA-protein docking. We searched the top-100 ranked structures for possible E2-erDNA complexes. The flexibility of the complex is not taken into account in ZDOCK. Therefore, we performed MD simulations to determine the predicted bound state. The highest-ranked docked structure was used as the starting configuration in MD simulations.

2.3. Molecular Dynamics Simulations

MD simulations are used to determine the bound configuration of E2-erDNA complex and estimate the strength of binding. All MD simulations are performed using NAMD 2.10 [20] and visualized using VMD 2.9.2 [21]. Throughout, we used the CHARMM36 force field [22, 23], and CHARMM27 force field for nucleic acids [24,25]. We used TIP3P water, with a time step of 2 fs, at a constant pressure (1 atm), and temperature (310 K). The temperature is below the expected melting temperature of the double stranded erDNA, which is approximately 325.6 K [26]. The E2-erDNA complex was solvated in a water box such that there is a layer of water 20 Å in each direction from the atom with the largest coordinate in that direction. The particle-mesh-Ewald (PME) algorithm was used for the electrostatics with a tolerance of 10^{-6}. The erDNA strand and E2 were initially held fixed to allow the water to equilibrate during the simulation period of 0.15 ns. Unbiased MD simulations were run for 40 ns to determine the stability of the bound complex. In these simulations no constraints are applied. Both the electrostatic and van der Waals non-bonded interaction energies are calculated using the NAMD Energy plugin available in NAMD2.10 [20].

We estimate the free energy of binding for the E2-erDNA complex using free energy perturbation (FEP) method. For FEP, the target is both annihilated and created in both a free and bound state. In other words, the transformation is performed bidirectionally and the forward and backward simulations are combined using the Bennett acceptance-ratio (BAR) estimator of the free energy which corresponds to the maximum likelihood value of the free energy. The ParseFEP plugin is used, with the Gram-charlier order set to 0 to compute the free energy difference between annihilation and creation simulations, and estimate the statistical error [27].

We build two different molecular systems; E2 in a water bath and E2 bound to erDNA, also in a water bath. To avoid any differences in the two simulations the dimensions of the simulation cell are identical for the free and bound state. In both cases a water box of 64 by 64 by 100 $Å^3$ is used. The annihilation is performed using 40 λ windows, differing by 0.025. For each window, 5000 fs of equilibration is performed, before ensemble averaging is turned on for a further 20,000 fs. A softcore potential is used to avoid explosively large energy values at each end of the λ scale when E2 is nearly annihilated. This scales down the electrostatic interactions from λ = 0.5 to λ = 1.0, and the van der Waals interactions from λ = 0 to λ = 1.0 for the annihilated molecule. To prevent E2 from moving away it is restrained to stay

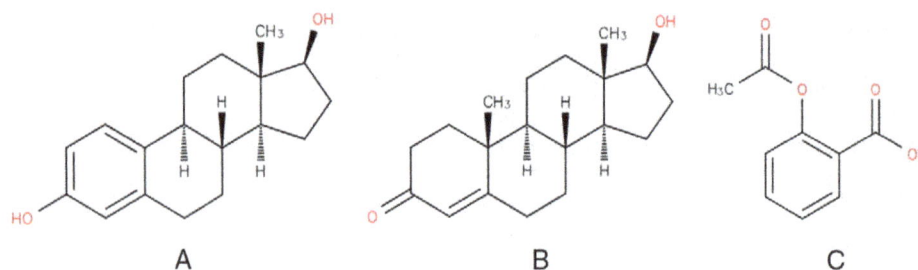

Fig. 2. Chemical structure of A) 17β–estradiol ($C_{18}H_{24}O_2$), B) testosterone ($C_{19}H_{28}O_2$), and C) aspirin ($C_9H_8O_4$). Image obtained from the Chemical Entities of Biological Interest (ChEBI) reference database (17β–estradiol, ChEBI ID: 16,469; testosterone, ChEBI ID: 17,347; aspirin, ChEBI ID: 15,365) (15).

within its normal fluctuating position from equilibration simulations. The positional restraint results in a loss of translational entropy, ΔG_{rest} equal to $-1/\beta\ln(c_0\Delta v)$, where β is equal to $1/k_BT$, Δv is the effective volume sampled by the target and c_0 is the usual standard concentration. The difference between the net free-energy changes for E2 in its free and bound states yields the binding free energy, to which the contribution due to the positional restraint is added, thus

$$\Delta G_{bind} = \Delta G_{free} - \Delta G_{bound} + \Delta G_{rest} \qquad (2)$$

We can then estimate the dissociation constant of target binding using the relation

$$K_d = 1/\exp(-\Delta G_{bind}/k_BT). \qquad (3)$$

The accuracy of MD simulations is dependent on the quality of sampling and the accuracy of the force-field [28]. Two force fields commonly used for simulations of nucleic acid-protein complexes are AMBER [29] and CHARMM [22–25]. The CHARMM force field has performed well with nucleic acid structural integrity due to its sophisticated atom-based smoothing of electrostatic forces [30]. Galindo-Murillo et al. [31] recently demonstrated the convergence and reproducibility of MD simulations using both AMBER and CHARMM nucleic acid force fields. They suggested that ~1 μs length simulation or longer are needed to converge the structural properties of free-DNA in solution, minus the two terminal base pairs at each end [31]. Due to the computational cost we were unable to run our simulations this long. However, investigations into several protein-DNA, ligand-DNA systems have demonstrated the feasibility of the MD simulation approach with smaller time scales [30,32]. For example, Mukherjee et al. [32] successfully used MD to provide detailed mechanistic insight into the intercalation of the anticancer drug duanomycin into DNA. Moreover, their results using only 7.5 ns MD simulations compare well with experimental results. Laughton and Harris [33] provide a good review relating to the computational simulation of DNA. As a control, we ran an additional simulation of the erDNA strand in the absence of E2 for 40 ns. We then compared the erDNA structures in the presence and absence of E2 to assess the effect of the force field and to determine the effect that E2 has on erDNA structure. As a further assessment of the chosen force field we also ran simulations of the erDNA structure with and without E2 using a recently developed CHARMM force field which was optimized for DNA [34].

To assess the importance of the ERE in the erDNA sequence in binding to E2 we ran two additional simulations. First, we used a DNA random sequence generator to scramble the ERE half site (www.faculty.

ucr.edu/~mmaduro/random.htm). This new sequence (rDNA) is illustrated in Table 2. Using 3D-DART we generated a 3D model of this randomised sequence [35]. Second, we examined the binding of testosterone to the original erDNA sequence. The atomic coordinates of testosterone were obtained from the Protein Data Bank (PDB) entry 2Q7I [36], and the chemical structure of testosterone is shown in Fig. 2B [18]. Chemicalize.org was used to obtain the atomic partial charges, January 2015, chemicalize.org and ChemAxon (http://www.chemaxon.com). As described above, we used ZDOCK and MD simulations to examine these additional bound complexes.

Spectroscopic studies have shown that aspirin binds to the minor groove of DNA, but does not intercalate into DNA [37]. Therefore, we also examine the binding of aspirin to the original erDNA sequence to provide a comparison with E2 and testosterone. The atomic coordinates of aspirin were obtained from the Protein Data Bank (PDB) entry 1TGM [38], and the chemical structure of aspirin is shown in Fig. 2C [18]. The atomic partial charges were taken from Jämbeck et al. [39].

Our simulations do not have explicit ions in the system rather a neutralizing background charge is applied [40]. We examine the effect of neutralizing background charge by simulating one of the bound complexes of E2 to erDNA with 34 explicit Na + ions. We find that the effect is negligible and therefore ions are not included in our system.

3. Results and Discussion

3.1. Molecular Docking

For the binding of E2 to erDNA, only two bound complexes are observed in the top-100 ranked structures from our rigid body docking simulations. These two complexes are bound at the location of the two palindromic half sites (Table 1) previously identified as important for binding with ER [7]. Specifically, in 67% of the complexes E2 is bound at G5 and T6, and in the remaining 33% of complexes E2 is bound at A14 and C15 (arrows shown in Table 1). These locations are identical due to DNA symmetry and therefore only one is included in subsequent MD simulations. The bound complex is shown in Fig. 3B.

For the binding of E2 to rDNA, there are two main bound complexes observed in the top-100 ranked structures. Specifically, E2 is bound to C8 and C9 in 51% of the complexes, and G12 and G13 in 48% of the complexes. Both of these complexes are included in subsequent MD simulations.

Two bound complexes are also observed in the top-100 ranked structures for the binding of testosterone to erDNA. As with E2 binding, these two complexes are bound at the location of the two palindromic half sites. Specifically, in 53% of the complexes testosterone is bound at G5, and in the remaining 47% of complexes testosterone is bound at C16. Again, due to symmetry only one of these complexes is included in subsequent MD simulations.

Similarly, two bound complexes are observed in the top-100 ranked structures for the binding of aspirin to erDNA at the location of the two palindromic half sites. Specifically, aspirin is bound at G4 and G5 in 67%

Table 2
The DNA sequence taken from PDB ID: 1HCQ (12) with the ERE palindromic half-site randomised. The two 6 base pair half sites are highlighted in grey.

	1	2	3	4	5	6	7	8	9	10	11	12	13	14	15	16	17	18	
5'-	C	C	T	C	A	A	G	C	C	A	G	G	C	T	T	G	A	G	-3'

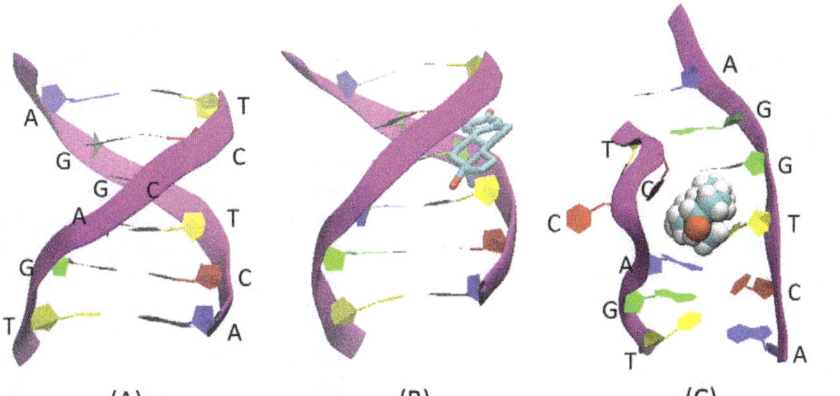

Fig. 3. Palindromic consensus half site with E2 (A) absent, (B) bound complex from ZDOCK, and (C) bound complex after 40 ns MD simulations. Note that only a portion of the DNA strand is shown, and is the location of the palindromic consensus half site. Colours are used to help identify backbone (purple), adenosine (blue), guanine (green), thymine (yellow) and cytosine (red). For clarity, in (B) E2 is shown in licorice with hydrogen atoms removed but as VdW spheres in (C).

of the complexes and in the remaining 33% is bound to C15 and C16. Therefore, due to symmetry we consider only one of these complexes in subsequent MD simulations.

3.2. MD Simulations

After 40 ns E2 remains bound to both the erDNA and one of the rDNA sequences. As shown in Fig. 3C, E2 is inserted into the erDNA strand at the centre of the palindromic half site. In agreement with the mass spectrometry results of Heger et al. [9], we find evidence for the plausibility of non-covalent bonding that is dominated by van der Waals interactions. As shown in Fig. 4, electrostatic interactions between E2 and the erDNA strand are approximately -4.47 ± 2.77 kcal/mol, whereas van

der Waals interactions are approximately -35.69 ± 3.98 kcal/mol. Our results provide detailed molecular insight into this interaction.

For the binding to erDNA, we find that E2 forms aromatic interactions with the thymine base on one strand and the cytosine base on the opposing strand, intercalating between bases G5 and T6. Fig. 5 illustrates the two hydrogen bonds which are formed between the base T6 and E2. A hydrogen bond is assumed to be formed if the donor-acceptor distance is <3.0 Å and the donor-hydrogen-acceptor angle is ≤30°. Both face-to-face and parallel displaced pi stacking is occurring between the bases T6 and C34 with E2, respectively. A pi-pi face-to-face interaction is assumed to be formed if the angle between two aromatic ring planes is less than 30° and the distance between the ring centroids is less than 4.4 Å (https://www.schrodinger.com/kb/1556). The

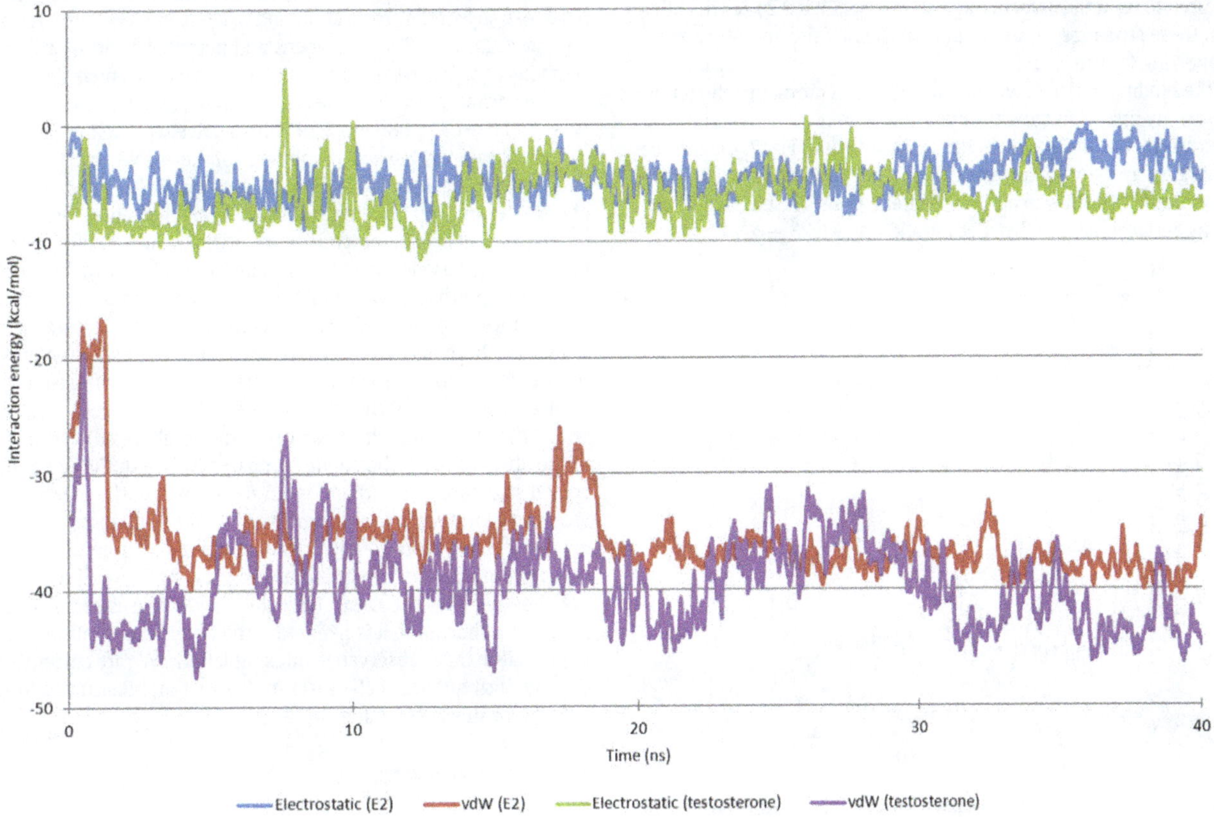

Fig. 4. Electrostatic and van der Waals (VdW) interaction energies between 17β-estradiol and testosterone, and the ds-erDNA. Moving average is displayed, averaging over 20 data points.

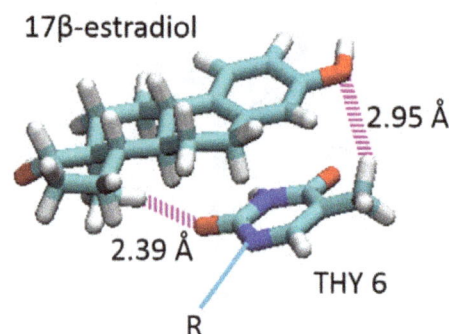

Fig. 5. Hydrogen bonding between base T6 and 17β-estradiol.

distance between T6 and E2 aromatic ring centroids is 3.98 Å with an angle of 19.4°, so that this interaction can be considered as a face-to-face pi-pi interaction. The interaction of E2 with C34 has a distance between aromatic ring centroids of 4.56 Å and an angle of 76.1°. These values are greater than the requirements for a face-to-face pi-pi interaction, but the offset distance of these aromatic rings is 1.09 Å, so that this interaction can be considered as a parallel displaced pi-pi interaction following the definition given by Hunter and Sanders [41] for an idealized pi atoms and provided in Fig. S1 of the Supporting Information. The Movie S1 in the Supporting Information illustrates E2 moving from its initial ZDOCK bound state to the intercalated state. The binding between G5 and T6 is fairly stable, as shown in Fig. 6 after approximately 2.5 ns there is little fluctuation of the distance between the centre of mass of E2 and the centre of the palindromic half site (1.49 ± 0.69 Å). Furthermore, one cytosine base swings away from the interior of the erDNA strand to accommodate the E2 molecule (see Fig. 3C). The RMSD between the erDNA strand in its bound form and the erDNA crystal structure is 11.68 Å, suggesting that there is considerable deviation from the original crystal structure. Furthermore, the RMSD between erDNA after 40 ns in the presence and absence of E2 is 8.21 Å. The differences in these structures in the vicinity of the half site are illustrated in the Supporting Information Fig. S2.

For the binding to rDNA, we find that E2 forms aromatic interactions with C8 on the first strand and the opposing guanine residues on the opposing strand (G11 and G12 in Table 2). As shown in Fig. 7, the interaction energies are similar magnitude to the binding to erDNA and binding is dominated by van der Waals interactions. The electrostatic interactions between E2 and the rDNA strand are −6.26 ± 2.64 kcal/mol,

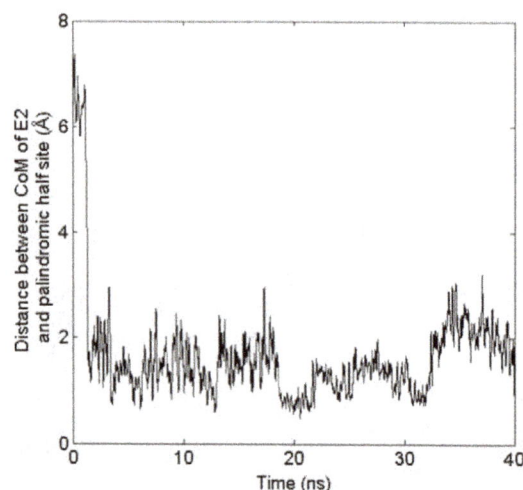

Fig. 6. The distance between the centre of mass (CoM) of E2 and the palindromic half site (located at G5/T6) as a function of simulation time. Moving average is displayed, averaging over 20 data points.

whereas the van der Waals interactions are −34.08 ± 3.19 kcal/mol. This suggests that binding is not specific to the ERE sequence.

We also find that after 40 ns testosterone remains bound to erDNA. Testosterone intercalates between bases G4 and G5. Again, van der Waals interactions dominate with interaction energies a similar magnitude to E2, as shown in Fig. 4. The electrostatic interaction energy between testosterone and erDNA is −6.15 ± 3.13 kcal/mol, whereas the van der Waals interaction energy is −33.36 ± 3.42 kcal/mol. This demonstrates that binding is not specific to the hormone E2, and that excess testosterone could have a similar negative impact on DNA.

In our simulations, the hormone (either E2 or testosterone) preferentially binds to the sequence GGT. These bases then stabilize the binding through aromatic interactions either on opposing strands, or in the case of rDNA between the two guanine bases. The anticancer drug daunomycin also intercalates into DNA [32] and has been shown to bind preferentially to the sequence A/T,C,G [42]. Another endocrine disruptor, bisphenol-A has been shown to intercalate between adjacent base pairs of DNA [15]. The presence of aromatic rings in these molecules renders them able to intercalate with the aromatic bases. An aromatic ring is also present in aspirin, but unlike E2 and testosterone we find that aspirin binds to the minor groove of erDNA and does not intercalate into the DNA strand as demonstrated in earlier spectroscopic studies [37]. Furthermore, aspirin does not stay bound to erDNA for longer than 0.5 ns. Prior to unbinding, the average electrostatic and van der Waals interaction energy over the first 0.5 ns is −11.03 ± 5.84 and −11.47 ± 2.75 kcal/mol, respectively. Fig. S3 in the Supporting Information illustrates the interaction energy between aspirin and erDNA over the 4 ns simulation.

As shown in Fig. 3 the binding of E2 has opened a gap in the DNA structure and disrupted the DNA structure such that the helical parameters have changed compared to the crystal structure [7]. It is well known that the binding of intercalators to DNA induces conformational changes in DNA structures including the opening of a gap between the flanking bases and an elongation and unwinding of the helical twist [43]. These structural changes affect the biological functions of DNA including the inhibition of transcription, replication, and DNA repair processes, thereby making intercalators potent mutagens and potential antitumor drugs [43]. Destabilization can trigger mutations leading to unwanted side effects [9].

The crystal structure of the erDNA duplex (1HCQ) has a mean helical twist and rise of 35.88° and 3.39 Å, respectively [7]. Using Curves+ [44, 45], we obtain local helical parameters of the erDNA structure in the vicinity of binding after 40 ns (i) in the presence of E2, (ii) in the presence of testosterone and (iii) in the absence of any hormone. We also examine the local helical parameters of the erDNA structure after 40 ns in the presence and absence of E2 for the alternative CHARMM force field [34]. In all cases (with and without E2/testosterone) the mean helical twist in the vicinity of binding decreases compared to the crystal structure. Thus, since both our control and bound complexes exhibited unwinding we are unable to comment on whether the binding of these hormones induces unwinding of the DNA duplex.

On the other hand, the mean rise of the equilibrated crystal structure in the absence of any hormone compares well with the unequilibrated crystal structure (3.39 Å) and is 3.5 Å. In contrast, the mean rise in the vicinity of binding in the presence of E2 and testosterone increases to 4.2 Å and 3.9 Å, respectively, demonstrating a lengthening of the DNA duplex compared to the equilibrated crystal structure. Since our results are showing that the mean helical twist decreases regardless of the presence of hormone it is likely that the force field is influencing these structural changes observed in our simulations. We are currently collaborating with quantum chemists to devise a more accurate force field specific for this DNA strand.

3.3. Effect on ER Binding

The ER protein is shown to interact with the central four base pairs of the 6 bp half site using four amino acid side chains on the surface of the

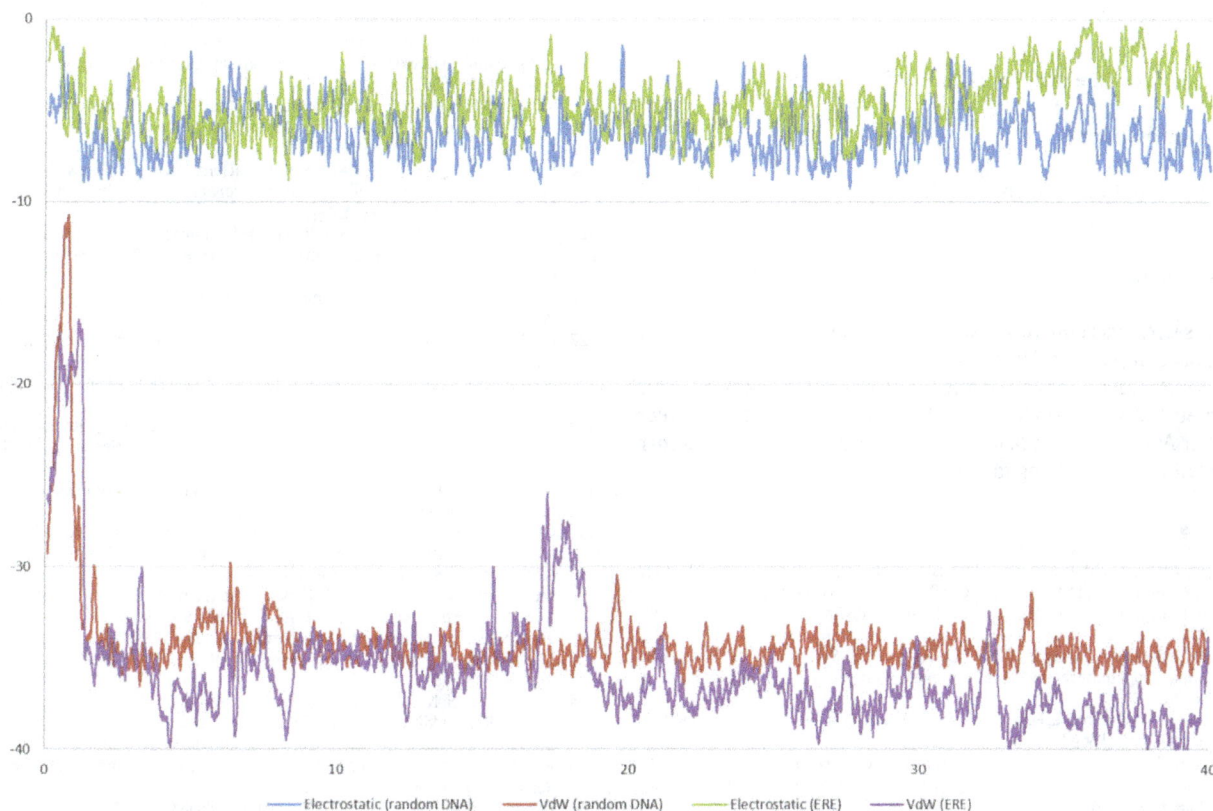

Fig. 7. Comparison of the electrostatic and van der Waals (VdW) interaction energies between 17β-estradiol and either the rDNA (random DNA) or erDNA (ERE) strand. Moving average is displayed, averaging over 20 data points.

recognition helix [7]. ER-α and ER-β bind cooperatively to DNA forming a dimer, as shown in Fig. 1. By forming this cooperative dimer, the protein measures both the spacing and helical repeat of its response element, thus greatly increasing the specificity of the interaction [7]. Side chains of the proteins make sequence specific hydrogen bonds to the phosphate backbone of the central 4 base pairs of the 6 bp half site. The dissociation constant of ER-α and ER-β binding to the consensus half site is approximately 0.6 nM and 1.5 nM, respectively [46,47], equivalent to a binding energy of between −12.1 and −12.7 kcal/mol. This binding is specific to the consensus half site as demonstrated by the fact that the dissociation constant of ER-α binding to plasmid DNA is 400-fold higher [47]. Using FEP we obtain a free energy of binding of E2 to erDNA -8.8 ± 1.2 kcal/mol, which corresponds to a dissociation constant of 388 nM.

The binding of E2 (or other hormones such as testosterone) may be sufficient to disturb the position of this very delicate conformational equilibrium. Typically, estrogens exert their biological effects through a direct interaction with the estrogen receptor which then activates the expression of genes encoding proteins with important biological functions. However, research has shown that unbound hormones can readily diffuse into cells. Oren et al. [48] demonstrated that steroid hormones, such as progesterone, testosterone and estradiol freely diffuse across biomembranes and that this diffusion is rapid, ~0.01 s to cross a 30 Å thick biomembrane. It is also possible for small molecules, such as hormones, to diffuse into the nucleus, either by passive diffusion [49] or through nuclear pore complexes [50]. Thus, excess E2 could bind to or near the response element prior to ER binding thus disrupting the DNA structure such that the protein spacing does not match the helical repeat of its response element. Alternatively, the portion of DNA structure with E2 bound may be recognized as an 'error' which could result in a deletion from the DNA sequence or a frameshift mutation

which would again affect the interaction of ER with DNA response element and interfere with signalling and regulation. Both of these possibilities could cause some serious unwanted side effects of E2 accumulation to normal hormone function. Intercalating drugs, such as ethidium bromide, have already been shown to inhibit the interaction of ER with DNA [51]. It is also possible that E2 could intercalate with DNA sequences other than the ERE, since we have demonstrated that E2 also remains bound to the randomised sequence (rDNA), thus damaging other cellular processes such as the activity of enzymes that bind DNA [51]. For example, some anticancer drugs (such as daunomycin) act by intercalating into DNA thus inhibiting the enzyme topoisomerase which stops DNA replication and leads to cell death [32].

4. Conclusions

We have demonstrated that E2 can bind to DNA directly and that binding not only occurs at the centre of the ERE half site but can also bind to random DNA sequences. E2 is shown to intercalate between base pairs, forming aromatic interactions with these base pairs. We demonstrate the intercalation results in a lengthening of the DNA duplex, but are unable to comment on the unwinding due to inaccuracies in the force field. In ongoing work we are working with colleagues to develop a more accurate force field to represent specific DNA sequences. We predict that this intercalation will alter the structure of the DNA duplex, and therefore have the potential to affect the biological functions of DNA including the inhibition of transcription, replication, and DNA repair processes [40]. Therefore, excess E2 has the potential to exert some serious side effects such as disruption of ER-DNA binding, DNA damage and possibly the initiation of cancer.

Supplementary data to this article can be found online at http://dx.doi.org/10.1016/j.csbj.2016.12.001.

Declaration of Interest

The authors declare that there is no conflict of interest.

Author Contributions

The manuscript was written through contribution of all authors. TAH conducted the simulations.

Acknowledgement

This research was undertaken with the assistance of resources from the National Computational Infrastructure (NCI), which is supported by the Australian Government. We gratefully acknowledge the support from the Australian Research Council through a Discovery Early Career Researcher Award. We thank Kenneth P. McNatty and Shalen Kumar for useful discussions leading to this work.

References

[1] Rodgers-Gray TP, Jobling S, Kelly C, Morris S, Brighty G, et al. Exposure of juvenile roach (*Rutilus rutilus*) to treated sewage effluent induces dose-dependent and persistent disruption in gonadal duct development. Environ Sci Technol 2001;35:462–70.

[2] Auriol M, Filali-Meknassi Y, Tyagi RD, Adams CD, Surampalli RY. Endocrine disrupting compounds removal from wastewater, a new challenge. Process Biochem 2006;41:525–39.

[3] Pacáková V, Loukotková L, Bosáková Z, Štulík K. Analysis for estrogens as environmental pollutants. J Sep Sci 2009;32:867–82.

[4] Diamanti-Kandarakis E, Bourguignon J-P, Giudice LC, Hauser R, Prins GS, et al. Endocrine-disrupting chemicals: an endocrine society scientific statement. Endocr Rev 2009;30:293–342.

[5] Basile T, Petrella A, Petrella M, Boghetich G, Petruzzelli V, et al. Review of endocrine-disrupting-compound removal technologies in water and wastewater treatment plants: an EU perspective. Ind Eng Chem Res 2011;50:8389–401.

[6] Gruber CJ, Gruber DM, Gruber IML, Wieser F, Huber JC. Anatomy of the estrogen response element. Trends Endocrinol Metab 2004;15:73–8.

[7] Schwabe JW, Chapman L, Finch JT, Rhodes D. The crystal structure of the estrogen receptor DNA-binding domain bound to DNA: how receptors discriminate between their response elements. Cell 1993;75:567–78.

[8] Caldon CE. Estrogen signalling and the DNA damage response in hormone dependent breast cancers. Front Oncol 2014;4(106).

[9] Heger Z, Guran R, Zitka O, Beklova M, Adam V, et al. *In vitro* interactions between 17β-estradiol and DNA result in formation of the hormone-DNA complexes. Int J Environ Res Public Health 2014;11:7725–39.

[10] Li L, Wang Q, Zhang Y, Niu Y, Yao X, et al. The molecular mechanism of bisphenol A (BPA) as an endocrine disruptor by interacting with nuclear receptors: insights from molecular dynamics (MD) simulations. PLoS One 2015;10, e0120330.

[11] le Maire A, Bourguet W, Balaguer P. A structural view of nuclear hormone receptor: endocrine disruptor interactions. Cell Mol Life Sci 2010;67:1219–37.

[12] Celik L, Lund JDD, Schiøtt B. Exploring interactions of endocrine-disrupting compounds with different conformations of the human estrogen receptor α ligand binding domain: a molecular docking study. Chem Res Toxicol 2008;21:2195–206.

[13] Atienzar FA, Billinghurst Z, Depledge MH. 4-n-Nonylphenol and 17-β estradiol may induce common DNA effects in developing barnacle larvae. Environ Pollut 2002;120:735–8.

[14] Roy D, Liehr JG. Estrogen, DNA damage and mutations. Mutat Res 1999;424:107–15.

[15] Zhang Y-L, Zhang X, Fei X-C, Wang S-L, Gao H-W. Binding of bisphenol A and acrylamide to BSA and DNA: insights into the comparative interactions of harmful chemicals with functional biomacromolecules. J Hazard Mater 2010;182:877–85.

[16] Chen R, Li L, Weng Z. ZDOCK: an initial-stage protein-docking algorithm. Proteins Struct Funct Genet 2003;52:80–7.

[17] Breton R, Housset D, Mazza C, Fontecilla-Camps JC. The structure of a complex of human 17beta-hydroxysteroid dehydrogenase with estradiol and NADP+ identifies two principal targets for the design of inhibitors. Structure 1996;4:905–15.

[18] Hastings J, de Matos P, Dekker A, Ennis M, Harsha B, et al. The ChEBI reference database and ontology for biologically relevant chemistry: enhancements for 2013. Nucleic Acids Res 2013;41:D456–63.

[19] Fanelli F, Ferrari S. Prediction of MEF2A-DNA interface by rigid body docking: a tool for fast estimation of protein mutational effects on DNA binding. J Struct Biol 2006;153:278–83.

[20] Phillips JC, Braun R, Wang W, Gumbart J, Tajkhorshid E, et al. Scalable molecular dynamics with NAMD. J Comput Chem 2005;26:1781–802.

[21] Humphrey W, Dalke A, Schulten K. VMD: visual molecular dynamics. J Mol Graph 1996;14:33–8.

[22] MacKerell Jr AD, Feig M, Brooks III CL. Extending the treatment of backbone energetics in protein force fields: limitations of gas-phase quantum mechanics in reproducing protein conformational distributions in molecular dynamics simulations. J Comput Chem 2004;25:1400–15.

[23] MacKerell Jr AD, Bashford D, Bellott M, Dunbrack Jr RL, Evanseck JD, et al. All-atom empirical potential for molecular modeling and dynamic studies of proteins. J Phys Chem B 1998;102:3586–616.

[24] Foloppe N, MacKerell Jr AD. All-atom empirical force field for nucleic acids: 1. Parameter optimization based on small molecule and condensed phase macromolecular target data. J Comput Chem 2000;21:86–104.

[25] MacKerell Jr AD, Banavali N. All-atom empirical force field for nucleic acids: 2. Application to molecular dynamics simulations of DNA and RNA in solution. J Comput Chem 2000;21:105–20.

[26] Kibbe WA. OligoCalc: an online oligonucleotide properties calculator. Nucleic Acids Res 2007;35:W43–6.

[27] Liu P, Dehez F, Cai W, Chipot C. A toolkit for the analysis of free-energy perturbation calculations. J Chem Theory Comput 2012;8:2606–26.

[28] Dans PD, Walther J, Gómez H, Orozco M. Multiscale simulation of DNA. Curr Opin Struct Biol 2016;37:29–45.

[29] Cornell WD, Cieplak P, CI B, Gould IR, Merz KM, et al. A second generation force field for the simulation of proteins, nucleic acids, and organic molecules. J Am Chem Soc 1995;117:5179–97.

[30] MacKerell Jr AD, Nilsson L. Molecular dynamics simulations of nucleic acid-protein complexes. Curr Opin Struct Biol 2008;18:194–9.

[31] Galindo-Murillo R, Roe DR, Cheatham III TE. Convergence and reproducibility in molecular dynamics simulations of the DNA duplex d(GCACGAACGAACGAACGC). Biochim Biophys Acta 2015;1850:1041–58.

[32] Mukherjee A, Lavery R, Bagchi B, Hynes JT. On the molecular mechanism of drug intercalation into DNA: a simulation study of the intercalation pathway, free energy, and DNA structural changes. J Am Chem Soc 2008;130:9747–55.

[33] Laughton CA, Harris SA. The atomistic simulation of DNA. WIREs Comput Mol Sci 2011;1:590–600.

[34] Hart K, Foloppe N, Baker CM, Denning EJ, Nilsson L, MacKerell Jr AD. Optimization of the CHARMM additive force field for DNA: improved treatment of the BI/BII conformational equilibrium. J Chem Theory Comput 2012;8:348–62.

[35] van Dijk M, Bonvin AMJJ. 3D-DART: a DNA structure modelling server. Nucleic Acids Res 2009;37:W235–9.

[36] Askew EB, Gampe RT, Stanley TB, Faggart JL, Wilson EM. Modulation of androgen receptor activation function 2 by testosterone and dihydrotestosterone. J Biol Chem 2007;282:25801–16.

[37] Bathaie SZ, Nikfarjam L, Rahmanpour R, Moosavi-Movahedi AA. Spectroscopic studies of the interaction of aspirin and its important metabolite, salicylate ion, with DNA, A·T and G·C rich sequences. Spectrochim Acta A 2010;77:1077–83.

[38] Singh RK, Ethayathulla AS, Jabeen T, Sharma S, Kaur P, Singh TP. Aspirin induces its anti-inflammatory effects through its specific binding to phospholipase A2: crystal structure of the complex formed between phospholipase A2 and aspirin at 1.9 angstroms resolution. J Drug Target 2005;13:113–9.

[39] Jämbeck JPM, Mocci F, Lyubartsev AP, Laaksonen A. Partial atomic charges and their impact on the free energy of solvation. J Comput Chem 2013;34:187–97.

[40] Hub JS, de Groot BL, Grubmüller H, Groenhof G. Quantifying artifacts in ewald simulations of inhomogeneous systems with a net charge. J Chem Theory Comput 2014;10:381–90.

[41] Hunter CA, Sanders JKM. The nature of π-π interactions. J Am Chem Soc 1990;112:5525–34.

[42] Chaires JB, Herrera JE, Waring MJ. Preferential binding of daunomycin to 5′A/TCG and 5′A/TGC sequences revealed by footprinting titration experiments. Biochemist 1990;29:6145–53.

[43] Gilad Y, Senderowitz H. Docking studies on DNA intercalators. J Chem Inf Model 2014;54:96–107.

[44] Lavery R, Moakher M, Maddocks JH, Petkeviciute D, Zakrzewska K. Conformational analysis of nucleic acids revisited: curves+. Nucleic Acids Res 2009;37:5917–29.

[45] Blanchet C, Pasi M, Zakrzewska K, Lavery R. CURVES + web server for analyzing and visualizing the helical, backbone and groove parameters of nucleic acid structures. Nucleic Acids Res 2011;39:W68–73.

[46] Tyulmenkov VV, Klinge CM. A mathematical approach to predict the affinity of estrogen receptors α and β binding to DNA. Mol Cell Endocrinol 2001;182:109–19.

[47] Peale Jr FV, Ludwig LB, Zain S, Hilf R, Bambara RA. Properties of a high-affinity DNA binding site for estrogen receptor. Proc Natl Acad Sci U S A 1988;85:1038–42.

[48] Oren I, Fleishman SJ, Kessel A, Ben-Tal N. Free diffusion of steroid hormones across biomembranes: a simplex search with implicit solvent model calculations. Biophys J 2004;87:768–79.

[49] Alberts B, Johnson A, Lewis J, et al. The transport of molecules between the nucleus and the cytosol. Molecular biology of the cell. 4th ed. New York: Garland Science; 2002.

[50] Hough LE, Dutta K, Sparks S, Temel DB, Kamal A, et al. The molecular mechanism of nuclear transport revealed by atomic-scale measurements. Elife 2015;4, e10027.

[51] André J, Pfeiffer A, Rochefort H. Inhibition of estrogen-receptor-DNA interaction by intercalating drugs. Biochemist 1976;15:2964–9.

12

The effects of shared information on semantic calculations in the gene ontology

Paul W. Bible[a],*, Hong-Wei Sun[b], Maria I. Morasso[c], Rasiah Loganantharaj[d], Lai Wei[a],*

[a]State Key Laboratory of Ophthalmology, Zhongshan Ophthalmic Center, Sun Yat-sen University, Guangzhou 510060, China
[b]Biodata Mining and Discovery Section, Office of Science and Technology, Intramural Research Program, National Institute of Arthritis and Musculoskeletal and Skin Diseases, Bethesda, Maryland
[c]Laboratory of Skin Biology, Intramural Research Program, National Institute of Arthritis and Musculoskeletal and Skin Diseases, Bethesda, Maryland
[d]Laboratory of Bioinformatics, Center for Advanced Computer Studies, University of Louisiana at Lafayette, Lafayette, Louisiana

ARTICLE INFO

Keywords:
Semantic similarity
Gene ontology
Function prediction
Machine learning
Protein–protein interaction
Gene expression

ABSTRACT

The structured vocabulary that describes gene function, the gene ontology (GO), serves as a powerful tool in biological research. One application of GO in computational biology calculates semantic similarity between two concepts to make inferences about the functional similarity of genes. A class of term similarity algorithms explicitly calculates the shared information (SI) between concepts then substitutes this calculation into traditional term similarity measures such as Resnik, Lin, and Jiang-Conrath. Alternative SI approaches, when combined with ontology choice and term similarity type, lead to many gene-to-gene similarity measures. No thorough investigation has been made into the behavior, complexity, and performance of semantic methods derived from distinct SI approaches. We apply bootstrapping to compare the generalized performance of 57 gene-to-gene semantic measures across six benchmarks. Considering the number of measures, we additionally evaluate whether these methods can be leveraged through ensemble machine learning to improve prediction performance. Results showed that the choice of ontology type most strongly influenced performance across all evaluations. Combining measures into an ensemble classifier reduces cross-validation error beyond any individual measure for protein interaction prediction. This improvement resulted from information gained through the combination of ontology types as ensemble methods within each GO type offered no improvement. These results demonstrate that multiple SI measures can be leveraged for machine learning tasks such as automated gene function prediction by incorporating methods from across the ontologies. To facilitate future research in this area, we developed the GO Graph Tool Kit (GGTK), an open source C++ library with Python interface (github.com/paulbible/ggtk).

1. Introduction

Researchers developed the gene ontology (GO) to provide a structured vocabulary that consistently describes the characteristics of genes and proteins across different organisms [1,2]. Specific GO terms in this vocabulary annotate proteins by specifying the biological processes in which they participate, their enzymatic and molecular functions, and their location within the cell. As a structured vocabulary, GO explicitly defines the relationships between terms using a directed acyclic graph (DAG). These relationships serve to clarify terminology, for example by identifying when one term may be a more specialized from of another. Three separate ontologies exist that provide a DAG of terms and relationships used to describe biological processes (BP), molecular functions (MF), and cellular components (CC). These term structures are not fixed. The Gene Ontology Consortium makes frequent updates to GO modifying the relationship structure and adding or removing terms to better reflect the current understanding of biological functions.

The annotation of gene products with GO terms provides a valuable resource allowing the comparison of functions both within and between separate organisms. For each annotation of a term to a protein, GO provides evidence codes that allow researchers to consider the methods that produced each annotation. The use of GO plays an important role in the analysis of high-throughput experiments

* Corresponding authors.
E-mail addresses: paul.bible@gmail.com (P.W. Bible), weil9@mail.sysu.edu.cn (L. Wei).

thanks in part to computational methods that utilize the rich domain knowledge encoded in GO annotations. The rigid, well-defined structure of GO proves to be an advantage that facilitates its integration into statistical and computational analyses. Methods of semantic similarity take advantage of this structure to quantify the similarity between the meaning of one term and another. Through semantic measures, the concept level knowledge stored in functional annotations provides the ability to quantify functional similarity between genes. Researchers have employed semantic methods for predicting protein–protein interactions [3–6], prioritizing host-pathogen interactions [7], and automated function prediction [8,9]. Increasingly, researchers apply computational methods to infer new GO annotations. These annotations receive the evidence code *inferred from electronic annotation* (IEA), and recent studies show that these predicted annotations are increasingly reliable [7,10].

Since the seminal works by Lord et al. [11,12], semantic similarity applications have become established tools in computational biology and bioinformatics. Many diverse methods exist for calculating semantic similarity between terms. The reviews of Refs. [13], [14], and [15] provide a thorough overview of semantic measures used in the Gene Ontology and other biomedical ontologies. The review by Pesquita et al. [13] categorizes semantic similarity at the term level into edge-based and node-based methods. Edge-based methods usually quantify semantic similarity using a function of the paths between two terms in the graph. Node-based methods use properties derived from the terms and often include operations on the shared ancestors or descendants of two terms. Lord et al. adapted three well studied information theoretic semantic measures, Resnik [16], Lin [17], and Jiang-Conrath [18], using a corpus-based calculation of term probability and information content (IC). These methods calculate the similarity between two concepts by operating on their shared and unique information. Lord et al. calculated shared information using the IC of the most informative common ancestor (MICA) between two terms. Information content based semantic methods have been extensively studied and research suggests that IC offers a superior conception of a term's specificity over methods based on graph depth [14,19].

Couto et al. [20] developed an alternative approach to shared information. While the MICA shared information considers only a single ancestor, the alternative approach, called the graph-based similarity measure (GraSM), considers multiple inheritance of ontology terms using a path counting method. GraSM calculates the mean IC of disjunctive common ancestors between two concepts. Couto et al. recognized the modularity of using alternative shared information methods and substituted GraSM shared information into Resnik, Lin, and Jiang-Conrath deriving three new term similarity algorithms. This method was shown to improve the performance of semantic similarity on accepted evaluation metrics such as correlation with sequence and domain similarity. The computational cost of path counting poses a significant challenge to the real-time calculation of GraSM. Zhang and Lai [21] proposed a faster GraSM alternative called exclusively inherited shared information that calculates a subset of the disjunctive common ancestors. The modular separation of shared information from term similarity leads to some interesting properties. Any new conception of shared information immediately implies the construction of three new term similarity measures. The success of GraSM and alternative conceptions of shared information clarifies the need for a thorough exploration of shared information in the biomedical ontologies.

Extending *term* similarity to *gene* similarity requires methods operating on term sets. For two genes, represented as term sets, the all-pairs term similarity is calculated then summarized by aggregation methods such as the *min* [22], *max* [23], *average* [11], or *best-match average* (BMA) [24]. See Ref. [13] for a detailed description of aggregation methods. Furthermore, genes are described by annotations from each of the independent ontologies (BP, MF, CC). The

choices for shared information calculation, term similarity algorithm, aggregation method, and ontology type lead to a combinatorial increase in the number of gene similarity measures. Considering the number of measure that can be constructed, we address the following questions in this work. Are the methods derived from alternative shared information calculations truly distinct or do they offer no statical difference in practical applications? If these methods are not distinct, it would imply that computationally intensive algorithms, such as GraSM, can be replaced by more efficient alternatives. Some gene similarity methods avoid explicit calculation of shared information, term similarity, and aggregation. Methods based on the simple yet powerful Jaccard set similarity, such as SimUI [25], SimGIC [19], SimDIC [26], and SimUIC [26], have been shown to perform well on a variety of tasks and are computationally easy to compute. Can methods derived from modular combination with shared information outperform these more efficient methods? If the shared information-based methods are truly distinct, can the large quantity of measures be leveraged for performance gains through ensemble integration techniques?

This work presents extensive and novel research on the understudied effects of shared information in semantic calculations in the gene ontology through robust evaluations of their performance on traditional and real-world tasks. As no method can determine *true* semantic similarity between genes [13], various methods for evaluation have been put forward to evaluate the performance of semantic similarity methods. Commonly used evaluations include correlation with sequence similarity [11,19], correlation with domain set similarity [13,20,27], correlations with gene expression [4,23,28], clustering genes into known pathways [3,5,21], and prediction of protein–protein interactions [3,6,29]. Based on these past works, we have constructed a new suite of six benchmarks to evaluate semantic similarity that provides evaluations for a broad range of use cases. These benchmarks are provided as lists of protein pairs and scores to facilitate use by other researchers. It is known that the GO annotations are incomplete and can suffer from bias and noise [30]. To address these issues, our evaluations use bootstrapping to provide robust statistical performance comparisons for each measure. Due to the large number of measures and the number of evaluations needed, no current tools addressed our simultaneous needs of speed and modularity. To achieve these goals, we have developed a new set of efficient, modular tools for working with GO graphs in C++, called the GO Graph Tool Kit (GGTK). With the aim of facilitating further research in the community, we provide an easy-to-use Python package that wraps the functionality of GGTK and release all GGTK code under the permissive BOOST License. GGTK will remain an ongoing open source project available at github.com/paulbible/ggtk.

2. Methods

2.1. Calculating information content and shared information

Understanding the structure of the GO DAG provides insight into how semantic similarity is derived at the term level. In GO graphs, each vertex or node represents a term or concept and each edge presents a relationship such as *is_a* or *part_of* [1]. A root node in GO represents the most general concept (e.g. *biological_process*), and all other concepts are considered descendants of this term. Information theoretic semantic measures rely on assigning a value of probability to a term. Lord et al. [11] proposed defining a term's probability as the number of times it occurs in a corpus of annotations divided by the number of occurrences for all terms. A term occurs if it or *any of its descendants* appear in the corpus. Eq. (1) shows the definition of probability where o_t is the number of occurrences for a term t. The information content (IC) follows, in Eq. (2), as the negative log of probability. As o_t includes all appearances of child terms, $P(root) = 1$

and any non-root term t satisfies $P(t) < P(root)$. These definitions ensure IC is monotonically increasing toward more specific terms.

$$P(t) = \frac{o_t}{o_{root}} \qquad (1)$$

$$IC(t) = -\log(P(t)) \qquad (2)$$

Using the above definitions Lord et al. used the IC calculation in adapting Resnik [16], Lin [17], and Jiang-Conrath [18] to GO. The original shared information approach calculates the set of common ancestors (CA) between two ontology terms (Eq. (3)), and finds the most informative common ancestor (MICA) or equivalently, after Resnik [16], the probability of the minimum subsumer (Eq. (4)). The shared information of these traditional term similarity measures is defined in Eq. (5).

$$CA(t_1, t_2) = Ancestors(t_1) \cap Ancestors(t_2) \qquad (3)$$

$$P_{ms}(t_1, t_2) = \min_{t \in CA(t_1, t_2)} \{P(t)\} \qquad (4)$$

$$SI_{MICA}(t_1, t_2) = IC\,(MICA(t_1, t_2))$$
$$= -\log(P_{ms}(t_1, t_2)) \qquad (5)$$

Using the above definitions, Resnik, Lin, and Jiang-Conrath semantic similarity (SS) can be re-written equivalently in terms of shared information (Eqs. (6), (7), and (8) respectively).

$$SS_{Resnik}(t_1, t_2) = SI(t_1, t_2) \qquad (6)$$

$$SS_{Lin}(t_1, t_2) = \frac{2 * SI(t_1, t_2)}{IC(t_1) + IC(t_2)} \qquad (7)$$

$$SS_{JC}(t_1, t_2) = 1 - (IC(t_1) + IC(t_2) - 2 * SI(t_1, t_2)) \qquad (8)$$

2.2. Implementation considerations affecting time complexity

Before describing the different shared information algorithms and their time complexity, we will explain the implementation of GO used within the Go Graph Tool Kit (GGTK). Varying reports on the time complexity of semantic similarity algorithms have been put forward due to implicit assumptions about the representation of the GO DAG and associated annotations. Here we clarify the graph representation and its implementation. GGTK reads the ontology files (go-basic.obo) provided by geneontology.org [1], and stores the graph structure for the three disjoint ontologies (BP, MF, CC) in memory as adjacency lists. Let n be the number of terms in the ontology. Probability and information content are calculated from annotations taken from UniProt-GOA [31]. GGTK reads GO annotations from file, indexes annotations by gene, and provides gene-to-annotation lookups in $O(1)$ time. Calculation of IC is performed after Lord et al. [11] and is accomplished in $O(n)$ time. Specifically, a depth first traversal visits every node in the GO graph. The number of annotation occurrences is calculated for each term starting with leaf nodes, moving to more general terms in the graph, and finishing with the root node. The number of cumulative occurrences for the root node is used to calculate term probability and IC after Eqs. (1) and (2) respectively. After the initial calculation, a function that maps a term to its IC is available as a map in which lookups are performed in $O(1)$ time. In this work, IC was calculated using human annotations (GOA Dec. 2016) including electronically inferred annotations and considering only is_a and part_of relationships.

The semantic similarity methods under study in this work all rely on the calculation of the set of common ancestors shared between two concepts. The DAG structure of GO complicates estimates of the size of this set with respect to the number of terms, n. For a tree based ontology, the size of the common ancestor sets is bounded by $O(\log n)$, but this complexity is not guaranteed for general DAGs. Using GGTK we compiled historical data on the average branching factor and number of ancestors of each term from 2006 to 2016. Fig. 1 shows summary data for the three ontologies BP, MF, and CC (considering only is_a and part_of relationships) as the graph topology evolved over 10 years. The number of nodes, mean branching factor (node degree), and mean ancestor number are growing over time. Fig. 1D shows that $\log n$ is an under estimate in the case of BP and CC but over estimates the average number of ancestors in MF. After Couto et al. [20], we will refer to the number of common ancestors as k in our complexity analysis where $|CA| = O(k)$. Based on our empirical analysis of the GO graph over the past 10 years, it appears that $O(k) \approx O(\log n)$ for current and past graphs; however, this relation may not hold in the future.

2.3. Shared information algorithms and their time complexity

Determining the shared information between two ontology terms is an area of ongoing research. In this work we address the effects of five unique shared information algorithms on their derived gene functional similarity measures. A thorough analysis of these methods must address their time complexity as well as qualitative features. In this section, we provide a brief description of the shared information algorithms analyzed in this work, describe their features, and address their time complexity. A more detailed description of the algorithms can be found in Appendix A. The source code for all methods is available at github.com/paulbible/ggtk. The following complexity bounds refer to the calculation of single-term-to-single-term shared information.

2.3.1. Common ancestor shared information

For purposes of evaluation, we developed a naive baseline algorithm called common ancestor shared information (CASI). CASI simply calculates the set of common ancestors and returns the mean IC as the shared information. This naive baseline serves as a useful tool for evaluating other shared information algorithms. The time complexity of CASI is $O(k)$. Calculating the common ancestor set is performed in $O(k)$ and using GGTK's $O(1)$ IC map allows the mean to be calculated in $O(k)$.

2.3.2. Most informative common ancestor shared information

The most informative common ancestor (MICA) shared information remains the most common shared information measure as it forms the basis of traditional term similarity methods. (Eq. (5), above). MICA shared information is equivalent to Resnik term similarity [11,16], and is defined as the IC of the MICA. Due to its simplicity, it is one of the most widely used shared information measures and forms the foundation for many more sophisticated semantic algorithms. The time complexity of MICA is $O(k)$. Construction of the common ancestor set takes $O(k)$ and finding the term with maximum IC in the set is also $O(k)$.

2.3.3. Couto et al. 2007, GraSM

Couto et al. devised GraSM [20] as an alternative to MICA. They observed that different paths in the ontology represent different interpretations of concepts. They reasoned that ancestors with multiple interpretations, designated disjunctive common ancestors (DCA), should factor into the shared information. Calculation of the DCA involves determining if a common ancestor has unique paths to each of the two terms that are separate from the other ancestors under consideration. For each common ancestor, t_a, paths

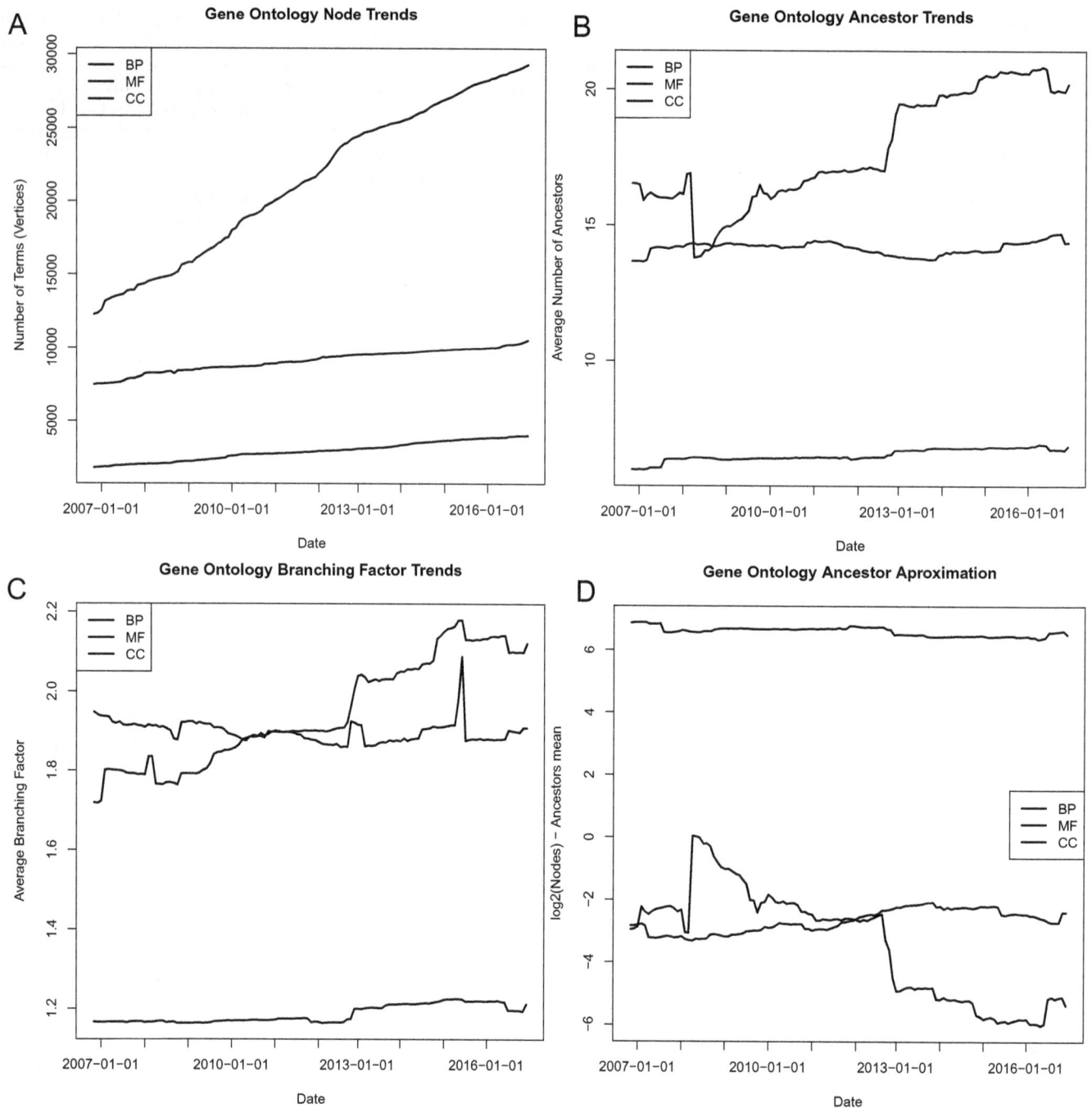

Fig. 1. Changes in GO graph structure (using *is_a* and *part_of* relationships) over time lead to variations in (A) number of nodes, (B) mean ancestor number, and (C) the mean branching factor for each term. Panel D shows the log of the number of terms minus the mean ancestor number.

are counted to determine if t_a is disjoint from another ancestor, say t_b. If the number paths from t_a to the two input terms under consideration is greater than or equal to the number of paths from t_b to the input terms, t_a must be disjoint and represent some unique interpretation of the shared information by virtue of having a unique path to the terms that does not pass though t_b. This process is repeated for all pairs of common ancestors. A detailed description of the GraSM algorithm is provided in Appendix A.3. Couto et al. reported the GraSM time complexity as $O(k^2)$, but their implementation uses a pre-computed path number map. The term-to-term path number map can be calculated for a one-time cost of $O(n^2)$, where n is number of terms in the ontology. The parallelized version of this

calculation can be performed by n separate topological sorts each taking $O(n)$ time. This 'one-time' cost could be quite large as $n \gg k$ and would need to be recalculated anytime other relationship edges were considered. To calculate GraSM shared information in real-time requires $O(k^3)$ operations. This result follows from the $O(k^2)$ calculations required to constructing the DCA and an added $O(k)$ cost for path counting at each step leading to a runtime that is cubic in the number of common ancestors.

2.3.4. Adjusted GraSM

Close study of the behavior of GraSM on real-world datasets leads to some unexpected results. In many cases, the DCA comprises a large proportion of the common ancestor set including the root node. An illustration of this behavior is provided in Supplemental Information Trace 1. GraSM is calculated by averaging the IC of DCA terms and may decrease as more shallow terms are included. In determining the membership of the DCA set, GraSM uses a *greater than or equal to* comparison on the number of paths. By changing the *greater than or equal to* ($\geq$) to a *strictly greater than* ($>$) in the path number comparison, the algorithm's behavior changes and fewer terms are included in the DCA set. This modified algorithm is referred to as adjusted GraSM (A-GraSM). See Supplemental Information Trace 1 and 2 for further details on A-GraSM. The time complexity of the adjusted algorithm is the same as GraSM, $O(k^3)$.

2.3.5. Semantic frontier

Zhang and Lai introduced the exclusively inherited common ancestors set as an alternative to the DCA [21] that calculated a subset of the DCA in linear time. Our group developed an efficient implementation of the algorithm by Zhang and Lai, called the semantic frontier (SF) algorithm. A simple analogy can help to explain the exclusively inherited common ancestor set. The common ancestor set can be considered a region of the GO graph. Terms that form the *semantic frontier* lie on the frontier of the territory formed by the common ancestors. Specifically, terms in the semantic frontier set are common ancestor terms that have an incoming edge leading from one of the input terms not shared by paths leading from the other. The average IC of this set is returned as the shared information. The SF implementation offers a reduced search space over the algorithm proposed by Zhang and Lai. The SF algorithm is based on the concept of a breadth first search (BFS) visitor. A BFS visitor performs actions when certain events in a BFS occur. A BFS is performed for each of the two input terms and the SF set is calculated by examining the visited edges entering the common ancestor set. A detailed description of the SF algorithm is available in Appendix A.5.

The time complexity of the SF algorithm is $O(k)$. SF calculates the common ancestor set in $O(k)$, each BFS visitor operates in $O(k)$, and checking for frontier edges takes $O(k)$ time. Zhang and Lai report the complexity of their algorithm as $O(n\log n)$ [21]. The greater complexity reported by Zhang and Lai may be the result of less efficient ancestor or IC access which GGTK provides in $O(k)$ and $O(1)$ time respectively. Based on the increase in GO graph size in past 10 years, our early exit and reduced search space features may yield further performance gains in the future.

2.4. From shared information to term similarity

Methods for calculating term similarity can be constructed from shared information calculations through substitution into Resnik, Lin, and Jiang-Conrath (see Eqs. (6), (7), and (8)). In order to construct term similarity calculators using these method, GGTK provides a shared information interface that allows the modular combination of shared information and term similarity algorithms. This combination allows multiple distinct term similarity measures to be constructed. The interface promotes extensibility allowing other researchers to construct shared information algorithms and immediately combine them with existing term similarity measures.

2.4.1. Gene similarity from all-pairs of terms aggregation

The calculation of gene functional similarity relies on methods that aggregate the all-pairs term similarity between gene annotations [6]. GGTK provides *max*, *average*, and *best-match average* (BMA) gene similarity aggregators. Research by Pesquita et al. [19] and others has suggested that BMA outperforms other methods of aggregation. Based on initial experiments using the Collaborative Evaluation of Semantic Similarity Measures (CESSM) online tool, we arrived at the same conclusion. We evaluated the gene similarity benchmarks

using BMA, but other aggregation methods may be useful in other contexts.

The number of gene-to-gene semantic similarity measures (N_{GSS}) explodes in a combinatorial fashion as described in Eq. (9) where $|si|$ is the number of shared information methods, $|ss|$ the number of semantic term similarity methods, $|a|$ the aggregation methods, and the final term, 3, represents the distinct ontologies, BP, MF, and CC. As new methods are devised to calculate term similarity, shared information, and aggregation of term similarity, the number of gene similarity measures grows combinatorially. This increasing makes evaluation a challenge, but offers a rich set of measures from which to choose. We explore whether these measures are significantly distinct in terms of performance and if gains can be achieved by using multiple measures.

$$N_{GSS} = |si| * |ss| * |a| * 3 \tag{9}$$

2.4.2. Jaccard-based gene similarity measures

Some gene similarity methods act on sets of terms without the need to explicitly calculate term-to-term similarity or employ aggregation of term similarity. To contrast the performance of shared information methods with other established approaches, we consider alternative gene similarity methods reported by the literature to perform well. Many of these approaches are variations of the Jaccard index [32] for set similarity. The Jaccard index calculates the ratio of the intersection to the union of sets. GGTK provides four Jaccard-based gene similarity measures. Gentleman [25] introduced a Jaccard-based gene similarity measure that calculates the ratio of shared ancestors between two terms to the union of each terms' ancestors. Let $\mathcal{A}_t$ be the set of all ancestors of term t (including t). The set $\mathcal{A}_t$ can be thought of as the induced subgraph of a term t in the ontology DAG. Eq. (10) describes the measure called SimUI, a similarity based on the union and intersection of term ancestors.

$$SimUI(t_1, t_2) = \frac{|\mathcal{A}_{t_1} \cap \mathcal{A}_{t_2}|}{|\mathcal{A}_{t_1} \cup \mathcal{A}_{t_2}|} \tag{10}$$

Pesquita et al. [19] developed an information content weighted version of SimUI called SimGIC, a graph-based information content similarity. SimGIC is defined in Eq. (11).

$$SimGIC(t_1, t_2) = \frac{\sum_{t \in \mathcal{A}_{t_1} \cap \mathcal{A}_{t_2}} IC(t)}{\sum_{t \in \mathcal{A}_{t_1} \cup \mathcal{A}_{t_2}} IC(t)} \tag{11}$$

Studies by Mazandu et al. [26] introduced two modified methods similar to SimGIC. These methods are called by those authors, SimDIC and SimUIC after Dice and *universal* indexes. These measures are defined in Eqs. (12) and (13).

$$SimDIC(t_1, t_2) = \frac{2 * \sum_{t \in \mathcal{A}_{t_1} \cap \mathcal{A}_{t_2}} IC(t)}{\sum_{t \in \mathcal{A}_{t_1}} IC(t) + \sum_{t \in \mathcal{A}_{t_2}} IC(t)} \tag{12}$$

$$SimUIC(t_1, t_2) = \frac{\sum_{t \in \mathcal{A}_{t_1} \cap \mathcal{A}_{t_2}} IC(t)}{\max\left\{\sum_{t \in \mathcal{A}_{t_1}} IC(t), \sum_{t \in \mathcal{A}_{t_2}} IC(t)\right\}} \tag{13}$$

As these methods operate on the gene-level term sets, they avoid the all-pairs-of-terms calculation and need no aggregation step making these Jaccard-based methods more efficient than the shared information methods. The linear time shared information algorithms showed acceptable speed for all the practical applications of this work despite being necessarily slower than these Jaccard-based methods. For this reason, their execution time was not measured.

2.4.3. Decoupling term similarity calculation from gene similarity

As some shared information methods are computationally inefficient, we wanted to decouple the calculation of shared information from the calculation of gene similarity. To achieve efficient calculation of gene similarity, GGTK provides capabilities to generate and import pre-computed term similarity matrices. By pre-computing term similarity, term similarity algorithms can operate on sets of terms and access the similarity of a pair of terms in $O(1)$ time. Although the memory cost would appear to be $O(n^2)$ in the worst case, an optimization mitigates this cost by calculating only similarity between terms that appear in a given corpus (rather than all terms in the ontology). The separation of terms in each of the disjoint ontologies (BP, MF, CC) provides further space savings. If no gene in the corpus has a particular function annotation there is no need to calculate its similarity to other terms. These optimizations greatly reduce the memory cost and allow for efficient calculation of the gene-to-gene functional similarity.

2.4.4. Measuring execution speed of shared information algorithms

The calculation of term similarity matrices also serves as a benchmark for measuring each algorithms' execution speed. In order to quantify the execution speed of the algorithms in each complexity class, the wall-clock time of each matrix calculation was recorded for each shared information algorithm using the Linux *time* command. In calculating the matrix, only those terms with annotations in the human corpus need to be analyzed. Using this optimization, the term similarity matrices were calculated processing 11,394 annotated terms for BP, 4149 for MF, and 1546 terms for CC. Taking advantage of symmetry ($Sim(A,B) = Sim(B,A)$), the resulting number of calculations equals the number of elements in the upper triangular portion of the term similarity matrix $\left(\frac{n(n-1)}{2} \right)$. Where the all-pairs-of-terms similarity is calculated, 64,905,921 pairs were processed for the BP, 8,605,026 for MF, and 1,194,285 for CC. The resulting wall-clock time for each of these calculations is reported as the execution time for each method.

2.5. Semantic similarity performance evaluations

Various performance evaluations have been put forward starting with Lord et al. [11] who measured the Pearson correlation between gene-to-gene semantic similarity scores and sequence similarity. Foundational approaches accepted in the literature measure the correlation between gene-to-gene semantic similarity and sequence, domain set, or expression profile similarity [14]. Perhaps more useful evaluations measure the predictive power of semantic similarity to discover protein–protein interactions [6,29] or to correctly cluster genes belonging to known pathways [21]. Machine learning evaluations such as these have more applications in the field.

Few studies have addressed the inherent uncertainty in both GO graph structure and the limited depth and breadth of gene annotations [29]. The machine learning community has long used bootstrapping to improve generalization and to overcome issues of noise [33]. To address the issue of uncertainty, we employ robust bootstrapping approaches that provide generalized measures of performance as well as performance distributions that allow direct statistical comparisons between similarity methods. In the following sections we describe the specific performance evaluations and datasets used to compare a gene similarity method with respect to the shared information algorithms. In this work we evaluate five shared information algorithms using three term similarity measures across the three ontologies of GO resulting in 45 distinct gene similarity measures (5 shared information methods * 3 term similarity methods * 3 ontology types). In addition, these methods are compared to 12 gene similarity methods derived from the Jaccard-based methods (4 Jaccard-based methods * 3 ontology types). The performance of 57 measures in total has been analyzed.

2.5.1. CESSM

The Collaborative Evaluation of Semantic Similarity Measures (CESSM) [34] is an online dataset and comparison tool used to evaluate gene similarity measures. Though CESSM has some issues [27], the community has accepted it as a useful but limited performance benchmark which can provide comparisons with 11 other measures. The CESSM annotation and test data was downloaded and used to evaluate each measure. The 45 shared information methods were submitted individually to the CESSM server and the results, once collected, were analyzed. CESSM is available at xldb.di.fc.ul.pt/tools/cessm/. CESSM was used to evaluate term aggregation techniques. BMA performed best on this benchmark. The aggregators *max* and *average* performed poorly and were not considered for subsequent analysis. This agrees with similar observations by Pesquita et al. [19] and others. This analysis motivated the choice of BMA and the results for each ontology type are available in Supplemental Information Figures S1–S3.

2.5.2. Relative reciprocal BLAST score

The relative reciprocal BLAST score (RRBS) is a measure of sequence similarity developed by Pesquita et al. [19] derived from the BLAST alignment tool [35]. Eq. (14) gives the RRBS definition between two sequences A and B [19]. All-pairs BLAST was performed using an e-value cut off of 1e−4 after [19] with the human protein dataset. Pesquita et al. noted the relationship between shared information and RRBS is non-linear. For this reason, we evaluated the non-linear Spearman's rank correlation, or Spearman's ρ, between gene similarity values and the sequence alignment derived RRBS. The performance distribution of the Spearman correlation was calculated by taking 1000 bootstrap samples of size 100,000 from the set of all RRBS scores calculated.

$$RRBS(A,B) = \frac{BLAST_{bitscore}(A,B) + BLAST_{bitscore}(B,A)}{BLAST_{bitscore}(A,A) + BLAST_{bitscore}(B,B)} \qquad (14)$$

2.5.3. Jaccard index set similarity of Pfam domains

The Jaccard index [32] measures the similarity between two sets as the size of their intersection divided by the size of their union. In general, the functional domains, rather than just sequence, dictate a protein's function. To this end, Pfam domains [36] have been used in various gene similarity measure evaluations [13,20,27]. Taking genes as sets of domains (D_A and D_B), a domain similarity method is derived (Eq. (15)). On the human protein dataset, Pfam domains were predicted using HMMER3 [37] with an e-value threshold of $3e^{-6}$. After removing gene isoforms, the all-pairs Jaccard Pfam similarity was calculated for the protein dataset. As above, performance distributions for each algorithm were generated by calculating the Spearman correlation between the gene similarity measure and Jaccard Pfam for 1000 bootstrap samples of size 100,000.

$$Sim_{Jaccard}(D_A, D_B) = \frac{|D_A \cap D_B|}{|D_A \cup D_B|} \qquad (15)$$

2.5.4. TF–IDF cosine similarity of Pfam domains

Term frequency –inverse document frequency (TF–IDF) is a technique in information retrieval that weights terms by their relative specificity [38]. Similarly, domains that appear frequently in a diverse set of proteins may have little influence on the protein's function. Song et al. [39] previously applied TF–IDF to protein domains. Eq. (16) shows the weight for a particular domain d belonging to a protein, where f_d is the frequency of that domain in the protein, N is the total number of proteins in the corpus, and n_d is the number of proteins having the domain. From this weight measure, proteins are represented as a vectors of domain weights (A and B)

and cosine similarity (Eq. (17)) represents the protein similarity. Performance distributions for this measure were calculated using Spearman correlation and bootstrapping as described above.

$$w_d = f_d * \log\left(\frac{N}{n_d}\right) \qquad (16)$$

$$Sim_{TF-IDF}(A, B) = \frac{A \cdot B}{\| A \| \| B \|} \qquad (17)$$

2.5.5. Gene expression across 79 human tissues

Highly correlated genes are often functionally related. Studies in the literature [4,23,28] have evaluated performance by measuring gene-to-gene semantic similarity correlation with gene expression correlation. The gene similarity measures were evaluated in terms of their correlation with microarray gene expression across 79 human tissues (NCBI GEO accession GSE1133). Probes from the human U133A array were mapped to their Refseq identifiers which were then mapped to Uniprot identifiers. Genes without any GO annotations in GOA were removed. After filtering, the all-pairs correlation of 5688 genes was calculated resulting in 16,173,828 unique correlation pairs. Both Pearson and Spearman gene correlations were calculated. As in previous literature [6], the absolute value of the correlation was calculated between expression pairs to attempt to detect a relationship either negative or positive. The semantic methods were then evaluated for their Spearman correlation to either the absolute Pearson or Spearman expression correlation. From these correlation pairs, 100 bootstrap samples were taken for each algorithm. Each sample had a size equal to 15% of the total number of pairs (>2.4 million).

2.5.6. Reactome pathway analysis

Uncovering pathway relationships between genes constitutes an important use case for semantic similarity algorithms that has been studied in the literature [3,5,21]. Using Reactome [40], we tested the performance of gene similarity methods in terms of their ability to partition sets of genes into known pathways by clustering. The variation of information (VI) criterion [41] was used to determine the agreement between known Reactome pathways and partitions of genes derived from hierarchical clustering using Ward's method [42] and gene-to-gene semantic similarity based distances (1 - similarity). Specifically, 100 datasets were generated from Reactome by randomly selecting 10 human pathways having between 10 and 150 proteins. These datasets were further processed replacing any overlapping pathways with non-overlapping pathways to remove any ambiguous assignments. For each gene similarity method, distance was calculated and the proteins were clustered into 10 groups. VI was calculated between the gene similarity based clustering and the known pathway assignments from Reactome. The 100 separate datasets were used to construct performance distributions.

2.5.7. Protein–protein interaction prediction

Several works in the literature [3,6,29] have used protein–protein interaction prediction to evaluate semantic similarity measures. Following previous methods, positive and negative datasets were generated from the Interologous Interaction Database (I2D) [43], and the semantic similarity measures were evaluated by their ability to distinguish interacting from non-interacting protein pairs. The I2D public version 2.3 was downloaded from http://ophid.utoronto.ca/ and used as the positive set of interacting proteins (228,847 human interactions). A negative dataset of equal size was generated after Guo et al. [3] by randomly choosing protein pairs resulting in a balanced dataset of 457,694 interactions. For each gene similarity method, 100 bootstrap samples of a size equal to 15% of the original dataset were used to calculate receiver operator characteristic (ROC) curves using the ROCR package [44] of the R programming language.

A ROC curve plots the true positive rate (TPR), the ratio of true predictions to all predictions made, against the false positive rate (FPR), the ratio of false predictions to all prediction, across a range of different thresholds. For the gene-to-gene semantic similarity measures, the ROC curve would be equivalent to sorting all the values of semantic similarity and counting the number of true and false predictions below each unique value for plotting. From these 100 samples, a performance distribution of the area under the ROC curve (AUC) values was used to statistically compare the performance of each semantic similarity measure. A classifier must have an AUC above 50% to be considered better than random guessing.

2.6. Isolating the effect of shared information from other factors

The choice of shared information, term similarity algorithm, and ontology type specify the gene-to-gene semantic similarity measure. To determine the effects of shared information on performance in the previous evaluations it is important to isolate the effects of other factors that influence the results.

2.6.1. Regression analysis of mean performance

Linear regression analysis on the bootstrapped mean performance values is used to examine the effects of shared information type, term similarity algorithm, and gene ontology type. A design matrix was created with each row representing a separate gene-to-gene semantic similarity method constructed through combining the different factors under study. Each column of the design matrix represents the different factors as categorical variables, with the ontology variable taking a value in {BP, MF, CC}, the term similarity variable taking a value in {Resnik, Lin, JC}, and the shared information variable taking a value in {casi, mica, grasm, agrasm, sf}. Using this design matrix, linear models were trained for each performance evaluation using the mean bootstrapped performance as the response variable. The models where creating using the lm function of the R programming language. The influence of each factor is reported as the negative log of the regression p-value taken from the analysis of variance of each fitted model using the anova function of the R programming language. Separate models were fitted for each of the six performance benchmarks and the relative influence of each factor is reported for all evaluations.

2.6.2. Performance ranking and statistical ties

Within each choice of term similarity and ontology type, the shared information algorithms were ranked based on statistical tests of their performance. For a specific choice of term similarity and ontology type, the shared information methods were sorted based on their mean performance and ranked using a statistical method operating on their performance distributions. A statistical tie in performance between two methods is determined if a t-test of the methods' performance distributions fails to reject the null hypothesis of equal means (p-value > 0.05, Welch's two sample t-test). The ranks are exhausted meaning that if two methods are tied for first place, method 1 and 2 receive the rank of 1, but the next best performer receives the rank of 3 (rather than 2). The overall ranks of all shared information methods are reported for every combination of term similarity and ontology type. The average shared information algorithm rank is reported for each ontology type and term similarity method.

2.7. Ensemble classifiers and cross-validation

With such a large number of gene similarity approaches, could they be combined through ensemble methods to improve performance on machine learning tasks? To answer this question, we applied majority voting to the task of protein-protein interaction prediction. A majority voting classifier makes a prediction based on the votes of a panel of other classifiers [45]. Although the area under the ROC curve provides a good estimate of a classifier's behavior across a range of thresholds, in practice, a single threshold must be chosen for a particular task. A process that maximized the F-score over a set of training data selected the threshold for an individual gene-to-gene semantic similarity method to be used as a classifier. The F-score represents the trade off between making true predictions and limiting false predictions. It is the harmonic mean of precision and recall and is defined in Eqs. (18), (19), and (20). These values were calculated using the ROCR [44] package in R. This training process was repeated for every individual method. From these individual classifiers, voting classifiers were constructed that predicted an interaction only if a majority of its constituent classifiers also predicted an interaction. Three ontology specific voting classifier were designed to include only methods in BP, MF, or CC categories. These classifiers helped to determine if any advantage could be derived from ensemble classifiers limited to a single GO type. A final voting predictor was constructed which consisted of all individual methods.

$$precision = \frac{true\ positives}{true\ positives + false\ positives} \qquad (18)$$

$$recall = \frac{true\ positives}{true\ positives + false\ negatives} \qquad (19)$$

$$F\text{-}score = 2 * \frac{precision * recall}{precision + recall} \qquad (20)$$

Ten-fold cross-validation was employed to test the performance of all individual classifiers as well as the four ensemble classifiers. Using the protein interaction dataset from Section 2.5.7, ten random partitions, or folds, were constructed each containing approximately 10% of the original 457,694 interactions. The classifiers were trained using nine folds and tested for the percent of misclassified instances (error rate) on the remaining fold. This process was repeated ten times. The classification error is reported for all 57 methods including the four voting classifiers.

3. Results

3.1. Execution time performance of the shared information algorithms

The all-pairs term similarity matrix for annotated terms in BP, MF, and CC was calculated for each shared information method and the execution times of these calculations were recorded. The wall-clock execution times for these calculations showed, as expected, that GraSM and A-GraSM methods are orders of magnitude slower than the other shared information methods. Fig. 2 shows the execution

times of all shared information methods combined with Resnik term similarity. The other term similarity measures showed comparable execution times. The complete list of execution times is available in Table S1. Both GraSM and A-GraSM have $O(k^3)$ time complexity, and their execution times dwarf the runtimes of the other algorithms. The execution time of the SF algorithm is greater than CASI or MICA. Both CASI and MICA proved to be the most efficient algorithms in terms of execution speed. The term similarity matrices for BP, MF, and CC required 64.91 million, 8.60 million, and 1.19 million term pair calculations respectively. Interestingly, despite requiring over 8 million more calculations, the MF processes completed faster than the CC processes. These results suggest that the topology of the GO graph plays an important role in determining the execution speed of semantic algorithms and that functions of the raw number of terms in an ontology may not accurately reflect their complexity. Fig. 1 supports this conclusion since the branching factor and average number of ancestors of the MF ontology is less than that of the CC ontology despite MF having more terms. The Jaccard-based methods do not require calculating term-to-term similarity so this evaluation does not apply. The Jaccard-based term-set level measures are known to be more efficient. These findings illustrate that the increased time complexity of the GraSM methods could be computationally prohibitive in some situations, and the problem may worsen as GO graph complexity grows.

3.2. RRBS: sequence similarity

Using the 519,892 protein pairs that passed selection, Spearman's ρ correlation between RRBS and each of the 45 shared information-based and 12 Jaccard-based gene similarity measures was evaluated by the bootstrapping method described in the Section 2.5.2. The correlation distributions of all measures are shown in Fig. 3. Supplemental Information Table S2 provides the mean performance and standard deviations for all measures. The measures are organized by term similarity type and ontology type. In general, BP methods perform best in terms of RRBS correlation. For Lin and Jiang-Conrath methods no dramatic differences are observed between shared information algorithms. For the Resnik methods, CASI and MICA lag behind GraSM, A-GraSM, and SF across all ontology types. With the CC ontology, the CASI and MICA Resnik methods show dramatically lower correlation than all other methods. Despite poor and average performance in the CC and MF ontologies respectively, Resnik BP methods using GraSM, A-GraSM, and SF show better correlation with RRBS than any other methods. Of the MF methods, Jiang-Conrath term similarity shows the lowest RRBS correlation among the term similarity algorithms. The Jaccard-based methods perform best out of the MF methods. The correlation of the MF Jaccard methods roughly equals the performance of the BP Jaccard methods. In the BP

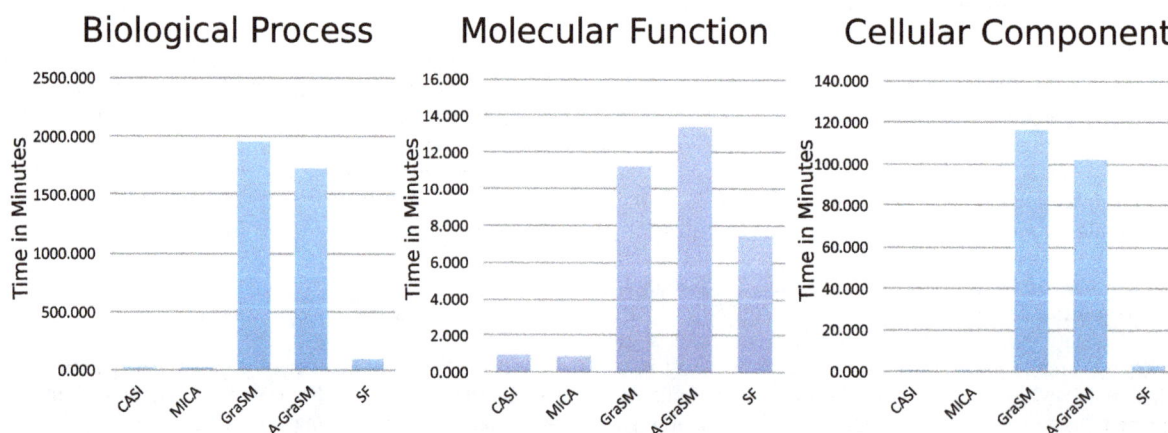

Fig. 2. The execution times for the all-pairs term similarity for Resnik term similarity show that GraSM and A-GraSM methods are slower than other shared information methods.

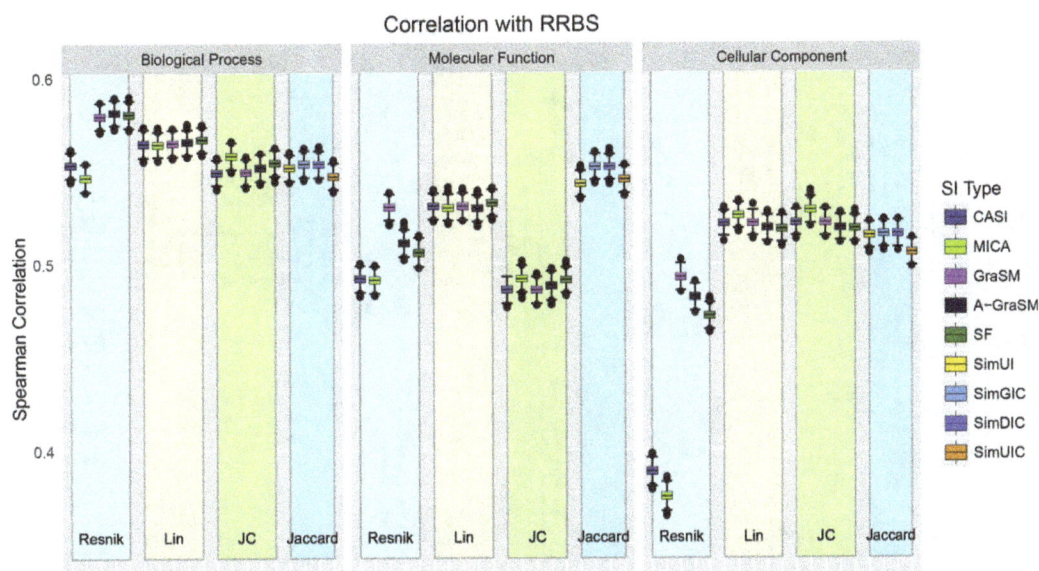

Fig. 3. Performance distributions for the BLAST-based RRBS benchmark for all 57 measures organized by term similarity algorithm and ontology type.

ontology, the Jaccard methods showed similar performance to the Jiang-Conrath methods but were lower than Lin and the best Resnik methods. These results show that BP ontology methods most closely correlate with RRBS similarity. Consistent with the results of Pesquita et al. [19], SimGIC performed well among the MF methods. In that work the authors did not consider BP or CC based methods citing the works Lord et al. [11,12] that report a loose correlation between gene semantic similarity and sequence similarity. Jain and Bader [4] analyzed the correlation between semantic methods and sequence similarity in all ontologies, and their results show the CC methods have a higher sequence correlation than MF or BP. From the results of Mazandu et al. [26], BP and CC methods appear to show better correlation with sequence than MF. As Pesquita et al. [19] demonstrated the relationship between semantic similarity and sequence similarity is not linear, making the Spearman correlation a more appropriate measure. The use of Spearman correlation over Pearson and using more recent annotation and GO graph data likely account for some of the differences with previous works.

Statistical comparisons between performance distributions examine the relative performance between the methods and establish a ranking. This approach to ranking is described in Section 2.6.2. Table 1 shows the rankings for all shared information methods. These data give insight into which shared information algorithms show statically significant differences in performance for each combination of term similarity and ontology type. SF shared information works best with BP and MF methods. CC methods work well with GraSM on the RRBS benchmark. GraSM and A-GraSM show good performance with Resnik. Lin based method showed CASI, GraSM, and SF as the best shared information methods. MICA shared information ranked first in all JC methods.

3.3. Jaccard index of Pfam domain sets

The all-pairs Pfam domain similarity was calculated with the Jaccard index resulting in 1,219,558 protein pairs. The performance distributions are shown in Fig. 4 depicting all 57 measures organized by term similarity type and ontology type. The complete listing of mean correlation performance and standard deviations can be found in Supplemental Information Table S3. As with the RRBS dataset, the shared information methods show smaller variability in performance compared to differences in ontology type and term similarity algorithm. The CASI and MICA methods combined with Resnik term similarity again perform poorly compared to the other methods, but these shared information method are comparable with GraSM, A-GraSM, and SF when combined with Lin and Jiang-Conrath. The CC ontology methods lag the performance of MF and BP in terms of correlation with Pfam Jaccard similarity. The Jaccard-based gene similarity methods perform best with MF annotations, with the best method being SimUI. These methods perform slightly better in MF than in BP. Overall MF Lin methods performed best, but BP Resnik methods combined with GraSM, A-GraSM, and SF outperformed most other methods. As the worst performing ontology group, the CC ontology may fail to encode detailed domain or structural information. Despite having a higher correlation than the CC ontology methods, the BP and MF correlations were lower than in the RRBS benchmark with the best methods only achieving a correlation near 0.35.

Table 2 shows the ranking of shared information methods for the Pfam Jaccard evaluations. Among the ontologies, SF performs best with BP and MF as in the RRBS evaluations. GraSM again shows the best performance for CC methods. Resnik, Lin and Jiang-Conrath

Table 1

Ranking of shared information for the BLAST-based RRBS benchmark against ontology type and term similarity type. Methods with greater correlation have lower rank. Bold font represents the best average rank for each category.

	Biological process				Molecular function				Cellular component				Mean by term similarity		
Method	Resnik	Lin	JC	Mean	Resnik	Lin	JC	Mean	Resnik	Lin	JC	Mean	Resnik	Lin	JC
CASI	4	3	5	4.00	4	2	4	3.33	4	2	2	2.67	4.00	**2.33**	3.67
MICA	5	5	1	3.67	5	4	1	3.33	5	1	1	2.33	5.00	3.33	**1.00**
GraSM	3	3	4	3.33	1	2	4	2.33	1	2	2	**1.67**	**1.67**	**2.33**	3.33
A-GraSM	1	2	3	2.00	2	5	3	3.33	2	4	4	3.33	**1.67**	3.67	3.33
SF	2	1	2	**1.67**	3	1	2	**2.00**	3	5	5	4.33	2.67	**2.33**	3.00

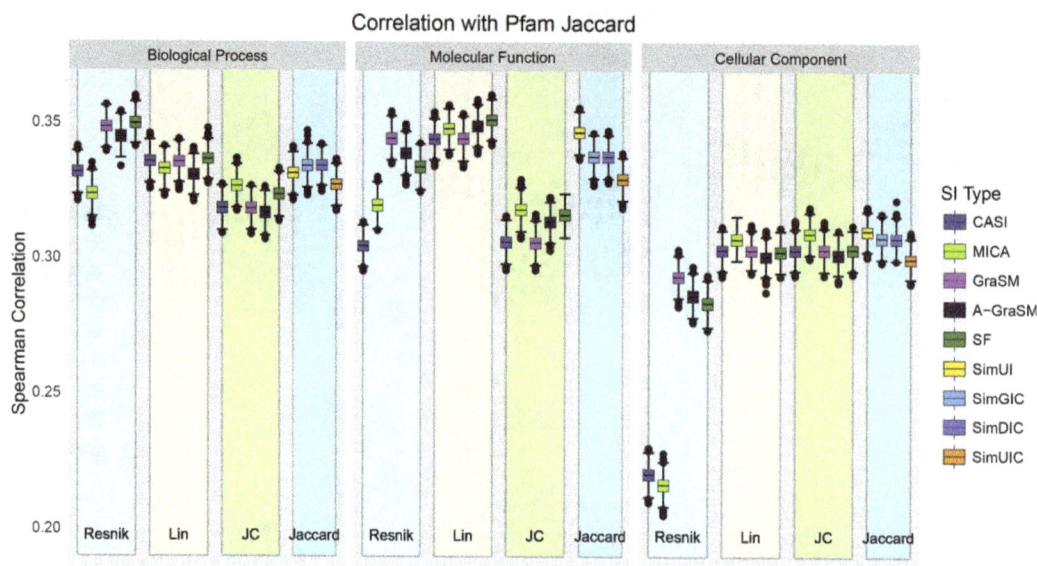

Fig. 4. Performance distributions for the Pfam Jaccard benchmark for all 57 measures organized by term similarity algorithm and ontology type.

method perform best with GraSM, SF, and MCIA respectively on the Pfam Jaccard dataset.

3.4. TF–IDF cosine similarity of Pfam domain sets

To correct for uninformative domains, we applied TF–IDF to proteins represented as sets of Pfam domains. With 1,219,558 unique protein pairs, evaluations were conducted using the bootstrapping method. Fig. 5 shows the distribution of Spearman correlations against TF–IF domain similarity (mean and standard deviation data are available in Supplemental Information Table S4). Again, shared information methods show less variability with the exception of Resnik methods using CASI and MICA. CASI and MICA Resnik methods are dramatically low in the CC ontology and show the lowest correlations of all measures. BP Resnik methods using GraSM, A-GraSM, and SF outperform all other methods in terms of TF–IDF correlation. The Jaccard-based methods perform similarly in BP and CC ontology, but these methods are among the worst performing measures in MF. The CC Jaccard-based methods show greater correlation with TF–IDF than the other CC methods. The BP methods preform best on average in this evaluations. These novels findings suggest that TF–IDF can be successfully applied to protein domain sets to correct for nonspecific domains and potentially uncover relationships in both biological processes and molecular functions.

The performance rankings of the shared information methods in the TF–IDF benchmark are shown in Table 3. With BP and CC methods in the TF–IDF benchmark, GraSM shows the best average ranking. SF shared information works best with MF methods in this dataset. Resnik methods work best with GraSM which is ranked first in all Resnik evaluations. Lin methods show the best correlations using

CASI and GraSM while Jiang-Conrath methods perform best with MICA and SF shared information.

3.5. Gene expression across 79 tissues

All 57 measures were evaluated based on their correlation with gene expression correlation. Fig. 6 shows the absolute gene expression correlation for all methods under study. The complete mean and standard deviation data are available in Supplementary Information Table S5 (Pearson gene correlation) and Table S6 (Spearman). BP Jiang-Conrath methods show the best performance at slightly greater than 0. All other methods show a negative correlation with expression correlation. These findings are consistent with the work of Xu et al. [6] which found that global gene expression correlates poorly with semantic similarity. Xu et al. observed a steady increase in semantic similarity as gene expression pairs are binned into sets of highly correlated gene sets. The same trend is confirmed in our data (Supplemental Figure S4). At such low levels of correlation, one might expect greater variability in performance due to a lack of any strong relationship. The data in Fig. 6 surprisingly shows tighter distributions than in other evaluations. The differences in performance between shared information types are minimal, with the exception of MICA.

The shared information rankings are provided in Table 4 for absolute gene expression. BP, MF, and CC methods work best with SF, MICA, and CASI shared information respectively for gene expression. CASI performs best with Resnik methods and A-GraSM with Lin methods. CASI, GraSM, and A-Grasm work well with Jiang-Conrath methods. The low and negative correlations associated with this benchmark makes the significance of this ranking less clear.

Table 2

Pfam Jaccard ranking of shared information against ontology type and term similarity type. Methods with greater correlation have lower rank. Bold font represents the best average rank for each category.

Method	Biological process				Molecular function				Cellular component				Mean by term similarity		
	Resnik	Lin	JC	Mean	Resnik	Lin	JC	Mean	Resnik	Lin	JC	Mean	Resnik	Lin	JC
CASI	4	2	3	3.00	5	4	4	4.33	4	2	4	3.33	4.33	2.67	3.67
MICA	5	4	1	3.33	4	3	1	2.67	5	1	1	2.33	4.67	2.67	**1.00**
GraSM	2	2	3	2.33	1	4	5	3.33	1	2	2	**1.67**	**1.33**	2.67	3.33
A-GraSM	3	5	5	4.33	2	2	3	2.33	2	5	5	4.00	2.33	4.00	4.33
SF	1	1	2	**1.33**	3	1	2	**2.00**	3	4	2	3.00	2.33	**2.00**	2.00

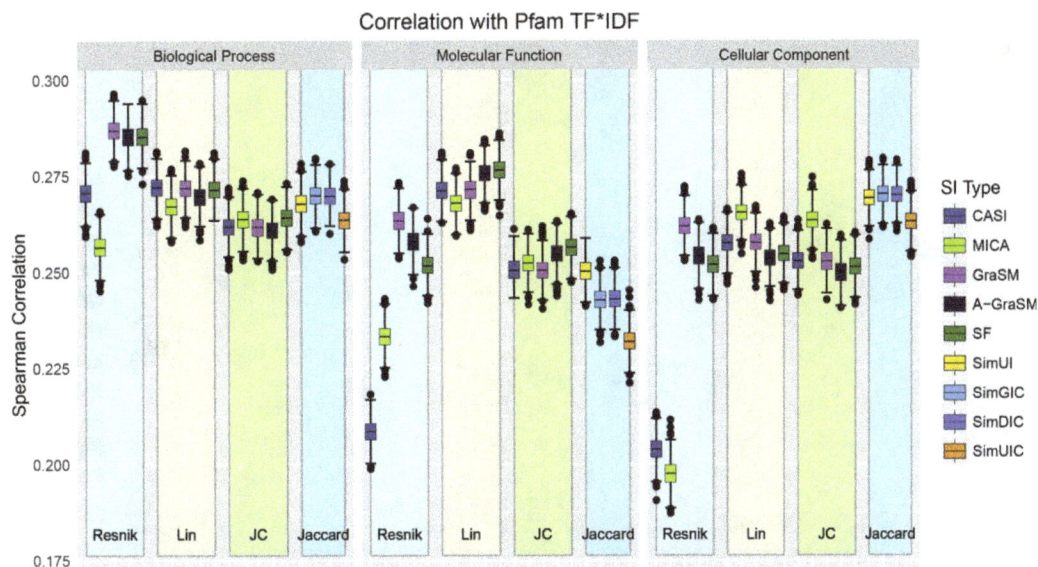

Fig. 5. Performance distributions for the Pfam TF–IDF benchmark for all 57 measures organized by term similarity algorithm and ontology type.

3.6. Reactome clustering

Using 100 randomly generated pathway datasets (see Section 2.5.6), the mean VI similarity (1 - normalized VI distance) was used to compare the closeness of semantic gene similarity derived clusters to their known Reactome assignments. Fig. 7 shows the VI similarity distribution for all 57 methods tested. The performance variability on this benchmark is much higher than in other evaluations. The BP methods outperform those of MF and CC, but within each ontology type all shared information based methods are essentially equivalent regardless of term similarity type. The Jaccard-based method tend to under perform the other methods. As the Reactome pathways were chosen randomly, the high variability in performance is expected. The success of BP methods confirms expectations since the BP ontology captures functional information most closely associated with biological pathways. The CC based methods out-perform MF method on average for this benchmark. The complete results are available in Supplemental Information Table S7.

Table 5 gives the performance rankings of the shared information methods for the Reactome benchmark. As the Reactome results show high variability, most methods could not be determined to be statistically different from one another. All shared information methods within each term similarity group proved to be equivalent in the MF ontology in terms of the VI similarity to Reactome pathway clusters. In BP, the CASI method performs poorly with Resnik. In the CC ontology, CASI and MICA perform poorly with MICA particular ill suited

to the Lin and Jiang-Conrath methods. Resnik methods show poor performance with CASI and MICA, while MICA Lin and Jiang-Conrath methods show poor performance reflecting issues in combination with the CC ontology.

3.7. Protein–protein interaction prediction

The predictive power of each gene similarity measure was evaluated on the human I2D interaction dataset. Performance distributions of the AUC (described in Section 2.5.7) appear in Fig. 8. Complete data for the mean and standard deviation of the AUC are available in Supplemental Information Table S8. This prediction-based evaluation showed heightened variability between shared information methods compared to other datasets, especially with the Jiang-Conrath methods. Within Jiang-Conrath term similarity, MICA shared information performed best. The shared information methods varied greatly within CC Reskin with the SF and A-GraSM versions giving the best AUC values of all methods. The Jaccard-based methods show the worst performance among MF methods and are superior to only the Jiang-Conrath methods in the BP and CC ontologies. BP and MF methods tend to under perform compared to CC methods with the exception of MF Jiang-Conrath. This result is at odds with older evaluations such as Guo et al. [3] that found BP methods are better predictors than CC methods. These differences are likely due to the smaller dataset used in their analysis as well as the changes in GO structure and annotations in recent years. Conflicting performance reports in the literature illustrate the need for

Table 3

Pfam TF–IDF ranking of shared information by term similarity type and ontology. Methods with greater correlation have lower rank. Bold font represents the best average rank for each category.

Method	Biological process				Molecular function				Cellular component				Mean by term similarity		
	Resnik	Lin	JC	Mean	Resnik	Lin	JC	Mean	Resnik	Lin	JC	Mean	Resnik	Lin	JC
CASI	4	1	3	2.67	5	3	4	4.00	4	2	2	2.67	4.33	**2.00**	3.00
MICA	5	5	2	4.00	4	5	3	4.00	5	1	1	2.33	4.67	3.67	**2.00**
GraSM	1	1	3	**1.67**	1	3	4	2.67	1	2	2	**1.67**	**1.00**	**2.00**	3.00
A-GraSM	2	4	5	3.67	2	2	2	2.00	2	5	5	4.00	2.00	3.67	4.00
SF	2	3	1	2.00	3	1	1	**1.67**	3	4	4	3.67	2.67	2.67	**2.00**

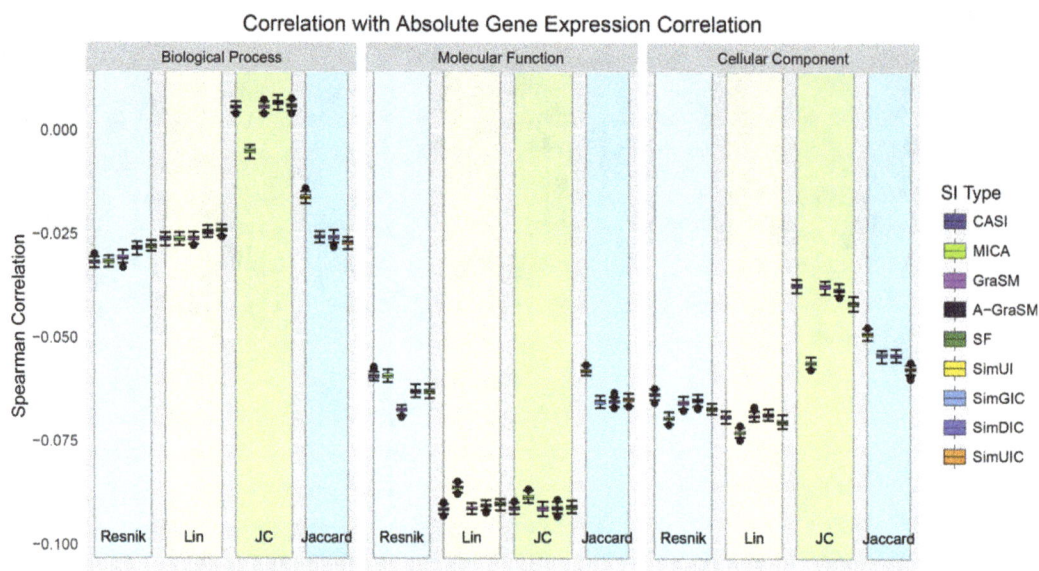

Fig. 6. Performance distributions for absolute gene expression correlation (Pearson) against all 57 measures organized by term similarity algorithm and ontology type.

transparent open source tools that can evaluate these methods on a level playing field using consistent ontologies and annotations.

The performance rankings for the protein–protein interaction prediction benchmark is shown in Table 6. Based on the distribution of the area under the ROC curve values, the shared information methods were statistically compared. MICA, A-GraSM, and SF shared information perform best for BP, MF, and CC ontologies respectively. GraSM and A-GraSM perform well for Resnik on this benchmark. CASI is the best performer for Lin term similarity, and MICA works best for JC across all the ontologies in these evaluations.

3.8. Analysis of factors affecting semantic similarity performance

Using the six performance benchmarks described in the previous section, regression models were trained to assess the influence of ontology type, term similarity type, and shared information type on the mean performance of the shared information methods. A design matrix of categorical variables was constructed and used to create a linear model of mean performance (described in Section 2.6.1). Fig. 9 presents the influence of each factor on each benchmark as the negative log of the regression p-value. This data demonstrates that the choice of ontology contributes most to the variability in performance followed by the choice of term similarity, and shared information contributes the least. All evaluations tested uphold this trend. The Reactome benchmark shows the strongest influence by ontology followed by the gene expression evaluations. Ontology type exerts the smallest influence on TF–IDF and PPI benchmarks; however, this effect is still much greater than the other two factors. The effects of

term similarity and shared information choice are small across all evaluations except PPI where the effects are negligible or zero. These results clearly show that the choice of ontology greatly effects the performance of gene similarity methods.

3.9. Leveraging semantic methods with majority voting

Using the protein–protein prediction dataset from the previous evaluations, we addressed the feasibility of combining semantic similarity methods to improve prediction. Using the 10-fold cross-validation method (described in Section 2.7), we analyzed the performance of all 57 measures in terms of their mean prediction error. For this evaluation, the percent of misclassified instances represents the classification error of each gene similarity semantic measure and voting classifier. Fig. 10 presents the cross-validation classification error for all methods organized by ontology type and term similarity method (lower is better). The ontology specific voting classifier performed equivalently with other methods within the ontology (the far right classifier within each ontology group in Fig. 10). Other single classifiers outperform the ontology specific voting predictor within each ontology. This result indicates that simple majority voting offers no advantage over the best methods within an ontology group; however, the combination of all semantic similarity methods substantially out performs even the best individual classifier. The failure of ontology specific voting predictors to confer any advantage for prediction suggests that variation among shared information or term similarity semantic methods alone cannot improve learning at least for this task. Analysis of the factors affecting semantic

Table 4

Absolute gene correlation benchmark ranking for shared information methods organized by term similarity type and ontology. Methods with greater correlation have lower rank. Bold font represents the best average rank for each category.

Method	Biological process				Molecular function				Cellular component				Mean by term similarity		
	Resnik	Lin	JC	Mean	Resnik	Lin	JC	Mean	Resnik	Lin	JC	Mean	Resnik	Lin	JC
CASI	4	3	3	3.33	1	5	3	3.00	1	2	1	**1.33**	**2.00**	3.33	**2.33**
MICA	4	3	5	4.00	1	1	1	**1.00**	5	5	5	5.00	3.33	3.00	3.67
GraSM	3	3	3	3.00	5	4	3	4.00	3	2	1	2.00	3.67	3.00	**2.33**
A-GraSM	2	2	1	1.67	3	3	3	3.00	2	1	3	2.00	2.33	**2.00**	**2.33**
SF	1	1	2	**1.33**	3	2	2	2.33	4	4	4	4.00	2.67	2.33	2.67

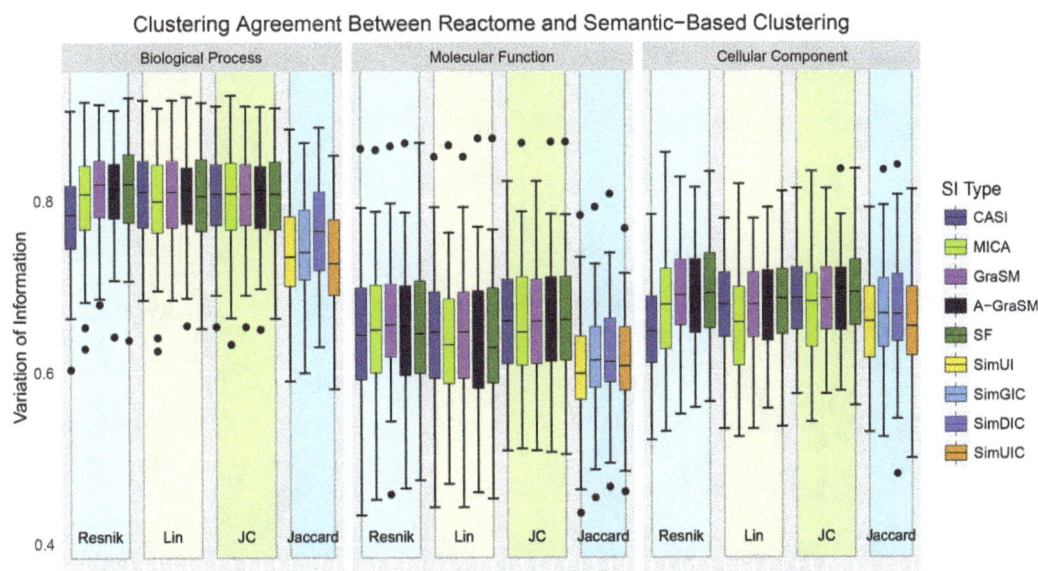

Fig. 7. Performance distributions for Reactome clustering compared to the 57 gene similarity semantic measures organized by term similarity algorithm and ontology type.

similarity performance of the previous section support this conclusion. The cross-validation evaluation show that BP methods tend to perform worse than all MF methods and worse than all but Jiang-Conrath CC methods. Interestingly, the MF methods perform best in cross-validation analysis while the CC methods tend to perform best in terms of the area under the ROC curve calculated through boot-strapping. These results indicate that the AUC metric may not totally capture the usefulness of a measure for classification.

4. Discussion

As semantic similarity measures have grown in popularity, a large number of measures have been developed making exhaustive performance assessment a growing challenge. IC based methods use measures of shared information to calculate term similarity and ultimately gene similarity. Recent works have devised alternative methods for calculating shared information. Through modular combination of shared information measures, term similarity calculations, and ontology choice, the number of semantic gene similarity measures explodes creating obvious challenges for performance evaluations. In a departure from previous works, we put forward robust methods to statistically compare semantic gene similarity measures in a manner that captures their generalized performance. We apply these methods to conduct a thorough investigation into the behavior of varying shared information measures. Given the large number of measure that can be created though modifications to the shared information, we considered the feasibility of leveraging these

measures in ensemble classifiers for prediction gains. By isolating the effects of ontology type, we determined that the shared information algorithms themselves could not be combined to improve protein-protein interaction prediction. Methods across the distinct ontologies can be leveraged to improve prediction.

Xu et al. [6] combined information from across ontologies to improve prediction; however, they incorporated an artificial root node connecting BP, MF, and CC into a single graph. Their research found that BP methods offered the best performance on a yeast dataset. Later Jain and Bader [4] also showed BP to out perform MF and CC methods in interaction prediction in yeast based on area under the ROC curve analysis both with and without the inclusion of electronically inferred annotations. While our work focused on human protein–protein interactions, the results resemble the more recent work of Yang et al. [29] in yeast where they found CC to perform better than other methods. Based on ROC analysis, our results showed CC methods performing best. The work by Yang et al. cites Collins et al. [46] in identifying that much of the ROC curve represents classification thresholds that would be useless in practice for protein interaction prediction. This results from thresholds that admit far too many false positive predictions to be useful. This issue became evident in our analysis when many of the MF predictors showed strong performance based on cross-validation evaluation but less competitive performance when only the area under the ROC curve was considered.

The work by Mazandu et al. [26] more closely relates to this study as they used human interactions to assess the predictive power

Table 5
Reactome based statistical ranking for shared information methods organized by term similarity type and ontology. Methods showing a higher VI similarity to Reactome pathways have lower rank. Bold font represents the best average rank for each category.

Method	Biological process				Molecular function				Cellular component				Mean by term similarity		
	Resnik	Lin	JC	Mean	Resnik	Lin	JC	Mean	Resnik	Lin	JC	Mean	Resnik	Lin	JC
CASI	5	1	1	2.33	1	1	1	**1.00**	5	1	1	2.33	3.67	**1.00**	**1.00**
MICA	1	1	1	**1.00**	1	1	1	**1.00**	4	5	5	4.67	2.00	2.33	2.33
GraSM	1	1	1	**1.00**	1	1	1	**1.00**	1	1	1	**1.00**	**1.00**	**1.00**	**1.00**
A-GraSM	1	1	1	**1.00**	1	1	1	**1.00**	1	1	1	**1.00**	**1.00**	**1.00**	**1.00**
SF	1	1	1	**1.00**	1	1	1	**1.00**	1	1	1	**1.00**	**1.00**	**1.00**	**1.00**

Area Under the ROC Curve for PPI Prediction

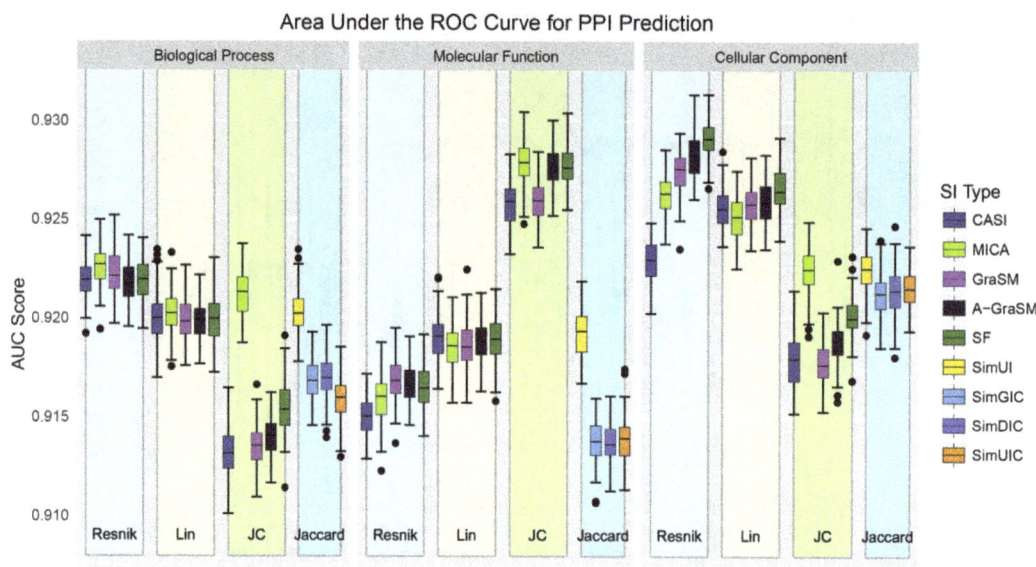

Fig. 8. Performance distributions for protein–protein interaction prediction by area under the ROC curve for the 57 gene similarity semantic measures organized by term similarity algorithm and ontology type.

Table 6
Protein interaction prediction ranking for shared information methods organized by term similarity type and ontology. Rankings were established by comparisons based on the area under the ROC curve calculated from bootstrap samples. Methods with greater AUC scores have lower rank. Bold font represents the best average rank for each category.

Method	Biological process				Molecular function				Cellular component				Mean by term similarity		
	Resnik	Lin	JC	Mean	Resnik	Lin	JC	Mean	Resnik	Lin	JC	Mean	Resnik	Lin	JC
CASI	3	1	4	2.67	5	1	4	3.33	5	2	4	3.67	4.33	**1.33**	4.00
MICA	1	1	1	**1.00**	4	4	1	3.00	4	5	1	3.33	3.00	3.33	**1.00**
GraSM	2	3	4	3.00	1	4	4	3.00	3	2	4	3.00	**2.00**	3.00	4.00
A-GraSM	3	3	3	3.00	1	1	1	**1.00**	2	2	3	2.33	**2.00**	2.00	2.33
SF	3	3	2	2.67	3	1	1	1.67	1	1	2	**1.33**	2.33	1.67	1.67

of semantic algorithms. The data set used by Mazandu et al. was comprised of a much smaller set of roughly 5000 curated interaction which were filtered to all contain BP and CC annotations. Citing their previous work [47], Mazandu et al. excluded MF based methods in their evaluation, and other work by Mazandu et al. [48] supported the exclusion of Jiang-Conrath methods. These caveats make a direct comparison to their work difficult. Mazandu et al. examined a large number of diverse measures which are out of score for this study of shared information methods. The shared methods among the two studies are the best match average (BMA) versions of

Resnik (called RBMA, by Mazandu et al.), Lin (LBMA), SimUI, SimGIC (AGIC), SimDIC (ADIC), and SimUIC (AUIC). The area under the ROC curve reported Mazandu et al. for the CC base methods is 0.9999656 (RBMA), 0.4853167 (LBMA), 0.8483416 (SimUI), 0.9173889 (AGIC), 0.8486233 (ADIC), and 0.9654985 (AUIC). The values reported for the BP verions of these methods are 0.9995277 (RBMA), 0.6194642 (LBMA), 0.9582268 (SimUI), 0.9689432 (AGIC), 0.9514534 (ADIC), and 0.9654985 (AUIC). The range of these values is consistent with those found in this study for the area under the ROC curve benchmark. All the methods tested achieved high performance using the most up-to-date annotations and GO graphs. Inconsistent with the reports of this study is the disparity between Resnik and Lin methods within the same ontology. Based on our evaluations, we would expect these methods to have scored more closely in the evaluations of Mazandu et al.

Difficulty in comparing the results of semantic similarity analysis in GO is a known problem [14]. In an excellent review, Mazandu et al. [15] describe two key challenges known as the *dataset issue*, where different tools use different version of GO or annotation datasets, and the *scaling issue* that results from tools making different assumption regarding normalization methods and other minor considerations such as root membership in ancestor sets etc. GGTK is an attempt to provide correct, transparent, and modular implementations of semantic algorithms where the dataset issue can be tackled easily and the assumptions that could lead to the scaling issue are clearly stated. The benchmarks used in this work are provided as simple lists of protein pairs and scores to facilitate comparisons by other researchers (available at github.com/paulbible/ggtk). This work demonstrates that many conclusions from the literature still

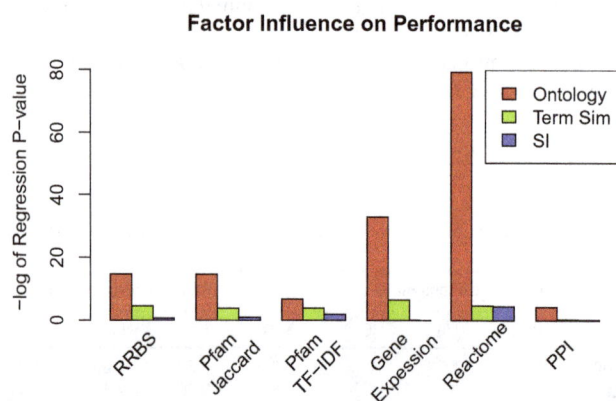

Fig. 9. The relative influence of ontology type, term similarity method, and shared information type (SI) on the mean performance across six evaluations.

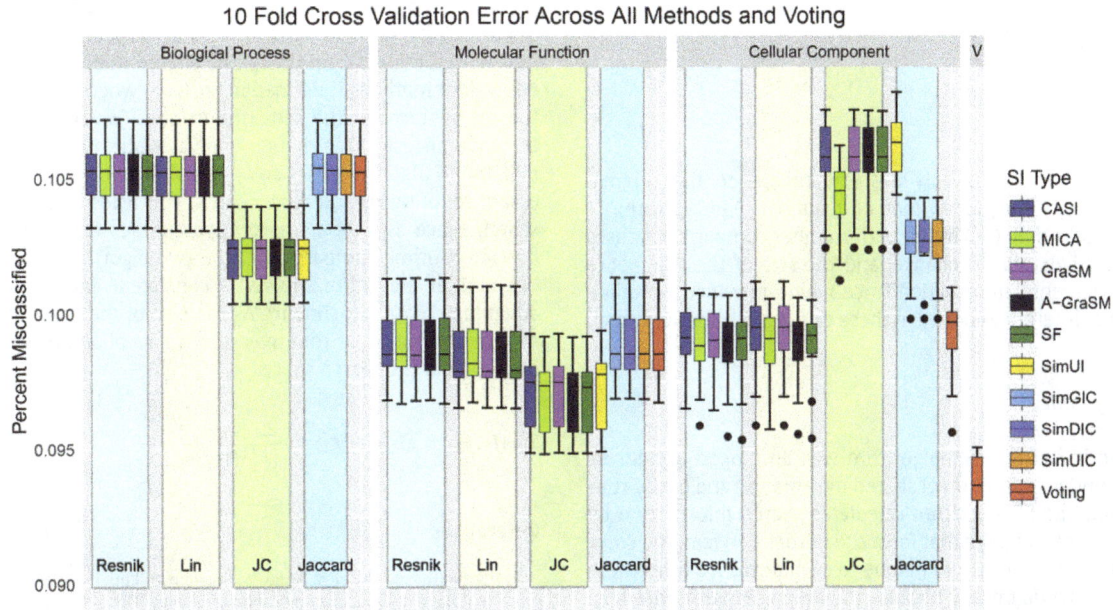

Fig. 10. The percent of misclassified samples for each method under study, trained as classifiers, and four voting predictors evaluated by 10-fold cross-validation. The voting predictor for an ontology type is presented as the last classifier within that ontology (red), and the voting predictor utilizing all semantic methods is presented at the far right.

hold despite changes in the structure of the ontologies and increases in the number of annotations. In opposition to previous reports, we find that MF methods performed best in predicting protein interactions using cross-validation. We found that an ensemble of gene similarity semantic methods could out perform any single method in interaction prediction, but this boost in performance results mainly from the integration of distinct information gained through the separate ontologies of BP, MF, and CC. Furthermore, we have developed an efficient C++ toolkit with an easy-to-use Python interface, GGTK, that will not only allow other researchers to take advantage of these findings but also facilitate their own research into semantic methods.

Acknowledgments

This work is supported by the National Basic Research Program of China 2015CB964601 and 2013CB967001, the Li Foundation Heritage Prize, the 1000 Young Talent Plan China, the National Natural Science Foundation of China 81570828, and the NSFC-Guangdong Fund for Application of Supercomputing using the Tianhe 2 Supercomputer to L. Wei. This study also utilized the high-performance computational capabilities of the Helix Systems at the National Institutes of Health, Bethesda, MD (http://helix.nih.gov). Additional support provided by the Intramural Research Program and the Office of Science and Technology of the National Institute of Arthritis and Musculoskeletal and Skin Diseases of the National Institutes of Health.

Appendix A. Detailed description of shared information algorithms

A.1. Common ancestor shared information

Common ancestor shared information (CASI) is a naive baseline algorithms that calculated the simple mean IC of all common ancestors. Eq. (A.2) defines CASI where $|CA(t_1, t_2)|$ is the number of shared ancestors.

$$SI_{CASI}(t_1, t_2) = \frac{\sum_{t \in CA(t_1, t_2)} IC(t)}{|CA(t_1, t_2)|} \qquad (A.1)$$

A.2. Most informative common ancestor shared information

Most informative common ancestor (MICA) shared information is equivalent to Resnik term similarity and is defined by the IC of minimum subsumer. MICA shared information is completely described in the main text above through Eqs. (3), (4), and (5). The MICA represents the most specific concept that contains both t_1 and t_2 as descendants.

A.3. Couto et al. 2007, GraSM

The graph-based semantic measure or GraSM was developed by Couto et al. [20]. GraSM calculates the set of disjunctive common ancestors (DCA) and computes their average as the shared information. Eq. (A.2) defines the disjunctive ancestor pairs for a term t where a_1 and a_2 are ancestor terms and *Paths* represents the number of unique paths from the ancestor to the descendant [49]. Couto et al. [20] give an algorithm for calculating if a pair of ancestor terms (a_1, a_2) are disjunctive in a descendant term t reproduced here in Algorithm 1. From this definition, the DCA set is defined in Eq. (A.3). Finally, the shared information is calculated as the mean IC of the set of $DCA(t_1, t_2)$ in Eq. (A.4).

$$DA(t) = \{(a_1, a_2) | (\exists p : p \in Paths(a_1, t) \wedge a_2 \notin p) \wedge$$
$$(\exists p : p \in Paths(a_2, t) \wedge a_1 \notin p)\} \qquad (A.2)$$

Algorithm 1. $DisjAnc(t, (a_1, a_2))$

Require: $IC(a_1) \leq IC(a_2)$
1: $nPaths = |Paths(a_1, a_2)|$
2: $nPaths_1 = |Paths(a_1, t)|$
3: $nPaths_2 = |Paths(a_2, t)|$
4: **return** $nPaths_1 \geq nPaths * nPaths_2$

$$DCA(t_1, t_2) = \{a_1 | a_1 \in CA(t_1, t_2) \wedge$$
$$\forall a_2 : (a_2 \in CA(t_1, t_2) \wedge IC(a_1) \leq IC(a_2) \wedge a_1 \neq a_2)$$
$$\implies (a_1, a_2) \in DA(t_1) \cup DA(t_2)\} \qquad (A.3)$$

$$SI_{GraSM}(t_1, t_2) = \frac{\sum_{t \in DCA(t_1, t_2)} IC(t)}{|DCA(t_1, t_2)|} \qquad (A.4)$$

A.4. Adjusted GraSM

Adjusted GraSM makes only a slight change to the original algorithm. Changing the *greater than or equal to* ($\geq$) in Algorithm 1 to a *strictly greater than* ($>$) in the path number comparison causes the algorithm's behavior to change, and the size of the DCA set is reduced. Supplemental Information Trace 1 and 2 provided a detailed example of the differences between these two algorithms on a small calculation.

A.5. Semantic frontier

The semantic frontier (SF) algorithm was developed to address the potential under estimation of shared information and costly runtime of GraSM. The SF algorithm calculates shared information by averaging the IC of concepts that form the border between the common ancestor set and ancestors unique to each term. We term these concepts the *semantic frontier* because they represent different borders to a shared territory. The SF calculation is outlined in Algorithms 2 and 3. The *target* function in Algorithm 3 returns the destination node of a directed edge. The SF algorithm is based on the concept of a breadth first search (BFS) visitor. A BFS visitor performs actions when certain events in a BFS occur. The SF algorithm constructs a BFS visitor that starts at a specific term in the GO graph and visits all ancestor terms on its way to the ontology root. Anytime a new edge is visited, it is added to the *visitedEdges* set for the starting term. Two sets are constructed in this way (*visitedEdgesT1* and *visitedEdgesT2*) for the terms of interest t_1 and t_2. Any term t that satisfies Eq. (A.5) is a member of the semantic frontier of t_1 and t_2.

Algorithm 2. *SemanticFrontier*(t_1, t_2)

```
1:  Anc = ConnomAncestors(t₁, t₂)
2:  visitedEdgesT1 = ∅
3:  visitedEdgesT2 = ∅
4:  IncidentEdges = An empty Map of vertex v to incident Edges e
5:  VisitedEdgeBFS(t₁, visitedEdgesT1, IncidentEdges)
6:  VisitedEdgeBFS(t₂, visitedEdgesT2, IncidentEdges)
7:  frontierTerms = ∅
8:  for a ∈ Anc do
9:      for e ∈ IncidentEdges[a] do
10:         if (e ∉ visitedEdgesT1) ∨ (e ∉ visitedEdgesT2) then
11:             frontierTerms = frontierTerms ∪ a
12:         end if
13:     end for
14: end for
15: return frontierTerms
```

Algorithm 3. *VisitedEdgesBFS*(t, *visitedEdges*, *IncidentEdges*)

```
1:  Construct a breadth first search visitor starting at node t
2:  On ExamineEdge(e):
3:      visitedEdges = visitedEdges ∪ e
4:      v = target(e)
5:      IncidentEdges[v] = IncidentEdges[v] ∪ e
```

$$t \in CA(t_1, t_2) \wedge \exists e \,|\, target(e) = t \wedge$$
$$((e \in visitedEdgesT1 \wedge e \notin visitedEdgesT2) \vee$$
$$(e \in visitedEdgesT2 \wedge e \notin visitedEdgesT1)) \qquad (A.5)$$

The SF algorithm is an efficient implementation of an algorithm developed by Zhang and Lai [21] that they call exclusively inherited shared information (EISI). The shared information of these two equivalent methods is defined in Eq. (A.6) where $SF(t_1, t_2)$ represents the set of terms in the semantic frontier (or exclusively inherited common ancestors). Although equivalent in runtime complexity to EISI, the SF algorithm uses an edge visitor to construct only the necessary set of terms and descendants. This leads to a reduction in the search space as well as more opportunities for early exit. A side-by-side runtime comparison of the two algorithms was conducted, but neither algorithm showed a consistent and repeatable speed advantage (data not shown). As the size of the GO graph continues to grow, the SF algorithm may offer more of an advantage through reduction of the search space.

$$SI_{SF}(t_1, t_2) = SI_{EISI}(t_1, t_2) = \frac{\sum_{t \in SF(t_1, t_2)} IC(t)}{|SF(t_1, t_2)|} \qquad (A.6)$$

References

[1] Ashburner M, Ball CA, Blake JA, Botstein D, Butler H, Cherry JM. et al. Gene ontology: tool for the unification of biology. The Gene Ontology Consortium.. Nat Genet 2000;25(1):25–9. http://dx.doi.org/10.1038/75556.

[2] Consortium GO. et al. Creating the gene ontology resource: design and implementation. Genome Res 2001;11(8):1425–33.

[3] Guo X, Liu R, Shriver CD, Hu H, Liebman MN. Assessing semantic similarity measures for the characterization of human regulatory pathways. Bioinformatics 2006;22(8):967–73.

[4] Jain S, Bader GD. An improved method for scoring protein–protein interactions using semantic similarity within the gene ontology. BMC Bioinf 2010;11(1):1.

[5] Wang JZ, Du Z, Payattakool R, Philip SY, Chen C-F. A new method to measure the semantic similarity of Go terms. Bioinformatics 2007;23(10):1274–81.

[6] Xu T, Du L, Zhou Y. Evaluation of Go-based functional similarity measures using *S. cerevisiae* protein interaction and expression profile data. BMC Bioinf 2008;9(1):1.

[7] Mulder NJ, Akinola RO, Mazandu GK, Rapanoel H. Using biological networks to improve our understanding of infectious diseases. Comput Struct Biotechnol J 2014;11(18):1–10.

[8] Vafaee F, Rosu D, Broackes-Carter F, Jurisica I. Novel semantic similarity measure improves an integrative approach to predicting gene functional associations. BMC Syst Biol 2013;7(1):22.

[9] Pandey G, Myers CL, Kumar V. Incorporating functional inter-relationships into protein function prediction algorithms. BMC Bioinf 2009;10(1):142.

[10] Jiang Y, Clark WT, Friedberg I, Radivojac P. The impact of incomplete knowledge on the evaluation of protein function prediction: a structured-output learning perspective. Bioinformatics 2014;30(17):i609–i616.

[11] Lord PW, Stevens RD, Brass A, Goble CA. Investigating semantic similarity measures across the gene ontology: the relationship between sequence and annotation. Bioinformatics 2003;19(10):1275–83.

[12] Lord PW, Stevens RD, Brass A, Goble CA. Semantic similarity measures as tools for exploring the gene ontology.. Pacific symposium on biocomputing. vol. 8. 2003. p. 601–12.

[13] Pesquita C, Faria D, Falcao AO, Lord P, Couto FM. Semantic similarity in biomedical ontologies. PLoS comput biol 2009;5(7). e1000443.

[14] Guzzi PH, Mina M, Guerra C, Cannataro M. Semantic similarity analysis of protein data: assessment with biological features and issues. Brief Bioinform 2012;13(5):569–85.

[15] Mazandu GK, Chimusa ER, Mulder NJ. Gene ontology semantic similarity tools: survey on features and challenges for biological knowledge discovery. Brief Bioinform 2016; bbw067.

[16] Resnik P. et al. Semantic similarity in a taxonomy: an information-based measure and its application to problems of ambiguity in natural language. J Artif Intell Res(JAIR) 1999;11:95–130.

[17] Lin D. An information-theoretic definition of similarity.. ICML. vol. 98. Citeseer.1998. p. 296–304.

[18] Jiang JJ, Conrath DW. Semantic similarity based on corpus statistics and lexical taxonomy. arXiv preprint cmp-lg/9709008.

[19] Pesquita C, Faria D, Bastos H, Ferreira AE, Falcão AO, Couto FM. Metrics for go based protein semantic similarity: a systematic evaluation. BMC bioinf 2008;9(5):1.

[20] Couto FM, Silva MJ, Coutinho PM. Measuring semantic similarity between gene ontology terms. Data Knowl Eng 2007;61(1):137–52.

[21] Zhang S-B, Lai J-H. Semantic similarity measurement between gene ontology terms based on exclusively inherited shared information. Gene 2015;558(1):108–17.

[22] Tao Y, Sam L, Li J, Friedman C, Lussier YA. Information theory applied to the sparse gene ontology annotation network to predict novel gene function. Bioinformatics 2007;23(13):i529–i538.

[23] Sevilla JL, Segura V, Podhorski A, Guruceaga E, Mato JM, Martinez-Cruz LA. et al. Correlation between gene expression and Go semantic similarity. IEEE/ACM Trans. Comput. Biol. Bioinform. 2005;2(4):330–8.

[24] Azuaje F, Wang H, Zheng H, Bodenreider O, Chesneau A. Predictive integration of gene ontology-driven similarity and functional interactions. Sixth IEEE International Conference on Data Mining-Workshops (ICDMW'06). IEEE.2006. p. 114–9.

[25] Gentleman R. Visualizing and distances using Go. :\ignorespaceshttp://www.bioconductor.org/docs/vignetteshtml.

[26] Mazandu GK, Mulder NJ. Information content-based gene ontology functional similarity measures: which one to use for a given biological data type? PloS one 2014;9(12). e113859.

[27] Xu Y, Guo M, Shi W, Liu X, Wang C. A novel insight into gene ontology semantic similarity. Genomics 2013;101(6):368–75.

[28] Wang H, Azuaje F, Bodenreider O, Dopazo J. Gene expression correlation and gene ontology-based similarity: an assessment of quantitative relationships. Computational intelligence in bioinformatics and computational biology, 2004. CIBCB'04. Proceedings of the 2004 IEEE symposium on. IEEE.2004. p. 25–31.

[29] Yang H, Nepusz T, Paccanaro A. Improving Go semantic similarity measures by exploring the ontology beneath the terms and modelling uncertainty. Bioinformatics 2012;28(10):1383–9.

[30] Thomas PD, Wood V, Mungall CJ, Lewis SE, Blake JA, Consortium GO. et al. On the use of gene ontology annotations to assess functional similarity among orthologs and paralogs: a short report. PLoS Comput Biol 2012;8(2). e1002386.

[31] Barrell D, Dimmer E, Huntley RP, Binns D, ODonovan C, Apweiler R. The GOA database in 2009—an integrated gene ontology annotation resource. Nucleic Acids Res 2009;37(suppl. 1):D396–D403.

[32] Jaccard P. The distribution of the flora in the alpine zone.. New Phytol 1912;11(2):37–50.

[33] Breiman L. Bagging predictors. Mach Learn 1996;24(2):123–40.

[34] Pesquita C, Pessoa D, Faria D, Francisco Couto. CESSM: Collaborative evaluation of semantic similarity measures. JB2009: Challenges Bioinform 2009;157:190.

[35] Altschul SF, Gish W, Miller W, Myers EW, Lipman DJ. Basic local alignment search tool.. J Mol Biol 1990;215(3):403–10. http://dx.doi.org/10.1006/jmbi.1990.9999.

[36] Finn RD, Bateman A, Clements J, Coggill P, Eberhardt RY, Eddy SR, Heger A, Hetherington K, Holm L, Mistry J. et al. Pfam: the protein families database. Nucleic Acids Res 2013;gkt1223.

[37] Eddy SR. et al. A new generation of homology search tools based on probabilistic inference. Genome Inform. vol. 23. 2009. p. 205–11.

[38] Salton G, Buckley C. Term-weighting approaches in automatic text retrieval. Inf Process Manag 1988;24(5):513–23.

[39] Song N, Sedgewick RD, Durand D. Domain architecture comparison for multidomain homology identification. J Comput Biol 2007;14(4):496–516.

[40] Croft D, Mundo AF, Haw R, Milacic M, Weiser J, Wu G. et al. The reactome pathway knowledgebase. Nucleic Acids Res 2014;42(D1):D472–D477.

[41] Marina Meilă. Comparing clusterings—an information based distance. J Multivar Anal 2007;98(5):873–95.

[42] Ward JH Jr,. Hierarchical grouping to optimize an objective function. J Am Stat Assoc 1963;58(301):236–44.

[43] Brown KR, Jurisica I. Unequal evolutionary conservation of human protein interactions in interologous networks. Genome Biol 2007;8(5):1.

[44] Sing T, Sander O, Beerenwinkel N, Lengauer T. ROCR: visualizing classifier performance in R. Bioinformatics 2005;21(20):7881. :\ignorespaceshttp://rocr.bioinf.mpi-sb.mpg.de.

[45] Dietterich TG. Ensemble methods in machine learning. International workshop on multiple classifier systems. Springer.2000. p. 1–15.

[46] Collins SR, Kemmeren P, Zhao X-C, Greenblatt JF, Spencer F, Holstege FC. et al. Toward a comprehensive atlas of the physical interactome of *Saccharomyces cerevisiae*. Mol Cell Proteomics 2007;6(3):439–50.

[47] Mazandu GK, Mulder NJ. A topology-based metric for measuring term similarity in the gene ontology. Adv Bioinforma 2012;

[48] Mazandu GK, Mulder NJ. Information content-based gene ontology semantic similarity approaches: toward a unified framework theory. Biomed Res Int 2013;

[49] Couto FM, Silva MJ. Disjunctive shared information between ontology concepts: application to gene ontology. J Biomed Semantics 2011;2(1):1.

13

Deep Assessment of Genomic Diversity in Cassava for Herbicide Tolerance and Starch Biosynthesis

Jorge Duitama [a,d,*], Lina Kafuri [b,c], Daniel Tello [b,c], Ana María Leiva [a], Bernhard Hofinger [b], Sneha Datta [b], Zaida Lentini [c], Ericson Aranzales [a], Bradley Till [b,1], Hernán Ceballos [a,1]

[a] Agrobiodiversity Research Area, International Center for Tropical Agriculture (CIAT), Cali, Colombia
[b] Plant Breeding and Genetics Laboratory, Joint FAO/IAEA Division, International Atomic Energy Agency, Seibersdorf, Austria
[c] Department of Biological Sciences, School of Natural Sciences, Universidad Icesi, Cali, Colombia
[d] Systems and Computing Engineering Department, Universidad de los Andes, Bogotá, Colombia

ARTICLE INFO

Keywords:
Cassava
Pooled targeted resequencing
Herbicide tolerance
Starch biosynthesis
SNP detection

ABSTRACT

Cassava is one of the most important food security crops in tropical countries, and a competitive resource for the starch, food, feed and ethanol industries. However, genomics research in this crop is much less developed compared to other economically important crops such as rice or maize. The International Center for Tropical Agriculture (CIAT) maintains the largest cassava germplasm collection in the world. Unfortunately, the genetic potential of this diversity for breeding programs remains underexploited due to the difficulties in phenotypic screening and lack of deep genomic information about the different accessions. A chromosome-level assembly of the cassava reference genome was released this year and only a handful of studies have been made, mainly to find quantitative trait loci (QTL) on breeding populations with limited variability. This work presents the results of pooled targeted resequencing of more than 1500 cassava accessions from the CIAT germplasm collection to obtain a dataset of more than 2000 variants within genes related to starch functional properties and herbicide tolerance. Results of twelve bioinformatic pipelines for variant detection in pooled samples were compared to ensure the quality of the variant calling process. Predictions of functional impact were performed using two separate methods to prioritize interesting variation for genotyping and cultivar selection. Targeted resequencing, either by pooled samples or by similar approaches such as Ecotilling or capture, emerges as a cost effective alternative to whole genome sequencing to identify interesting alleles of genes related to relevant traits within large germplasm collections.

1. Introduction

Cassava is one of the most important crops in the tropics, surpassed only by maize and rice [1], and it is usually grown by poor farmers living in marginal and submarginal lands of the tropics [2]. It provides staple food for over 700 million people in Africa (51%), Asia (29%) and South America (20%) [3], being their main source of carbohydrates, in part due to its capacity to produce more energy per hectare than other crops [4,5]. Cassava is also preferred among other crops in these areas because it keeps competitive yields under poor soils, drought, acidic conditions, high air temperatures and evapotranspiration, pests, and diseases [6–8]. In marginal areas where grain crops often fail, cassava can strive, allowing farmers to harvest it when needed [9,10].

In addition to human and animal consumption, cassava has great potential as a source of industrial starch [11]. In fact, cassava is the second most important source of starch worldwide. In the last two decades, cassava production has increased mainly owing to its superior starch quality; which is used primarily in food-processing, paper, glue, textiles, and pharmaceutical industries or occasionally for ethanol production [8]. Therefore one important goal of cassava breeding programs is to develop new varieties with high starch content [12] and with variation in its starch functional properties [13,14]. The biosynthesis of starch involves the production of amylose and amylopectin molecules, which is catalyzed by a series of enzymes (Fig. 1). The synthesis of amylose is catalyzed by the *GBSSI* (Granule bound starch synthase) enzyme [15]. Mutations that knock out this protein are known as *waxy* mutations, because the resulting starches lack amylose [16]. There is a whole complex of enzymes involved in the synthesis of amylopectin: four soluble starch synthases (*SSI, SSII, SSIII* and *SSIV*), two types of starch branching enzymes (*SBEI* and *SBEII*), the Glucan Water Dikinase (*GWD*), and various debranching enzymes and kinases [17]. The SS and the SBE enzymes contribute glucose units to the main chain, and mediate the cleavage

* Corresponding author at: Cra 1 Este No 19A - 40, Bogotá, Colombia.
  E-mail address: ja.duitama@uniandes.edu.co (J. Duitama).
  [1] These authors contributed equally to this work and should be considered joint last authors.

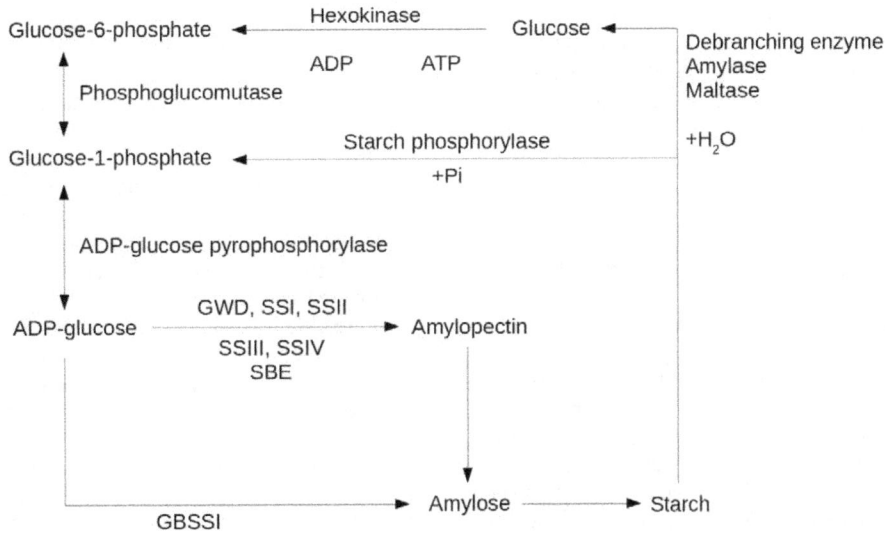

Fig. 1. Metabolic reactions related to starch biosynthesis. Arrows indicate reactions catalyzed by the enzymes listed close to the corresponding arrow.

and branch formation of the amylopectin units [18]. Alteration in SBE activity affects the number of and size distribution of amylopectin branches [17]. It is hard to determine the exact role of each isoform of the soluble starch synthases in this process due to their different gene expression, which depends on both genotypic and environmental variations [18]. GWD controls the overall rate of starch breakdown with a central rate limiting role in starch breakdown machinery and downstream starch synthesis [19]. Plants lacking this protein accumulate abnormally high levels of starch [20].

Another central goal in cassava breeding is the development of herbicide-tolerant cultivars, because the use of herbicides is an effective mechanism to control weeds, reducing labor and alleviating problems of soil erosion associated with mechanical weeding [21]. Studies on the impact of introducing herbicide resistance cassava in Colombia estimated production cost savings between 15% and 25% [22]. Additionally, the positive environmental effects which reduce tillage would bring for increased sustainability of the crop on marginal lands [23].

Resistance to two types of herbicides, inhibiting amino acid biosynthesis, has been commercially exploited in different crops and was targeted in this study. The first group of herbicides (imidazolinones, sulfonylureas, triazolopyrimidine, pyrimidinyl-thiobenzoates, and sulphonyl-aminocarbonyl-triazolinone), interact with the enzymes Acetohydroxyacid synthase (AHAS) and acetolactate synthase (ALS) [24,25]. AHAS has an important role during the synthesis of branched chain amino acids such as valine, leucine, and isoleucine, which are important for the synthesis of several proteins [24]. However, variations in just one amino acid in the binding site of AHAS enzymes can lead to a change in their quaternary structure, blocking herbicide binding and conferring tolerance in the plant. At least five naturally occurring mutations in AHAS, leading to resistance, have been reported in different plant species [24]. The second class of herbicides also affecting amino acid synthesis is the PPT (L-phosphinothricin), also known as glufosinate, and act on the glutamine synthase enzyme (GS). GS synthesizes glutamine and is very important in the regulation of the nitrogen metabolism [26,27]. With the development of transgenic technology, studies established a protocol of using somatic cotyledons as explants for the transformation of cassava [28] successfully transformed a herbicide-resistance gene into the cotyledons of cassava Per 183 by the Agrobacterium mediated method [21]. However, the development of transgenic herbicide-resistant cassava faces regulatory problems that have restricted the adoption of the technology in Africa (with the exception of South Africa).

CIAT holds in trust the largest global germplasm collection of cassava and other *Manihot* species (more than 6000 accessions). The *in vitro* collection at CIAT was initiated in 1978 soon after the technology for slow growth *in vitro* became available [29]. The germplasm collection is a valuable asset and the main repository of genetic variability of cassava. Advanced materials developed from it were the sources of amylose free starch mutations [14]. Although these discoveries provided important proof of the value of the collection, it also highlighted the limited exploration and exploitation of its genetic variability. This work also highlighted how time consuming and inefficient it is to expose useful recessive traits by conventional self-pollination methods. A recent partial screening of the collection allowed discovering varieties carrying two mutations responsible for improved starch quality traits [30]. These findings are encouraging to explore cost-effective alternatives to screen the germplasm collection in search for useful mutations for agronomically relevant traits.

In recent years, the development of high throughput sequencing technologies led to major progress in the understanding of genomic variation in plants, increasing the number of sequenced genomes [31]. However, despite the economic importance of cassava, studies of its genomic diversity are much less complete, compared to other crops such as rice, wheat or maize. Up-to-date the largest study of genomic variability in cassava, which includes 1280 accessions, is based on 402 single nucleotide polymorphisms (SNPs) scattered across the genome [32]. Although a draft cassava genome was assembled and made available in 2012 [33], a chromosome-level assembly was only achieved in 2016 [34]. In the meantime, genotyping by sequencing (GBS) has been a commonly used alternative to obtain dense datasets of genome-wide SNP markers [35]. These SNPs have been used to develop saturated genetic maps for breeding populations, genetic mapping of traits [36–38], and markers for fingerprinting [39]. More recently they have been used to perform a Genome-wide Association Study (GWAS) to identify loci related to resistance to the Cassava mosaic disease [40]. Although GBS is an efficient technique to screen markers and gather information across the genome, it does not allow the study and discovery of variability within specific genes. Sequencing of RNA has also been used as an alternative to identify expressed variation across thousands of genes [41]. However, the cost per sample of this technique is still prohibitive for large numbers of samples. For this reason, targeted resequencing remains an alternative approach to study genetic variability in specific loci.

In this study, we performed pooled targeted resequencing of DNA from 1667 cassava accessions to detect rare SNPs in specific genes associated with the starch biosynthesis pathway and with herbicide resistance. Selected accessions represent about one fourth of the entire collection and include landraces from the most important regions of cassava production in Latin America. We combined the results of 7 variant calling tools applied to aligned reads obtained with two different algorithms to develop a dataset of more than 2000 SNPs within the genes of interest. These SNPs can be prioritized and validated for allele mining and efficient identification of mutated genes in accessions within the cassava germplasm collection.

2. Results

2.1. Targeted Pooled Sequencing of the Cassava Germplasm Bank

DNA was extracted from a total of 1728 accessions from the germplasm collection (Supplementary Table 1). In general, DNA quality was good, with 70% of the samples showing clear shaped bands without significant smearing (Supplementary figure 1). Only 61 samples were discarded due to low DNA concentration. For pooled resequencing two possible methods to normalize DNA concentration across samples were evaluated: the use of paramagnetic beads and visual concentration determination using agarose gels (see Section 4.2). Owing to inconsistencies observed when using beads, a 96 well agarose gel system was adopted. Based on a literature review and on blast searches of the cassava reference genome [34], a total of 6 genes related to herbicide tolerance and 8 genes related to starch biosynthesis were chosen for this study (Supplementary Table 2). To capture the exonic regions of the targeted genes, a total of 121 primer pairs having an expected amplicon length of 600 bp, were designed (Supplementary Table 3). This resulted in an expected total length of 72 kbp of DNA sequence targeted in the assay. To assess the quality of these primers, PCR assays were performed

on one of the pooled samples. Only 18 primers failed to amplify, 13 of them located within the gene GWD (Supplementary figure 2).

Amplicon products for each pool were sent to the high throughput sequencing Illumina MiSeq instrument available at the Plant Breeding and Genetics Laboratory from the International Atomic Energy Agency (IAEA) in Seibersdorf, Austria. After one 2×300 paired-end sequencing run, around 2.5 million fragments were obtained for each pool. Assuming that these fragments are evenly distributed across the targeted regions, this raw sequencing production represents a expected read depth of around $20,000\times$ per targeted base pair within each pool. Reads were trimmed to 240 bp for the first read and to 170 bp for the second read to remove low quality ends. Alignment of the trimmed reads to the reference genome yielded an overall alignment rate of 97%, with 89% of the fragments aligning to unique locations and with the expected distance and orientation (Fig. 2a). Even requiring a stringent reciprocal overlapping of 90% between each aligned fragment and a targeted region, 91% of the total fragments could be reliably assigned to a single region defined by one primer pair (Supplementary Table 3). This percentage represents the capture success rate of the experiment. Moreover, fragments within each pool were assigned more or less evenly to the targeted regions for which primer amplification was successful (Fig. 2b). Besides 17 of the 18 primers for which amplification failed, only five additional primers had less than 20 reads assigned within each pool. Except for the case of pool 7, more than half of the regions had more than 20,000 fragments assigned within each pool. Pool 7 had only 38 regions with this minimum read depth because about 600,000 fewer fragments were sequenced for this pool. In principle, each fragment assigned to a region represents one read of the entire region.

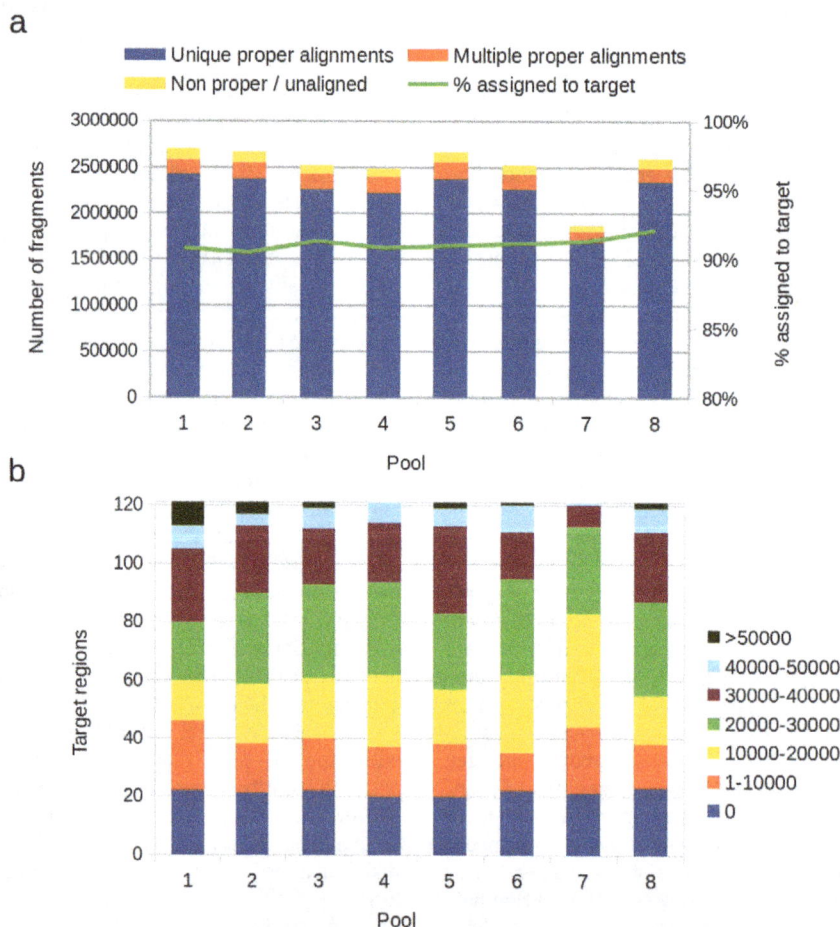

Fig. 2. Read alignment statistics per pool. a) Number of fragments sequenced as paired-end reads for each pool. Counts are discriminated as number of fragments aligning with the expected distance and orientation (proper pair) to a unique region of the genome, fragments aligning as a proper pair to multiple regions and fragments not aligned or not aligned as a proper pair. The line indicates the percentage of fragments that could be uniquely assigned to a targeted region defined by the coordinates of its corresponding primer pair. b) Distribution of the number of fragments assigned to each target region within each pool.

However, the initial trimming performed on each read reduced the sequenced portion of its corresponding region, leaving uncovered the central parts of some of the regions (Supplementary figure 4).

2.2. Comparison of Tools for SNP Discovery in Pooled Data

The number of fragments assigned to each region is tightly related to the total read depth available within each particular locus to assess the presence of non-reference alleles, call variation, and estimate relative allele frequencies based the number of reads supporting each allele. Theoretically, if 10,000 fragments are assigned to one region within one pool, the minor allele of a biallelic variant with a frequency of 0.01 within the samples included in the pool should be observed in about 100 reads. Because about 200 samples were included in each pool, heterozygous variants present in only one sample would have a minor allele frequency (MAF) of $1/400 = 0.0025$ within one pool. Although in this experiment some of these variants would have enough read support be detected, it becomes increasingly difficult to separate the support of true alleles with low frequency from sequencing errors.

To identify sites with evidence of variation within the pools, we combined the results of 12 previously published bioinformatic pipelines

designed to discover single nucleotide polymorphisms (SNPs) and in some cases small indels. The pipelines are the combination of 2 read alignment tools, Bowtie2 [42] and the Burrows-Wheeler Aligner (BWA) [43] with 7 variant discovery programs: Freebayes [44], the Genome Analysis Toolkit (GATK) [45], the Next Generation Sequencing Experience Platform (NGSEP) [46], Samtools [47], SNVer [48], VarScan [49] and VipR [50]. From these tools, SNVer and VipR were particularly designed to identify variation in pools. Because Freebayes and GATK presented problems or were not compatible with bowtie2 alignments, we only ran these tools using as input BWA alignments. On average 1350 variants (1270 SNPs) were predicted within each pool, being SNVer on BWA alignments the pipeline reporting the smallest number of SNPs (294) and VipR on bowtie2 alignments the pipeline reporting the largest number (4354) (Fig. 3a). The average number of indels was 80. VipR and SNVer were not able to detect any indel and VarScan detected indels only from bowtie2 alignments.

Merging the variants predicted by the different pipelines, a raw dataset of 7925 variants was obtained, including 7348 biallelic SNPs, 258 biallelic indels and 319 multiallelic variants. Reads supporting each allele of each variant within each pool were counted following the genotyping step of the NGSEP pipeline and allele frequencies were

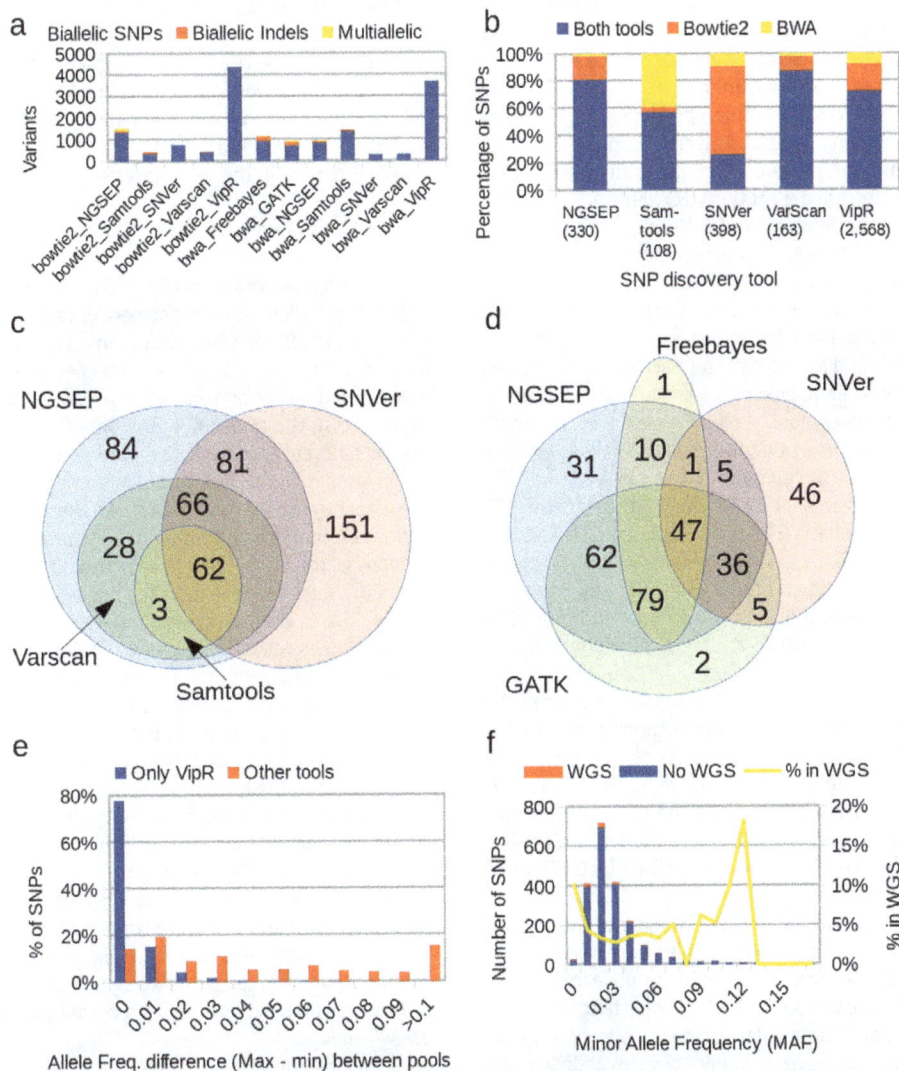

Fig. 3. Comparison of variant calls with different pipelines. a) Number of total variants detected by each variant caller; b) Comparison of number of SNPs called by each SNP discovery tool on alignments obtained with bowtie2 and with BWA; c) Comparison of number of SNPs called between different SNP calling tools on bowtie2 alignments; d) Comparison of number of SNPs called between different SNP calling tools on BWA alignments; e) Distribution of differences in predicted alternative allele frequency between pools for the curated dataset of SNPs; f) Distribution of minor allele frequency for SNPs identified only by VipR discriminating SNPs found in a dataset of variants obtained from WGS data. The line indicates the percentage of such SNPs within each category.

estimated from these counts. About 70% of the raw variants are located within the targeted regions. At first sight, this percentage looks inconsistent with the capture success rate of 91% reported above. The explanation for this outcome is that variants outside targeted regions are called from the few reads falling away from targeted regions and then the total read depth of those variants is much lower than that of the variants within the targeted regions (Supplementary figure 3). The raw variants were filtered by minimum read depth, number of pools in which the variant is observed, and minimum alternative allele frequency. To differentiate true rare SNPs from sequencing errors, the number of errors for each raw SNP was estimated as the average between the third and the fourth smallest allele read depth. Then, the ratio between the read depth of the allele with the second count and the estimated number of sequencing errors was calculated and the SNP was filtered out if this ratio was less than 5. This filtering procedure yielded a curated dataset of 2614 SNPs (Supplementary Table 4). Estimated allele frequencies for curated SNPs were adjusted taking into account read counts of the two predicted alleles. Contrasting the raw calls obtained using each tool during the discovery step with this filtered dataset, we found that 80% of the SNPs in the final set were discovered only by VipR and only 46 SNPs were reported by tools different than VipR. The filters reduced the number of SNPs called by each method to about half in the case of vipR and SNVer, and up to 1 over 10 in the case of Samtools. Samtools only reported 108 of the filtered SNPs with only one SNP not shared by other tools. SNVer and NGSEP were the second and third tools reporting more SNPs within this dataset with 398 and 330 SNPs respectively. The SNPs contributed by the same discovery tool using different read alignment methods were compared to assess the consistency of each method relative to the input alignments (Fig. 3b). Although Varscan only called a total of 163 SNPs, 87% of them were consistently called from bowtie2 and BWA alignments. 80% of the SNPs called by NGSEP were consistent across alignment tools. The smallest percentage of intersection (25.6%) was reported by SNVer. With the exception of Samtools, the other tools reported more SNPs using bowtie2 alignments than BWA alignments.

In absence of a gold-standard to perform a formal quality assessment of the variants predicted by different pipelines, we also calculated the intersections between SNP discovery tools, excluding vipR (Fig. 3c and d). Starting from alignments built using bowtie2, Varscan calls every SNP called by Samtools, and NGSEP calls every SNP called by Varscan or by Samtools. NGSEP and SNVer share 209 SNPs, which represents the 58% of the SNPs called by SNVer and the 64% of the SNPs called by NGSEP. Starting from BWA alignments the sharing between the same 4 tools remains consistent, with the exception of one SNP called by Samtools, which is not called by any other tool (including vipR) and four SNPs called by samtools, NGSEP and SNVer and not called by Varscan. Every SNP called by Varscan is also called by NGSEP. GATK and Freebayes were added to the comparison performed starting from BWA alignments. 47 SNPs were identified by the four methods and 117 additional SNPs were called by three out of four methods. The number of shared SNPs between NGSEP and SNVer (89) still represents 63% of the total SNPs called by SNVer. However, in this case the same number only represents 33% of the SNPs called by NGSEP. From the 182 SNPs called by NGSEP and not called by SNVer, 83% are called either by GATK or by Freebayes.

We also investigated the consistency of allele frequency estimations between pools, taking into account that the samples were pooled without information of population structure and hence the allele frequencies of variants should be stable across pools. Fig. 3e shows that the differences between the largest and the smallest predicted allele frequency for each variant are generally small, having only 213 cases of differences larger than 0.05 and 78 cases of differences larger than 0.1. Because the set of SNPs identified in this study is largely dominated by the SNPs only identified by VipR, this comparison was performed independently for the SNPs predicted only by VipR and for the SNPs predicted by at least one of the other tools. As expected, the subset of variants only called

by vipR consists on SNPs with low MAF (Fig. 3f). Overall, this result indicates that the predictions are stable, especially for the SNPs with high MAF in which large errors on the prediction of allele frequencies could be expected. The largest difference was observed in the SNP located at 27,238,423 of chromosome 3. Whereas the alternative allele (Guanine) is predominant in pool 4 with 27,876 reads supporting this allele and only 989 reads supporting the alternative allele (Adenine), in pool 8 the alternative allele is supported by only 6 reads, which is much smaller than the read support of the reference allele (13,028) and it is even smaller than the read counts for cytosine and thymine (9 and 13 respectively). Read counts in the other pools are relatively balanced between the reference and the alternative allele.

Looking for further evidence to assess the precision of the SNP calling procedure, we compared the SNPs predicted in this work with the SNPs identified from an analysis of whole genome sequencing (WGS) data from 58 cassava varieties [34]. Due to the reduced number of samples, it would be expected that most SNPs with low MAF would not be observed in the WGS panel. However, to the best of our knowledge, this is the only publicly available dataset of SNPs aligned to the current cassava reference genome. A total of 350 SNPs (13.4%) appear in the two datasets (Supplementary Table 4). Whereas 54.3% (272) of the variants called by at least one of the other tools appear in the WGS dataset, only 3% (78) of the variants predicted only by vipR appear in the WGS dataset. However, these 78 SNPs are not skewed toward the highest MAF ranges within the subset of VipR SNPs, as it would be the case if the SNPs in the lower MAF ranges would be mostly false positives. The SNPs present in the WGS dataset are well distributed across the different ranges of MAF and in particular 10% of the SNPs with MAF less than 0.01 appear in the WGS dataset.

2.3. Functional Characterization of Variants within Targeted Genes

Functional annotations of the dataset of filtered SNPs using both NGSEP and SNPeff were performed, obtaining 317 synonymous, 1037 missense and 59 non sense mutations (Fig. 4a). At first sight, the number of missense mutations looks unexpectedly high. However, this can be explained by the accumulation of rare mutations over the varieties sequenced in the pools. Keeping only variants called by at least one method different than VipR, the number of missense mutations (91) becomes similar to the number of synonymous mutations (84). Fig. 4a shows that the percentage of rare variants reduces to 35% and that synonymous mutations and mutations in introns tend to have larger allele frequencies than non-synonymous mutations. Fig. 4b shows the distribution of mutations in coding regions per gene. The AHAS genes accumulate 55% of the mutations and seem to have larger SNP density than the genes related to amylose content, even after normalizing by the length of the covered exonic regions. Within the SS family, *SSIII* and *SSIV* show a larger SNP density and for *SSIV* in particular the number of synonymous mutations (3) is much smaller than the number of missense mutations (11). Six of these missense mutations have a predicted MAF larger than 0.1. The number of non-sense mutations reduced to only seven. Interestingly, two of these mutations, which modify the codons 141 and 143 at exon 4 of the gene GWD showed alternative allele frequencies close to 0.5 and to 0.25 respectively over the 8 pools. Read counts indicate that in almost all pools the alternative alleles of both mutations were supported by over 3000 reads and that the number was always 5-fold higher than the number of reads supporting other alternative allele. Three additional mutations with MAFs larger than 0.15 are located close to the end of the *SSIII* and the *AHAS4* genes.

Unfortunately vipR and SNVer, which were the two software packages implementing models for pooled sequencing data, were not designed to call small indels. Combining results of the other tools, 4 small indels were identified within coding regions of the sequenced genes (Supplementary Table 5). One of these indels, located within the gene SBE was a missense 3 bp deletion, which removes a lysine

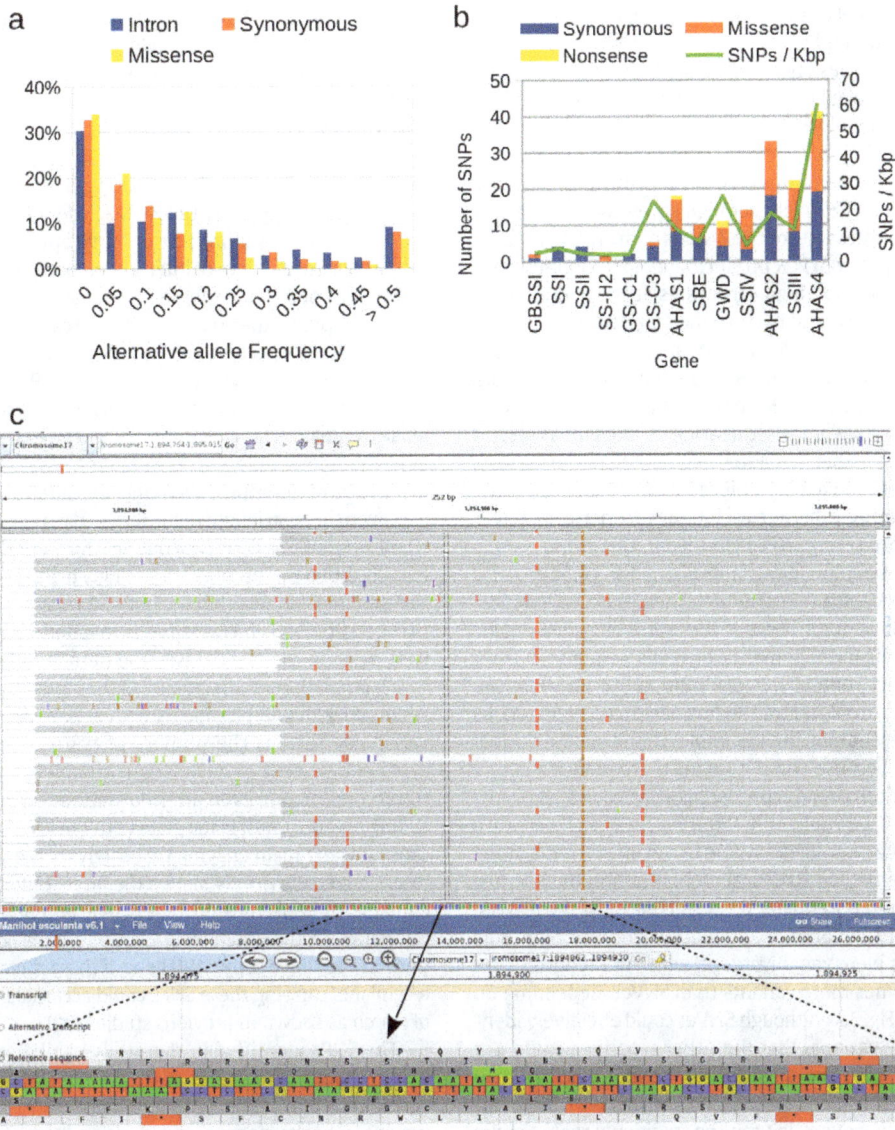

Fig. 4. Functional analysis of variants. a) Distribution of alternative allele frequencies observed over the 8 pools for the dataset obtained removing SNPs that were called only by vipR. b) Distribution of SNPs within coding regions of the genes sequenced in this study. The line represents the number of SNPs per kilo base pair c) Reads supporting a 1 bp deletion changing the open reading frame to generate an early stop codon in the allele of the AHAS gene at chromosome 17. The upper panel is a visualization using the integrative genomics viewer (IGV) of the reads spanning the region (gray rectangles). Colors different than gray indicate base calls different than the reference allele. The highlighted column shows reads reporting a 1 bp deletion. The lower panel shows a view of the JBrowse visualizer available in phytozome of the highlighted subregion, including the nucleotide sequence and the six possible amino acid translations. The arrow indicates the location of the frameshift deletion.

amino acid. The three remaining indel mutations are all 1 bp deletions located at the AHAS 4 gene located at chromosome 17 (Fig. 4c). The three mutations are predicted to change the open reading frame of the gene, which is likely to produce an early stop codon. Predicted allele frequencies based on read counts indicate that these mutations are present in about 15% of the sequenced cultivars.

3. Discussion

The recent releases of chromosome-level assemblies for different plants and the continuous reduction in sequencing costs allows research in staple crops such as cassava to enter the post-genomic era in which comprehensive characterization of genomic diversity across complete genebank collections becomes a feasible task [51]. However, because whole genome sequencing (WGS) costs are still in the order of $500 per sample for cassava, cost-effective sequencing alternatives are preferred for different applications. Genotype by Sequencing (GBS), which recently became the method of choice for applications such as

construction of genetic maps, population structure and association mapping, has as main disadvantage that it does not allow to obtain complete sequencing of any single gene. Because the objective in this work was to perform allele mining over the CIAT germplasm collection for genes already known to be related to starch content and herbicide tolerance, we decided to implement a targeted sequencing approach based on PCR assays guided by carefully selected primers. This strategy allowed maximizing the power of high throughput sequencing (HTS) to obtain accurate information of variability across more than 1500 varieties from the germplasm collection. To the best of our knowledge, this study is up-to-date the sequencing effort involving the largest number of samples in cassava.

The targeted sequencing strategy followed in this experiment indeed revealed a large amount of variants at different allele frequencies within the targeted genes. A comparison with the SNPs identified by whole genome sequencing of 58 African varieties (Bredeson, 2016) served as validation of the variants with high Minor Allele Frequency (MAF) but also showed that sequencing a limited number of varieties does not allow identification of a large amount of genetic variation that could be potentially relevant for breeding purposes. The consistency in predictions of allele frequencies observed across the eight pools suggests that the method employed for DNA normalization and the bioinformatic analysis were generally effective and hence they can be used for future pooled sequencing experiments. The main drawback that we could observe using the pooled targeted sequencing approach was a reduction of the regions effectively sequenced by the experiment due to the increased error rates toward the 3' ends of the reads. Because reads are directly sequenced from PCR products and not randomly sampled within the targeted regions, high error rates at the 3' end of the reads will accumulate at the central parts of the targeted regions, producing a large amount of false positives. If reads are trimmed to prevent this effect, central parts of some of the targeted regions are lost. In future experiments, amplicon lengths of PCR products should be reduced to take into account the error rate of the sequencing instrument. A second drawback of this approach is that individual genotyping of the variants revealed by the experiment can not be achieved within the experiment. We are currently evaluating different techniques to perform direct genotyping of the most promising SNPs identified in this work.

The most commonly used tools for variants discovery (NGSEP, GATK, Samtools, Freebayes and Varscan) are not designed to detect low frequency variants in pooled samples, because they were designed to perform variants discovery from alignments of reads sequenced from individual samples. Hence, one of the assumptions to improve the genotyping quality in these tools is that the two alleles in heterozygous sites will have even representation in the sample. This is not the normal case for pooled samples because population allele frequencies determine the relative proportion of read counts supporting each allele within variant sites. However, we could only find two additional software tools (VipR and SNVer) that would be feasible to run on current aligned HTS reads and that implemented statistical models to find the low frequency variants that could potentially be extracted from these data. An initial comparison of the variants obtained with these two tools showed that their results were very divergent, with VipR reporting between five and twelve times more variants than SNVer, depending on the read alignment tool (Fig. 3a). Although SNVer could effectively identify some low frequency variants that the other pipelines could not identify, these variants were not consistently identified across read alignment tools. Moreover, SNVer missed some variants with large frequency that could be discovered even with the traditional tools. On the other hand, manual examination of the read counts for some of the raw SNPs with low frequency alternative nucleotides predicted by VipR showed that these counts were almost the same as the read counts supporting the other two nucleotides, which were likely to be produced by sequencing errors. Regarding other types of variation, VipR and SNVer were not designed to call variants beyond SNPs. Finally, the output VCF format provided by both tools was largely outdated, which made us feel reluctant of the sustainability of these tools over time. In this scenario, we considered a good alternative to try all the options that we had available, and compare the variants obtained using the different pipelines. As expected, the commonly used tools for variants discovery reported between 4 and 13 times less variants than VipR. A comparison between them was consistent with a previous benchmark that we performed using GBS data, in which NGSEP identifies more SNPs than the other tools [52]. In this case, a possible reason for this difference is that Samtools, GATK and Freebayes were designed to analyze WGS data of human samples. Hence, the models implemented in these tools include filters of balance between read alignment strands, which are not adequate for analysis of reads taken from region-specific PCR products. It is worth to clarify that in the absence of a gold-standard dataset, the comparison presented in this manuscript is not a formal benchmark between methods but a survey of the available alternatives performed from a user perspective. We believe that the results presented in this survey would be helpful for other researchers performing pooled resequencing experiments and also that improved methods for

variants discovery in pooled samples could be developed to take full advantage of the data generated by similar experiments.

The final outcome of the comparison between pipelines for variants discovery and the filtering process, including the filtering of variants in which the minor allele could not be clearly separated from sequencing errors, is a dataset of 2614 SNPs within the targeted genes (Supplementary Table 4). Despite of the filtering procedure, close to 80% of these variants are still SNPs with low MAF identified only by VipR. Although we could follow a more conservative approach and report only SNPs called by a certain type of intersection between the tools, this would remove most of the rare mutations that are actually interesting for follow up genotyping experiments. For this reason, we decided to retain the union of the SNPs identified by the different tools after performing the filters described above. However, each SNP is reported with functional annotations, intersection with SNPs obtained from WGS data, predicted allele frequencies, raw read counts and pipelines that reported each variant. This allows different researchers to use common excel filters to select the most appropriate variants for different follow up experiments.

Given the total length of the targeted region, the SNPs identified in this study amount to a density of one SNP for each 26 base pairs. Although we initially found this number surprisingly high, the latest release of the 3000 rice genomes project [53] includes 32 million SNPs for a 400 Mega base pair genome, which corresponds to a density of one SNP for each 12.5 base pairs. In the rice dataset, the number of variants is also increased by accumulation of rare alleles as the sample size increased. Individual genotyping should provide us with a more accurate measure of genetic variability such as the number of pairwise differences per kbp. The AHAS genes seem to have larger variability than the genes related to starch production, even after normalization by the covered portion of coding regions. GBSSI is the gene with the lowest variability, probably because it is the main enzyme that catalyzes the reaction to produce amylose. Conversely AHAS4 shows the largest number of variants and also contains three frameshift indels that potentially produce silencing of this paralog. Other interesting variants are the non-sense mutations identified in the single copy GWD gene. If these mutations have a silencing effect, plants carrying these SNPs could accumulate abnormally high levels of starch as shown in previous studies [20].

The SNPs identified in this study can be prioritized based on read evidence and predictions of functional consequences, and then they can be tested in a direct genotyping platform. We are currently exploring different alternatives to perform individual genotyping, not only for validation but also to identify varieties with rare alleles that could exhibit interesting characteristics for the traits of interest that then could be selected as new sources of genetic variability for the cassava breeding program. The publication of the SNPs identified in this experiment is helpful to encourage other groups to perform individual genotyping of these SNPs in their own germplasm collections, accelerating the discovery of varieties with improved phenotypes. Moreover, the genetic variation that we could identify in the CIAT collection, within genes that a-priori could be thought as completely conserved, is also encouraging to try alternative cost-efficient techniques such as multi-dimensional pooled EcoTILLING [54] in future experiments. Although EcoTILLING is in principle a more expensive technique because it requires the design of a tridimensional pooling strategy in which each sample is included in three different pools, it allows direct identification of samples carrying rare alleles. Based on the results of this experiment, we believe that improved methods for targeted resequencing, such as those used in this study, will provide cost-effective valuable information to accelerate breeding cycles through the use of molecular techniques.

4. Methods

4.1. DNA Extraction

DNA was extracted from a total of 1728 accessions from the germplasm collection at CIAT. The DNA was isolated by using 1 g of cassava leaf tissue grounded with liquid nitrogen in 15 mL tubes using the CTAB method. Thereafter, 3 mL of the prewarmed extraction buffer was added (100 mM tris HCl (pH 8), 20 mM EDTA (pH 8), 2 M NaCl, 2% CTAB (w/v), 2% PVP) to each sample and they were mixed. The samples were incubated at 65 °C for 1 h with frequent swirling. An equal volume of phenol: chloroform: isoamyl alcohol (25:24:1) was added to each sample and mixed gently for 30 min. The samples were centrifuged at 3000 rpm for 30 min at room temperature. Approximately 2 mL of the supernatant was transferred to a new tube. The supernatant was precipitated with 1/1 volume of isopropanol and was incubated for 30 min at 4 °C. The precipitated nucleic acids were collected and washed twice with 70% ethanol. The obtained nucleic acid pellet was air-dried until the ethanol was evaporated and dissolved in 200 uL of TE buffer (10 mM tris-HCl pH 8, 1 mM EDTA pH 8). The nucleic acid dissolved in TE buffer was treated with ribonuclease A (RNase A, 10 mg/mL) and incubated at 37 °C for 30 min. The quality of extracted DNA was stained with SYBR safe (Invitrogen) and visualized by agarose gel electrophoresis (1%). The purity of the DNA was estimated by spectrophotometry, which estimates A260/280 and A260/230 ratio. After this, the dried samples were packed to be shipped to the Plant Breeding and Genetics Laboratory in Austria.

4.2. Determination of DNA Quality and Quantity, and Sample Pooling

Once the DNA samples arrived to the Plant Breeding and Genetics Laboratory in Austria for processing and sequencing, were centrifuged and then hydrated by the addition of 100 uL (water). Samples were incubated at room temperature for 10 min followed by a short vortex and an additional 5 min incubation to ensure that DNA was completely in solution. Samples were stored at 4 °C for a minimum of 24 h prior to additional processing.

To ensure even sequencing coverage of all DNA samples in a pool, methods were evaluated to normalize DNA concentrations. Experiments employing paramagnetic bead-based purification systems (e.g. MagQuantTM) yielded inconsistent concentrations, possibly due to variations of input DNA (data not shown). Therefore a system using 96 well gels and image based quantification was employed [55]. Briefly, 12.5 μL of DNA from each tube was transferred to a well in a 96 well plate to facilitate liquid handling. 5 μL of DNA was loaded onto 96 well E-gels® 2%. Five microliters lambda DNA standards diluted to specific concentrations (3, 4.5, 6.8, 10.1, 15.2, 22.8, 34.2, 51.3 ng/μL) in the last column of the gel. Samples were electrophoresed, the gel photographed and concentrations determined with the aid of the image analysis program ImageJ. Samples' concentrations were adjusted, samples pooled together and the final concentration of each of 8 pools was adjusted to 3.57 ng/μL for PCR.

4.3. Primer Design and PCR Performance

A total of 121 primer pairs were designed for the exonic regions of genes related to herbicide tolerance (AHAS1, AHAS2, AHAS3, AHAS4, GS-C1 and GS-C3), and starch biosynthesis, (GWD, GBSSI, SS-H2, SSI, SSII, SSIII, SSIV and SBE). Primer3 [56] was used to design primers with a length between 25 and 30 bp, with a Tm between 65 °C and 72 °C, with an optimal of 70 °C, to amplify fragments between 550 and 650 bp. The TaKaRa Ex Taq® polymerase was used to perform the PCR using 17.85 ng of pooled DNA according to manufacturer's recommendations. Amplification was performed as follows: The initial denaturing cycle was 2 min at 95 °C, followed by 8 cycles of denaturing at 94 °C for 20 s, annealing at 65 °C for 30 s and extension at 72 °C for 1 min. The last cycle extension was held for an extra 5 min, followed by holding at 8 °C. The concentration of PCR products was determined using 96 well E-gels® 1%. PCR products produced from the same DNA were pooled together such that 8 samples of pooled PCR products deriving from the 8 DNA pools created.

4.4. Sequencing

Illumina library preparation was performed using the TruSeq® Nano DNA Library Prep (version 15041110 Rev. D) with minor modification. Briefly, the first normalization and fragmentation steps were not performed and library preparation began with the first bead-based cleanup step. All other steps were followed according to the protocol. Dual indexes were used. Quantification was performed using Qubit fluorometry. Libraries were normalized to 4 nM and pooled together. The concentration of this pool was further checked, adjusted, and the pool denatured and diluted to 17.5 pM according to the Illumina protocol. Samples were sequenced on an Illumina MiSeq using 2 × 300 Paired End version 3 chemistry. Fastqc [57] was used to perform an initial quality assessment of the raw reads. The reads did not pass the base quality filter after 240 bp in the first read and after 170 bp of the second read. Accordingly, reads were trimmed to these lengths.

4.5. Read Alignment

The reference genome Manihot esculenta v6.1 was downloaded from the webpage of Phytozome 11 [58], including the corresponding GFF3 file with gene functional annotations. Two different tools were used to align reads to the reference genome: bowtie2-2.2.5 [42] and BWA 0.7.12-r1039 [43]. The alignment using bowtie2-2.2.5 was made according to the documentation, indexing the cassava reference genome first. The program was run with default parameters, except for the maximum number of alignments per read, which was set to 3, the minimum fragment length to 0 and the maximum fragment length to 800. Picard-2.2.4 [59] was used to sort the BAM files. BWA 0.7.12-r1039 was also used to align reads to the reference genome according to the documentation. The program was executed with the default parameters, setting the bandwidth for banded alignment to 600. Samtools 1.3.1 was used to convert the SAM files into BAM files, to sort them and index them. Visualization of read alignments was performed using the Integrative Genomics Viewer (IGV) [60].

4.6. SNP Discovery

Seven variant callers were combined with the two read alignment tools to obtain twelve different pipelines. The procedure for each pipeline is briefly described below.

4.6.1. Freebayes
Freebayes v1.0.2-33-gdbb6160 [44] was executed only from BAM files generated by BWA, according to the documentation available in the website. Samtools-1.3.1 was used to merge the VCF file obtained from each pool and create a final VCF file containing the information of the eight samples. This variant caller could not be executed using files obtained with bowtie2.

4.6.2. GATK
To run GATK 3.5-0-g36282e4 [45] a Sequence Dictionary had to be created using picard 2.2.4, as well as indexing the reference genome using samtools-1.3.1. The Haplotype Caller option was run to obtain the SNPs present in each sample, with the default parameters, except for read downsampling, which was set to 0. At the end, eight VCF files were obtained, one per sample, with all the information about the SNPs present in each of them. This was followed by the Merge Variants option available in this program to obtain a final VCF with the SNP information of all the samples. It's important to mention, that GATK is only compatible with files obtained from BWA, so it was not possible to use this variant caller with the alignment information obtained with bowtie2.

4.6.3. NGSEP

The NGSEP-3.0.1 [46] pipeline was used to discover SNPs and indels. This pipeline was executed with default parameters, except for the maximum number of alignments allowed to start at the same reference site, which was set to 0. The options to find repetitive regions, CNV, large indels and inversions were turned off during the variants discovery and the genotyping steps of the pipeline. Because NGSEP is compatible with bowtie2 and BWA, the pipeline was run with the files obtained with these two alignment programs, with the same parameters mentioned above.

4.6.4. Samtools

The variant calling was performed according to the documentation (version 1.3.1) [47]. Mpileup files were generated and the multi allelic variant caller option was used to detect SNPs. At the end of this process, eight VCF files with the SNP information of each sample were obtained, and the program was used to merge them to obtain a final VCF with the information of all the SNPs present. Because Samtools is compatible with alignment files obtained with bowtie2 and BWA, the same pipeline was run using the different alignment files.

4.6.5. SNVer

SNVer-0.5.3 [48] was executed according to the documentation available. To run this variant caller, a file with five columns that contained the sample name information, number of haploids per pool, number of samples, minimum quality and maximum base quality values, respectively had to be created. At the end, a final VCF file with the information of all the samples was obtained. Because SNVer is compatible with bowtie2 and BWA, this pipeline was run with the information obtained with these two alignments tools.

4.6.6. VarScan

To run VarScan v2.3.9 [49], the documentation available was followed. Mpileup files had to be created first using Samtools. With these mpileup files one of the tools available on the VarScan folder was used to detect the SNPs present in each sample, so at the end of this process eight VCF files with the SNP information were obtained. These files were merged using Samtools to obtain a final VCF file. Because VarScan is compatible with bowtie2 and BWA, this pipeline was run with the files obtained with these two alignment tools.

4.6.7. VipR

This program was executed according to the documentation available (version 0.0.16) [50]. First mpileup files had to be created with Samtools, using the parameters recommended for the documentation. These mpileup files had to be converted into a vipR files. Then, an R script was run following the documentation, setting the number of haploids to 536, corresponding to the biggest pool created in the experiment. At the end, a final VCF file with all the SNP information of each sample was obtained. Because VipR is compatible with bowtie2 and BWA, this pipeline was run with the files obtained with these two alignment tools.

4.7. Downstream Analysis

At the end 12 VCF files were obtained as a result of the combination of alignment files made with bowtie2 and BWA and the 7 variant callers. With these 12 VCF files the NGSEP pipeline was used to do the genotyping, first merging the variants present in all the VCF files, and then running the genotyping process with default parameters, except for the maximum number of alignments allowed to start at the same reference site, which was set to 0. This was done with the BAM files for each read alignment tool, generating two final VCF files.

The functional annotation was performed using NGSEP and SNPeff [61], having the GFF3 cassava file as a reference. NGSEP was also used to filter this final file, removing the variants embedded in indels first,

and then filtering to keep biallelic SNPs with a read depth of $10000\times$ or more and those which were present in at least two pools. A custom script written in java was used to filter variants in which the read count of the minor allele is less than five times the read count of the average between the read counts of the third and the fourth allele. Custom scripts were also written to calculate statistics related to the coverage of genes and primers.

Acknowledgements

The financial support from COLCIENCIAS-Colombia (Project code 223670048777 – Contract 393-2015, with resources from World and Inter-American Development Banks) and the technical monitoring by Cesar Augusto Trujillo Beltran were fundamental for the completion of the research described in this article. We also thanks Luis Augusto Becerra for his general supervision of the work of Ana Maria Leiva.

References

[1] FAO. Why cassava? Food and Agriculture Organization of the United Nations; 2008 [Available at: http://www.fao.org/ag/agp/agpc/gcds/index_en.html. Accessed 2016 Dec 22].

[2] Aerni P. Mobilizing Science and Technology for development: the case of the Cassava Biotechnology Network (CBN). AgBioforum 2006;9(1):1–14.

[3] Food and Agriculture Organization of the United Nations. Statistics Division; 2017 [Available at: http://www.fao.org/faostat/en. Accessed 2017 Jan 27].

[4] Batista de Souza CR. Genetic and genomic studies of cassava. Genes Genomes Genomics 2007;1(2):157–66 [Available at: http://www.globalsciencebooks.info/Online/GSBOnline/images/0712/GGG_1(2)/GGG_1(2)157-166o.pdf. Accessed 2016 Dec 22].

[5] Montagnac JA, Davis CR, Tanumihardjo SA. Nutritional value of cassava for use as staple food and recent advances for improvement. Compr Rev Food Sci Food Saf 2009;8(3):181–94. http://dx.doi.org/10.1111/j.1541-4337.2009.00077.x.

[6] Burns AE, Gleadow J, Cliff J, Zacarias A, Cavagnaro T. Cassava: the drought, war and famine crop in a changing world. Sustainability 2010;2:3572–607. http://dx.doi.org/10.3390/su2113572.

[7] El-Sharkawy MA. International research on cassava photosynthesis, productivity, eco-physiology, and responses to environmental stresses in the tropics. Photosynthetica 2006;44(4):481–512.

[8] FAO. Save and grow: cassava. A guide to sustainable production intensification; 2013 [Available at: http://www.fao.org/3/a-i3278e/index.html. Accessed 2016 Dec 22].

[9] Ceballos H, Iglesias CA, Pérez JC, Dixon AGO. Cassava breeding: opportunities and challenges. Plant Mol Biol 2004;56(4):503–16. http://dx.doi.org/10.1007/s11103-004-5010-5.

[10] Pérez JC, Lenis JI, Calle F, Morante N, Sánchez T, et al. Genetic variability of root peel thickness and its influence in extractable starch from cassava (Manihot esculenta Crantz) roots. Plant Breed 2011;130(6):688–93. http://dx.doi.org/10.1111/j.1439-0523.2011.01873.x.

[11] Da G, Dufour D, Giraldo A, Moreno M, Tran T, et al. Cottage level cassava starch processing systems in Colombia and Vietnam. Food Bioprocess Technol 2013;6(8):2213–22. http://dx.doi.org/10.1007/s11947-012-0810-0.

[12] Kunkeaw S, Yoocha T, Sraphet S, Boonchanawiwat A, Boonseng O, et al. Construction of a genetic linkage map using simple sequence repeat markers from expressed sequence tags for cassava (Manihot esculenta Crantz). Mol Breed 2011;27(1):67–75. http://dx.doi.org/10.1007/s11032-010-9414-4.

[13] Ceballos H, Hershey C, Becerra-López-Lavalle LA. New approaches to cassava breeding. Plant Breed Rev 2012;36:427–504. http://dx.doi.org/10.1002/9781118358566.ch6.

[14] Morante N, Ceballos H, Sánchez T, Rolland-Sabaté A, Calle F, et al. Discovery of new spontaneous sources of amylose-free cassava starch and analysis of their structure and techno-functional properties. Food Hydrocoll 2016;56:383–95. http://dx.doi.org/10.1016/j.foodhyd.2015.12.025.

[15] Buléon A, Colonna P, Planchot V, Ball S. Starch granules: structure and biosynthesis. Int J Biol Macromol 1998;23(2):85–112.

[16] Jobling S. Improving starch for food and industrial applications. Curr Opin Plant Biol 2004;7(2):210–8. http://dx.doi.org/10.1016/j.pbi.2003.12.001.

[17] Brummell DA, Watson LM, Zhou J, Mckenzie MJ, Hallett IC, et al. Overexpression of starch branching enzyme II increases short-chain branching of amylopectin and alters the physicochemical properties of starch from potato tuber. BMC Biotechnol 2015;15:28. http://dx.doi.org/10.1186/s12896-015-0143-y.

[18] Martin C, Smith A. Starch biosynthesis. Plant Cell 1995;7(7):971–85. http://dx.doi.org/10.1105/tpc.7.7.971.

[19] Zeeman S, Smith SM, Smith AM. The breakdown of starches in leaves. New Phytol 2004;163(2):247–61. http://dx.doi.org/10.1111/j.1469-8137.2004.01101.x.

[20] Skeffington AW, Graf A, Duxbury Z, Gruissem W, Smith AM. Glucan, water dikinase exerts little control over starch degradation in Arabidopsis leaves at night. Plant Physiol 2014;165(2):866–79. http://dx.doi.org/10.1104/pp.114.237016.

[21] Taylor N, Chavarriaga P, Raemakers K, Siritunga D, Zhang P. Development and application of transgenic technologies in cassava. Plant Mol Biol 2004;56(4):671–88. http://dx.doi.org/10.1007/s11103-004-4872-x.

[22] Pachico D, Rivas L. A preliminary comparison of the potential welfare and employ-ment effects of herbicide tolerant, high yielding, of mechanized cassava in different markets in Colombia. In: Fauquet CM, Taylor NJ, editors. Cassava: an ancient crop for modern timesProceedings of the 5th International Meeting of the Cassava Biotech-nology Network (4–9 November 2001, St. Louis, MO); 2003.

[23] Ceballos H, Ramírez J, Bellotti A, Jarvis A, Alvarez E. Adaptation of cassava to chang-ing climates. In: Yadav SS, Redden RJ, Hatfield JL, Lotze-Campen H, Hall AE, editors. Crop adaptation to climate change. Oxford, UK: Wiley-Blackwell; 2011. http://dx. doi.org/10.1002/9780470960929.ch28.

[24] Cobb AH, Reade JPH. Herbicides and plant physiology. 2nd ed. Chichester, UK: Wiley-Blackwell; 2010[286 pp.].

[25] Tan S, Evans RR, Dahmer ML, Singh BK, Shaner DL. Imidazolinone-tolerant crops: history, current status and future. Pest Manag Sci 2005;61(3):246–57. http://dx. doi.org/10.1002/ps.993.

[26] Betti M, García-Calderón M, Pérez-Delgado CM, Credali A, Estivill G, et al. Glutamine synthetase in legumes: recent advances in enzyme structure and functional geno-mics. Int J Mol Sci 2012;13(7):7994–8024. http://dx.doi.org/10.3390/ijms13077994.

[27] De Block M, Botterman J, Vandewiele M, Dockx J, Thoen C, et al. Engineering herbicide resistance in plants by expression of a detoxifying enzyme. EMBO J 1987;6(9):2513–8.

[28] Sarria R, Torres E, Angel F, Chavarriaga P, Roca WM. Transgenic plants of cassava (Manihot esculenta) with resistance to Basta obtained by Agrobacterium-mediated transformation. Plant Cell Rep 2000;19(4):339–44. http://dx.doi.org/10.1007/s002990050737.

[29] Hershey C, Debouck D. A global conservation strategy for cassava (Manihot esculenta) and wild Manihot species. Cali, Colombia: Centro Internacional de Agricultura Tropical (CIAT); 2010 [Available at: https://www.croptrust.org/wp-con-tent/uploads/2014/12/cassava-strategy.pdf, Accessed 2016 Dec 22].

[30] Sánchez T, Salcedo E, Ceballos H, Dufour D, Mafla G, et al. Screening of starch quality traits in cassava (Manihot esculenta Crantz). Stata J 2009;61(5):12–9. http://dx.doi. org/10.1002/star.200990027.

[31] Goodwin S, McPherson JD, McCombie WR. Coming of age: ten years of next-generation sequencing technologies. Nat Rev Genet 2016;17:333–51. http://dx.doi. org/10.1038/nrg.2016.49.

[32] de Oliveira EJ, Ferreira CF, da Silva SV, de Jesus ON, Oliveira GAF, da Silva MS. Poten-tial of SNP markers for the characterization of Brazilian cassava germplasm. Theor Appl Genet 2014;127(6):1423–40. http://dx.doi.org/10.1007/s00122-014-2309-8 [PMID:24737135].

[33] Prochnik S, Marri PR, Desany B, Rabinowicz PD, Kodira C, et al. The cassava genome: current progress, future directions. Trop Plant Biol 2012;5(1):88–94. http://dx.doi. org/10.1007/s12042-011-9088-z [PMID:PMC3322327].

[34] Bredeson J, Lyons JB, Prochnik SE, Wu GA, Ha CM, et al. Sequencing wild and cultivated cassava and related species reveals extensive interspecific hybridization and genetic di-versity. Nat Biotechnol 2016;34:562–70. http://dx.doi.org/10.1038/nbt.3535.

[35] Elshire RJ, Glaubitz JC, Sun Q, Poland JA, Kawamoto K, et al. A robust, simple genotyping-by-sequencing (GBS) approach for high diversity species. PLoS One 2011;6(5):e19379. http://dx.doi.org/10.1371/journal.pone.0019379.

[36] International Cassava Genetic Map Consortium (ICGMC). High-resolution linkage map and chromosome-scale genome assembly for cassava (Manihot esculenta Crantz) from 10 populations. G3 2014;5(1):133–44. http://dx.doi.org/10.1534/g3. 114.015008 [PMID: PMC4291464].

[37] Rabbi I, Hamblin M, Gedil M, Kulakow P, Ferguson M, et al. Genetic mapping using genotyping-by-sequencing in the clonally propagated cassava. Crop Sci 2014; 54(4):1384–96. http://dx.doi.org/10.2135/cropsci2013.07.0482.

[38] Soto JC, Ortiz JF, Perlaza-Jiménez L, Vásquez AX, Lopez-Lavalle LAB, et al. A genetic map of cassava (Manihot esculenta Crantz) with integrated physical mapping of immunity-related genes. BMC Genomics 2015;16:190. http://dx.doi.org/10.1186/s12864-015-1397-4.

[39] Rabbi IY, Kulakow PA, Manu-Aduening JA, Dankyi AA, Asibuo JY, et al. Tracking crop varieties using genotyping-by-sequencing markers: a case study using cassava (Manihot esculenta Crantz). BMC Genet 2015;16:115. http://dx.doi.org/10.1186/s12863-015-0273-1.

[40] Wolfe MD, Rabbi IY, Egesi C, Hamblin M, Kawuki R, et al. Genome-wide association and prediction reveals genetic architecture of cassava mosaic disease resistance and prospects for rapid genetic improvement. Plant Genome 2016;9(2). http://dx.doi. org/10.3835/plantgenome2015.11.0118.

[41] Pootakham W, Shearman JR, Ruang-areerate P, Sonthirod C, Sangsrakru D, et al. Large-scale SNP discovery through RNA sequencing and SNP genotyping by targeted enrich-ment sequencing in cassava (Manihot esculenta Crantz). PLoS One 2014;9(12): e116028. http://dx.doi.org/10.1371/journal.pone.0116028 [PMID: PMC4281258].

[42] Langmead B, Salzberg S. Fast gapped-read alignment with Bowtie 2. Nat Methods 2012;9:357–9. http://dx.doi.org/10.1038/nmeth.1923.

[43] Li H, Durbin R. Fast and accurate long-read alignment with burrows-wheeler trans-form. Bioinformatics 2010;26(5):589–95. http://dx.doi.org/10.1093/bioinformatics/btp698 [PMID: 19451168].

[44] Garrison E, Marth G. Haplotype-based variant detection from short-read sequenc-ing; 2012 [Available at: https://arxiv.org/abs/1207.3907. Accessed 2016 Dec 22]

[45] McKenna A, Hanna M, Banks E, Sivachenko A, Cibulskis K, et al. The genome analysis toolkit: a MapReduce framework for analyzing next-generation DNA sequencing data. Genome Res 2010;20:1297–303. http://dx.doi.org/10.1101/gr.107524.110.

[46] Duitama J, Quintero JC, Cruz DF, Quintero C, Hubmann G, et al. An integrated frame-work for discovery and genotyping of genomic variants from high-throughput se-quencing experiments. Nucleic Acids Res 2014;42(6):e44. http://dx.doi.org/10. 1093/nar/gkt1381.

[47] Li H, Handsaker B, Wysoker A, Fennell T, Ruan J, et al. The sequence alignment/map (SAM) format and SAMtools. Bioinformatics 2009;25(16):2078–9. http://dx.doi.org/ 10.1093/bioinformatics/btp352 [PMID: 19505943].

[48] Wei Z, Wang W, Hu P, Lyon GJ, Hakonarson H. SNVer: a statistical tool for variant calling in analysis of pooled or individual next-generation sequencing data. Nucleic Acids Res 2011;39(19):e132. http://dx.doi.org/10.1093/nar/gkr599 [PMID: 21813454].

[49] Koboldt DC, Zhang Q, Larson DE, Shen D, McLellan MD, et al. VarScan 2: somatic mutation and copy number alteration discovery in cancer by exome sequencing. Genome Res 2012;22(3):568–76. http://dx.doi.org/10.1101/gr.129684.111.

[50] Altmann A, Weber P, Quast C, Rex-Haffner M, Binder EB, Muller-Myhsok B. vipR: variant identification in pooled DNA using R. Bioinformatics 2011;27(13):177–84. http://dx.doi.org/10.1093/bioinformatics/btr205.

[51] The 3000 rice genomes project. The 3000 rice genomes project. GigaScience 2014;3: 7. http://dx.doi.org/10.1186/2047-217X-3-7.

[52] Perea C, De La Hoz JF, Cruz DF, Lobaton JD, Izquierdo P, et al. Analysis of genotype by sequencing (GBS) data with NGSEP. BMC Genomics 2016;17(Suppl. 5):498. http:// dx.doi.org/10.1186/s12864-016-2827-7.

[53] Mansueto L, Fuentes RR, Borja FN, Detras J, Abriol-Santos JM, et al. Rice SNP-seek da-tabase update: new SNPs, indels, and queries. Nucleic Acids Res 2016. http://dx.doi. org/10.1093/nar/gkw1135 [in press].

[54] Comai L, Young K, Till BJ, Reynolds SH, Greene EA, et al. Efficient discovery of DNA polymorphisms in natural populations by Ecotilling. Plant J 2004;37(5):778–86. http://dx.doi.org/10.1111/j.0960-7412.2003.01999.x.

[55] Huynh OA, Jankowicz-Cieslak J, Saraye B, Hofinger B, Till BJ. Low-cost methods for DNA extraction and quantification. In: Jankowicz-Cieslak J, Tai TH, Kumlehn JK, Till BJ, editors. Biotechnologies for plant mutation breeding. Biotechnologies for plant mutation breedingSpringer; 2017. p. 227–39.

[56] Koressaar T, Remm M. Enhancements and modifications of primer design program Primer3. Bioinformatics 2007;23(10):1289–91. http://dx.doi.org/10.1093/bioinfor-matics/btm091.

[57] FastQC. A quality control tool for high throughput sequence data; 2017 [Available at: http://www.bioinformatics.babraham.ac.uk/projects/fastqc/. Accessed 2017 Jan 27].

[58] Phytozome v12.0, 2017, [Available at: https://phytozome.jgi.doe.gov/pz/portal.html. Accessed 2017 Jan 27].

[59] Picard tools - by Broad Institute, 2017, [Available at: http://broadinstitute.github.io/ picard/. Accessed 2017 Jan 27].

[60] Robinson JT, Thorvaldsdóttir H, Winckler W, Guttman M, Lander ES. Integrative ge-nomics viewer. Nat Biotechnol 2011;29:24–6. http://dx.doi.org/10.1038/nbt.1754.

[61] Cingolani P, Platts A, Wang LL, Coon M, Nguyen T, et al. A program for annotating and predicting the effects of single nucleotide polymorphisms, SnpEff: SNPs in the genome of Drosophila melanogaster strain w1118; iso-2; iso-3. Flying 2012;6(2): 80–92. http://dx.doi.org/10.4161/fly.19695 [PMID: 22728672].

14

Protein Structure Classification and Loop Modeling Using Multiple Ramachandran Distributions ☆

Seyed Morteza Najibi[a], Mehdi Maadooliat[b,e], Lan Zhou[c], Jianhua Z. Huang[c], Xin Gao[d,*]

[a]Department of Statistics, College of Sciences, Shiraz University, Shiraz, Iran
[b]Department of Mathematics, Statistics and Computer Science, Marquette University, WI 53201-1881, USA
[c]Department of Statistics, Texas A&M University, TX 77843-3143, USA
[d]Computational Bioscience Research Center (CBRC), Computer, Electrical and Mathematical Sciences and Engineering Division, King Abdullah University of Science and Technology (KAUST), Thuwal 23955-6900, Saudi Arabia
[e]Center for Human Genetics, Marshfield Clinic Research Institute, Marshfield, WI 54449, USA

ARTICLE INFO

Keywords:
Bivariate splines
Log-spline density estimation
Protein structure
Ramachandran distribution
Roughness penalty
Trigonometric B-spline
Protein classification
SCOP

ABSTRACT

Recently, the study of protein structures using angular representations has attracted much attention among structural biologists. The main challenge is how to efficiently model the continuous conformational space of the protein structures based on the differences and similarities between different Ramachandran plots. Despite the presence of statistical methods for modeling angular data of proteins, there is still a substantial need for more sophisticated and faster statistical tools to model the large-scale circular datasets. To address this need, we have developed a nonparametric method for collective estimation of multiple bivariate density functions for a collection of populations of protein backbone angles. The proposed method takes into account the circular nature of the angular data using trigonometric spline which is more efficient compared to existing methods. This collective density estimation approach is widely applicable when there is a need to estimate multiple density functions from different populations with common features. Moreover, the coefficients of adaptive basis expansion for the fitted densities provide a low-dimensional representation that is useful for visualization, clustering, and classification of the densities. The proposed method provides a novel and unique perspective to two important and challenging problems in protein structure research: structure-based protein classification and angular-sampling-based protein loop structure prediction.

1. Introduction

Proteins are large biomolecules or macromolecules that perform a vast array of functions for the biological processes within the cell of organisms. A protein is a linear chain of amino acids, each of which is composed of an amino group ($-NH_2$), a central carbon atom (C_α), a carboxyl group ($-COOH$), and a side-chain group that is attached to C_α and is specific to each amino acid. Depending on the amino acid sequence (different amino acids have different biochemical properties) and interactions with their environment, proteins

fold into a three-dimensional structure, which allows them to interact with other proteins and molecules to perform their function. Hence, an important topic in the field of structural biology is the determination of the three-dimensional (3D) structure of a protein. In a protein, each amino acid is called a residue and the chain of carbon, nitrogen and oxygen atoms are referred to as the backbone. While the side-chain structures determine local structures and interactions of the amino acids of the protein, the backbone structure determines the overall shape of the protein and is the focus of much research.

The backbone conformation of proteins can be represented equivalently by Cartesian coordinates of carbon, nitrogen and oxygen atoms, or the backbone dihedral angles (ϕ, ψ), and ω, with the assumption of standard bond lengths and angles. Moreover, the global folds of proteins can be equivalently represented by either the Cartesian coordinates of C_α traces or the 2 pseudo-angles (θ, τ) between the two consecutive planes formed by 4 successive C_α. The Ramachandran plot, a scatter plot of ϕ vs. ψ, can reflect

☆ The first two authors, Najibi and Maadooliat, made equal contributions to the paper.
 * Corresponding author.
   E-mail addresses: mor.najibi@gmail.com (S.M. Najibi), mehdi@mscs.mu.edu (M. Maadooliat), lzhou@stat.tamu.edu (L. Zhou), jianhua@stat.tamu.edu (J.Z. Huang), xin.gao@kaust.edu.sa (X. Gao).

the allowed regions of conformational space available to protein chains. By analogy to Ramachandran's concept of dihedral angles, the pseudo-Ramachandran plot, a scatter plot of θ vs. τ, can provide a distinctive classification of protein structures and largely contribute to different applications [1].

In the development of protein tools over the last two decades, the angular representation of proteins and Ramachandran plots have been applied in various protein structure-related problems, such as protein structural model checking [2–4], structure prediction [5–9], model quality assessment [10–12], prediction server ranking [13, 14], protein structure alignment [15, 16], free energy function learning [17–19], molecular dynamics simulation [20], empirical energy functions [21] and classification functions such as backbone-dependent rotamer library [22, 23].

Since the seminal work of Ramachandran et al. [24], the two-dimensional histogram of Ramachandran plot has been commonly used to determine accessible regions and validate new protein structures [2, 3]. The histogram is a rough non-parametric density estimation where the number of parameters is equal to the number of data points. Furthermore, because of the circular nature of the protein angles, the traditional parametric or non-parametric density estimation methods cannot be used for estimating Ramachandran distributions. In the last decade, novel parametric and non-parametric methods have been introduced to address this problem. The parametric methods propose to use directional distributions such as von Mises distribution or short Fourier series that are naturally designed for periodic data [25–29]. On the other hand, the non-parametric techniques use kernel density estimates with periodic kernels, Dirichlet process with boundary modification, or a mixture of directional distributions [30–32].

Depending on the purpose of the study, one may produce Ramachandran plots based on residues associated with some specific amino acids, and/or some specific structural elements. In some cases, the number of residues (data points) is too small, and that makes it challenging to obtain reliable bivariate densities using techniques that estimate each Ramachandran distribution separately. An intuitive solution to this problem is to borrow information from a group of Ramachandran plots that has some common features. To this end, Lennox et al. [33] proposed a hierarchical Dirichlet process technique based on bivariate von Mises distributions that can simultaneously model angle pairs at multiple sequence positions. This method is typically used for predicting highly variable loop and turn regions. Ting et al. [34] and Joo et al. [35] also used this technique with further modification to produce near-native loop structures. In another approach, Maadooliat et al. [36] proposed a penalized spline collective density estimator (PSCDE) to represent the log-densities based on some shared basis functions. This method showed some significant improvements for loop modeling of the hard cases in a benchmark dataset where existing methods do not work well [36].

Comparing to other competitive approaches, PSCDE is more efficient in estimating the densities in the sparse regions by incorporating the shared information among the distributions. In this technique, the bivariate log-densities are represented using a common set of basis functions. Each log-density has its own coefficient vector in the basis expansion, and it can be used for clustering and classification of the densities. Furthermore, using a common set of basis functions significantly reduces the number of parameters to be estimated. This method has been applied to estimate the neighbor-dependent Ramachandran distributions to make the angular-sampling-based protein structure prediction more accurate. In this paper, we make an innovative and constructive development over the PSCDE method.

The PSCDE method is constructed based on Bernstein-Bézier spline basis functions defined over triangles to estimate the log-densities in a complex domain [36]. In simple words, in PSCDE,

we artificially extended the constraints of the adjacent triangles to the triangles in boundaries in order to estimate the densities in a two-dimensional circular domain. Here, we propose an alternative approach that uses the tensor product of trigonometric B-spline basis to handle the angular nature of the data. The main advantage of the proposed method is that there is no need to implement any further constraints to take into account the continuity and circularity of the data since the new bases are trigonometric functions that are smooth and intrinsically periodic. Another improvement in the proposed procedure is on selecting the smoothing parameter. In the existing PSCDE procedure, the tuning parameter is selected using the Akaike Information Criterion. Therefore a grid search is needed to choose the optimal tuning parameter and that could become time-consuming, especially if different tuning parameters are used for different basis functions. Following Schellhase and Kauermann [37], we propose to update the smoothing parameter within the Newton–Raphson iterative procedure that is used for the density estimation.

The PSCDE method is originally applied to the protein loop modeling problem. Here, we focus on a new application and use an extension of PSCDE to the protein structure classification problem. There is a large literature on the classification of the protein structures in the Protein Data Bank (PDB) [38–40]; because a good classification can reveal the evolutionary relationship between the proteins and step toward understanding the protein functions. While a vast majority of the literature deals with the protein classification in a pairwise structural comparison framework, the proposed estimated densities can be used as an alternative technique based on angular representation for the structural classification.

Specifically, the estimated angular density corresponding to a protein structure has a basis expansion whose coefficients can be used as an input to a clustering algorithm. Furthermore, most of the existing techniques for protein classification are using sequence and/or 3D structure comparison to classify the proteins based on some (dis)similarity scores obtained after pairwise alignments. The proposed method is an alignment-free procedure that provides a vector of coefficients (i.e. features), associated with each structure (density), that can be directly used to classify the proteins.

We also applied the proposed method to the loop modeling problem and compared the result with the other methods in the online supplementary. In this application, we trained the neighbor-dependent distributions of the backbone dihedral angles (i.e., neighbor-dependent Ramachandran distributions) using the new collective density estimation approach and fed the results into the Rosetta loop modeling procedure to study the accuracy and efficiency of the Rosetta server in predicting the loop regions. The main concern of using the neighbor-dependent Ramachandran distributions is that we are partitioning the data into smaller groups, some partitions may end up with a limited number of observations, and therefore we may lose accuracy in estimating the Ramachandran distributions due to the data sparsity. The proposed collective estimation procedure can overcome this difficulty and thereby improve the accuracy of the estimated densities. We encourage the interested readers to read the online supplementary materials for the implementation of the proposed method on loop-modeling application.

The rest of the paper is organized as follows. Section 2 introduces the penalized spline collectively density estimator procedure based on the new trigonometric basis functions to incorporate the circular nature of data. Section 3 presents the protein structure classification problem and the implementation of the new procedure for this application. Section 4 concludes the paper with a discussion. A web-based toolbox is also introduced in the Appendix to illustrate the advantages of the proposed technique. This toolbox can be used further by the research community to obtain the collective estimation of Ramachandran distributions for any other related application (e.g. backbone-dependent rotamer library [22, 23]).

2. Collective Estimation of Multiple Probability Density Functions

In this section, we review and extend a procedure for estimating the multiple probability density functions, known as the PSCDE [36]. Suppose that we observe data from m bivariate probability distributions with the density functions $f_i, i = 1, \cdots, m$. We assume that each log-density can be represented by a set of common basis functions. Therefore we write each log-density function as

$$\log\{f_i(\boldsymbol{x})\} = \omega_i(\boldsymbol{x}) + c_i, \tag{1}$$

where $\omega_i(\boldsymbol{x})$ is a linear combination of the basis functions $\{\phi_k, k = 1, \ldots, K\}$ such that

$$\omega_i(\boldsymbol{x}) = \sum_{k=1}^{K} \phi_k(\boldsymbol{x})\alpha_{ik} \quad \forall i = 1, \ldots, m, \tag{2}$$

and c_i is a normalizing constant ($c_i = -\log \int \exp \omega_i(\boldsymbol{x})d\boldsymbol{x}$) to ensure that each f_i is a valid density function. In our setting, the value of K and the basis functions (ϕ_k's) are not pre-specified and will be determined based on data. We assume that ϕ_k's fall in a low-dimensional subspace of a function space spanned by a rich family of fixed basis functions, $\{b_\ell(\boldsymbol{x}), \ell = 1, \ldots, L\}, (L \gg K)$, such that

$$\phi_k(\boldsymbol{x}) = \sum_{\ell=1}^{L} b_\ell(\boldsymbol{x})\theta_{\ell k}.$$

This framework provides a common set of basis functions to represent the log-densities. Also, each density in this model is represented with a set of coefficients $\alpha_{ik}, k = 1, \ldots, K$, which can be used as an excellent feature for comparison, assessment and classification of the densities. Furthermore, similar to the scree plot in principal component analysis (PCA), one may plot the sum of square of the component coefficients ($g(k) = \sum_i \alpha_{ik}^2$) as a function of component index, to select number of significant components, K, e.g. see Figs. 1A and 2A.

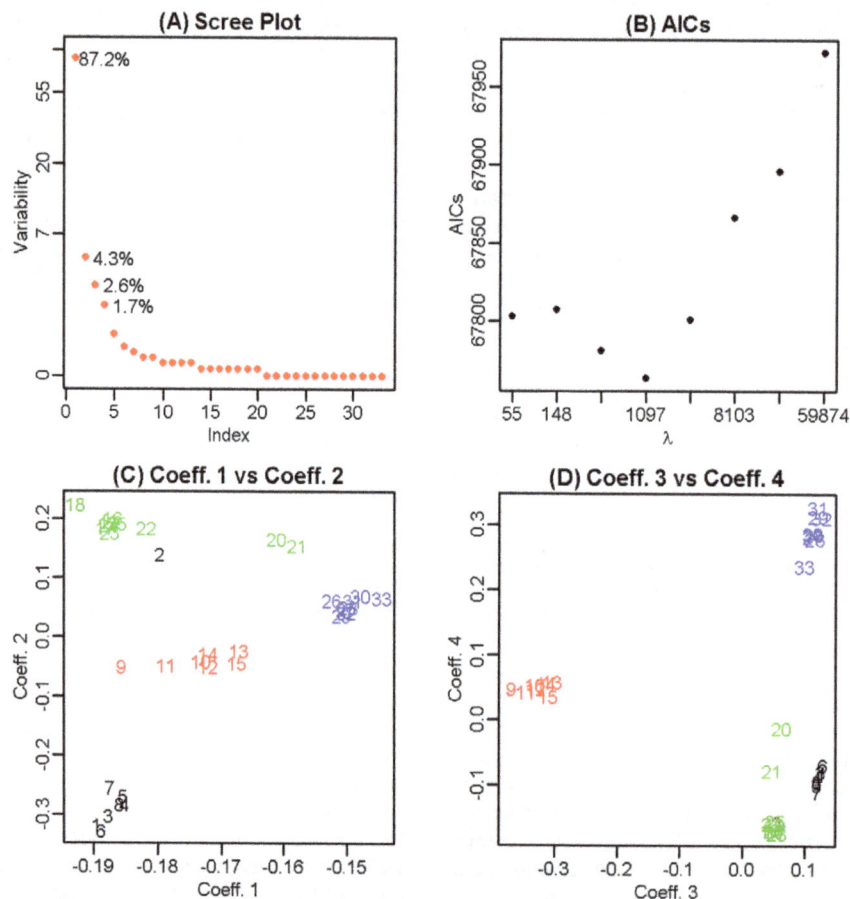

Fig. 1. A classification task with 33 domains from four *Species* of the same protein class, separated at the bottom of SCOP hierarchy with PSCDE approach [36]. (A) The scree plot with numbers showing the percentage of variability explained by the leading components; (B) the AIC plot; (C) the scatter plot of coefficients 1 vs 2; and (D) the scatter plot of coefficients 3 vs 4.

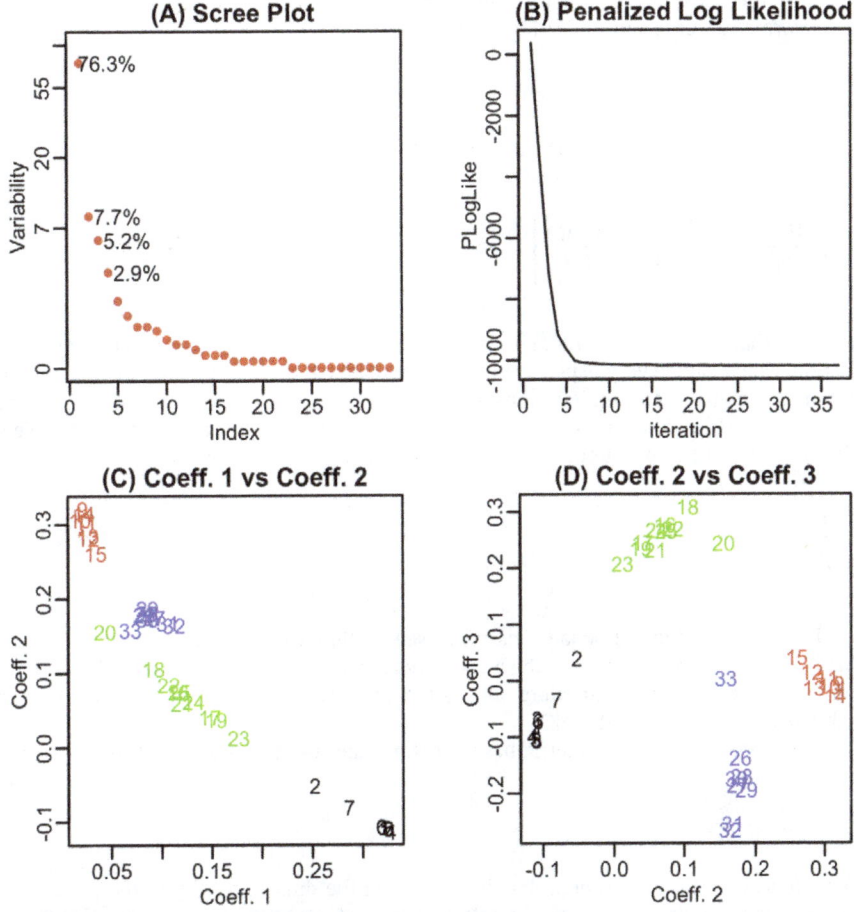

Fig. 2. A classification task with 33 domains from four *Species* of the same protein class, separated at the bottom of SCOP hierarchy with PSCDE(T) approach. (A) The scree plot with numbers showing the percentage of variability explained by the leading components; (B) the trace of the penalized log-likelihood function; (C) the scatter plot of coefficients 1 vs 2; and (D) the scatter plot of coefficients 2 vs 3.

Here, we use the tensor product technique to construct bivariate trigonometric splines that are smooth and intrinsically periodic in one or two directions. The details on how to construct the basis functions are given in Section 2.1. To further simplify the presentation, let $\phi(\boldsymbol{x}) = (\phi_1(\boldsymbol{x}), \phi_2(\boldsymbol{x}), \ldots, \phi_K(\boldsymbol{x}))^\top$, $\boldsymbol{\alpha}_i = (\alpha_{i1}, \alpha_{i2}, \ldots, \alpha_{iK})^\top$, $\mathbf{b}(\boldsymbol{x}) = (b_1(\boldsymbol{x}), b_2(\boldsymbol{x}), \ldots, b_L(\boldsymbol{x}))^\top$, $\boldsymbol{\theta}_k = (\theta_{1k}, \theta_{2k}, \ldots, \theta_{Lk})^\mathrm{T}$ and $\boldsymbol{\Theta} = (\boldsymbol{\theta}_1, \boldsymbol{\theta}_2, \ldots, \boldsymbol{\theta}_K)$, then $\omega_i(\boldsymbol{x})$ given in Eq. (2) can be written as

$$\omega_i(\boldsymbol{x}) = \phi(\boldsymbol{x})^\top \boldsymbol{\alpha}_i = \mathbf{b}(\boldsymbol{x})^\top \boldsymbol{\Theta} \boldsymbol{\alpha}_i, \quad i = 1, \ldots, m. \tag{3}$$

If we evaluate the densities on common regular grids ($\boldsymbol{x}_j, j = 1, \cdots, n$) in the circular plane, we may further simplify the presentation of the densities in an $n \times m$ matrix: $\boldsymbol{\Omega} = \{\omega_i(\boldsymbol{x}_j)\}^\top$. Specifically, let $\mathbf{B} = (\mathbf{b}(\boldsymbol{x}_1), \mathbf{b}(\boldsymbol{x}_2), \ldots, \mathbf{b}(\boldsymbol{x}_n))^\top$, and $\boldsymbol{A} = (\boldsymbol{\alpha}_1, \boldsymbol{\alpha}_2, \ldots, \boldsymbol{\alpha}_m)^\top$, then Eq. (3) can be written in the matrix form, $\boldsymbol{\Omega} = \mathbf{B}\boldsymbol{\Theta}\boldsymbol{A}^\mathrm{T}$, where the parameters to be estimated are $(\boldsymbol{\Theta}, \boldsymbol{A})$. To address the identifiability issue raised by the product of two matrices $(\boldsymbol{\Theta}, \boldsymbol{A})$, we follow the remedy given in [36] based on the singular value decomposition (SVD) technique.

Now, by assuming observations $x_{ij}, j = 1, \ldots, n_i$ from the ith group, $i = 1, \ldots, m$, the log-likelihood function has the following form:

$$\ell(\boldsymbol{\Theta}, \boldsymbol{A}) = \sum_{i=1}^{m} \sum_{j=1}^{n_i} \{\omega_i(x_{ij}) + c_i\}. \tag{4}$$

To obtain smooth densities, the parameters can be estimated by introducing the roughness penalty [41] and minimizing the penalized likelihood criterion:

$$-2\, \ell(\boldsymbol{\Theta}, \boldsymbol{A}) + \lambda \, \mathrm{trace}(\boldsymbol{\Theta}^\top \boldsymbol{D} \boldsymbol{\Theta}), \tag{5}$$

where $\boldsymbol{D}$ penalizes wiggliness (induces smoothness) and $\lambda > 0$ is the tuning parameter. We then use the alternating blockwise Newton–Raphson algorithm in Maadooliat et al. [36] to minimize the penalized likelihood function.

There are different well-known methods to select the tuning parameter. A commonly used technique is to choose the tuning parameter, λ, that minimizes the Akaike Information Criterion (AIC) [42]:

$$\mathrm{AIC}(\lambda) = -2\ \ell(\boldsymbol{\Theta}, \boldsymbol{A}) + 2\ \mathrm{df}(\lambda),$$

where $\ell(\boldsymbol{\Theta}, \boldsymbol{A})$ is the log likelihood function and $\mathrm{df}(\lambda)$ is the degrees of freedom, defined as:

$$\mathrm{df}(\lambda) = \sum_{k=1}^{K} \mathrm{trace}\left\{ \left[\frac{\partial^2 \ell(\boldsymbol{\Theta}, \boldsymbol{A})}{\partial \theta_k \partial \theta_k^\top} + \lambda \boldsymbol{D} \right]^{-1} \left[\frac{\partial^2 \ell(\boldsymbol{\Theta}, \boldsymbol{A})}{\partial \theta_k \partial \theta_k^\top} \right] \right\}.$$

Selecting the tuning parameter that minimizes the AIC, requires training the model for different values of λ's and then pick the one that minimizes the criterion function, which can be very expensive in time. Instead, we present an alternative procedure that updates the value of the tuning parameter within the Newton–Raphson iterations. This idea has been used in generalized mixture model to iteratively update the smoothing parameter [43]. Schellhase and Kauermann [37] extended this approach for density estimation. We borrow their formulation, and use the parameter estimates in the ith step to update the tuning parameter, $\hat{\lambda}_{i+1}$, through

$$\hat{\lambda}_{i+1}^{-1} = \frac{\mathrm{trace}\left(\hat{\boldsymbol{\Theta}}_i^\top \boldsymbol{D} \hat{\boldsymbol{\Theta}}_i \right)}{\mathrm{df}\left(\hat{\lambda}_i \right) - (a-1)}, \tag{6}$$

where a is the order of the differences used in the penalty matrix $\boldsymbol{D}$ (see Section 2.1). From what we have seen in the implementation of the new procedure, updating the tuning parameter within the Newton–Raphson iterations, on average, does not increase the number of the iterations required to converge. Therefore the new procedure obtains the final result p times faster than the older procedure, where p is the number of λ's used in the grid search to minimize the AIC.

In the following subsection, we obtain the trigonometric basis functions and the penalty matrix that has been used in minimizing the penalized likelihood function (Eq. (5)).

2.1. Basis Functions and the Penalty Matrix

There are a variety of basis functions that can come in handy depend on the dimensionality of the problem and the data structure. In this context, the circular nature of the protein angles is an obstacle that prevents us from using the standard B-spline functions. Maadooliat et al. [36] proposed to use bivariate spline functions over triangulations, and they artificially extended the constraints for two adjacent triangles [44] to the triangles in boundaries. Triangulation is a sophisticated procedure that works perfectly for complex geometries with unbalanced observations over irregular grid points. For Ramachandran plot, we evaluate the densities over regular grid points in a smooth rectangular plane that is obtained by unfolding a simple manifold (torus or sphere), and it is better if we can avoid such sophisticated procedure. Furthermore, extending the triangulation technique beyond the bivariate case, and implementing the PSCDE via triangulations in higher dimensions is not straightforward.

A frequently used basis functions for Euclidean space is the tensor product of standard B-spline functions which is appealing and very easy to use in the real world applications [45]. With some small alteration, the tensor product of trigonometric spline can be defined by sin and cos functions which are smooth and naturally periodic functions [46]. Moreover, this method can be easily applied to higher dimensional density estimation.

We need to develop rich set of basis functions $\{b_\ell(\boldsymbol{X}), \ell = 1, \ldots, L\}$, that is required for estimating the Ramachandran or pseudo-Ramachandran distributions, over the support set (Ω or Ω') which can be defined as

$$\Omega = \{-\pi \le \phi \le \pi \quad \text{and} \quad -\pi \le \psi \le \pi\} \quad \text{or} \quad \Omega' = \{-\pi \le \theta \le \pi \quad \text{and} \quad 0 \le \tau \le \pi\}. \tag{7}$$

From a geometric point of view, Ω resembles the surface of a torus with some fixed minor/major radiuses and Ω' represents the surface of a sphere with fixed radius. In fact, the existing parametric models take into account the topology and develop a parametric framework on surfaces of a torus or sphere with some fixed radiuses to model the bivariate densities [32, 47, 48]. In contrast, non-parametric methods use either a periodic kernel or some boundary modification technique to address this issue.

Here we present the tensor product of two sets of trigonometric basis functions and construct the bivariate bases that can be used to represent the space for two dihedral angles (ϕ, ψ) defined over Ω. One may proceed with a similar procedure based on the Kronecker product of a trigonometric spline and a standard B-spline to obtain the bivariate basis representation for the pair of dihedral, planar angles (θ, τ) defined over Ω'.

A univariate normalized trigonometric spline with κ knots, $(x_1, x_2, \cdots, x_\kappa)$, and order of ν, can be represented recursively as a periodic spline on a circle; see Schumaker [49, ch. 8] for details. In specific, for every ϕ within the interval $[x_i, x_{i+\nu}]$ the spline functions are defined as

$$S_i^1(\phi) = \begin{cases} 1 & x_i \le \phi \le x_{i+1} \\ 0 & \text{o.w.} \end{cases},$$

$$S_i^\nu(\phi) = \frac{\sin\left(\frac{\phi - x_i}{2}\right)}{\sin\left(\frac{x_{i+\nu-1} - x_i}{2}\right)} S_i^{\nu-1}(\phi) + \frac{\sin\left(\frac{x_{i+\nu} - \phi}{2}\right)}{\sin\left(\frac{x_{i+\nu} - x_{i+1}}{2}\right)} S_{i+1}^{\nu-1}(\phi). \tag{8}$$

The same methodology should be used to create basis functions for dihedral angle, ψ. The main advantage of using these linearly independent basis functions over the standard B-spline choice is that the continuity of the tangent plane for any smooth function on surface of a sphere is the result of the former one. Therefore, there is no need to introduce any periodic constraints for the trigonometric spline functions (for more details see Schumaker and Traas [50]), due to the fact that each piece lies in $span(\mathscr{F}_m)$, where:

$$\mathscr{F}_m = \begin{cases} \{\cos(\phi/2), \sin(\phi/2), \ldots, \cos((2q-1)\phi/2), \sin((2q-1)\phi/2)\} & \text{if } \nu = 2q, \\ \{\cos(\phi), \sin(\phi), \ldots, \cos(q\phi), \sin(q\phi)\} & \text{if } \nu = 2q-1. \end{cases}$$

In matrix form, we denote $\boldsymbol{B}_\phi$ and $\boldsymbol{B}_\psi$ to be the matrices that represent the trigonometric basis functions associated to ϕ and ψ directions with ranks M and N respectively. The matrix $\boldsymbol{B}$ that represents the bivariate spline basis functions can be then obtained from the Kronecker product of $\boldsymbol{B}_\phi$ and $\boldsymbol{B}_\psi$:

$$\boldsymbol{B} = \boldsymbol{B}_\phi \otimes \boldsymbol{B}_\psi,$$

where the symbol $\otimes$ is used to represent the Kronecker product.

It should be noted that the number of knots, κ, directly influence the smoothness of the estimated functions. The smaller κ results smooth, but biased estimates. While increasing κ will reduce the bias, but it will consequently increase the variability and therefore, we end up with some rough estimates. It is customary to have a large number of knots in the model and control the smoothness of estimates by introducing a roughness penalty into the likelihood function, to control the bias-variance tradeoff. Here, we monitor the roughness of the estimated functions by using difference penalty [51] to achieve the appropriate level of smoothness. In a nutshell, the variability is controlled through a difference function of order a, Δ_a, where $\Delta_1 \boldsymbol{\theta}_k := \boldsymbol{\theta}_k - \boldsymbol{\theta}_{k-1}$, and Δ_a is obtained recursively. For example, the second order difference function, Δ_2, has the following form:

$$\Delta_2 \boldsymbol{\theta}_k := \Delta_1 \Delta_1 \boldsymbol{\theta}_k = \boldsymbol{\theta}_k - 2\boldsymbol{\theta}_{k-1} + \boldsymbol{\theta}_{k-2}.$$

We may write the difference functions Δ_a into a matrix form, $\boldsymbol{L}_a$. For example, for $a = 1$ we have

$$\boldsymbol{L}_1 = \begin{bmatrix} 1 & -1 & 0 & \cdots & 0 \\ 0 & 1 & -1 & \ddots & 0 \\ \vdots & \ddots & \ddots & \ddots & 0 \\ 0 & \cdots & 0 & 1 & -1 \end{bmatrix}_{(M-1 \times M)}$$

The positive definite penalty matrix used to control the smoothness in the ϕ direction is defined as D^ϕ, and it has the following quadratic form: $D^\phi = \boldsymbol{L}_a^\top \boldsymbol{L}_a$. Now, we may use the tensor product technique to derive the penalty matrix for the bivariate domain, (ϕ, ψ), as the following:

$$\boldsymbol{D} = \left[\boldsymbol{I}_N \otimes D^\phi + D^\psi \otimes \boldsymbol{I}_M \right], \tag{9}$$

where $\boldsymbol{D}^\phi = \left(\boldsymbol{L}_a^\phi \right)^\top \boldsymbol{L}_a^\phi$ and $\boldsymbol{D}^\psi = \left(\boldsymbol{L}_a^\psi \right)^\top \boldsymbol{L}_a^\psi$.

We now have the required tools to proceed with the estimation procedure. The minimization of the penalized likelihood function (Eq. (5)) can be obtained through the Newton–Raphson algorithm, of which the details can be found in [36]. After convergence, the densities can be obtained using Eq. (1). From now on, we refer to our new procedure that uses the trigonometric basis expansion in PSCDE as PSCDE(T).

3. Application: Protein Structure Classification

In this section, we introduce an application of collective density estimation in protein structural comparison. To evaluate the proposed method, we designed four protein clustering tasks from the Structural Classification of Proteins (SCOP) database, and then try to cluster the proteins in each task without knowing their labels in the SCOP tree. The final clustering result of PSCDE(T) is compared with seven competitive approaches using two external measures (the descriptions are given in Sections 3.3 and 3.4), where SCOP labels are used as the gold standard. Since the class labels were not used, this is a clustering or unsupervised learning problem.

3.1. Structural Classification of Proteins

The Structural Classification of Proteins is a widely used database that stores the results of classification of known protein structures and is available at http://scop.mrc-lmb.cam.ac.uk/scop/. The SCOP has been constructed manually by visual inspection and comparison

of structures. Since manual inspection and classification is time-consuming and subjective, automated classification methods have been developed in the past two decades, including alignment-based methods [52–54], alignment-free methods [55], and consensus methods [56, 57]. However, it is well acknowledged that a reliable automatic protein classification method is not yet available, partly due to the fact that most of the existing methods depend on distance-based similarity measures and are biased by sequence alignments [55, 58]. In this section, we report the results from some experiments of using the SCOP database as a benchmark to evaluate the potential use of angular distributions for automatic protein structure classification. In contrast to the existing protein structure classification methods, our method is completely alignment-free and does not depend on sequence similarity or distance-based measures, thus provides a unique perspective to the problem.

In the SCOP database, protein domains are classified hierarchically according to their sequential, structural and functional relationship. From top to bottom, the SCOP hierarchy comprises the following seven levels: *Class, Fold, Superfamily, Family, Protein, Species,* and

Domain. The *Domain* level lists the individual protein domains of known structures. We refer to Murzin et al. [38] and Andreeva et al. [40] for more details regarding the description of the SCOP hierarchy and how the database is organized.

3.2. Task Designs

To evaluate the performance of PSCDE(T) in different datasets, we designed four SCOP tasks with "Easy", "Somewhat Hard", "Hard" and "Challenging" level of difficulty, that we call them SCOP.1 to SCOP.4, respectively:

1. SCOP.1 (Easy Task): In this task, we considered an easy protein classification. The goal is to classify 63 protein domains that were randomly selected from three remote Protein *Classes* in SCOP. The constituents of the collection of protein domains and the details of this SCOP tree are available in the online supplementary materials.
2. SCOP.2 (Somewhat Hard Task): We considered a protein classification task for which 33 domains were extracted from four *Species* under the same *Protein* subclass that belongs to the "all-alpha protein" *Class*. The constituents of the collection of domains and the details of the SCOP tree involving these domains are available in the online supplementary materials. This classification task is considered somewhat harder than the easy task, because the domains are very similar both sequentially and structurally—they are very close in the SCOP tree and depart only at the bottom (i.e., the *Species* level) of the SCOP hierarchy.
3. SCOP.3 (Hard Task): We considered a protein classification task for which 40 protein chains were randomly selected from three different *Fold/Superfamily* levels, where all chains belong to the "Alpha and beta proteins (a+b)" *Class*. The constituents of the collection of domains and the details of the SCOP tree involving these domains are available in the online supplementary materials. This classification task is considered harder than the SCOP.2, because the similarities within a group of chains branched out from a specific *Superfamily* level is not as strong as branching out at a specific *Species* levels. This task can be used to evaluate different methods in detecting the remote homology relationship at the *Superfamily* level.
4. SCOP.4 (Challenging Task): Fischer et al. [59] provided a challenging benchmark to assess the performance of a fold recognition method in an objective, unbiased and thorough way. We have selected 26 protein chains from their benchmark in the "All beta proteins" *Class* within three different *Folds*. This classification task is considered the hardest task in this paper, which is also indicated in [59].

After choosing the protein domains from the SCOP database, the complete information of the proteins were obtained from the Protein Data Bank (PDB). The PDB record of each protein structure contains its 3D atomic coordinates, secondary structure assignments, as well as atomic connectivity. While different types of dihedral/planar angles can be obtained using the atomic coordinates, we used the R package PRESS [60] to derive the (θ, τ) angles from the PDB files for each task. We observed that θ angles are within the range $(75, 165)$ and τ's are within $(-180, 180)$.

3.3. Protein Classification Approaches and Distance Matrices

Due to the tree based structure of the SCOP database, we use the agglomerative hierarchical clustering technique to group the protein structures. In order to do this, we need to feed in a pairwise (dis)similarity matrix as an input to the clustering algorithm. In this subsection, we illustrate how to obtain such (dis)similarity matrices to compare five non-density based and three density based approaches, respectively.

Since clustering cannot be directly performed on 3D protein structures, a protein structure or sequence comparison algorithm is usually applied to generate (dis)similarity scores between any pair of structures and such scores are then used for clustering [61]. We considered five such algorithms that cover a broad spectrum of existing methods:

- Needleman–Wunsch (NW) algorithm for global sequence alignment [62], with implementation available in the R package Biostrings;
- Smith–Waterman (SW) algorithm for local sequence alignment [63], with implementation available in the R package Biostrings;
- TM-align [64], available at http://zhanglab.ccmb.med.umich.edu/TM-align/;
- Yakusa [65], available at http://bioserv.rpbs.jussieu.fr/Yakusa/download/index.html;
- Dali [66], available at http://ekhidna.biocenter.helsinki.fi/dali_lite/downloads/v3/.

The first two methods are based on sequence comparison, and the other three methods are based on structure comparison. After we apply these five algorithms, we follow Sam et al. [61] to transform the similarity matrices to distance matrices.

We also considered three density based approaches: Kernel Density Estimator (KDE), PSCDE and PSCDE(T) for protein classification. We used Symmetric Kullback–Leibler Divergence (SKLD) between Ramachandran distributions to obtain pairwise distance matrices between proteins [14]. In the KDE, we used Gaussian kernel density estimation with slight modification to consider the angular structure of the data to obtain an estimate of each density separately [14].

In the PSCDE(T) method, we initialized the algorithm with the cubic B-spline basis functions with 5 degrees of freedom in the θ direction and the cubic trigonometric B-spline basis functions with 15 degrees of freedom in the τ direction. The final tensor product basis functions are obtained and evaluated over 90 gird points in each direction. Furthermore, we selected the number of common basis to be equal to the number of classes in the gold standard associated to each task (four common basis for SCOP.2, and three common basis for the remaining three tasks). In general, one may use scree plot based on the initial estimates (obtained by mapping the kernel density estimators to the column space of the basis expansion) or other approaches available in the literature to select the number of common basis. After estimating the parameters $(\mathbf{A}, \mathbf{\Theta})$ using the Newton–Raphson algorithm, the densities can be obtained using Eq. (1). The PSCDE results can be obtained similarly. In order to have comparable initial basis functions for PSCDE, we partitioned the (θ, τ) domain to 64 similar right triangles with cubic bivariate B-spline basis functions over each triangle (see [36] for more details).

The distance matrices obtained for the above eight approaches: NW, SW, TM-align, Yakusa, Dali, KDE, PSCDE and PSCDE(T) are used as an input to the hierarchical clustering algorithm, implemented in the hclust function with option {method="ward.D"} in the R package stats to obtain dendrograms [67] (e.g. see Fig. 3). In order to obtain the clusters, we cut the dendrograms of all eight approaches into the number of the original clusters in the SCOP database. To evaluate the performance of the proposed method in discovering the correct label (gold standard), we used two external measures that are commonly used in the clustering evaluation literature and discussed in Section 3.4.

3.4. External Evaluation Measures

Consider A and B be two clusterings of a dataset consisting of N records. Let A cluster the data in r clusters and define a_i as the size of

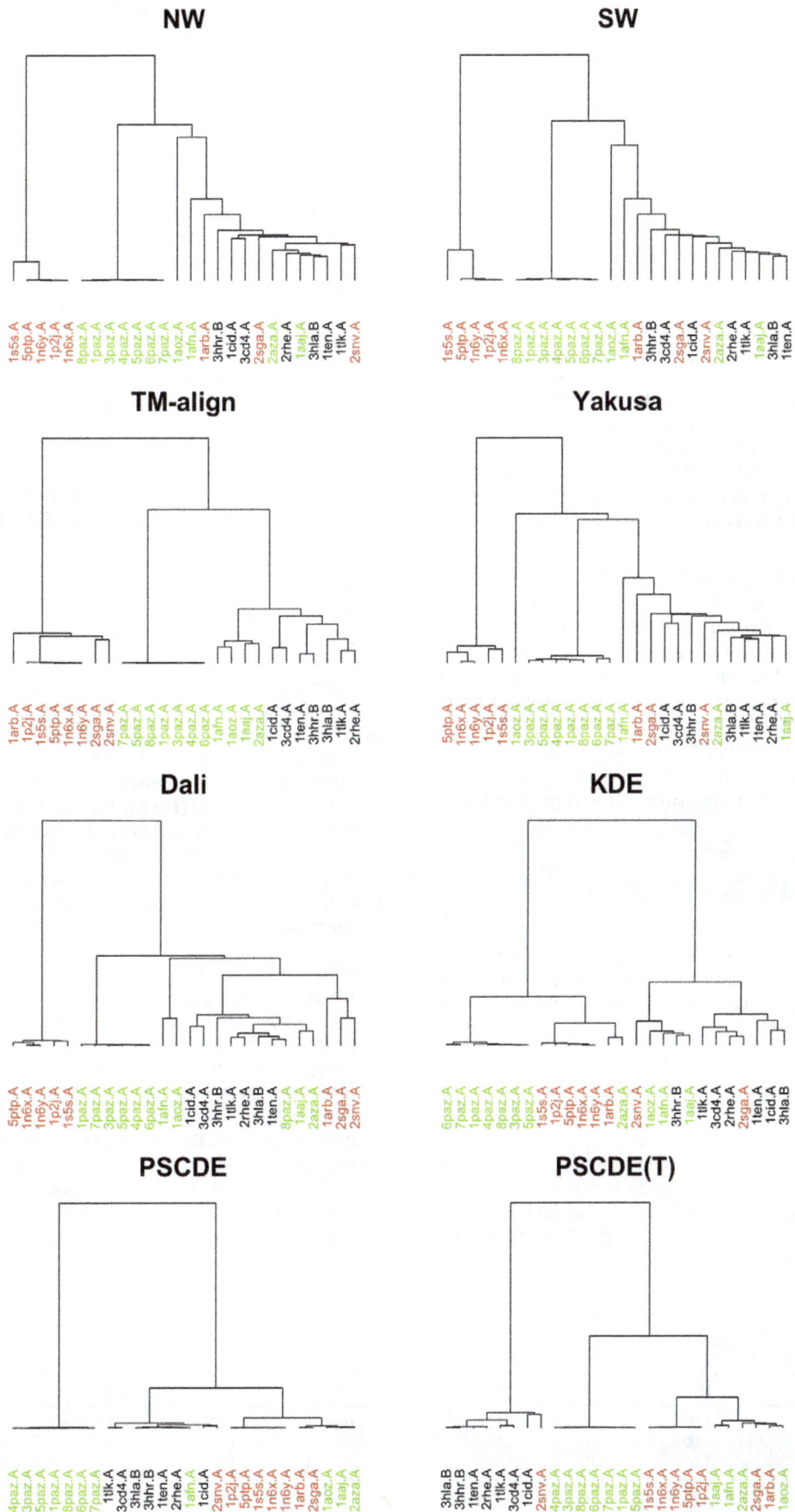

Fig. 3. Dendrograms from hierarchical clustering for SCOP.4 task.

Table 1
$r \times c$ contingency table M relating to two clustering A and B.

		B				
		b_1	...	b_j	...	b_c
A	a_1	n_{11}	...	.	...	n_{1c}
	$\vdots$	$\vdots$		$\vdots$		$\vdots$
	a_i	.		n_{ij}		.
	$\vdots$	$\vdots$		$\vdots$		$\vdots$
	a_r	n_{r1}	...	.	...	n_{rc}

cluster $i = 1, \ldots, r$, and let B cluster the data in c clusters of size b_j for each cluster $j = 1, \ldots, c$ (Note that, in our comparison $r = c$). Given that A and B are partitions of the same data it is possible to count the elements that belong both to cluster i and j. Let n_{ij} denote the number of records shared between cluster i and j. The overlap between two clusterings can be represented in matrix form by a $r \times c$ contingency table M such as the one in Table 1. We refer to $a_i = \sum_j n_{ij}$ as the row marginals and to $b_j = \sum_i n_{ij}$ as the column marginals.

Here, we have used two external measures as follows:

1. Normalized Mutual Information (NMI): In the information theory, the mutual information of two random variables is a measure of the mutual dependence between the two variables. The concept of mutual information is intricately linked to that of entropy of a random variable. The entropy in clustering is defined as the expected value of its information content if it is seen as a random variable. We can therefore define entropy for clustering A and B as $H(A) = -\sum_{i=1}^{r} \frac{a_i}{N} \log \frac{a_i}{N}$ and $H(B) = -\sum_{j=1}^{c} \frac{b_j}{N} \log \frac{b_j}{N}$, respectively. Formally, the mutual information of two clusterings [68] can be defined as

$$MI(A, B) = \sum_{i=1}^{r} \sum_{j=1}^{c} \frac{n_{ij}}{N} \log \frac{n_{ij} N}{a_i b_j}.$$

The mutual information has many possible upper bounds that might be used to obtain the Normalized Mutual Information. Here, we have used $\max\{H(A), H(B)\}$ to normalize the MI as follows:

$$NMI = \frac{MI}{\max\{H(A), H(B)\}}. \tag{10}$$

2. Adjusted Rand Index (ARI): The Rand index in data clustering is a measure of the similarity between two data clusterings. The adjusted Rand Index (ARI) is defined to adjust the chance grouping of elements [69]. ARI is related to the

accuracy but is applicable even when class labels are not used. It is defined as

$$ARI = \frac{\sum_{ij} \binom{n_{ij}}{2} - \left[\sum_i \binom{a_i}{2} \sum_j \binom{b_j}{2}\right] / \binom{N}{2}}{\frac{1}{2}\left[\sum_i \binom{a_i}{2} + \sum_j \binom{b_j}{2}\right] - \left[\sum_i \binom{a_i}{2} \sum_j \binom{b_j}{2}\right] / \binom{N}{2}}. \tag{11}$$

Both NMI and ARI indices change between 0 to 1, with 0 indicating that the two clusters do not agree on any pairs and 1 indicating that the clusters are exactly the same. In the following subsection, we present and compare the clustering results for the eight approaches given in Section 3.3 considering the original SCOP labels as the gold standard in four different tasks given in Section 3.4 using the above external measures. We should point out that we do not use the class labels when applying the clustering algorithms and only use the class labels for evaluation of the clustering results.

3.5. Results

We implemented the eight approaches introduced in Section 3.3 (NW, SW, TM-align, Yakusa, Dali, KDE, PSCDE and PSCDE(T)) on four design tasks, given in Section 3.2 as SCOP.1 to SCOP.4, and obtained the external measures between the eight different clusters and the gold standard for each task (based on the SCOP tree). The results are presented in Table 2.

Furthermore, Table 3 compares the running time and number of iterations needed to run PSCDE and PSCDE(T) in a personal Macintosh computer with 2.5 GHz Intel Core i5 and 10 GB memory. The running time of PSCDE(T) is clearly faster than PSCDE. This is due to the fact that PSCDE(T) updates the tuning parameter (λ) within each iteration while PSCDE runs independent Newton–Raphson iterations for each tuning parameters separately and pick the one that minimizes the AIC [36] (see e.g. Fig. 1B). It is worth to note that the web-application is approximately 3 times faster due to the higher performance of the Shiny servers.

In the remaining of this section, we emphasize some important outcomes of each task and refer the readers to the online supplementary materials for further details.

SCOP.1 (Easy Task): As it is expected, Table 2 confirms that all of the eight competing approaches do a great job in this easy task. This can be seen in the dendrogram given in the online supplementary materials (Fig. S.2) as well. The height of the vertical lines, indicates the degree of difference between branches. The longer the line, the greater the difference.
SCOP.2 (Somewhat Hard Task): The results shown in Table 2 confirm that PSCDE and PSCDE(T) methods are again competitive with the other methods on this clustering task. The PSCDE procedure mislabels only one of the structures (same as "TM-align" method), while the other six methods give perfect classification. The dendrogram associated to this task is also given

Table 2
Comparing the clustering performance of eight approaches (NW, SW, TM-align, Yakusa, Dali, KDE, PSCDE and PSCDE(T)) on four different tasks: "Easy", "Somewhat Hard", "Hard" and "Challenging" (SCOP.1–SCOP.4) based on Normalized Mutual Information (NMI) and Adjusted Rand Index (ARI).

Task	Measure	NW	SW	TM-align	Yakusa	Dali	KDE	PSCDE	PSCDE(T)
SCOP.1	NMI	1.00	1.00	1.00	1.00	1.00	1.00	1.00	1.00
	ARI	1.00	1.00	1.00	1.00	1.00	1.00	1.00	1.00
SCOP.2	NMI	1.00	1.00	0.93	1.00	1.00	1.00	0.93	1.00
	ARI	1.00	1.00	0.91	1.00	1.00	1.00	0.91	1.00
SCOP.3	NMI	0.47	0.32	1.00	0.86	1.00	0.87	1.00	1.00
	ARI	0.34	0.19	1.00	0.86	1.00	0.86	1.00	1.00
SCOP.4	NMI	0.48	0.48	0.71	0.29	0.44	0.39	0.56	0.64
	ARI	0.30	0.30	0.60	0.17	0.23	0.30	0.47	0.51

Table 3
Running time and number of iterations to achieve the final results of PSCDE and PSCDE(T) in a personal computer.

Method	PSCDE(T)		PSCDE	
	Time (min)	Iterations	Time (min)	Iterations
SCOP.1	1.24	19	11.49	176
SCOP.2	1.39	37	14.85	324
SCOP.3	1.00	18	15.19	174
SCOP.4	0.12	7	8.52	112

in the online supplementary materials. The height of the vertical lines in Fig. S.4 suggests that our angular density-based method provides a competitive result in clear separation of the four clusters.

The difference between the results of PSCDE and PSCDE(T) in SCOP.2 makes it interesting to further compare the associated results in more details and illustrate their properties. Fig. 1 presents the results from applying the PSCDE method. The scree plot (Fig. 1A) indicates that four components can represent most of the variability among the angular densities. The AIC plot (Fig. 1B) shows a clear minimum of the AIC corresponding to the selected penalty parameter. The scatter plots (Fig. 1C and D) of coefficients in the fitted exponential family densities show that no single coefficient can separate the four classes. Neither any pairs of the coefficients can provide a good separation, but all coefficients together give some good separation. However, one of the proteins (indicated as number 2 with black color) is mislabeled and is closer to the green color cluster.

In a similar framework, Fig. 2 presents the results from applying the proposed PSCDE(T) method. Similarly, the scree plot (Fig. 2A) indicates that four components represent most of the variability among the angular densities. The penalized log-likelihood versus iterations (Fig. 2B) shows that the convergence is achieved after 37 iterations. Although, the scatter plots (Fig. 2C) and D) of coefficients in the fitted exponential family densities show that no single coefficient can separate the four classes, but the coefficients 2 and 3 can separate three classes and with coefficients 1 together give a perfect separation of four classes.

By comparing the results of Figs. 1 and 2 some interesting observations were obtained. Although the information (energy) in components 1 is less (76.3%) in PSCDE(T), it does a good job in separating the classes. Furthermore, instead of running the PSCDE for 8 different tuning parameters (324 Newton–Raphson iterations that took 14.85 min to run) and then pick the optimal one that minimizes the AIC, the proposed PSCDE(T) gives even better estimates in one run (37 iterations in 1.39 min). Note that, in PSCDE(T) the tuning parameter gets updated within the Newton–Raphson iterations, which leads to obtaining the results almost 8 (number of different tuning parameters used in PSCDE) times faster than PSCDE procedure.

SCOP.3 (Hard Task): The results shown in Table 2 indicate that TM-align, Dali, PSCDE and PSCDE(T) provided the clustering results that are in total agreement with the gold standard of SCOP.3 task. While the results of Yakusa and KDE are somehow acceptable, the performances of NW and SW are poor for this clustering task. Similar to the previous two tasks, the associated dendrogram to SCOP.3 is presented in the online supplementary material (Fig. S.6). Fig. S.6 also confirms that the first two methods (NW and SW), which are motivated from pairwise sequence alignment, produced unacceptable hierarchical clustering results. While (i) Yakusa and KDE results are somehow

acceptable; (ii) TM-align and Dali have no mislabeling in this case; but clearly PSCDE and PSCDE(T) have the longest vertical lines among the respective dendrograms, indicating the highest degree of difference (separation) between the branches.

SCOP.4 (Challenging Task): The results shown in Table 2 clearly indicate that five of the approaches (NW, SW, Yakusa, Dali and KDE) failed to produce acceptable results (NMI < 0.50 and ARI ≤ 0.30), while PSCDE(T) and TM-align produced the external measures (NMI and ARI) greater than 0.50. Fig. 3 provides the dendrograms for all eight approaches and confirms that TM-align and PSCDE(T) have the longest vertical lines among the respective dendrograms with acceptable degree of separation (compared with the other six approaches). It is worth to mention that TM-align incorporates an optimal alignment of the whole 3D structures, while PSCDE(T) is only a summary statistics and ignores many aspects of the protein structure.

4. Discussion

This paper develops an extension to a recent technique for collective estimation of multiple bivariate densities. The proposed method develops a new set of bivariate spline functions, using a tensor product approach, which can replace the bivariate B-spline functions (based on triangulation) implemented in PSCDE. The construction of the new bivariate basis function is simpler, more appealing, and can be easily extended to handle cases with more than two dimensions. While PSCDE handles the circular nature of the angular data with some artificial constraints (that extend the notion of adjacent triangles to the triangles in boundaries), the proposed method simply uses the trigonometric spline functions, that are naturally periodic. Another advantage of the new procedure is to speed up the process by updating the smoothness parameter within the Newton–Raphson iterations and avoid a grid search over the space of smoothing parameter, λ, which could be very expensive in time.

The estimated coefficients of the basis expansion based on PSCDE(T) provide a low-dimensional representation of the densities that can be used for visualization and clustering the densities. In general, the PSCDE(T) algorithm is faster, more appealing and interpretable in comparison to the previous approach, PSCDE.

We have applied the proposed method to four protein structural comparison tasks with different levels of difficulties. The results of these tasks show that PSCDE(T) is a new competitive method compared with existing approaches. Furthermore, the last two tasks illustrate that the PSCDE(T) can improve the efficiency of the estimated densities by borrowing strength across distributions while the non-collective estimation method of KDE does not have such ability. This improvement directly influenced the efficiency of clustering in the last two harder tasks.

We also used this method in estimating the neighbor-dependent Ramachandran distributions (the results are given in online supplementary materials), and fed those estimates into Rosetta for loop modeling application. The ultimate results showed that PSCDE(T) is competitive with other similar methods and occasionally improve the results for some hard cases. We also included, in our web application tool, the corresponding input file that contains the 800 neighbor-dependent Ramachandran densities. This can be used by the scientific community to test the quality and applicability of PSCDE(T) approach in loop modeling or any other applications that use the neighbor-dependent Ramachandran distributions (e.g. backbone-dependent rotamer library [22, 23]).

In summary, since the angular density is only a summary statistics and ignores many aspects of the protein structure, we do not expect that it always gives the best results in an arbitrary dataset. This new methodology can be used independently or as a supplement to the existing methods.

Acknowledgment

We are grateful to Professor Roland L. Dunbrack for providing the data set for the neighbor-dependent Ramachandran distribution application, and to Amelie Stein for help with the implementation of Rosetta. Part of Maadooliat's work was done during his sabbatical leave at Marshfield Clinic Research Institute. Maadooliat is greatly appreciative of the support he received from the Center for Human Genetics at Marshfield Clinic Research Institute. The research reported in this publication was supported by the King Abdullah University of Science and Technology (KAUST) Office of Sponsored Research (OSR) under Award No. URF/1/1976-04.

References

[1] Oldfield TJ, Hubbard RE. Analysis of $C\alpha$ geometry in protein structures. Proteins 1994;18(4):324–37.

[2] Laskowski R, MacArthur MW, Moss D, Thornton JM. Procheck: a program to check the stereochemical quality of protein structures. J Appl Crystallogr 1993;26:283–91.

[3] Hooft RWW, Sander C, Vriend G. Objectively judging the quality of a protein structure from a Ramachandran plot. Comput Appl Biosci: CABIOS 1997;13(4):425–30.

[4] Davis IW, Murray LW, Richardson JS, Richardson DC. Molprobity: structure validation and all-atom contact analysis for nucleic acids and their complexes. Nucleic Acids Res 2004;32(Web Server issue):W615–W619.

[5] Simons KT, Bonneau R, Ruczinski I, Baker D. Ab initio protein structure prediction of CASP III targets using ROSETTA. Proteins 1999;37(Suppl 3):171–6.

[6] Hamelryck T, Kent JT, Krogh A. Sampling realistic protein conformations using local structural bias. PLoS Comput Biol 2006;2(9):e131.

[7] Boomsma W, Mardia KV, Taylor CC, Ferkinghoff-Borg J, Krogh A, Hamelryck T. A generative, probabilistic model of local protein structure. Proc Natl Acad Sci USA 2008;105(26):8932–7.

[8] Zhao F, Peng J, Debartolo J, Freed KF, Sosnick TR, Xu J. A probabilistic and continuous model of protein conformational space for template-free modeling. J Comput Biol 2010;17(6):783–98.

[9] Rohl CA, Strauss CEM, Misura KMS, Baker D. Protein structure prediction using Rosetta. Methods Enzymol 2004;383:66–93.

[10] Benkert P, Tosatto SCE, Schomburg D. Qmean: a comprehensive scoring function for model quality assessment. Proteins 2008;71(1):261–77.

[11] Gao X, Xu J, Li SC, Li M. Predicting local quality of a sequence-structure alignment. J Bioinforma Comput Biol 2009;7(5):789–810.

[12] Archie J, Karplus K. Applying undertaker cost functions to model quality assessment. Proteins 2009;75(3):550–5.

[13] Qiu J, Sheffler W, Baker D, Noble WS. Ranking predicted protein structures with support vector regression. Proteins 2008;71(3):1175–82.

[14] Maadooliat M, Gao X, Huang JZ. Assessing protein conformational sampling methods based on bivariate lag-distributions of backbone angles. Brief Bioinform 2013;14(6):724–36.

[15] Miao X, Waddell PJ, Valafar H. Tali: local alignment of protein structures using backbone torsion angles. J Bioinforma Comput Biol 2008;6(1):163–81.

[16] Challis CJ, Schmidler SC. A stochastic evolutionary model for protein structure alignment and phylogeny. Mol Biol Evol 2012;29(11):3575–87.

[17] Mu Y, Nguyen PH, Stock G. Energy landscape of a small peptide revealed by dihedral angle principal component analysis. Proteins 2005;58(1):45–52.

[18] Altis A, Otten M, Nguyen PH, Hegger R, Stock G. Construction of the free energy landscape of biomolecules via dihedral angle principal component analysis. J Chem Phys 2008;128(24):245102.

[19] Riccardi L, Nguyen PH, Stock G. Free-energy landscape of RNA hairpins constructed via dihedral angle principal component analysis. J Phys Chem B 2009;113(52):16660–8.

[20] Altis A, Nguyen PH, Hegger R, Stock G. Dihedral angle principal component analysis of molecular dynamics simulations. J Chem Phys 2007;126(24):244111.

[21] Buck M, Bouguet-Bonnet S, Pastor RW, MacKerell AD Jr,. Importance of the CMAP correction to the CHARMM22 protein force field: dynamics of hen lysozyme. Biom J 2006;90(4):L36–L38.

[22] Bhuyan MSI, Gao X. A protein-dependent side-chain rotamer library. BMC Bioinforma 2011;12(Suppl 14):S10. 1–12.

[23] Shapovalov MV, Dunbrack RL Jr,. A smoothed backbone-dependent rotamer library for proteins derived from adaptive kernel density estimates and regressions. Structure 2011;19(6):844–58.

[24] Ramachandran GN, Ramakrishnan C, Sasisekharan V. Stereochemistry of polypeptide chain configurations. J Mol Biol 1963;7:95–9.

[25] Mardia KV. Statistics of directional data. J R Stat Soc Ser B Methodol 1975;37:349–93.

[26] Rivest LP. A distribution for dependent unit vectors. Comput Stand: Theory Methods 1988;17:461–83.

[27] Singh H, Hnizdo V, Demchuk E. Probabilistic model for two dependent circular variables. Biometrika 2002;89:719–23.

[28] Mardia KV, Taylor CC, Subramaniam GK. Protein bioinformatics and mixtures of bivariate von Mises distributions for angular data. Biometrics 2007;63:505–12.

[29] Pertsemlidis A, Zelinka J, Fondon JW, Henderson RK, Otwinowski Z. Bayesian statistical studies of the Ramachandran distribution. Stat Appl Genet Mol Biol 2005;4(1):1–18.

[30] Dahl DB, Bohannan Z, Mo Q, Vannucci M, Tsai JW. Assessing side-chain perturbations of the protein backbone: a knowledge based classification of residue ramachandran space. J Mol Biol 2008;378:749–58.

[31] Dunbrack RL, Cohen FE. Bayesian statistical analysis of protein side-chain rotamer preferences. Protein Sci 1997;6(8):1661–81.

[32] Lennox KP, Dahl DB, Vannucci M, Tsai JW. Density estimation for protein conformation angles using a bivariate von Mises distribution and Bayesian nonparametrics. J Am Stat Assoc 2009;104:586–96.

[33] Lennox KP, Dahl DB, Vannucci M, Day R, Tsai JW. A Dirichlet process mixture of hidden Markov models for protein structure prediction. Ann Appl Stat 2010;4(2):916–42.

[34] Ting D, Wang G, Shapovalov M, Mitra R, Jordan MI, Dunbrack RL Jr,. Neighbor-dependent Ramachandran probability distributions of amino acids developed from a hierarchical Dirichlet process model. PLoS Comput Biol 2010;6(4):e1000763.

[35] Joo H, Chavan AG, Day R, Lennox KP, Sukhanov P, Dahl DB. et al. Near-native protein loop sampling using nonparametric density estimation accommodating sparcity. PLoS Comput Biol 2011;7(10):e1002234.

[36] Maadooliat M, Zhou L, Najibi SM, Gao X, Huang JZ. Collective estimation of multiple bivariate density functions with application to angular-sampling-based protein loop modeling. J Am Stat Assoc 2016;111(513):43–56.

[37] Schellhase C, Kauermann G. Density estimation and comparison with a penalized mixture approach. Comput Stat 2012;27(4):757–77.

[38] Murzin AG, Brenner SE, Hubbard T, Chothia C. SCOP: a structural classification of proteins database for the investigation of sequences and structures. J Mol Biol 1995;247(4):536–40.

[39] Orengo CA, Michie AD, Jones S, Jones DT, Swindells MB, Thornton JM. CATJ — a hierarchic classification of protein domain structures. Structure 1997;5(8):1093–108.

[40] Andreeva A, Howorth D, Chandonia J-M, Brenner SE, Hubbard TJP, Chothia C. et al. Data growth and its impact on the scop database: new developments. Nucleic Acids Res 2008;36(Database issue):D419–D425.

[41] Green P, Silverman B. Nonparametric regression and generalized linear models: a roughness penalty approach. Chapman & Hall/CRC.; 1994.

[42] Akaike H. A new look at the statistical model identification. IEEE Trans Autom Control 1974;19(6):716–23.

[43] Schall R. Estimation in generalized linear models with random effects. Biometrika 1991;78(4):719–27.

[44] Lai M, Schumaker L. Spline functions on triangulations. Number v. 13 in encyclopedia of mathematics and its applications. Cambridge University Press. 2007. 9780521875929.

[45] De Boor C. A practical guide to splines. vol. 27. Springer-Verlag New York; 1978.

[46] Lyche T, Winther R. A stable recurrence relation for trigonometric-splines. J Approx Theory 1979;25(3):266–79.

[47] Singh H, Hnizdo V, Demchuk E. Probabilistic model for two dependent circular variables. Biometrika 2002;89(3):719–23.

[48] Mardia KV, Taylor CC, Subramaniam GK. Protein bioinformatics and mixtures of bivariate von Mises distributions for angular data. Biometrics 2007;63(2):505–12.

[49] Schumaker LL. Spline functions: basic theory. New York: Wiley; 1981.

[50] Schumaker LL, Traas C. Fitting scattered data on spherelike surfaces using tensor products of trigonometric and polynomial splines. Numer Math 1991;60(1):133–44.

[51] Eilers PH, Marx BD. Flexible smoothing with b-splines and penalties. Stat Sci 1996;89–102.

[52] Gough J, Karplus K, Hughey R, Chothia C. Assignment of homology to genome sequences using a library of hidden Markov models that represent all proteins of known structure. J Mol Biol 2001;313(4):903–19.

[53] Getz G, Starovolsky A, Domany E. F2CS: FSSP to CATH and SCOP prediction server. Bioinformatics 2004;20(13):2150–2.

[54] Cui X, Gao X. K-nearest uphill clustering in the protein structure space. Neurocomputing 2017;220:52–9.

[55] Rogen P, Fain B. Automatic classification of protein structure by using Gauss integrals. Proc Natl Acad Sci 2003;100(1):119–24.

[56] Cheek S, Qi Y, Krishna SS, Kinch L, Grishin N. SCOPmap: automated assignment of protein structures to evolutionary superfamilies. BMC Bioinf 2004;5:197:1–25.

[57] Camoglu O, Can T, Singh AK, Wang Y-F. Decision tree based information integration for automated protein classification. J Bioinforma Comput Biol 2005;3(3):717–42.

[58] Koehl P. Protein structure similarities. Curr Opin Struct Biol 2001;11(3):348–53.

[59] Fischer D, Elofsson A, Rice D, Eisenberg D. Assessing the performance of fold recognition methods by means of a comprehensive benchmark. Pac Symp Biocomput 1996;300–18.

[60] Huang Y, Bonett S, Kloczkowski A, Jernigan R, Wu Z. P.R.E.S.S. — an R-package for exploring residual-level protein structural statistics. J Bioinforma Comput Biol 2012;10(3):1242007.

[61] Sam V, Tai C-H, Garnier J, Gibrat J-F, Lee B, Munson P. Towards an automatic classification of protein structural domains based on structural similarity. BMC Bioinformat 2008;9:74(1):1–18.

[62] Needleman SB, Wunsch CD. A general method applicable to the search for similarities in the amino acid sequence of two proteins. J Mol Biol 1970;48(3):443–53.

[63] Smith TF, Waterman MS. Identification of common molecular subsequences. J Mol Biol 1981;147(1):195–7.

[64] Zhang Y, Skolnick J. TM-align: a protein structure alignment algorithm based on the TM-score. Nucleic Acids Res 2005;33(7):2302–9.

[65] Carpentier M, Brouillet S, Pothier J. Yakusa: a fast structural database scanning method. Proteins 2005;61(1):137–51.

[66] Holm L, Rosenström P. Dali server: conservation mapping in 3D. Nucleic Acids Res 2010;38(Suppl 2):W545–W549.

[67] Core Team R. R: a language and environment for statistical computing. Vienna, Austria: R Foundation for Statistical Computing; 2016. URL https://www.R-project.org/.

[68] Strehl A, Ghosh J. Cluster ensembles — a knowledge reuse framework for combining multiple partitions. J Mach Learn Res March 2003;3:583–617. ISSN 1532-4435.

[69] Kuncheva LI, Hadjitodorov ST. Using diversity in cluster ensembles. 2004 IEEE International Conference on Systems, Man and Cybernetics (IEEE Cat. No.04CH37583). vol. 2. Oct 2004. p. 1214–9.

15

BOG: R-package for Bacterium and virus analysis of Orthologous Groups

Jincheol Park [a], Cenny Taslim [b], Shili Lin [c,*]

[a] *Department of Statistics, Keimyung University, South Korea*
[b] *Ohio State University Medical Center, USA*
[c] *Department of Statistics, State University, USA*

ARTICLE INFO

Keywords:
Bacterium and virus analysis
Clusters of Orthologous Groups
Hypergeometric test
Mann–Whitney Rank Sum test
Gene set enrichment analysis
Tabular and graphical visualization

ABSTRACT

BOG (Bacterium and virus analysis of Orthologous Groups) is a package for identifying groups of differentially regulated genes in the light of gene functions for various virus and bacteria genomes. It is designed to identify Clusters of Orthologous Groups (COGs) that are enriched among genes that have gone through significant changes under different conditions. This would contribute to the detection of pathogens, an important scientific research area of relevance in uncovering bioterrorism, among others. Particular statistical analyses include hypergeometric, Mann–Whitney rank sum, and gene set enrichment. Results from the analyses are organized and presented in tabular and graphical forms for ease of understanding and dissemination of results. BOG is implemented as an R-package, which is available from CRAN or can be downloaded from http://www.stat.osu.edu/~statgen/SOFTWARE/BOG/.

1. Introduction

BOG (Bacterium and virus analysis of Orthologous Groups) is an R-package for identifying groups of differentially regulated genes in the light of gene functions for various virus and bacteria genomes. BOG can be useful in transcriptional profiling of virulent pathogens taking into account of functional categories, an important scientific research area of relevance to detection of bioterrorism. For example, in human host, the concentration of free iron available to bacterium controls the pathogen growth. Effective strategies for adaptation to this altered environmental conditions and, subsequently, the acquisition of iron, are vital to the survival of most bacterial pathogens. Many pathogens undergo significant changes in their gene and protein expression to adapt to growth in iron limiting conditions, including *Bacillus anthracis*, the causative agent of anthrax, a highly virulent pathogen that has been used in recent history as a biological weapon [3]. BOG may also be applicable to studies of marine ecosystems. An example is the study of how hydrostatic pressure may impact the transcriptome of a deep-sea indigenous organism, *Desulfovibrio hydrothermalis* [1]. Such a study is critical in understanding the marine ecosystems, especially those of the deep sea, which represent a major volume of the biosphere. Other examples include bacterial biofilms, important for the study of resistance to antibiotics [6], and *Brassica napus*, an important oil crop [4].

For the type of studies discussed above, the typical first step is to profile the entire transcriptome to identify genes that are differentially expressed (DE) under different conditions (e.g. iron depleted vs. iron replenished in *B. anthracis*, or in situ hydrostatic pressure vs. atmospheric pressure in *D. hydrothermalis*). For this task, many software packages are available, including DEseq [2], EdgeR [8], Cufflinks [14], and DIME [5]. However, finding the set of DE genes is typically not the end goal. Rather, the interest is to find Clusters of Orthologous Groups (COGs) that are enriched (i.e. over-represented) among the DE genes identified in the first step. This, the second, step is essential for providing new insights into the underlying molecular mechanisms linked to the adaptation of a bacterium or a virus from a native to a perturbed condition. Despite the critical importance of this task, studies of this nature are largely descriptive rather than inferential. Pie charts and bar graphs are often the only tools used to visually depict COGs having a larger share of the DE genes, which are then interpreted as indication of enrichment [3,1,4,6]. However, this does not take into account the sizes of COGs, which can be problematic as a larger share of the DE genes may not be that unusual if the corresponding COG also contains more genes. Further, the descriptive nature of the methods does not lead to conclusions that are based on proper evaluation of scientific evidence. Despite an abundance of software for finding DE genes, to the best of our knowledge, there is no computational tool/software currently available for identifying COGs that are significantly enriched with DE genes. Although such an analysis is similar to finding gene ontology (GO) functional categories that are significantly enriched, a software package for such a purpose, such as GOTM [15], is not directly applicable to finding COGs that are over represented among DE genes. Hence, we

* Corresponding author at: Department of Statistics, The Ohio State University, 1958 Neil Avenue, Columbus, OH 43210-1247, USA.

  E-mail address: shili@stat.ohio-state.edu (S. Lin).

believe that it is of value for a software package like BOG that is capable of quick and accurate identification of COGs that are over-represented among differentially expressed genes through rigorous statistical tests.

BOG consists of three modules: (optional) DIME processing, analysis, and output modules (Fig. 1(a)). More specifically, after reading in a raw input data set, BOG performs a differential analysis through a mixture ensemble procedure and computes local fdr as a differential score for each gene using the DIME software (http://cran.r-project.org/web/packages/DIME/) [11,12]. If the input data are already (adjusted) p-values rather than raw data, then BOG will skip the DIME preprocessing step. The scores (either calculated or as input) are delivered to the analysis module, which performs three alternative statistical tests to identify COGs that are over represented among the differentially expressed genes: hypergeometric, Mann–Whitney, and gene set enrichment analysis. The analysis results will then be delivered to the output module for tabular and graphical presentation for ease of understanding and dissemination of results.

2. Statistical tests in the analysis module

Suppose we have a list of genes $\mathscr{G} = \{g_1, ..., g_N\}$ in an experiment; their associated memberships with a set of known orthologous groups (M) are denoted by $\mathscr{M} = \{m(g_1), ..., m(g_N)\} : m(g_i) \in M\}$. We also attach to each gene a differential score $s(g_i)$ (local fdr or p-values): $\mathscr{S} =$

$\{s(g_1), ..., s(g_n)\}$, which are either obtained directly from user's input or computed by DIME. For each orthologous group $m \in M$ with the corresponding gene set $\mathscr{G}_m = \{g_i : m(g_i) = m\}$, we denote its size by $n_m = ||\mathscr{G}_m||$. In the following, we describe each of the three analysis methods.

2.1. Hypegeometric (HG)

We let K be the number of genes that are deemed to be differentially expressed under two conditions, that is, $K = \sum_i^N I\{s(g_i) < s^*\}$, where s^* is a preset threshold (default is set to be 0.05 on BOG but can be changed by user) and $I\{\cdot\}$ is the usual indicator function taking the value of 1 or 0. For each orthologous group $m \in M$, under the null hypothesis that this group is not over-represented among the set of differentially expressed genes, the test statistic $T_{HG} = \sum_{g_i \in \mathscr{G}_m} I\{s(g_i) < s^*\}$ follows the HG distribution $H(K, N, n_m)$. The null hypothesis is rejected if the associated p-value is small, that is, T is much larger than what one would expect under the HG distribution.

2.2. Mann–Whitney Rank Sum (RANK)

To avoid the need to preset a "significance" threshold (which is somewhat arbitrary), we consider all genes by using their rankings based on their differential scores. Specifically, for each gene $g_i \in \mathscr{G}, i = 1, ..., N$, we assign it a ranking $r(g_i) \equiv r\{s(g_i)\}$ such that a gene with a

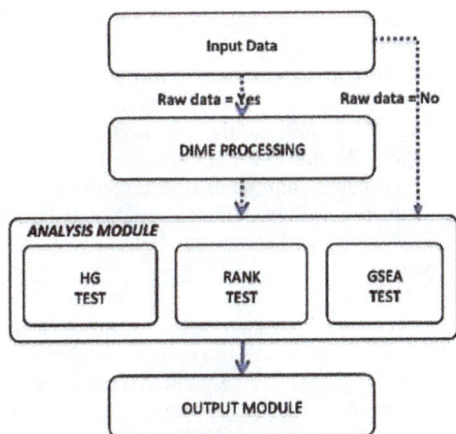

(a) Modules and processing flow

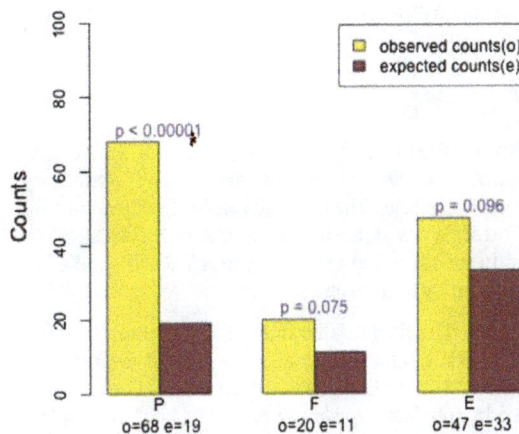

(b) HG Test

COG	P-value	P-value(adjusted)
P	<0.00001	<0.00001
N	0.0002	0.0039
F	0.0147	0.2651
...	...	...

(c) RANK Test

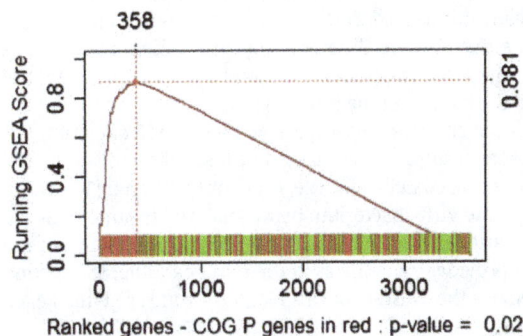

Ranked genes - COG P genes in red : p-value = 0.02

(d) GSEA Test

Fig. 1. Flowchart and sample outputs. (a) The flowchart depicts the three sequential modules that made up BOG. (b) COGs with adjusted p-value < 0.1 from the hypergeometric test. For each COG (P, F and E), the left bar represents the observed number of differentially expressed genes identified, while the right bar is for the expected number according to the size of the COG. The p-values indicated are adjusted p-value taking into account of multiple testing. (c) Tabular outcome from the Mann–Whitney rank test. The middle column gives raw p-values, while the last column provides adjusted p-values taking multiple testing into consideration. (d) An example GSEA scoring path for the "P" category. One can see the maximum score is reached at 358 genes, with the majority of the genes in the top 358 coming from the "P" category (in red). The p-value is adjusted for multiple testing.

smaller score will be assigned a higher rank (large number). For each orthologous group $m \in M$, we compute the test statistic $T_{RANK} = \sum_{g_i \in \mathscr{G}_m} r(g_i)$. Under the null hypothesis that this group is not over-represented among the set of differentially expressed genes, the expected value of T_{RANK} is $n_m(N + 1)/2$. If the observed statistic is significantly larger than this expected value, then this orthologous group is deemed over-represented.

2.3. Gene set enrichment analysis (GSEA)

Instead of basing on correlations as in the original GSEA [9], the modified GSEA in this paper uses rankings of the scores for all genes, like the RANK test. As such, there is no need to preset a threshold of significance. However, unlike the RANK test, evidence of over representation of a COG is evaluated in a sequential manner. More specifically, let $\tilde{\mathscr{G}} = \{\tilde{g}_1, ..., \tilde{g}_N : \tilde{g}_i \in \mathscr{G}\}$ be the ordered set of genes such that $r(\tilde{g}_1) \geq ... \geq r(\tilde{g}_N)$. Recall that a smaller score will receive a higher ranking value. For each orthologous group $m \in M$, we evaluate the sequences of the P_i^+ and P_i^- values, $i = 1, \cdots N$:

$$P_i^+ (\mathscr{G}_m) = \sum_{k \in K_m(i)} \frac{r(\tilde{g}_k)}{r_m}, P_i^- (\mathscr{G}_m) = \sum_{k \in K_m(i), k \leq i} \frac{1}{N - n_m}, \quad (1)$$

where $K_m(i) = \{k \leq i : \tilde{g}_k \in \mathscr{G}_m\}$, $r_m = \sum_{\tilde{g}k \in \mathscr{G}_m} r(\tilde{g}k)$. We define the GSEA statistic as $T_{GSEA} = \max_i \{P_i^+ (\mathscr{G}_m) - P_i^- (\mathscr{G}_m)\}$. Its associated p-value for evaluating evidence of over representation of differential expression of genes in m is determined by a permutation test by randomly permuting the N gene labels.

3. BOG software package

We briefly describe the main functions and data input. More details, especially on control parameters of functions, are available in the documentation of the BOG package. The package can be downloaded from CRAN or from http://www.stat.osu.edu/ statgen/SOFTWARE/BOG/. BOG is the flagship function that performs the HG, RANK, and GSEA tests. It takes two primary arguments:

- data: BOG accepts a data file (R dataframe) of two columns. The first column is the geneIDs (characters) and the second is numerical measures for the corresponding genes, which has two possible options controlled by the data. type argument: (1) "data", normalized "differences" of gene expressions between two comparison groups, (2) "pval", (adjusted) p-values for each gene if differential analysis is carried out beforehand. Under option (1), BOG assumes that the data are already normalized. Further, "difference" is in a broad sense, which can either be log-difference or just difference without performing log-transformation first, depending on the preference of the user and the context of the problem [10].
- cog_file: This can either be a user specified input file (R dataframe) or simply the specification of the name of one of the built-in COGs: anthracis, brucella, coxiella, difficile, ecoli, or francisella [13]. More specifically, if the virus/bacterium being analyzed is not one of the six build-in varieties, then a file with two columns is required: the first column provides gene IDs as in the input data file; the second column specifies the Clusters of Orthologous Groups to which each gene belongs.

The output module receives results from the analysis module and summarizes them in a tabular format with three columns: COG, p-value, and adjusted p-value, for each of the tests performed. A user can display the table by running the command printHG, printRANK, or printGSEA. Further, BOG provides several graphical functions for visualizing the results, including hgplot and gseaplot.

4. Example

To demonstrate the use of BOG, we analyze a set of gene expression levels of *B. anthracis* grown in iron depleted media (0μM iron concentration) and iron replenished media (30μM iron concentration) at the four hour time point after treatment [3].

To identify genes whose expressions are altered when iron is depleted, we took the average difference of normalized gene expression values at 0 μM vs. 30 μM after 4 h of treatment (each with four replicates). We first ran DIME to analyze the data and obtain the local fdr value for each of the genes. This list of fdr value was then saved as input to BOG and made available in the BOG package as input file *anthracis_iron*. We chose to demonstrate our example in a "piecemeal" fashion to facilitate greater understanding. We ran the following command with the BOG main function to analyze over representation:

```
bog <- BOG(data = "anthracis_iron", data.type = "pval",
cog.file = "anthracis", hg.thresh = 0.01, gsea = TRUE).
```

The output in *bog* is then processed using various function in the Output model and the results are presented in Fig. 1(b–d). Output from the HG test, summarized using hgplot(bog), is visualized in Fig. 1(b) for COGs with (adjusted) p-value < 0.1. From the results, we can see that "P" (inorganic ion transport and metabolism) is the most significant COG. The results from the RANK test are being summarized in a tabular form (Fig. 1(c)) using the command printRANK(bog), which shows that COG "P" is also returned as the most significant. Finally, we demonstrate the GSEA-path for category "P" in Fig. 1(d) by using the command gseaplot(bog, " P "), from which one can see that this category is being selected as over-represented among genes that are differentially expressed in iron depleted condition against iron repleted one. The consistent results from all three tests are reassuring. More importantly, this finding is also consistent with current understanding of the science, as the significant increase in ion transport mechanism and some aspects of metabolism is a clear indication of adaptation to growth under iron depleted condition [3].

5. Discussion

We develop an R package (BOG) for identification of Clusters of Orthologous Groups in bacteria and viruses that are enriched among genes that have gone through significant changes under different conditions. Three tests are available to provide user with greater choices. Hypergeometric and Mann–Whitney rank tests are computationally efficient, although note that the hypergeometric test requires the specification of a "significance" threshold. On the other hand, the gene set enrichment analysis based on fdr instead of correlation as in [9] does not need the specification of a threshold, but it is computationally intensive. Therefore, the package provides user with the flexibility of whether to run the GSEA option. As we demonstrated through application to the *B. anthracis* example, all three tests consistently identified the same category as the most enriched gene set. For convenience, we use gene expression as our example data type, although BOG is also applicable to other high-throughput data, including DNA-protein binding and methylation data. For the initial step of finding DE genes, we use DIME as the default in BOG, although this can be replaced by any other package including those mentioned in Section 1. The software is written in such a way that the step for finding DE genes can be performed using a user-desired software before calling BOG to identify COGs that are enriched among the set of DE genes. As such, BOG is directly applicable to all examples discussed in Section 1; the set of DE genes or the rankings can be used as input to BOG to formally test which BOGs are enriched in addition to simple descriptive statistics/graphs used therein. In addition to its intended use in detection of pathogens, BOG might also find applications in analyzing gut microbiota community compositions, a subject with recent surge of interests, as such compositions may be related to obesity and other health conditions [7]. For instance, in an analysis of 16S rRNA gene from a study of obese and lean individuals, one may first detect taxa, at a particular taxonomic rank (e.g. species), that have significantly different proportions among these two groups of individuals. Then BOG can be called to identify categories at a higher taxonomic rank (e.g. family) that are significantly enriched.

Acknowledgment

This work was supported in part by the National Science Foundation under Agreement No. 0931642, DMS-1042946, and DMS-1220772, and by the Bisa Research Grant of Keimyung University in 2014.

References

[1] Amrani A, Bergon A, Holota H, Tamburini C, Garel M, Ollivier B, et al. Transcriptomics reveal several gene expression patterns in the piezophile *Desulfovibrio hydrothermalis* in response to hydrostatic pressure. PLoS One 2014;9:e106831. http://dx.doi.org/10.1371/journal.pone.0106831.

[2] Anders S, Huber W. Differential expression analysis for sequence count data. Genome Biol 2010;11:R106 [URL http://genomebiology.com/2010/11/10/R106].

[3] Carlson PE, Carr KA, Janes BK, Anderson EC, Hanna PC. Transcriptional profiling of *Bacillus anthracis* Sterne (34f2) during iron starvation. PLoS One 2009;4:e6988.

[4] Huang J-Y, Jie Z-J, Wang L-J, Yan X-H, Wei W-H. Analysis of the differential expression of the genes related to *Brassica napus* seed development. Mol Biol Rep 2011;38:1055–61.

[5] Khalili A, Huang T, Lin S. A robust unified approach to analyzing methylation and gene expression data. Comput Stat Data Anal 2009;53:1701–10.

[6] Qin N, Tan X, Jiao Y, Liu L, Zhao W, Yang S, et al. RNA-Seq-based transcriptome analysis of methicillin-resistant *Staphylococcus aureus* biofilm inhibition by ursolic acid and resveratrol. Sci Rep 2014;4:5467 [URL http://d360prx.biomed.cas.cz:2062/srep/2014/140627/srep05467/full/srep05467.html].

[7] Ridaura V, Faith J, Rey F, Cheng J, Duncan A, Kau A, et al. Gut microbiota from twins discordant for 55 obesity modulate metabolism in mice. Science 2013;341.

[8] Robinson MD, McCarthy DJ, Smyth GK. edgeR: a bioconductor package for differential expression analysis of digital gene expression data. Bioinformatics 2009;26:139–40.

[9] Subramanian A, Tamayo P, Mootha VK, Mukherjee S, Ebert BL, Gillette MA, et al. Gene set enrichment analysis: a knowledge-based approach for interpreting genome-wide expression profiles. Proc Natl Acad Sci U S A 2005;102:15545–50.

[10] Taslim C, Huang K, Huang T, Lin S. Analyzing ChIP-seq data: preprocessing, normalization, differential identification and binding pattern characterization. Methods Mol Biol 2012;802:275–91.

[11] Taslim C, Huang T, Lin S. DIME: R-package for identifying differential ChIP-seq based on an ensemble of mixture models. Bioinformatics 2011;27:1569–70.

[12] Taslim C, Lin S. A mixture modeling framework for differential analysis of high-throughput data. Comput Math Methods Med 2014 [Artical ID 758718, 9 pages].

[13] Tatusov RL, Fedorova ND, Jackson JD, Jacobs AR, Kiryutin B, Koonin EV, et al. The COG database: an updated version includes eukaryotes. BMC Bioinf 2003;4:41.

[14] Trapnell C, Roberts A, Goff L, Pertea G, Kim D, Kelley DR, et al. Differential gene and transcript expression analysis of RNA-seq experiments with TopHat and Cufflinks. Nat Protoc 2012;7:562–78 [URL http://www.pubmedcentral.nih.gov/articlerender.fcgi?artid=333 4321&tool=pmcentrez&rendertype=abstract].

[15] Zhang B, Schmoyer D, Kirov S, Snoddy J. GOTree Machine (GOTM): a web-based platform for interpreting sets of interesting genes using gene ontology hierarchies. BMC Bioinf 2004;5:16.

16

A Graph Based Framework to Model Virus Integration Sites

Raffaele Fronza [a],*, Alessandro Vasciaveo [a,b], Alfredo Benso [b], Manfred Schmidt [a]

[a] Department of Translational Oncology, National Center for Tumor Diseases and German Cancer Research Center, Im Neuenheimer Feld 581, 69120 Heidelberg, Germany
[b] Department of Control and Computer Engineering, Politecnico di Torino, Corso Duca degli Abruzzi 24, 10129 Torino, Italy

ARTICLE INFO

Keywords:
Gene therapy
Systems biology
Genomics
Insertional mutagenesis

ABSTRACT

With next generation sequencing thousands of virus and viral vector integration genome targets are now under investigation to uncover specific integration preferences and to define clusters of integration, termed common integration sites (CIS), that may allow to assess gene therapy safety or to detect disease related genomic features such as oncogenes.

Here, we addressed the challenge to: 1) define the notion of CIS on graph models, 2) demonstrate that the structure of CIS enters in the category of scale-free networks and 3) show that our network approach analyzes CIS dynamically in an integrated systems biology framework using the Retroviral Transposon Tagged Cancer Gene Database (RTCGD) as a testing dataset.

1. Introduction

Viral vector integration is a process exploited in gene therapy (GT) to correct defective cells of an individual and to drive the health status from the pathological condition to a normal one [1–6]. As consequence of this perturbation, i.e. if vectors integrate into cellular genome positions where the expression of an important gene is dysregulated, the affected cell may step from the primary illness state to a secondary state. Thus, insertional mutagenesis is a potential risk that may accompany vector integration events [7–11].

Therefore, large insertional mutagenesis screenings are used to assess the safety of the treatment in clinical GT, to design safer GT protocols and to discover new disease (i.e. cancer) candidate genes [12–16].

The central role in integration site (IS) analyses is given in the assessment of the genome-wide integration profile and the identification of integration clusters that could alter gene expression. The definition of these clusters or common integration sites (CIS) is not standardized and usually based on accumulation of IS that are unlikely to occur by chance and statistically significant different compared to a random *in silico* control. A regular interpretation (Standard Windows Method, SWM) is founded on the number of integrations in a predefined genomic window, that classifies CIS as follows: a) 2 IS are within 30 Kb or b) 3 IS within 50 Kb or c) 4 IS within 100 Kb or d) ≥5 IS within 200 Kb [17,18]. It is obvious that this historical definition can be used only as a first approximation for discovery of biologically (and clinically)

relevant CIS, because the results are highly dependent from the size of the IS dataset [19]. Even if methods not constrained on a predefined set of fixed windows were released [20–23], the data importation together with CIS generation and analysis remain tedious tasks. The static tabular and text oriented nature of the CIS representation requires extensive processing steps that can involve custom programming, format exchanging and manual interpretation of the results. These computational difficulties reduce the analysis capability of standard life science labs and strictly rely on elevated bioinformatics skills.

We hypothesized that in the next generation sequencing area only systems biology approaches may be able to dissect biologically (and clinically) relevant CIS. Here, we developed a new CIS construction framework using an approach based on graphs. This approach has numerous advantages: 1) the resulting CIS are represented by networks, 2) graph theory can be used to infer characteristics and properties of the integration process (i.e. the node degree distribution of the networks can be linked to the randomness of the integration process, and 3) a large repertoire of IS can be imported and parsed without any prior constraint (except for the maximal distance between two IS).

More in detail, the graph model allows an easy structural organization of the annotations in the Gene Atmosphere (GA) (e.g. protein coding atmosphere, post-transcriptional atmosphere, etc.) by using different layers of node categories while performing the enrichment as described in Paragraph 2.6. The implementation of this model in software tools, which allow networks visualization, provides a broader overview of the data at a glance. An example of this feature is depicted in Fig. 3 where the CIS network is disposed on a plane by applying a force-directed layout. Furthermore, with the availability of Hi-C data, this

---

* Corresponding author.
  E-mail address: raffaele.fronza@nct-heidelberg.de (R. Fronza).

framework is ready to embed relations among CIS in the spatial organization of the genome, enriching the modeling to a multidimensional level [24] (i.e. moving from a linear genomic vicinity modeling to a topological genomic modeling). Another interesting feature which emerges using this graph model is the capability of assessing biological properties by exploiting topological characteristics of the network. For example, the scale-free distribution of a set of CIS can be used to establish if the dataset under analysis contains genomic regions enriched in IS, an observation that is a prerequisite in order to properly recognize CIS.

Recent approaches to the identification of hot-spots try to take into account the size of the IS's dataset and the prior knowledge about vector integration preferences [23]. The implementation of this graph model on a normal computer machine could be greedy of computational resources when dealing with very huge IS datasets (i.e. millions of nodes). To our knowledge, there are no IS datasets in literature big enough that cannot be easily represented by our model. Enhanced with annotated genomic data, this model could be easily extended in order to drive the identification of CIS exploiting the information contained in the annotations. It is our intention to evaluate this possibility in future. One of the main differences between our model and the statistical frameworks used to the identification of CIS is that here the statistical method is applied after the CIS identification leaving to the user of the model the ability to give a biological meaning to the CIS (e.g. the CIS is not excluded a priori by the statistical method). With the complex annotation feature, as described in Paragraph 2.6, our model is able to perform a many-to-many mapping against genomic features (i.e. genes) and integration sites while other methods just perform a one-to-one mapping (i.e. one gene, one IS). In this way, our model provides a more refined granularity, when it is enhanced with complex annotation, allowing the simultaneous representation of different genomic features in the model (i.e. transcriptional elements, protein coding genes, etc.), that other models do not allow.

A Cytoscape [25] draft prototype plugin was developed to test the framework of this paper.

2. Results and Discussion

2.1. CIS Definition

First of all we define what a common integration site is. A set of n IS in the database is represented as the set of n vertices V of the graph G. Then, for each couple of vertices v_i and v_j ($i, j = 1, 2, ..., n$ $i \neq j$) we add an edge e_{ij} if the distance between the corresponding IS is below a threshold T_H of 50 Kbp. A weight w_{ij} is associated with the edge e_{ij} and represents the distance between the corresponding IS. The default value of 50 Kbp was selected using the maximal influence window size where a causal relation is found between an insertion event and gene expression [22]. De Jong showed that the presence of viral integration is correlated with the local amount of gene expression and that 50 Kbp is an upper bound on which the presence of IS can be linked with gene expression. At the end of this process, we obtain the undirected weighted graph $G = (V, e)$ as abstract representation of all the distance relations in the IS dataset. The graph G is composed by a set of unconnected subgraphs (Connected Components, CC). Each CC is the natural graph representation of a CIS in which the order is represented by the number of vertices.

2.2. Integration Process and Node Degree Distribution

The non-random character of virus and viral vector integration suggests the existence of sub genomic regions that are preferentially targeted. As many complex biological systems where many components interact together, also the viral integration process derives from intricate functional interactions that involve viral and host proteins/DNA. The behaviors of complex systems are captured by a characteristic of

the network that is called scale-free property [26–28]. This property depends on the distribution of the nodes degree. The node degree is the number of edges that connect a node with the neighbors. The degree distributions of several networks follow a power law, precisely defined with the functional $d(k) = ak^{-\gamma}$, where $d(k)$ is the degree distribution, $k = 0,1,2,...$ is the node degree, a is the normalization constant and γ is the degree exponent. In scale-free network the exponent is usually less than three ($\gamma < 3$), whereas in random networks $\gamma \geq 3$.

To prove that the mechanism of viral CIS or hot-spot (HS) formation is embedded via scale-free property into the network representation, we developed a series of synthetic transfection experiments that consisted of placing a fixed number of integrations on human genome carrying a random number of artificial hot-spots. The integrations were divided in two subsets: 1) IS placed on a simulated genome with hot-spots (IS_{SYN}) and 2) IS randomly placed on a genome without hot-spots (IS_{RAND}). The scale-free property of CIS networks found in IS_{RAND} and IS_{SYN} was then verified using the Cytoscape "Network Analysis" plugin.

We further verified the presence of a HS driven mechanism on six datasets: five in which we expected a scale free behavior (LV [1], HIV [29], GV1 [2], GV2 [16] and RTCGD [12]); and one from an adeno-associated viral (AAV) vector study [30] where we expected a random integration profile. In Fig. 1 the degree distributions of the groups of the experimental IS sets are plotted. The richness in integration sites of the datasets is: ~1000 IS_{SYN} (g), ~15,000 IS_{RAND} (e), ~4000 IS_{LV1} (b), ~2000 IS_{AAV} (a), ~35,000 IS_{HIV} (d), ~15,000 IS_{GV1} (h), ~800 IS_{GV2} (c), and ~8800 IS_{RTCGD} (f). All the experimental and synthetic sets, except for the AAV and RAND set, have a log–log degree distribution that follows a power law with gamma exponent $\gamma < 3$. Only two datasets, the random dataset IS_{RAND} ($\gamma = 3.6$) and IS_{AAV} ($\gamma = 4.8$) have no scale-free degree distribution. This last finding is in line with our and other published studies that did not attribute to AAV any HS driven integration pattern [30,31]. From a practical point of view and as a first result of our graph modeling, the node degree distribution in a network that represent integration events indicate the presence of an accumulation process driven by genomic hotspots.

Réka [26] demonstrated that complex systems that display a high degree of error tolerance (robustness) are represented by scale-free networks. An incomplete IS dataset can be seen as the result of a process that remove IS from the complete basin of integrations present in a sample, due to unavoidable experimental subsampling. Recalling the robustness property for scale free networks we can prove that genomic hot-spots are identified even within an incomplete set of experimental IS.

2.3. General Structure of the CIS Pool and RTCGD Dataset

The Retroviral Transposon tagged Cancer Gene Database (RTCGD; http://variation.osu.edu/rtcgd/, [12]) was used as test case for our graph model.

The RTCGD dataset has been first analyzed in order to compare two general CIS properties, the order and the dimension of the 10 biggest CIS identified by our framework and the SWM. Fig. 2(A) shows the general structure and shape of all the CIS with order bigger than 9 as they appear analyzing RTCGD integration.

5110 IS are selected by the CIS construction tool as belonging to reputed CIS and 4035 compose CIS with p-value < 0.05 (see Appendix A Table 1 in [41]). How the p-value is computed per CIS is explained in the Paragraph 3.6. The CIS order goes from 2 to 82 (in Fig. 2(A) CIS from order 2 to order 8 are not shown). RTCGD data contains 2910 IS and the CIS falls in the same range order. No statistical model is applied in order to test the CIS significance.

In the 10 biggest CIS the order and dimension of 3 of them (*myc*, *ahi* and *rasgrp1*) were returned identical by the analysis performed using our model and RTCGD and other 4 CIS (*gif1*, *lvis1*, *pim1* and *notch1*) were comparable (difference in the order is less than 10; see Table 1).

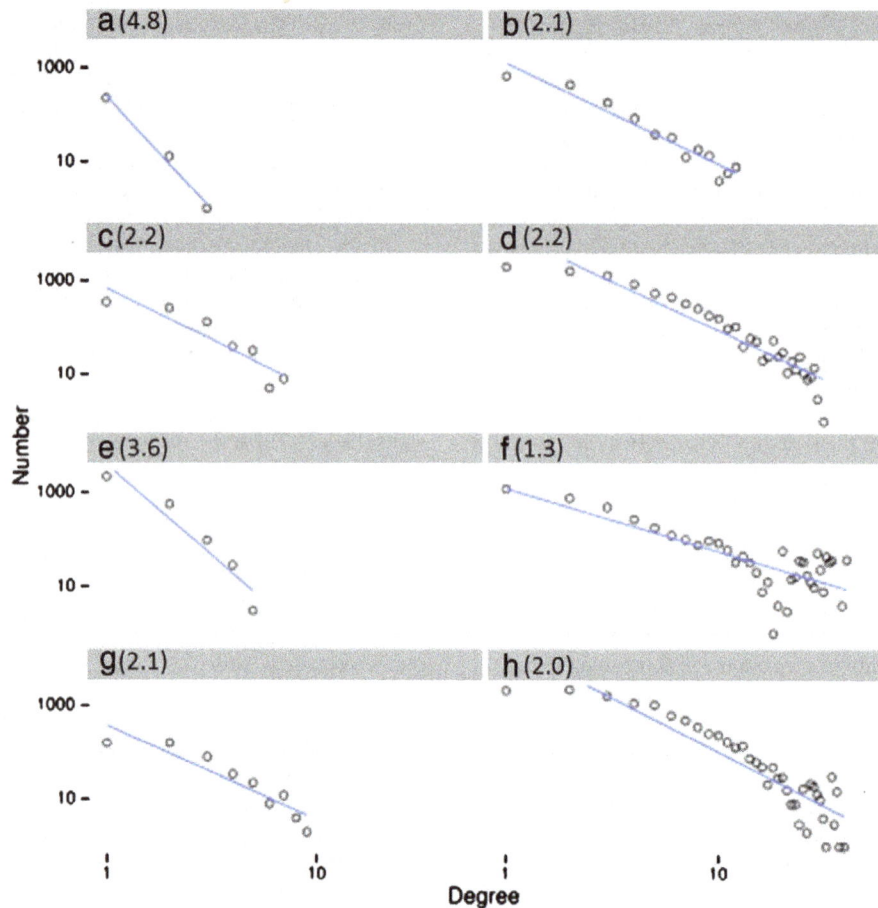

Fig. 1. Degree distributions for datasets used to test scale free properties of CIS networks. The degree distribution of a scale-free network follows the power law $d(k) = ak^{-\gamma}$, with $\gamma < 3$. This distribution is interpolated as a straight line on log–log plot.

For the remaining common integration sites the RTCGD ones were fragmented in two separated CIS by using our model (*sox4* and *meis1*) or vice versa (*fgf3*). Fig. 2(B) shows the structure of these CIS. A box surrounding two CIS indicates that the CIS are combined in RTCGD.

The CIS discovered using our graph model have a wide spectrum of dimensions imposing only the threshold distance between two IS. As shown in Table 1 the high order CIS are generally more compact than the CIS retrieved by a SWM. This behavior can easily explained by the construction rules that SWM imposes to fill a region of 200 Kbp in CIS with order bigger than four.

As a further point of discussion we would like to make some observation on the structure of CIS in sets of IS that contains many IS that accumulate in genomic hotspots that cover several megabases. The IS_{HIV} set of IS contains more than 35,000 objects and produces a typical integration profile with dense accumulation regions. The biggest CIS is found on chromosome 16 at position 89,730,888 and spans 0.69 Mb. The graph that represents this hotspot is shown in Fig. 3 and is strongly anisotropic. In order to arrange the nodes in this manner we used a force-direct layout embedded in Cytoscape. Accumulations of nodes (strongly connected) indicate an area in the CIS rich in integration. Regions in the network where few links (weakly connected, 3e) are present represent areas that act as interface between two IS dense areas. The CIS in the figure might be dissected in at least 4 independent subregions that cover 300 (3a), 162 (3b), 32 (3c) and 47 (3d) kilobases and where different genomic elements are targeted. In Fig. 4 the networks structure of a region of 1.2 Mb on chromosome 3 is shown. Five independent CIS of order 25–68 are detected. Only one of them 4II is isotropic and does not show subregions of accumulation. Three CIS, 4I, 4II and 4IV, contain at least two subregions whereas in 4V, 46 IS cover

more or less uniformly a region of about 200 Kb. To conclude this paragraph, the networks can be visually represented to be easily examined by eye or is possible to use graph algorithms that automatically extract topological structures related with specific hotspot features.

2.4. Shannon Index and Dataset Comparison

Entropy [32] is the most elementary concept of information theory and is widely used in many fields in order to measure the complexity of a defined system. We can introduce a level of complexity to the CIS analysis combining the IS that belong to different investigational classes like tumors or vectors type. A hotspot can consequently be composed by a mixture of IS that derive from different sources. In order to measure the complexity of one CIS we computed the entropy value, absolute and relative. Low entropy values are found in CIS where IS come from few tumor source (exactly one source if normalized entropy $NE = 0$). If $NE = 1$ all the tumor sources contribute equally to the CIS. To determine significant tumor hotspots we extracted the set of CIS that belong to 10% of the upper (highly heterogeneous) and lower (highly homogeneous) tail of the entropy distribution. The NE distribution for 203 CIS (order > 4; 2129 IS) is comprised between 0 and 0.65 (median = 0.4). 23 CIS (201 IS, 9.4%) with $NE < 0.1515$ and 18 CIS (198 IS, 9.3%) with $NE > 0.5375$ were selected as homo and heterogeneous CIS respectively. The largest homogeneous CIS was found on chromosome 15: 98795118 (Wnt1 gene) and contains 28/29 IS that are associated with mammary tumor. The most heterogeneous CIS was found at the position chr2:173665 (*cebpb* gene) and contains 23 IS. It is originated from 6 different tumors: T cell, B cell, myeloid, lymphoma, brain tumor and HS. On

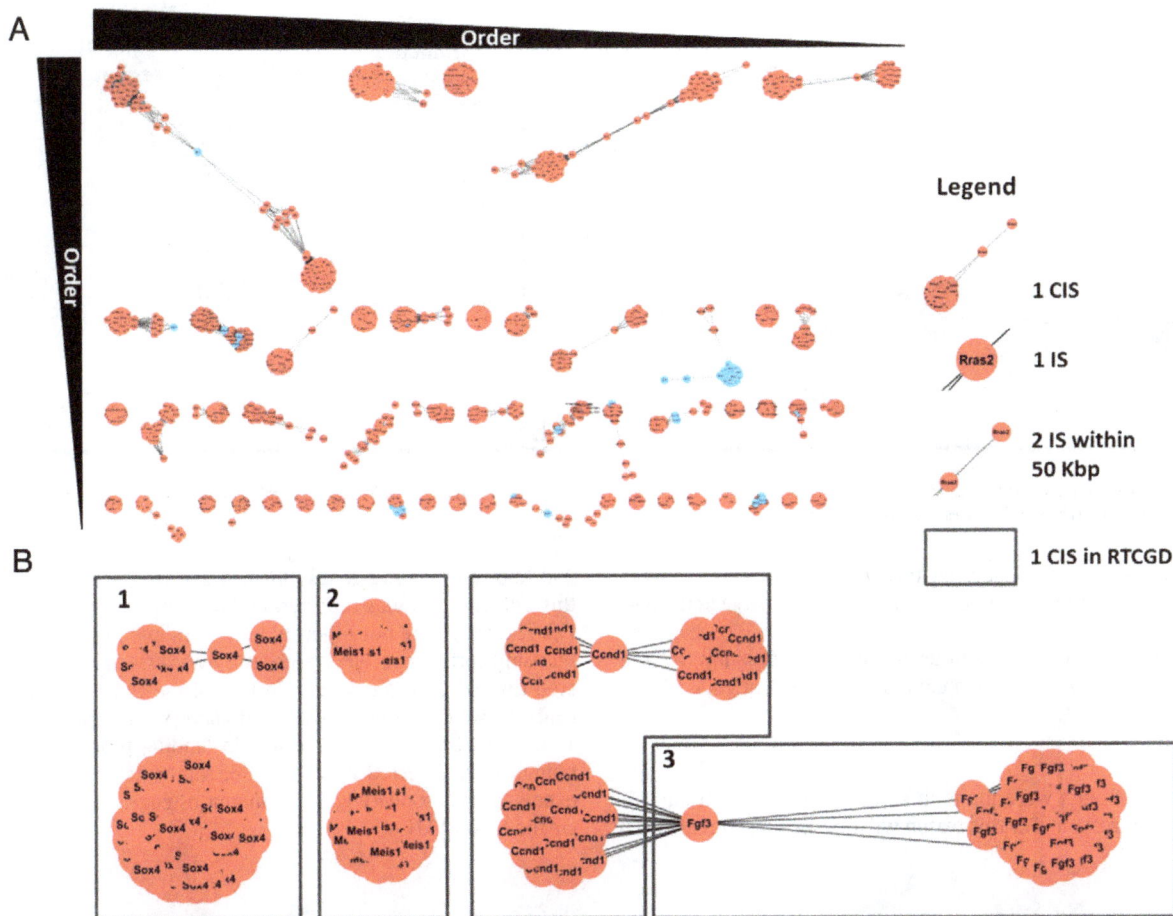

Fig. 2. A) CIS representation on graphs of RTCGD. A CIS is represented as a network of nodes that symbolize a single integration event. An edge between two nodes indicates that the two corresponding IS are within a range of 50 Kbp. B) Comparison between three high order CIS in graph representation and original RTCGD.

average the CIS order and dimension were 11 and 99.5 Kbp respectively in the heterogeneous set and 9 and 58 Kbp in homogeneous CIS.

2.5. Literature Analysis

A detailed literature exploration of the list of genes returned in the homogeneous set was performed in order to detect potential candidates to further experimental tasks. The probability that a homogeneous CIS appears by chance was assessed on a multinomial model. The complete list of homogeneous CIS is given in Table 2. We divided the genes present in this list in 3 collections concordantly with the CIS homogeneity: 1) in the "mammary" group those found in "mammary tumor", 2) in the

Table 1
RTCGD CIS order and dimension comparison.
Only the 10 biggest CIS are considered. In parenthesis the order or the dimension of the CIS that complete the corresponding CIS found applying the dual method.

CIS name	Graph CIS		RTCGD	
	Order	Dimension	Order	Dimension
Gif1	81	122,521	82	197,286
Myc	78	234,094	78	234,094
Sox4	65 (10)	51,755 (130,119)	75	235,775
Ahi	61	297,769	61	297,769
Pim1	35 (9)	12,131 (81,543)	43	144,748
Meis1	26 (13)	9,894 (46,027)	38	154,229
Lvis1	42	118,556	38	89,966
Fgf3	53	143,856	38 (33)	95,144 (182,060)
Rasgrp1	36	86,755	36	86,755
Notch1	36	92,014	34 (2)	90,385 (1,629)

"sarcoma" group those found in "sarcoma" and 3) combining in the "Leukemia" collection those found in "B cell", "T cell" or "myeloid". The analysis was performed using ProteinQuest, a literature mining tool [33,34].

All the genes found specifically for mammary tumor were already well known as contributing to mammary neoplasia. In the sarcoma group genes, *tmem74* and *rabgap1l*, only *rabgap1l* was previously linked with tumors [35], but not with sarcomas, while in the leukemic group *usp49*, *smsg6* and *zpf423* were not linked with blood diseases.

2.6. Complex Annotation

The previous investigation is directly based on the gene annotation provided by RTCGD, annotation found using the Next Gene Approach (NGA). This approach just annotates an IS with the name of the proximal genetic element, that habitually is a protein coding genes. This procedure implicitly force a many-to-one relation between IS and a gene element. We think that this is a true analysis limitation because it is not possible to assign multiple genes to the same IS. Moreover, due to the variety of the gene types (i.e. protein coding, miRNA or lincRNA), to the complexity of genetic loci (i.e. introns can contain non coding genes) or simply to the high density of gene elements in few kilo bases, the NGA should be limited only to a raw estimation of the genes involved in the viral integration. In our network model we decided to introduce a new category of nodes that represent gene elements. These auxiliary nodes added to the network symbolize genes found within a certain distance (50 Kbp by default) to any IS. With this simple addition to the network structure we naturally introduced a many-to-many relation between integration sites and genomic elements (referred as Transcriptional Elements (TE) in the model). The group of the genes found in one CIS network is defined as the gene

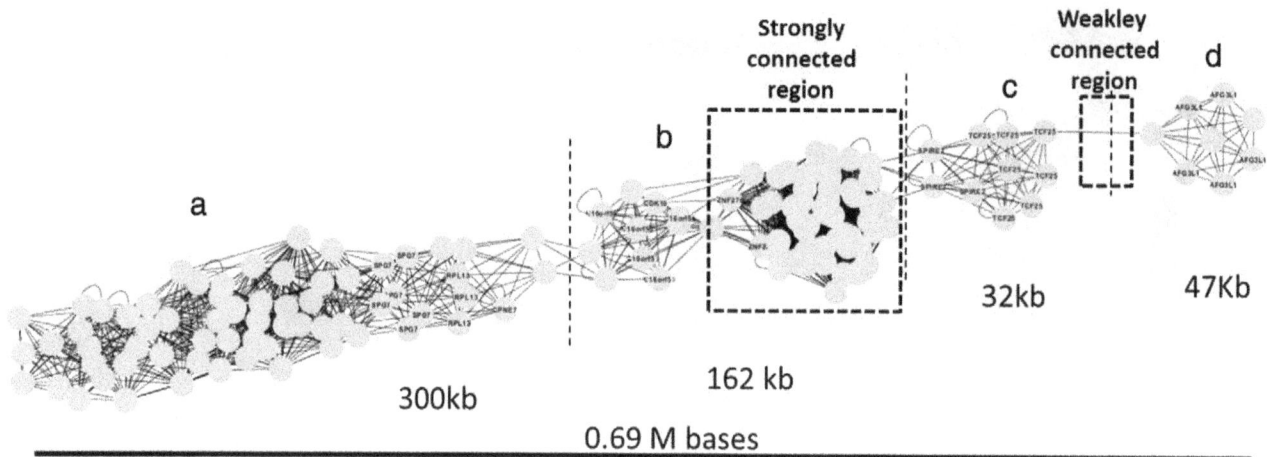

Fig. 3. The network representation of the biggest CIS (0.69 Mbases) in IS_{HIV} (chromosome 16: 89,730,888; ~35,000 IS). The structure contains some regions with many (a, b, c and d) and few (e) integration sites. Labels in the nodes correspond to genes in the original IS annotation.

atmosphere (GA) of the CIS. Applying this new network layer, we were able to notice many other gene elements that did not emerge in the previous literature analysis.

In the following analysis we focus only on genes found in GA (see Appendix A Table 2 in [41]) that are not present in the original RTCGD annotation.

2.6.1. Sarcomas

No gene elements were found in sarcoma CIS Fig. 5. AI and AII using 50 Kbp threshold. For this reason we increased the threshold to

200 Kbp. Two coding genes, *tmem74* and *trhr*, constitute the GA of this CIS. The first was reported in RTCGD and used to annotate the CIS. The other one not reported in the original list, the thyrotropin-releasing hormone receptor gene (*trhr*), was associated with differentiated thyroid cancer risk [36]. In CIS AII the composition of the GA contains two coding genes (*rabgap1l*, *cacybp*), 1 miRNA (mir1927) and 1 snoRNA (GM23528). The calcyclin-binding protein (*cacybp*) was ubiquitously detected in all kinds of tumor tissues and was highly expressed in nasopharyngeal carcinoma, osteogenic sarcoma, and pancreatic cancer [37].

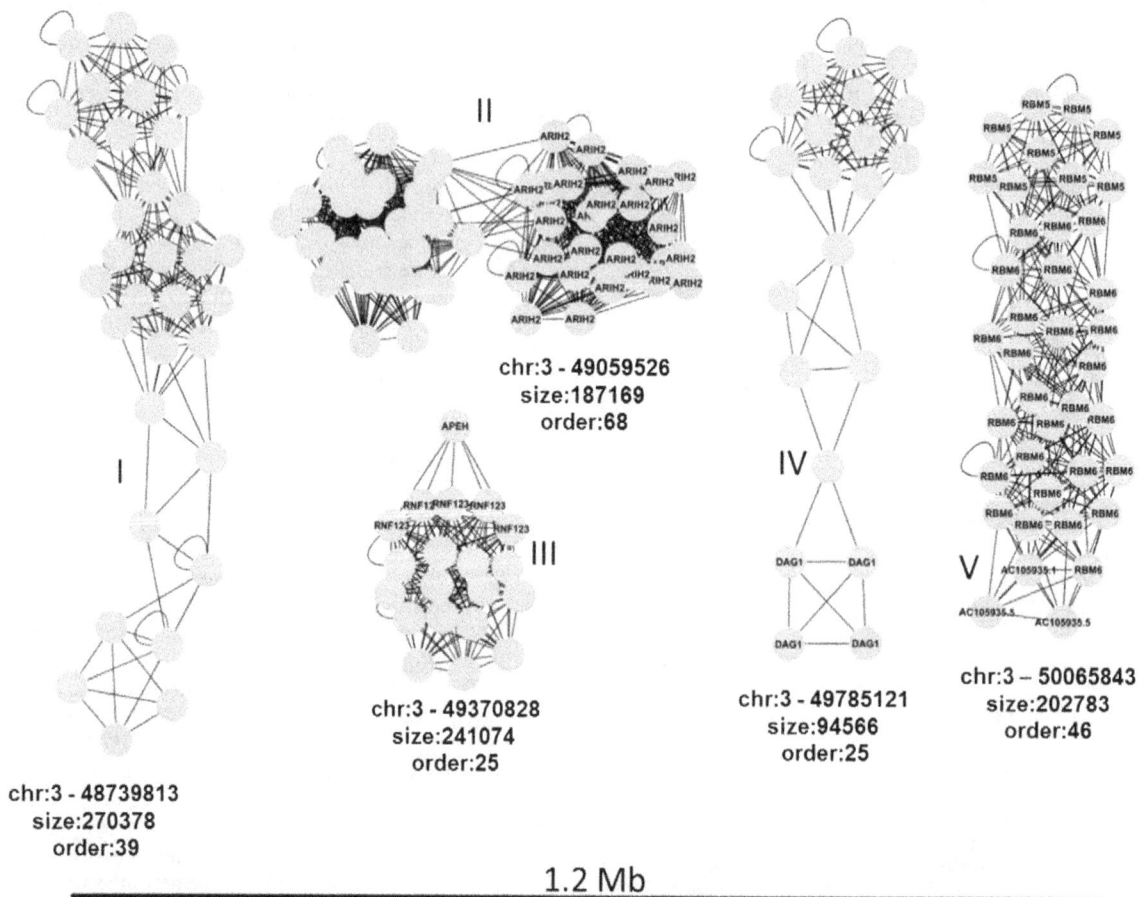

Fig. 4. 5 independent CIS of order 25–68 are detected in a IS dense region of 1.2 Mbases (chromosome 3:49,000,000–50,000,000). Three of them (I, II and IV) are composed by at least two sub regions. Labels in the nodes correspond to genes in the original IS annotation.

Table 2
Genes found in homogeneous CIS (normalized entropy, NE < 0.1515).

Chr	Position	Dimension	Order	Tumor type	Gene	Entropy	p-Value
15	98795118	65490	29	Mammary_tumor	Wnt1	0.055	<0.001
3	97979620	120117	20	Bcell	Notch2	0.146	<0.001
11	103817471	61740	12	Mammary_tumor	Wnt3	0.000	<0.001
14	115040166	9438	12	Tcell	Mirn17	0.106	<0.001
11	59292730	16451	10	Mammary_tumor	Wnt3a	0.000	<0.001
1	160495222	156990	9	Sarcoma	Rabgap1l	0.129	<0.001
4	3916043	43556	9	Myeloid	Plag1	0.129	<0.001
7	25114306	15181	9	Bcell	Pou2f2	0.000	<0.001
4	154549687	57802	8	Bcell	Prdm16	0.139	0.002
11	74982305	88850	8	Bcell	Smg6	0.000	<0.001
13	37813516	74223	7	Bcell	Rreb1	0.139	0.002
15	43989356	57412	8	Sarcoma	Tmem74	0.000	<0.001
12	86876152	126384	7	Bcell	6430527G18Rik	0.151	0.004
20	12087680	59992	7	Bcell	Bcor	0.151	0.004
16	30038826	74290	6	Bcell	Hes1	0.000	0.001
17	45556480	2280	6	Bcell	Nfkbie	0.000	0.001
2	31079892	40607	5	Bcell	Fnbp1	0.000	0.003
6	125321059	37420	5	Bcell	Ltbr	0.000	0.003
8	87959778	25262	5	Bcell	Zfp423	0.000	0.003
15	80537828	36086	5	Tcell	Grap2	0.000	<0.001
16	24166711	82290	5	Bcell	Bcl6	0.000	0.003
16	39882896	180	5	Mammary_tumor	ENSMUSG00000068293	0.000	<0.001
17	47754686	85300	5	Bcell	Usp49	0.000	0.003

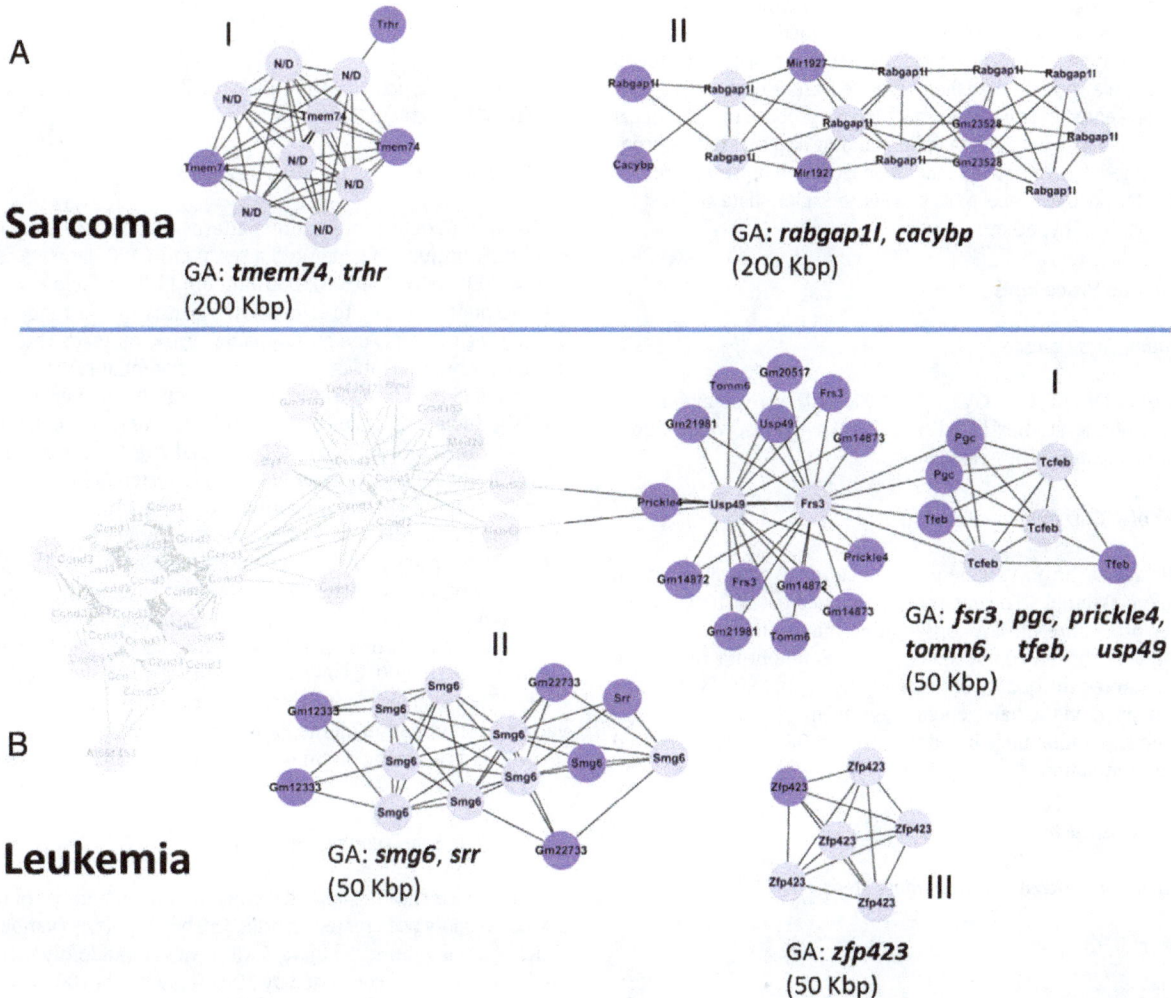

Fig. 5. Gene atmosphere of the homogeneous CIS associated only with A) Sarcomas or B) Leukemia but not linked with the same disease in the literature analysis. Purple nodes: genes, light blue: insertion sites.

2.6.2. Leukemia

As displayed in Fig. 5.BI two CIS fall in the same locus. The two twin CIS are falsely connected for the reason that IS belonging to distinct CIS are coupled over two common genes (*med20* and *usp49*). The GA of BI contains 6 coding genes *fsr3, pgc, prickle4, tomm6, tfeb, usp49*. Interestingly this block of mouse genes was found and described in a human large linkage disequilibrium block, which contains CCND3, BYSL, TRFP, USP49, C6ofr49, FRS3, and PGC [38]. This block contains somatic alterations that correlate with breast cancer prognosis and survival in humans. In CIS BII the GA contains *smg6* and *srr*. We didn't find any verified involvement of *srr* in tumor development. In CIS BIII we found complete concordance between our graph model complex annotation and RTCGD because only *zfp423* compose the GA.

Therefore, implementing an annotation strategy based on a single simple network construction rule, we were able to explore several potential tumor related genes both coding and not coding that are not described in the original dataset.

We would like to briefly illustrate a pathway analysis that we performed on 1421 unique genes found in the GA (see Appendix A Table 2 in [41]). Processes associated with the chromatin rearrangement and the ribosome structure emerge clearly from the data. Because all the IS in RTCGD are strongly correlated with the appearance of tumors in mice, the likelihood that genes in the neighborhood of an IS are in a significant pathway is in fact high. Consequently, we do not discuss further this result.

In conclusion in this work was our interest to show that the network representation of integration sites hotspots is a convincing framework since it links topological structures of the graphs with biological features of the CIS. To demonstrate that the complex interactions between viral and cellular components during the viral integration originates the IS accumulation, we showed that the graphs representing putative hotspots are represented as scale-free networks. We propose that the visual modeling introduces a better data understanding and that the incorporation of the model in frameworks that manage graph representation of biological data, like Cytoscape, makes easier to explore data and test alternative functional hypothesis.

3. Experimental Procedures

3.1. Integration Sites Datasets

Six datasets, LV, HIV, GV1, GV2, AAV and RTCGD were obtained from the authors of the studies [1,2,12,16,29,30] whereas two random datasets were created as described below.

3.2. Liftover of RTCGD from mm9 to mm10

RTCGD dataset (http://variation.osu.edu/rtcgd/) is actually based on mm9 genome assembly. The dataset was updated to mm10 assembly via liftover procedure [39]. A conservative minimum ratio of bases remapping was chosen (0.95) along with unique output regions in order to guarantee the quality of the mapping. From 8807 IS only two were not mapped on the new genome assembly.

The gene annotation table based on Ensembl GRCm38.p2 was used for complex annotation

3.3. Network Construction

A graph is constructed following these steps:

CIS Layer:

1. Order the Insertion Site Dataset D by Insertion Position (location)

2. For each IS i in the ordered Insertion Site Dataset D do

2.1. Add a node v_i, annotating the location of the IS i to the graph G_D

2.2. For each node v_j in the graph G_D do

2.2.1. If the location of v_i is at a distance smaller than the threshold T_H from the node v_j, with $i \neq j$

  a) Connect the v_i and v_j with the edge e_{ij} and weight w_{ij}, add e_{ij} to graph G_D

  b) Otherwise continue from 2.2 (next j)

2.3. Continue from 2 (next i)

GA Layer:

1. Order the Transcriptional Element Dataset D_{TE} by Transcription Start Site (location)

2. For each TE i in the ordered Transcriptional Element Dataset D_{TE} do

2.1. Add a node v_i, annotating the location of the TE i to the graph G_D (previously created and ordered by location)

2.2. For each node v_j in the graph G_D do

2.2.1. If the location of v_i is at a distance smaller than the threshold T_H from the node v_j, with $i \neq j$

  a) Connect the v_i and v_j with the edge e_{ij} and weight w_{ij}, add e_{ij} to graph G_D

  b) Otherwise continue from 2.2 (next j)

2.3. Continue from 2 (next i)

Scale free structure of graphs is tested using NetworkAnalyzer plugin [40] directly in Cytoscape.

3.4. Random Datasets and Random CIS Statistics

In order to detect integration patterns that deviate from a random even distribution we generated a set of $m = 100$ datasets composed of $l = 500, 2000, 5000, 10,000$ random IS. Two factors were taken into account in order to avoid some biases in the random model: 1) only regions reported as mappable by 100 bp reads (http://moma.ki.au.dk/cgi-bin/hgTables) are chosen as potential IS areas and 2) the fraction of IS per chromosomes is shaped on the frequency detected in the LV1 studio. From the created random datasets we extracted (connection threshold 50 Kbp) and grouped the CIS by the order $i = 2,..,10$. In the random dataset we never detected CIS with order larger than 9. On these groups we computed 1) the CIS rarity distributions (see Paragraph 3.6), and 2) the maximal number of random CIS of order i found in any of the m datasets of numerosity l. Except for random CIS of order $i = 2$, where R_{l2} uniformly distributed over the threshold distance, the distribution of R used to compute the p-value and loglikelihood was approximated by a continuous normal distributions of mean $\mu_{\theta rnd}$ (26,000 bases) and standard deviation $\sigma_{\theta rnd}$ (12,000 bases). These values were chosen from the smallest mean and biggest variance found within all the R_{li}. Also for the experimental CIS we approximated the rarity distribution using mean and variance found in LV1 samples.

3.5. Synthetic Transfection Experiment

An in silico chromosome that contains a fixed number of randomly placed hotspots was created. In brief, 159 hot-spots with various dimensions (from few hundred bp to 5 Mbp) where randomly marked on a linear chromosome composed by 100 Mbp. A set of 100, 500 and 1000 insertion sites were simulated on this chromosome. A random process then assigns a fraction of IS randomly in the hotspots (H_{is}) and a fraction on the complete chromosome (H_{rnd}). We decided to use a ratio H_{is}:

H_{rnd} = 1:3 between random IS and hotspot IS in order to simulate an experiment with a low signal to noise ratio.

3.6. Statistical Model, p-Value and Log-Likelihood Ratio Test

For each CIS a p-value and a log likelihood ratio for a 2-class problem, with full specified null (θ_{rnd}) and alternative (θ_{ex}) hypotheses, are returned.

Two classes of CIS are extracted from the data: CIS_{RAND}, the CIS found in the simulation experiment without hot-spots (IS_{RAND}) and CIS_{EXP}, the experimental CIS found in one clinical study [1]. The simulation experiment was repeated 100 times (connection threshold 50 Kbp) in order to obtain a measurement of the CIS frequency.

The rarity R measures the IS compactness in a CIS and the rarity distributions were computed for both CIS classes. The rarity for dataset l and order i R_{li} is expressed as $\frac{dimension\ of\ the\ CIS\ in\ datatset\ l}{i-1}$. Except in random CIS of order 2, where R uniformly distributed over the threshold distance, the rarity was approximated by two continuous normal distributions of mean ($\mu_{\theta rnd}, \mu_{\theta ex}$) and variance ($\sigma^2_{\theta rnd}, \sigma^2_{\theta ex}$).

Two prior probabilities for the hypothesis, $p(\theta_{ex})$ and $p(\theta_{rnd})$, were calculated from the number of experimental (N_{ex}) and expected (N_{rnd}) CIS. The expected number of random CIS, grouped by order, was equated to the maximal CIS frequency found in the random experiments for CIS of same order. Then the priori for the experimental class is $p(\theta_{ex}) = N_{ex}/(N_{ex} + N_{rnd})$ and $p(\theta_{rnd}) = 1 - p(\theta_{ex})$.

The probability of an error, interpreting a CIS as not random (rejection of θ_{rnd}) given the CIS rarity x, is

$$p(\theta_{rnd}|x) = p(x|\theta_{rnd})p(\theta_{rnd})$$
$$= p(\theta_{rnd})\left(2\pi\sigma^2_{\theta rnd}\right)^{-\frac{1}{2}}\int_{-\infty}^{x} \exp\left(-0.5\left(\frac{x-\mu_{\theta rnd}}{\sigma_{\theta rnd}}\right)^2\right)dx.$$

We returned also the log-likelihood ratio $\Lambda(x)$ in order to indicate the CIS class more probably given the rarity x.

For a 2-class problem the class θ_{ex} is chosen when $p(\theta_{ex}|x) > p(\theta_{rnd}|x)$. By the Bayes theorem it is equivalent to write $p(x|\theta_{ex})p(\theta_{ex}) > p(x|\theta_{rnd})p(\theta_{rnd})$. Rearranging this relation we can write

$$\Lambda(x) = \log_2 \frac{p(x|\theta_{ex})p(\theta_{ex})}{p(x|\theta_{rnd})p(\theta_{rnd})}$$

obtaining the decision rule that if $\Lambda(x) > 0$ the hypothesis θ_{rnd} is rejected. The higher is $\Lambda(x)$, the better the is hypothesis θ_{ex}.

This probability $p(\theta_{rnd}|x)$ and the log-likelihood $\Lambda(x)$ score are associated to any CIS and returned to the user.

3.7. Integration Sites Analysis Software and Availability of Supporting Data

Cytoscape 2.8 was used as prototyping environment to perform all the network examination. Supporting material as the prototype plugin for Cytoscape 2.8 and tutorials are available on request.

Abbreviations

CIS	Common Insertion Sites
GA	Gene Atmosphere
GT	Gene Therapy
HS	Hot Spot
IS	Insertion Sites
NGA	Next Gene Annotation
SWM	Standard Window Method

Competing Interests

The authors declare that there is no conflict of interest.

Acknowledgments

We would like to gratefully acknowledge Irene Gil-Farina for assistance with the paper preparation.

References

[1] Cartier N, Hacein-Bey-Abina S, Bartholomae CC, Veres G, Schmidt M, Kutschera I, et al. Hematopoietic stem cell gene therapy with a lentiviral vector in X-linked adrenoleukodystrophy. Science 2009;326:818–23. http://dx.doi.org/10.1126/science.1171242.

[2] Boztug K, Schmidt M, Schwarzer A, Banerjee PP, Díez IA, Dewey RA, et al. Stem-cell gene therapy for the Wiskott–Aldrich syndrome. N Engl J Med 2010;363:1918–27. http://dx.doi.org/10.1056/NEJMoa1003548.

[3] Aiuti A, Biasco L, Scaramuzza S, Ferrua F, Cicalese MP, Baricordi C, et al. Lentiviral hematopoietic stem cell gene therapy in patients with Wiskott–Aldrich syndrome. Science 2013;341:1233151. http://dx.doi.org/10.1126/science.1233151.

[4] Cavazzana-Calvo M, Hacein-Bey S, de Saint Basile G, Gross F, Yvon E, Nusbaum P, et al. Gene therapy of human severe combined immunodeficiency (SCID)-X1 disease. Science 2000;288:669–72.

[5] Biffi A, Montini E, Lorioli L, Cesani M, Fumagalli F, Plati T, et al. Lentiviral hematopoietic stem cell gene therapy benefits metachromatic leukodystrophy. Science 2013;341:1233158. http://dx.doi.org/10.1126/science.1233158.

[6] Cavazzana-Calvo M, Payen E, Negre O, Wang G, Hehir K, Fusil F, et al. Transfusion independence and HMGA2 activation after gene therapy of human β-thalassaemia. Nature 2010;467:318–22. http://dx.doi.org/10.1038/nature09328.

[7] Hacein-Bey-Abina S, Von Kalle C, Schmidt M, McCormack MP, Wulffraat N, Leboulch P, et al. LMO2-associated clonal T cell proliferation in two patients after gene therapy for SCID-X1. Science 2003;302:415–9. http://dx.doi.org/10.1126/science.1088547.

[8] Hacein-Bey-Abina S, Garrigue A, Wang GP, Soulier J, Lim A, Morillon E, et al. Insertional oncogenesis in 4 patients after retrovirus-mediated gene therapy of SCID-X1. J Clin Invest 2008;118:3132–42. http://dx.doi.org/10.1172/JCI35700.

[9] Howe SJ, Mansour MR, Schwarzwaelder K, Bartholomae C, Hubank M, Kempski H, et al. Insertional mutagenesis combined with acquired somatic mutations causes leukemogenesis following gene therapy of SCID-X1 patients. J Clin Invest 2008;118:3143–50. http://dx.doi.org/10.1172/JCI35798.

[10] Stein S, Ott MG, Schultze-Strasser S, Jauch A, Burwinkel B, Kinner A, et al. Genomic instability and myelodysplasia with monosomy 7 consequent to EVI1 activation after gene therapy for chronic granulomatous disease. Nat Med 2010;16:198–204. http://dx.doi.org/10.1038/nm.2088.

[11] Deichmann A, Brugman MH, Bartholomae CC, Schwarzwaelder K, Verstegen MMA, Howe SJ, et al. Insertion sites in engrafted cells cluster within a limited repertoire of genomic areas after gammaretroviral vector gene therapy. Mol Ther 2011;19:2031–9. http://dx.doi.org/10.1038/mt.2011.178.

[12] Akagi K, Suzuki T, Stephens RM, Jenkins NA, Copeland NG. RTCGD: retroviral tagged cancer gene database. Nucleic Acids Res 2004;32:D523–7. http://dx.doi.org/10.1093/nar/gkh013.

[13] Bokhoven M, Stephen SL, Knight S, Gevers EF, Robinson IC, Takeuchi Y, et al. Insertional gene activation by lentiviral and gammaretroviral vectors. J Virol 2009;83:283–94. http://dx.doi.org/10.1128/JVI.01865-08.

[14] Montini E, Cesana D, Schmidt M, Sanvito F, Ponzoni M, Bartholomae C, et al. Hematopoietic stem cell gene transfer in a tumor-prone mouse model uncovers low genotoxicity of lentiviral vector integration. Nat Biotechnol 2006;24:687–96. http://dx.doi.org/10.1038/nbt1216.

[15] Montini E, Cesana D, Schmidt M, Sanvito F, Bartholomae CC, Ranzani M, et al. The genotoxic potential of retroviral vectors is strongly modulated by vector design and integration site selection in a mouse model of HSC gene therapy. J Clin Invest 2009;119:964–75. http://dx.doi.org/10.1172/JCI37630.

[16] Stein S, Scholz S, Schwäble J, Sadat MA, Modlich U, Schultze-Strasser S, et al. From bench to bedside: preclinical evaluation of a self-inactivating gammaretroviral vector for the gene therapy of X-linked chronic granulomatous disease. Hum Gene Ther Clin Dev 2013;24:86–98. http://dx.doi.org/10.1089/humc.2013.019.

[17] Shen H, Suzuki T, Munroe DJ, Stewart C, Rasmussen L, Gilbert DJ, et al. Common sites of retroviral integration in mouse hematopoietic tumors identified by high-throughput, single nucleotide polymorphism-based mapping and bacterial artificial chromosome hybridization. J Virol 2003;77:1584–8.

[18] Abel U, Deichmann A, Nowrouzi A, Gabriel R, Bartholomae CC, Glimm H, et al. Analyzing the number of common integration sites of viral vectors—new methods and computer programs. PLoS One 2011;6, e24247. http://dx.doi.org/10.1371/journal.pone.0024247.

[19] Wu X, Luke BT, Burgess SM. Redefining the common insertion site, vol. 344; 2006.

[20] De Ridder J, Uren A, Kool J, Reinders M, Wessels L. Detecting statistically significant common insertion sites in retroviral insertional mutagenesis screens. PLoS Comput Biol 2006;2, e166. http://dx.doi.org/10.1371/journal.pcbi.0020166.

[21] Presson AP, Kim N, Xiaofei Y, Chen IS, Kim S. Methodology and software to detect viral integration site hot-spots. BMC Bioinf 2011;12:367. http://dx.doi.org/10.1186/1471-2105-12-367.

[22] De Jong J, de Ridder J, van der Weyden L, Sun N, van Uitert M, Berns A, et al. Computational identification of insertional mutagenesis targets for cancer gene discovery. Nucleic Acids Res 2011;39, e105. http://dx.doi.org/10.1093/nar/gkr447.

[23] Bergemann TL, Starr TK, Yu H, Steinbach M, Erdmann J, Chen Y, et al. New methods for finding common insertion sites and co-occurring common insertion sites in transposon- and virus-based genetic screens. Nucleic Acids Res 2012;40:3822–33. http://dx.doi.org/10.1093/nar/gkr1295.

[24] Babaei S, Akhtar W, de Jong J, Reinders M, de Ridder J. 3D hotspots of recurrent ret-roviral insertions reveal long-range interactions with cancer genes. Nat Commun 2015;6:6381. http://dx.doi.org/10.1038/ncomms7381.

[25] Shannon P, Markiel A, Ozier O, Baliga NS, Wang JT, Ramage D, et al. Cytoscape: a soft-ware environment for integrated models of biomolecular interaction networks. Ge-nome Res 2003;13:2498–504. http://dx.doi.org/10.1101/gr.1239303.

[26] Albert R, Jeong H, Barabasi A. Error and attack tolerance of complex networks. Na-ture 2000;406:378–82. http://dx.doi.org/10.1038/35019019.

[27] Albert R. Scale-free networks in cell biology. J Cell Sci 2005;118:4947–57. http://dx.doi.org/10.1242/jcs.02714.

[28] Bonifazi P, Goldin M, Picardo M a, Jorquera I, Cattani a, Bianconi G, et al. GABAergic hub neurons orchestrate synchrony in developing hippocampal networks. Science 2009;326:1419–24. http://dx.doi.org/10.1126/science.1175509.

[29] Wang GP, Ciuffi A, Leipzig J, Berry CC, Bushman FD. HIV integration site selection: analysis by massively parallel pyrosequencing reveals association with epigenetic modifications. Genome Res 2007;17:1186–94. http://dx.doi.org/10.1101/gr.6286907.

[30] Kaeppel C, Beattie SG, Fronza R, van Logtenstein R, Salmon F, Schmidt S, et al. A largely random AAV integration profile after LPLD gene therapy. Nat Med 2013; 19:889–91. http://dx.doi.org/10.1038/nm.3230.

[31] Pañeda A, Lopez-franco E, Kaeppel C, Unzu C, Sampedro A, Mauleon I, et al. Safety and liver transduction efficacy of rAAV5-cohPBGD in non-human primates: a poten-tial therapy for Acute Intermitent Porphyria. Hum Gene Ther 2013;1–37. http://dx.doi.org/10.1089/hum.2013.166.

[32] Shannon CE. A mathematical theory of communication. Bell Syst Tech J 1948;27: 379–423 [623–56].

[33] Giordano M, Natale M, Cornaz M, Ruffino A, Bonino D, Bucci EM. iMole, a web based image retrieval system from biomedical literature. Electrophoresis 2013;34:1965–8. http://dx.doi.org/10.1002/elps.201300085.

[34] Benso A, Cornale P, Di Carlo S, Politano G, Savino A. Reducing the complexity of com-plex gene coexpression networks by coupling multiweighted labeling with topolog-ical analysis. Biomed Res Int 2013;2013:676328. http://dx.doi.org/10.1155/2013/676328.

[35] Roberti MC, La Starza R, Surace C, Sirleto P, Pinto RM, Pierini V, et al. RABGAP1L gene rearrangement resulting from a der(Y)t(Y;1)(q12;q25) in acute myeloid leukemia arising in a child with Klinefelter syndrome. Virchows Arch 2009;454:311–6. http://dx.doi.org/10.1007/s00428-009-0732-z.

[36] Akdi A, Pérez G, Pastor S, Castell J, Biarnés J, Marcos R, et al. Common variants of the thyroglobulin gene are associated with differentiated thyroid cancer risk. Thyroid 2011;21:519–25. http://dx.doi.org/10.1089/thy.2010.0384.

[37] Zhai H, Shi Y, Jin H, Li Y, Lu Y, Chen X, et al. Expression of calcyclin-binding protein/Siah-1 interacting protein in normal and malignant human tissues: an immunohis-tochemical survey. J Histochem Cytochem 2008;56:765–72. http://dx.doi.org/10.1369/jhc.2008.950519.

[38] Azzato EM, Driver KE, Lesueur F, Shah M, Greenberg D, Easton DF, et al. Effects of common germline genetic variation in cell cycle control genes on breast cancer sur-vival: results from a population-based cohort. Breast Cancer Res 2008;10:R47. http://dx.doi.org/10.1186/bcr2100.

[39] Hinrichs AS, Karolchik D, Baertsch R, Barber GP, Bejerano G, Clawson H, et al. The UCSC Genome Browser Database: update 2006. Nucleic Acids Res 2006;34: D590–8. http://dx.doi.org/10.1093/nar/gkj144.

[40] Assenov Y, Ramírez F, Schelhorn S-E, Lengauer T, Albrecht M. Computing topological parameters of biological networks. Bioinformatics 2008;24:282–4. http://dx.doi.org/10.1093/bioinformatics/btm554.

[41] Vasciaveo Alessandro, et al. Common Integration Sites of published datasets identi-fied using a graph based framework. Computational and Structural Biotechnology Journal 2015;14:87–90.

17

Prediction of anticancer peptides against MCF-7 breast cancer cells from the peptidomes of *Achatina fulica* mucus fractions

Teerasak E-kobon [a], Pennapa Thongararm [b], Sittiruk Roytrakul [c], Ladda Meesuk [d], Pramote Chumnanpuen [b,*]

[a] *Department of Genetics, Faculty of Science, Kasetsart University, Bangkok 10900, Thailand*
[b] *Department of Zoology, Faculty of Science, Kasetsart University, Bangkok 10900, Thailand*
[c] *National Center for Genetic Engineering and Biotechnology, Thailand Science Park, Pathum Thani 12120, Thailand*
[d] *Faculty of Dentistry, Thammasat University, Pathum Thani 12120, Thailand*

ARTICLE INFO

Keywords:
Peptidomics
Cytotoxic peptides
Achatina fulica
Breast cancer
Bioinformatics prediction
Snail mucus

ABSTRACT

Several reports have shown antimicrobial and anticancer activities of mucous glycoproteins extracted from the giant African snail *Achatina fulica*. Anticancer properties of the snail mucous peptides remain incompletely revealed. The aim of this study was to predict anticancer peptides from *A. fulica* mucus. Two of HPLC-separated mucous fractions (F2 and F5) showed in vitro cytotoxicity against the breast cancer cell line (MCF-7) and normal epithelium cell line (Vero). According to the mass spectrometric analysis, 404 and 424 peptides from the F2 and F5 fractions were identified. Our comprehensive bioinformatics workflow predicted 16 putative cationic and amphipathic anticancer peptides with diverse structures from these two peptidome data. These peptides would be promising molecules for new anti-breast cancer drug development.

1. Introduction

Breast cancer is one of the most common diseases in women globally [1]. Several factors make women at high risk of the breast cancer [2]. Early detection and the use of radiation therapy, surgery, and chemotherapeutic drugs including selective estrogen receptor modulators (SERMs) and aromatase inhibitors can reduce invasive breast cancer. However, the patients remain traumatized by the unfavorable side effects [3,4]. The search for target-specific and less side-effect cancer therapy is still undergoing.

Anticancer peptides have been proved to be effective small molecules (<50 amino acids) that can act specifically against cancerous cells by either membranolytic mechanism or disruption of mitochondria [5]. The net negative charge of the cancer membrane is an important factor for peptides' selectivity and toxicity [6], as compared to the typically zwitterionic property of non-cancerous eukaryotic membranes. Amphiphilicity levels and hydrophobic arc size allow penetration of these peptides through the cancerous cell membranes and lead to destabilization of the membrane integrity [7,8]. For example, pleurocidin-like peptides (NRCs) identified from fish could kill breast cancer cells and human mammary epithelial cells by causing membrane

damage with subtle harm to human fibroblasts [9]. These cell-penetrating peptides could be used as cancer-specific drug delivery. For example, the non-specific cell-penetrating anticancer peptide buforin IIb was modified to enhance the cancer specificity with no effects on normal cells [10]. This cancer-specific peptide derivative was successfully used to deliver apoptosis-induced antibody into the cancer cells. Distinctively, a peptide SA12 could induce apoptosis on SKBr-3 breast cancer cells by the mitochondrial pathway [11]. Taken together, these physicochemical properties and experimentally validated information were used to develop bioinformatic programs for anticancer peptide prediction and design. AntiCP predicts anticancer peptides by using amino acid composition and binary profiles to develop support vector machine models (SVM) [12]. Another SVM-based program, ACPP, particularly screens for anticancer peptides that contain apoptotic domain [13]. In this regard, these prediction tools will assist high-throughput screening for anticancer peptides from complex peptidomes of an array of natural products.

Giant African snails (*Achatina fulica*) are invasive animals that seriously cause damages to agricultural and ornamental plants worldwide. Only one antimicrobial peptide, namely Mytimacin-AF, was identified from the mucus of *A. fulica* [14]. Mytimacin-AF (9.7 kDa) was a novel cysteine-rich peptide that could inhibit the growth of both fungi and bacteria with little hemolytic effect on human red blood cells. However, we hypothesized that the anticancer peptides from the mucus of *A. fulica* may not be completely revealed. Thus, this study aimed to predict putative anticancer peptides from the most effective HPLC-

* Corresponding author at: Department of Zoology, Faculty of Science, Kasetsart University, 50 Ngam Wong Wan Road, Chatuchak, Bangkok 10900, Thailand.

E-mail address: pramote.c@ku.ac.th (P. Chumnanpuen).

separated mucous fractions against the breast cancer cell line MCF-7 using mass spectrometric and bioinformatic analysis methods. Our results provide alternative high-throughput screening methods to identify potential anticancer peptides from nearly a thousand peptides within the snail mucus for further validation.

2. Experimental procedure

2.1. Cell culture

The breast cancer cell line MCF-7 and the kidney epithelial cell line Vero used in this study were kindly provided by the Department of Biochemistry, Faculty of Medicine, Chiangmai University, Thailand and the Genome Institute, National Center for Genetic Engineering and Biotechnology (BIOTEC), Thailand. The cells were cultured and passaged in Dulbecco's Modification of Eagle's Medium (DMEM, Gibco-RBL, Life Technologies, NY) supplemented with 10% Fetal Bovine Serum (FBS, Hyclone, Thermo Fisher Scientific Inc., USA), 1% Penicillin–Streptomycin (PAA, Laboratories GmbH, Austria) and 1% Amphotericin B (PAA, Laboratories GmbH, Austria). The cells were maintained at 37 °C in 95% relative humidified atmosphere containing 5% CO_2. Cell growth was measured under a light microscope and 80% confluence of the cells was used in all experiments.

2.2. Separation of A. fulica mucus by HPLC

The snail mucus samples were collected from adult A. fulica by intermittent irritation in an ultrasonicating bath at 30 °C sporadically. The crude mucous samples were separated by ZORBAX 300SB-4.6 × 150 mm C18 column, 5 μm, (Agilent, Palo Alto, CA) with Agilent® 1200 system using methanol–water (50:50) with 0.1% trifluoroacetic acid (adjusted from [15]) as mobile phase and the flow rate was 0.30 ml/min. Numbers of the HPLC peaks were used to determined numbers of the fractions. Six HPLC-separated mucous fractions were collected manually and named as F1, F2, F3, F4, F5 and F6 fractions. All HPLC fractions and the crude mucus were concentrated by freeze-drying at − 100 °C and kept at − 20 °C until use.

2.3. Determination of cytotoxicity of the mucous fractions by MTT assay

Cell viability count was performed using 3-(4,5-dimethylthiazol-2-yl)-2,5-diphenyltetrazolium bromide (MTT) assay [16]. Cells were seeded at 2×10^4 cells per well (200 μl/well) in 96-well tissue culture plates and allowed cells to adhere for 24 h at 37 °C in the CO_2 incubator. The culture medium was then replaced with 200 μl/well of the fresh medium for the control group and 200 μl/well of the fresh medium containing the same concentration (1000 μg/ml) of the crude mucus or the six HPLC-separated fractions. After 72 h incubation, 50 μl/well of tetrazolium bromide salt solution (2 mg/ml of stock in phosphate buffered saline, PBS) was added into 150 μl of the cell suspension. Four hours before completion, the reaction mixture was carefully taken out and 200 μl/well of dimethyl sulfoxide or DMSO (Sigma, USA) was added to each well before the addition of 25 μl/well of Sorensen's glycine buffer (Research Organics, USA). The optical densities (OD) were measured at 570 nm using microplate reader (Tecan Sunrise, Switzerland). Finally, the highest effective anti-breast cancer fraction with the lowest percentage of cell viability was then selected for further analysis.

Cytotoxicity of the mucous fractions against the MCF-7 and Vero cells was compared by slightly modified the above-described method due to the limited quantity of the fractions. The cells were seeded at 4×10^3 cells per well in 96-well tissue culture plates and allowed cells to adhere for 24 h at 37 °C in the CO_2 incubator. The culture medium was then replaced with 100 μl/well of the fresh medium for the control group and 100 μl/well of the fresh medium containing three concentrations (1, 10 and 100 μg/ml) of the crude mucus, the F2, and F5 fractions. After 24 h incubation, 25 μl/well of tetrazolium bromide salt solution

(5 mg/ml of stock in PBS) was added to the cell suspension. Four hours before completion, the reaction mixture was carefully taken out, and 100 μl/well of DMSO was added to each well. The optical densities were measured at 570 nm.

2.4. Statistical analysis of the MTT assay

The results were presented as mean ± sem (standard error of mean) or mean ± sd (standard deviation). The parameters were analyzed with one-way analysis of variance (One-way ANOVA) followed by Sidak's multiple comparisons test. Statistical analysis was conducted with Graphpad Prism version 6.0 for Windows (Graphpad software, San Diego, California, USA). Significant levels were considered at $p < 0.05$ and highly significant level at $p < 0.01$ comparing with control group.

2.5. Mass spectrometric analysis of the selected cytotoxic fractions

2.5.1. Sample fractionation
Individual selected fractions were fractionated based on their molecular size using Macrosep® 3 K, 10 K and 50 K Omega centrifugal devices (Pall Life Sciences, USA) into four sub-fractions: lower than 3 kDa, between 3 and 10 kDa, between 10 and 50 kDa, and larger than 50 kDa sub-fractions. These sub-fractions were mixed well with two volumes of cold acetone and incubated overnight at − 20 °C. The mixture was centrifuged at 10,000 ×g for 15 min, and the supernatant was discarded. The pellet was freeze-dried and stored at − 80 °C before use.

2.5.2. Determination of protein/peptide concentration by lowry method
The pellets were resuspended in 0.15% Sodium Deoxycholic acid (DOC) or 0.5% SDS and determined protein concentration by Lowry method [17]. The absorbance at 750 nm (OD_{750}) was measured, and the protein concentration was calculated using the standard curve, plotted between OD_{750} on Y-axis and BSA concentration (μg/ml) on X-axis.

2.5.3. In-solution digestion
Each protein sub-fractions were hydrolyzed by trypsin at an enzyme to the protein ratio of 1:50 at 37 °C for 24 h, except the lower than 3 kDa sub-fraction. The peptides were dried by vacuum centrifuge and kept at − 80 °C for further mass spectrometric analysis.

2.5.4. HCTultra LC–MS analysis of the peptidomes
Peptide solutions were analyzed using an HCTultra PTM Discovery System (Bruker Daltonics Ltd., U.K.) coupled to an UltiMate 3000 LC System (Dionex Ltd., U.K.). Peptides were separated on a nanocolumn (PepSwift monolithic column 100 μm i.d. × 50 mm). Eluent A was 0.1% formic acid, and eluent B was 80% acetonitrile in water containing 0.1% formic acid. Peptide separation was achieved with a linear gradient from 10% to 70% B for 13 min at a flow rate of 300 nl/min, including a regeneration step at 90% B and an equilibration step at 10% B, one run took 20 min. Peptide fragment mass spectra were acquired in data-dependent AutoMS (2) mode with a scan range of 300–1500 m/z, 3 averages, and up to 5 precursor ions selected from the MS scan 50–3000 m/z.

2.5.5. Identification of peptide sequences
The MS/MS data from LC–MS were submitted to database search using the Mascot software (Matrix Science, London, U.K., [18]). The peptide sequence data was searched against the NCBI database for protein identification. Database interrogation was; taxonomy (other metazoans); enzyme (trypsin); variable modifications (carbamidomethyl, oxidation of methionine residues); mass values (monoisotopic); protein mass (unrestricted); peptide mass tolerance (1 Da); fragment mass tolerance (± 0.4 Da), peptide charge state (1 +, 2 + and 3 +), max missed cleavages (1) and instrument = ESI-TRAP. Proteins considered as identified proteins had at least one peptides with an individual Mascot score corresponding to $p < 0.05$. The resultant peptides with

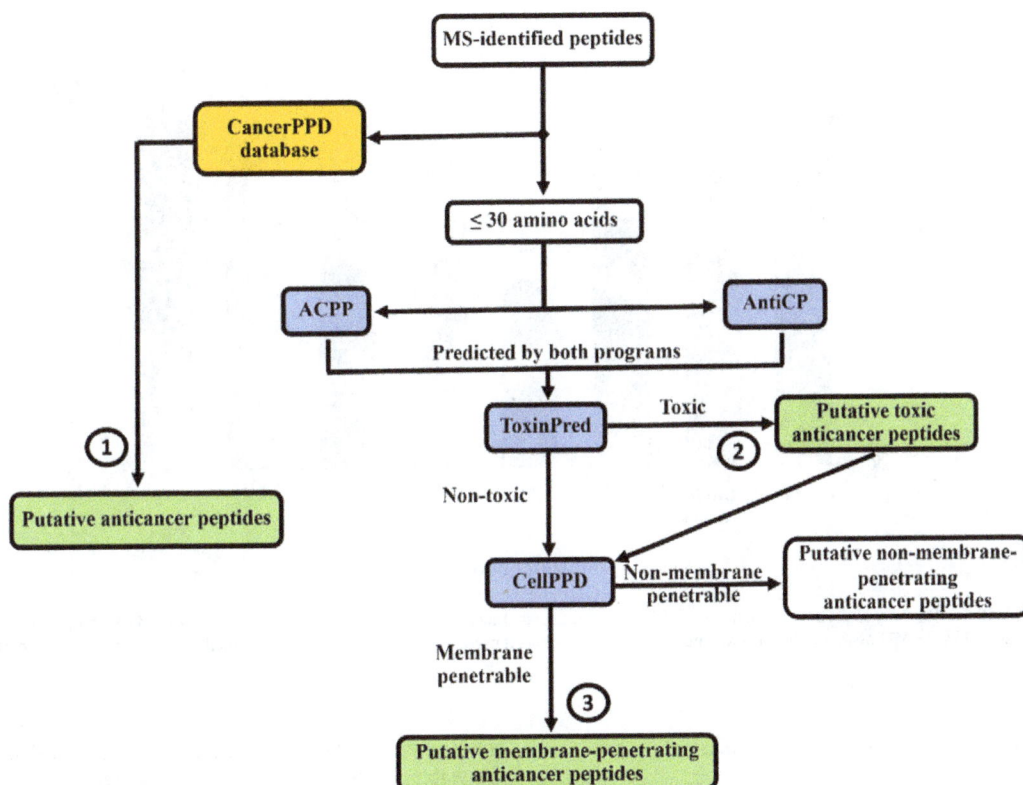

Fig. 1. Workflow for bioinformatic prediction of putative anticancer, cytotoxic and membrane-penetrating peptides (labeled with numbers 1, 2 and 3) obtained from peptidomic analysis of *A. fulica* mucous fractions. Details were described in the text. CancerPPD (yellow box) was a database of anticancer peptides. ACPP, AntiCP, ToxinPred and CellPPD were the bioinformatics prediction programs (blue boxes).

the rank of ion matches (pep_rank) equivalent to one were considered significance.

2.6. Bioinformatic prediction of anticancer peptides

The identified peptides from the fractions F2 and F5 were then blasted against a database of anticancer peptides (cancerPPD, [19]). These peptides were screened for putative anticancer peptides by using consensus prediction. Each peptide was submitted for the prediction of two programs; ACPP [13] and AntiCP [12] for anticancer peptide prediction based on amino acid composition, conserved features and physicochemical properties. The hydrophobicity, hydrophilicity,

amphiphaticity and hydrophathicity scores were computed by AntiCP and their means were statistically compared between the anticancer and non-anticancer peptides by using an R package for independent two sample t-test. Peptides positively predicted by both programs were considered putative anticancer peptides. These peptides were checked for toxicity by submission to ToxinPred for the prediction of toxic peptides [20]. The predicted anticancer peptides were subjected to CellPPD for prediction of cell penetrating properties [21]. These peptides were classified into putative membrane-penetrating and non-membrane-penetrating anticancer peptides (Fig. 1). If the number of predicted anticancer peptides was very high, the cutoff score (from 0.5–1.0) was then applied to narrow the list down to a short list of those with the high prediction scores. Structures of these putative anticancer peptides were then modeled by the PEP-FOLD program version 1.5 [22]. Structural similarity was used to classify these predicted peptide structures.

3. Results and discussion

The *A. fulica* mucus was successfully collected by intermittent irritation and separated into six fractions by the C18-reverse phase HPLC system using methanol–water mobile phase. All six fractions were manually collected and named as F1, F2, F3, F4, F5 and F6 fractions as shown in Fig. 2. The fraction was scanned between 210 to 290 nm which was a suitable range for peptide bond and aromatic amino acid absorptions. The sample showed the highest signal when scanning at 220 nm, but revealed two distinguished peaks at the absorbance of 210, 230, 280 and 290 nm. Thus, these two peaks were collected in the F3 and F4 fractions. The lowest signal peak was collected in the F2 fraction. Other three moderate signal peaks were collected in the F1, F5 and F6 fractions.

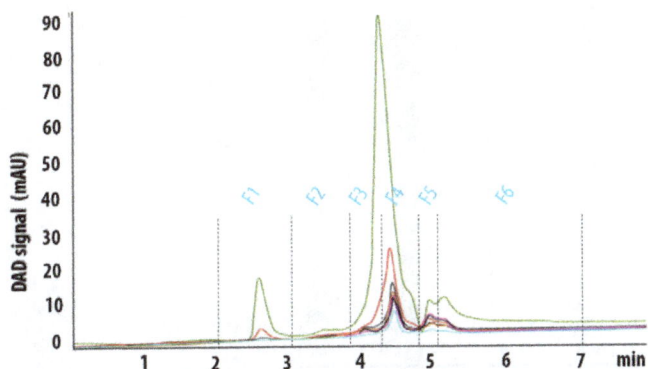

Fig. 2. HPLC chromatogram shows retention times (minutes) of the six separated fractions of the *A. fulica* mucus. The signal according to the diode array detector (DAD) is represented in milli-arbitary unit (mAU). The fractions were scanned at different ultraviolet wavelengths; 210 (pink), 220 (green), 230 (purple), 250 (red), 280 (brown green) and 290 (blue) nm.

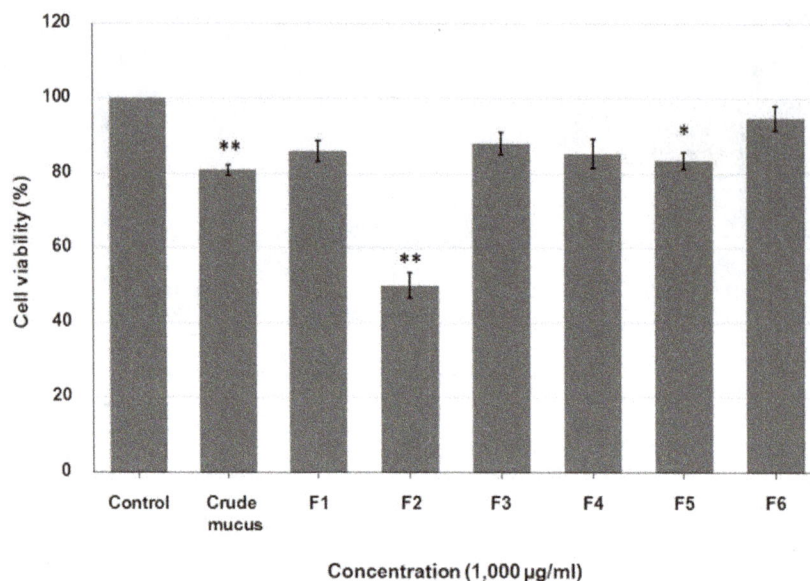

Fig. 3. Percentages of the MCF-7 cell viability after 72 h treatment with *A. fulica* crude mucus and six HPLC-separated fractions (F1, F2, F3, F4, F5 and F6) at the concentration of 1000 µg/ml. The cell viability was measured by the MTT assay. Control was the untreated cells. Error bars represented the SEM, * and ** represented levels of significance at $p < 0.05$ and $p < 0.01$.

3.1. Selection of highly cytotoxic mucous fractions against MCF-7 cells

Cytotoxicity of the *A. fulica* crude mucus and six HPLC-separated fractions (F1, F2, F3, F4, F5 and F6) on the MCF-7 cells were evaluated by the MTT assay to measure cytotoxic effect on cell viability. Results showed that fractions F1, F3, F4 and F6 did not significantly induce cell death ($p > 0.05$) comparing to the control group. The treatments of these four fractions resulted in more than 85% of the viable cells. Oppositely, the crude mucus, fractions F2 and F5 significantly had cytotoxic effects to the MCF-7 cells ($p < 0.05$) comparing to the control group. The F2 fraction showed the highest cytotoxic effect ($50.03 \pm 3.38\%$ of viable cells) followed by crude mucus and the F5 fraction (80.91 ± 1.38 and $83.61 \pm 2.26\%$, respectively) (Fig. 3).

Cytotoxicity against the MCF-7 and Vero cells was further compared by varying the concentration of the crude mucus, the F2 and F5 fractions to 1, 10 and 100 µg/ml. Results in Fig. 4 showed that the increased

concentration of crude mucus and the F2 fraction significantly reduced the viability of both cell types to 72–84% ($p < 0.05$). The F5 fraction at 100 µg/ml significantly decreased the viability for both Vero and MCF7 cells but showed no different reduction at 1 and 10 µg/ml. The crude mucus and these two fractions significantly suppressed cell viability of the Vero cells more than those of the MCF-7 cells ($p < 0.05$), except the crude mucus at 100 µg/ml. The Vero cells were insignificantly affected by these three fractions at 100 µg/ml ($p < 0.05$), whereas the viability of the MCF-7 cells were significantly interfered by the crude mucus and the F5 fraction more than the F2. Although this experiment was not conducted at the concentration of 1000 µg/ml as described in the previous section (Fig. 3) due to limited amount of the HPLC-separated fractions. The percentages of cell viability from these two independent experiments was unable to compare. However, the authors observed similar pattern of the cell viability reduction between the crude mucus and the F5 fraction. For the F2 fraction, the cytotoxic activity could

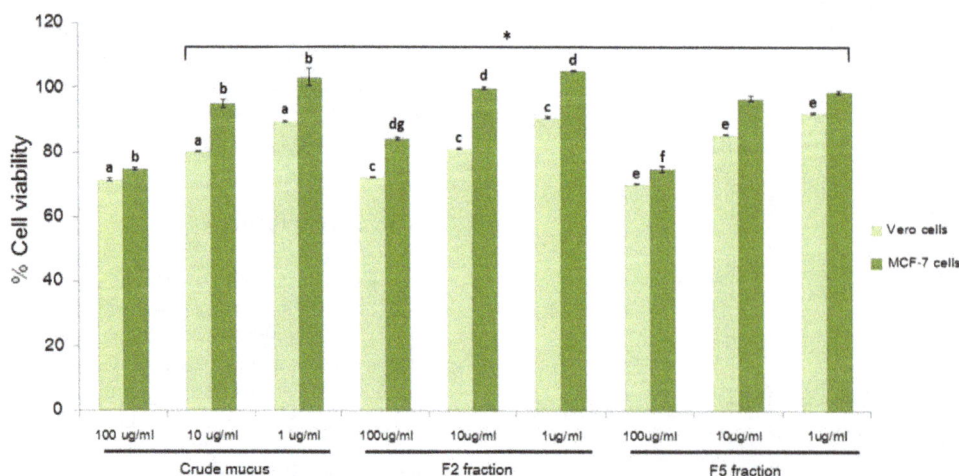

Fig. 4. Percentages of the Vero and MCF-7 cell viabilities after treatment with *A. fulica* crude mucus and the F2 and F5 fractions at the concentration of 1, 10 and 100 µg/ml. The cell viability was measured by the MTT assay. Control was the untreated cells. Error bars represented the SD, * represented statistical difference between the viability of Vero and MCF-7 cells at the significance of $p < 0.05$. a, b, c, d, e and f: significant difference between concentrations of the crude mucus (a and b), the F2 (c and d) and F5 (e and f) fractions, respectively, $p < 0.05$. g represented the significant difference ($p < 0.05$) of the F2 fraction from the crude mucus and the F5 fraction at 100 µg/ml.

Table 1
Details of putative anticancer peptides predicted from the F2 and F5 fractions of the *A. fulica* mucus by using four bioinformatics programs. Each fraction consisted of four sub-fractions: <3 kDa, 3–10 kDa, 10–50 kDa and >50 kDa. Predictive scores and amino acid sequence for each peptide were also shown. These peptides were structurally classified into five groups including random coiled, a single helix, helix-consisted loop, β-sheet-consisted loop and short peptide. The SVM scores above the threshold of 0.90 are highlighted in red. The 16 significantly putative anticancer peptides are showed in bold amino acid sequences.

ID	Fraction	Subfraction	Amino acid sequence	No of residues	ACPP	AntiCP	CellPPD	ToxinPred	ACPP	AntiCP	CellPPD	ToxinPred	Structural type
0	F2	< 3kDa	DTPRCCR	7	+	+	-	+	0.70	1.24	-0.23	0.46	Short peptide
1			GGPIAAPEASK	11	+	+	-	-	0.94	0.39	-0.38	-1.25	Random coiled
2			LGGIVVVVVSRR	12	+	+	-	-	0.62	0.53	-0.62	-1.10	Random coiled
3			GLAWGALLYLGAALADALGK	20	+	+	-	-	0.83	0.54	-0.22	-0.92	Single helix
4			GIGSAAGAAVSLVYMLPK	18	+	+	-	-	0.70	0.85	-0.32	-1.79	Single helix
5			NAIATTTQK	9	+	+	-	-	0.88	0.50	-0.19	-0.70	Beta–sheet consisted loop
6			GIKINPGAFIQTISVIKVR	19	+	+	-	-	0.78	0.19	-0.21	-0.99	Beta–sheet consisted loop
7			HNGCGSNGSTK	11	+	+	-	-	0.66	0.59	-0.53	-0.83	Beta–sheet consisted loop
8			VKHLIGNF	8	+	+	-	-	0.75	0.55	-0.28	-0.93	Short peptide
9			GGAMIMK	7	+	+	-	-	0.61	0.38	-0.25	-0.72	Short peptide
10			VVPVCTK	7	+	+	-	-	0.52	0.41	-0.26	-1.28	Short peptide
11	F2	3–10 kDa	GGPIAAPEASK	11	+	+	-	-	0.88	0.39	-0.38	-1.25	Random coiled
12			NAIATTTQK	9	+	+	-	-	0.78	0.50	-0.19	-0.70	Random coiled
13			CVGLGGRGC	9	+	+	-	-	0.72	0.93	-0.58	-0.62	Random coiled
14			GLAWGALLYLGAALADALGK	20	+	+	-	-	0.83	0.54	-0.22	-0.92	Single helix
15			GIGSAAGAAVSLVYMLPK	18	+	+	-	-	0.70	0.85	-0.32	-1.79	Single helix
16			MGIPRLGLVGIMGIPRLGLVGIQCK	25	+	+	-	-	0.53	0.80	-0.34	-1.50	Helix–consisted loop
17			VKHLIGNF	8	+	+	-	-	0.51	0.55	-0.28	-0.93	Short peptide
18	F2	10–50 kDa	KLERAAGSK	9	+	+	+	-	0.64	0.04	0.06	-1.01	Helix–consisted loop
19			GYAAGIK	7	+	+	-	-	0.88	1.62	-0.24	-1.07	Short peptide
20			HANGGVLK	8	+	+	-	-	0.58	1.16	-0.27	-1.08	Short peptide
21	F2	> 50 kDa	GYAAGNK	7	+	+	-	-	0.90	1.18	-0.24	-1.00	Short peptide
22			AAAIHVSK	8	+	+	-	-	0.50	0.73	-0.24	-1.25	Short peptide
23	F5	< 3 kDa	HALLIIFNASKK	12	+	+	-	-	0.94	0.93	-0.34	-1.07	Random coiled
24			VCKALIPGLIPLSFGHGLEPK	21	+	+	-	-	0.71	0.92	-0.29	-0.90	Random coiled
25			HLIKAKGSD	9	+	+	+	-	0.78	0.82	0.17	-0.41	Random coiled
26			RNAGLAKLGSSLLGAAKSLMGK	22	+	+	-	-	0.86	0.97	-0.20	-0.45	Single helix
27			IVASTMKIIK	10	+	+	-	-	0.81	0.06	-0.32	-0.60	Single helix
28			LAVVGILGLGLLASIAALMRMISYK	25	+	+	-	-	0.91	0.42	-0.43	-1.75	Helix–consisted loop
29			GGGTMGNAGGVGAAK	15	+	+	-	-	0.77	0.83	-0.60	-0.62	Beta–sheet consisted loop
30			HAILLITKGIFK	12	+	+	-	-	0.88	1.60	-0.11	-1.83	Beta–sheet Consisted loop
31			AGWRHAGS	8	+	+	-	-	0.65	1.02	-0.24	-0.73	Short peptide
32			HKGCAMTA	8	+	+	-	-	0.77	1.05	-0.26	-1.01	Short peptide
33			AGAGMHE	7	+	+	-	-	0.69	0.32	-0.29	-0.64	Short peptide
34	F5	3–10 kDa	VKGAPVKTK	9	+	+	-	-	0.95	0.55	-0.17	-1.02	Random coiled
35			FSKGISKTGPK	11	+	+	-	-	0.87	0.11	-0.18	-1.10	Random coiled
36			AKATKPDAK	9	+	+	+	-	0.56	0.11	0.14	-0.32	Random coiled
37			VAGAVTSAK	9	+	+	-	-	0.52	0.52	-0.62	-0.73	Random coiled
38			HSIKNFYLIAKPATKNGR	18	+	+	+	-	0.66	0.06	0.02	-1.40	Helix–consisted loop
39			RGFNVIIK	8	+	+	-	-	0.65	0.76	-0.25	-0.75	Short peptide
40			GCGNS	5	+	+	-	-	0.56	1.15	-0.27	-0.66	Short peptide
41	F5	> 50 kDa	YGGKFVAIK	9	+	+	-	-	0.70	1.84	-0.18	-0.88	Random coiled
42			GNIAILKIMVK	11	+	+	+	-	0.62	0.82	0.02	-1.24	Single helix

substantially depend on its concentration. These results leaded to the hypothesis that peptides or proteins within the F2 and F5 fractions could have different cytotoxic mechanisms against the cells. Modification of these peptides or proteins could enhance specificity to the MCF-7 cells. Thus, the F2 and F5 fractions were selected for peptidomic analysis.

3.2. Peptidomic identification of the selected mucous fractions

The fractions F2 and F5 showed in vitro inhibitive effects on the MCF-7 and Vero cells. These two fractions were then further separated by molecular masses into four sub-fractions: lesser than 3 kDa, between 3 kDa and 10 kDa, between 10 kDa and 50 kDa and larger than 50 kDa. While the <3 kDa sub-fraction was directly subjected to the LC–MS/MS analysis, the other three sub-fractions were tryptic digested in solution prior mass spectrometric analysis. The MASCOT search of the results against the NCBI database did not give significant matched proteins, despite 424 and 404 peptides were detected. Limited genomic and proteomic information of gastropods in the NCBI database caused this unsuccessful protein identification. If the database did not contain the unknown peptides, it should at least find the closest homologue. This

finding was similar to an annotation of a non-model gastropod *Nerita melanotragus* transcriptome that only 15–18% of the contigs were assigned with a putative function [23]. However, amino acid sequences of these peptides were further analyzed by peptidomics. The majority of these peptides (82–84%) were less than 10 kDa. These peptides were blasted against the anticancer peptide database CancerPPD [19] to find similarities to any known anticancer peptides, but no significant hits were found.

Four bioinformatic programs; ACPP, AntiCP, CellPPD and ToxinPred, were then used to predict putative anticancer peptides from the mass spectrometric-detected peptides using our designed workflow in Fig. 1. Forty-three putative anticancer peptides were firstly predicted from the F2 and F5 fractions by both ACPP and AntiCP (Table 1) with the positive prediction scores.

Of these 43 peptides, 23 (5.4%) peptides were from the F2 fraction and 20 (4.9%) peptides were from the F5 fractions. The majority of the putative anticancer peptides of the F2 and F5 fractions were less than 10 kDa with the range of sizes between 5 to 25 amino acids. Analysis of physicochemical properties of the 43 putative anticancer peptides by AntiCP showed a higher average score of hydrophobicity ($p < 0.01$) and a lower average score of hydrophilicity ($p < 0.01$) in compared to

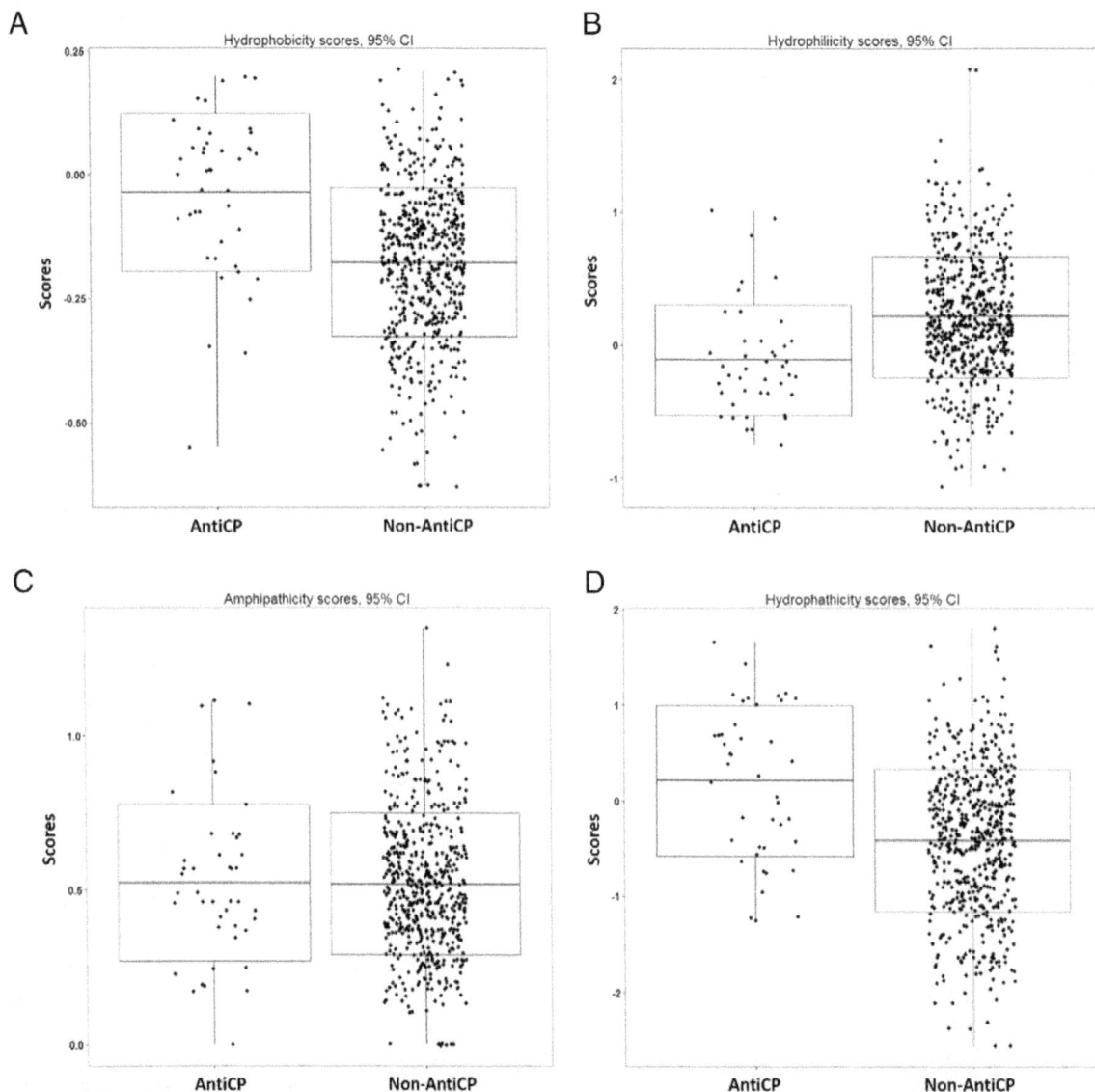

Fig. 5. Physicochemical properties of the putative anticancer peptides compared with the non-anticancer peptides from fractions F2 and F5 of *A. fulica* mucus. Hydrophobicity (A), hydrophilicity (B), amphipaticity (C) and hydrophathicity (D) scores of these peptides were computed by AntiCP. The boxplots were drawn by R program. The scores were shown in the black dots. The uppermost and lowermost ends represented maximum and minimum scores. The middle line indicated means and the two paralleled lines showed standard deviations.

the average scores of 785 non-anticancer peptides from both HPLC-fractions (Fig. 5A and B). These putative cationic anticancer peptides revealed a greater score of hydropathicity (p < 0.01) meaning that they were more hydrophobic (Fig. 5D). Positive scores of amphipathicity were averagely similar to the non-anticancer peptides (Fig. 5C). This amphiphatic property indicated the ability of these peptides to bind and penetrate the breast cancer cell membrane [7,8]. However, this property was not clearly separated between the two groups in this study.

Five of these peptides from both F2 and F5 fractions had cell penetrating ability as predicted by CellPPD and only one peptide from the F2 fraction was toxigenic as predicted by ToxinPred. The CellPPD and ToxinPred programs helped predicting specific characteristics of these peptides. The results could partially explain the cytotoxic mechanisms of the F2 and F5 fractions against the Vero and MCF-7 cells. More complicated situations could be that different peptides worked in combination or the cytotoxicity was also concentration-dependent. The predictions could select candidate peptides for further experimental validations.

Structures of these 43 peptides were simple and could be classified into five structural categories: random coiled, a single helix, helix-consisted loop, β-sheet-consisted loop and short peptide (Fig. 6). The short putative anticancer peptides (less than 9 amino acid residues) were predominantly observed in the F2 fraction, whereas the F5 fraction contained more of the random coiled peptides. Similar observation [24] was shown in the classification of antimicrobial peptides from the ADP database that the majority of the antimicrobial peptides had no known structure (40%). Notably, most of the known structures of these peptides were β-sheet structure with disulfide bonds and helical structure.

As the predictions relied on the support-vector-machine (SVM) score for confident determination of the anticancer peptides by ACPP and AntiCP programs, the threshold of 0.9 was optimally set for the putative anticancer peptide candidates. Applying this stringent criterion narrowed the result down to 16 significantly putative cationic anticancer peptides (Table 1 and Fig. 7). Four short-length (No 0, DTPRCCR; No 19, GYAAGIK; No 20, HANGGVLK and No 21,

GYAAGNK) and two random-coiled (No 1, GGPIAAPEASK and No 13, CVGLGGRGC) peptides were from the F2 fraction, while ten peptides were from the F5 fraction which contained one single helix (No 26, RNAGLAKLGSSLLGAAKSLMGK), one helix-consisted loop (No 28, LAVVGILGLGLLASIAALMRMISYK), one β-sheet-consisted loop (No 30, HAILLITKGIFK), four random-coiled (No 23, HALLIIFNASKK; No 24, VCKALIPGLIPLSFGHGLEPK; No 34, VKGAPVKTK and No 41, YGGKFVAIK) and three short peptides (No 31, AGWRHAGS; No 32, HKGCAMTA and No 40, GCGNS) (Fig. 6). Of these 16 peptides, one random-coiled peptide (No 23) had scores beyond 0.9 from both programs. A recent review had shown that the alpha-helical cationic anticancer peptides were promising anticancer agents with unique mechanisms of cytoplasmic membrane disruption leading to necrosis and apoptotic induction by interruption of mitochondrial membrane [25]. Alteration of amino acid residues of these predicted peptides to improve their net charge, hydrophobicity and helicity could enhance their specific binding to the cancer cells [8]. Therefore our findings could be further optimized by a de novo peptide design and modification.

In discussion, this study examined cytotoxicity of the crude A. fulica mucus and six HPLC-separated fractions against the breast cancer cell line MCF-7 by using the MTT assay and found that the F2 and F5 fractions significantly induced cell death. These fractions also affected the normal Vero cells. As the MTT assay measured cell viability by colorimetric assessing the activity of NADH-dependent oxidoreductase, this study showed the anticancer activity of the A. fulica mucus and the cytotoxic compounds were potentially located in the F2 and F5 fractions. By scanning each HPLC-separated fractions at 210–290 nm, the absorbance at 220 and 280 nm of proteins was clearly observed. This was supported by the protein assay and the mass spectrometric analyses of the F2 and F5 fractions. Therefore, we ascertained that these fractions contained peptides. Previous study suggested that the snail mucus contained complex mixture of proteoglycans, glycosaminoglycans, glycoproteins, hyarulonic acid, small peptides and metal ions [26]. As the crude mucus samples were not processed through enzymatic digestion or heat treatment, the proteoglycan, glycosaminoglycan and hyarulonic

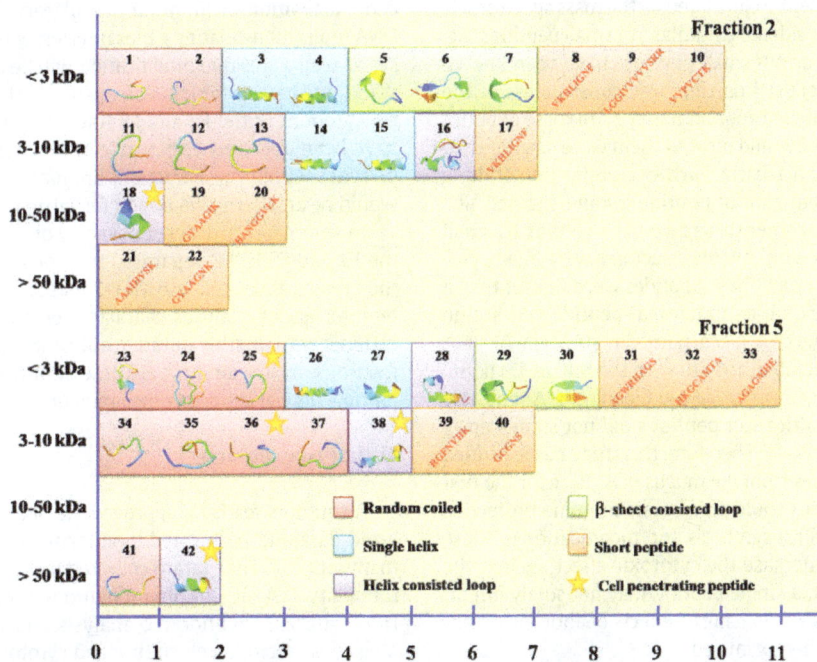

Fig. 6. Structural predictions of the putative anticancer peptides obtained from the F2 and F5 fractions of the A. fulica mucus. These two fractions were separated into four sub-fractions: <3 kDa, 3–10 kDa, 10–50 kDa and >50 kDa. The peptide contents of these sub-fractions were analyzed by protein assay and mass spectrometry. All sub-fractions were trypsinized except the <3 kDa sub-fraction. Putative anticancer peptides and their structures were predicted by bioinformatic programs and shown in this figure. The structures of these peptides were classified into five categories: random coiled, a single helix, helix-consisted loop, β-sheet-consisted loop and short peptide. Putative peptides predicted from the F2 fraction were displayed on the top half, while those of the F5 fraction were shown on the bottom half. The asterisks indicated the cell-penetrating peptides.

```
No 0:    D  T  P  R  C  C  R
No 1:    L  G  G  I  V  V  V  V  V  S  R  R
No 13:   C  V  G  L  G  G  R  G  C
No 19:   G  Y  A  A  G  I  K
No 20:   H  A  N  G  G  V  L  K
No 21:   G  Y  A  A  G  N  K
No 23:   H  A  L  L  I  I  F  N  A  S  K  K
No 24:   V  C  K  A  L  I  P  G  L  I  P  L  S  F  G  H  G  L  E  P  K
No 26:   R  N  A  G  L  A  K  L  G  S  S  L  L  G  A  A  K  S  L  M  G  K
No 28:   L  A  V  V  G  I  L  G  L  G  L  L  A  S  I  A  A  L  M  R  M  I  S  Y  K
No 30:   H  A  I  L  L  I  T  K  G  I  F  K
No 31:   A  G  W  R  H  A  G  S
No 32:   H  K  G  C  A  M  T  A
No 34:   V  K  G  A  P  V  K  T  K
No 40:   G  C  G  N  S
No 41:   Y  G  G  K  F  V  A  I  K
```

- Positively-charged amino acid
- Negatively-charged amino acid
- Polar uncharged amino acid
- Non-polar amino acid

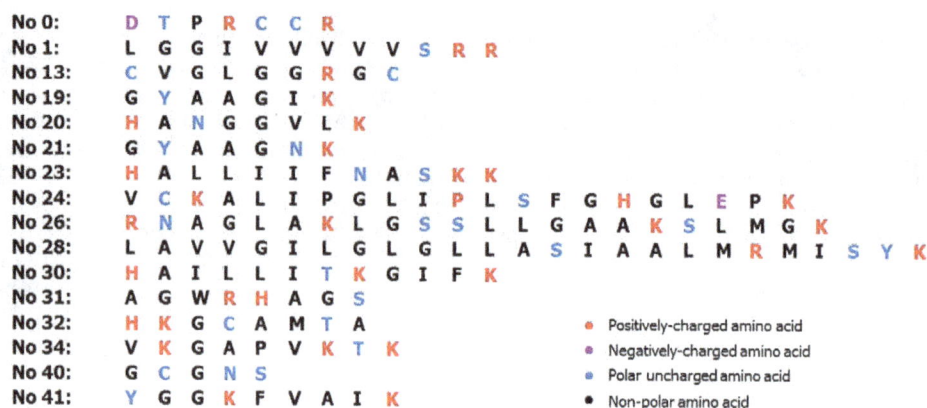

Fig. 7. Amino acid constituents of 16 significantly predicted putative anticancer peptides from the F2 and F5 fractions of the *A. fulica* mucus. Colors represented amino acid properties as positively-charged, negatively-charged, polar and non-polar.

acid contents of the mucus could almost be precipitated after centrifugation, yielding only small soluble peptides, glycoproteins and other small compounds.

The cytotoxic effect of the peptide/protein contents within the F2 and F5 fractions was initially hypothesized as the result of a previously characterized glycoprotein enzyme of *A. fulica* mucus, named achacin. Kanzawa et al. found that the purified achacin from *A. fulica* mucus could induce death of HeLa cells at the IC_{50} of 10 µg/ml [27]. Achacin (59 kDa) is an L-amino acid oxidase which catalyzes oxidative deamination of L-amino acids to α-keto acids, hydrogen per oxide (H_2O_2) and ammonia (NH_3) [28]. The production of H_2O_2 could induce apoptosis by causing membrane blebbling. Achacin could also reduce amount of free amino acids and triggered the caspase pathway causing chromatin condensation and DNA fragmentation [27,29]. Although an amino acid sequence of achacin is available in the protein database, proteomic analyses of these two fractions did not detect any peptide fragments of achacin. Therefore, other small peptides could be the cause of these cytotoxic effects.

The peptidomic contents of the F2 and F5 fractions were subfractioned by molecular sizes before elucidated by the mass spectrometric analysis which revealed considerable numbers of small peptides. Little information about these various small peptides have been known because the majority have focused on the large constituents of the *A. fulica* mucus. This study predicted approximately 5% of these peptides were putative anticancer peptides and most of them were smaller than 10 kDa, considerably tinier than achacin. Further classification of these anticancer peptides after prediction of peptide folding showed five structural groups. Bringing together these two parts so that 16 small cationic amphipathic peptides with variable structures were finally predicted from the F2 and F5 fractions. These peptides were shorter than a previously reported cysteine-rich antimicrobial peptide (80 amino acids, 9.7 kDa) from the mucus of *A. fulica*, named mytimacin-AF [14]. Mytimacin-AF showed antimicrobial activity with the MIC of 1.9 µg/ml and little hemolytic activity against human red blood cells. Amino acid sequences of our predicted anticancer peptides did not significantly share similarity with mytimacin-AF. Therefore, this study has identified the smallest anticancer peptides from the mucus of *A. fulica* for the first time. These anticancer peptides could play innate immune protective roles in the *A. fulica*. Biochemical synthesis and modification of these peptides are in progress to increase the cytotoxic effect against the breast cancer cells. Moreover, a single peptide may not solely inhibit the growth of the MCF-7 cells, combinatorial effects of anticancer peptides and proteins will be further examined.

Several molluscan small bioactive peptides have been previously reported since 1996 in *Mytilus* sp., *Crassostrea* sp., *Ruditapes philippinarum*, *Biomphalaria glabrata*, *Argopecten irradians*, *Bathymodiolus azoricus*, *Chlamis farreri*, *Haliotis* sp., *Mercenaria mercenaria*, *Littorina littorea*, *Hyriopsis cumingii*, *Venerupis philippinarum* and *A. fulica* [30], but only a few peptides were identified from gastropod species such as antimicrobial peptides in *B. glabrata*, littorein in *L. littorea* and Mytimacin-AF in *A. fulica* [14]. Most of these peptides were small cationic cysteine-rich amphipathic molecules. The 16 predicted anticancer peptides in this study were also small cationic molecules with amphipathic properties. A proline-rich antimicrobial peptide from the Chilean scallop *A. purpuratus* was an example of non-cysteine-rich bioactive peptides [30]. The anticancer peptides are potential targets over other chemicals for new therapeutic agents due to their low molecular masses, simple structures, specific cytotoxicity to cancer cells over normal cells, various routes of administration, and lower the risk of drug resistance induction [31]. These peptides have diverse mechanisms of action as reviewed by Mulder et al. [31]. The cationic anticancer peptides can bind to the negatively-charged membrane of the cancer cells and disrupt the membrane stability and fluidity, such as pore formation by the carpet model or barrel-stave model, or increase of calcium ion influx, leading to cell death. Some of these peptides may inhibit cancer cell growth by modification of lysosomal membrane, enhancement of proteasome activity, induction of mitochondrial pathway of apoptosis by the caspase cascade, activation of immune modulatory pathway, and inhibition of DNA replication-relating genes interfering the cell cycle. Detailed analyses as well as residue modification will be conducted for better understanding the cytotoxic mechanisms of these 16 peptides. However, tiny amount of the F2 and F5 fractions as well as their sub-fractions have been obtained, exhaustive experimental screening of the active peptides within these fractions are hardly difficult. Synthetic peptides would be an alternative option for further analysis.

In summary, anticancer property of small peptide contents within the F2 and F5 HPLC-separated fractions from *A. fulica* mucus against the breast cancer cell line MCF-7 was shown. Mass spectrometric and bioinformatics analyses characterized and predicted relatively small cationic amphipathic putative anticancer peptide candidates in these fractions, for the first time. These peptides will be promising targets for new anticancer drug development.

Author contributions

All authors read and approved the final manuscript. T. E-kobon analyzed data and performed bioinformatics predictions and wrote the manuscript; P. Thongaram collected mucus fraction and conducted cytotoxicity test; P. Chumnanpuen designed experiments, performed HPLC analysis, peptidomic analysis and revised the manuscript; L. Meesuk assisted in cell culture and cytotoxicity test; S. Roytrakul performed LC–MS/MS analysis and assisted in cytotoxicity test.

Acknowledgments

This work was supported by Kasetsart University Research and Development Institute (KURDI, grant number KURDI-2.56), Faculty of Science, Kasetsart University, (grant number KURDI-02-2556) the Development and Promotion of Science and Technology Talent Project (DPST, grant number DPST-SC-KU-013-2556), and Do Day Dream Co., Ltd. (grant number SNAILWHITE-01-2558). The Central Laboratory of Department of Zoology facilitated mucus extraction and separation. J. Sapkaew and internship students provided assistance with mucus sample preparation. Assoc. Prof. Ratana Banjerdpongchai at the Department of Biochemistry, Faculty of Medicine, Chiang Mai University kindly provided the MCF-7 cell lines. Prof. Sittichai Koontongkaew and laboratory members at the Faculty of Dentistry, Thammasat University provided assistance with the cell line culture and cytotoxicity assays.

References

[1] Schnitt SJ. Classification and prognosis of invasive breast cancer: from morphology to molecular taxonomy. Mod Pathol 2010;23(Suppl. 2):S60–4.
[2] Goldstein NSZ, C. R. Risk factors and risk assessment. Early Diagn Treat Cancer Ser: Breast Cancer 2011:55–69.
[3] Mantyh PW, Clohisy DR, Koltzenburg M, Hunt SP. Molecular mechanisms of cancer pain. Nat Rev Cancer 2002;2:201–9.
[4] Ghafoor A, Jemal A, Ward E, Cokkinides V, Smith R, Thun M. Trends in breast cancer by race and ethnicity. CA Cancer J Clin 2003;53:342–55.
[5] Harris F, Dennison SR, Singh J, Phoenix DA. On the selectivity and efficacy of defense peptides with respect to cancer cells. Med Res Rev 2013;33:190–234.
[6] Schweizer F. Cationic amphiphilic peptides with cancer-selective toxicity. Eur J Pharmacol 2009;625:190–4.
[7] Dennison SR, Whittaker M, Harris F, Phoenix DA. Anticancer α-helical peptides and structure/function relationships underpinning their interactions with tumour cell membranes. Curr Protein Pept Sci 2006;7:487–99.
[8] Y-b Huang, Wang X-f, Wang H-y, Liu Y, Chen Y. Studies on mechanism of action of anticancer peptides by modulation of hydrophobicity within a defined structural framework. Mol Cancer Ther 2011;10:416–26.
[9] Hilchie AL, Doucette CD, Pinto DM, Patrzykat A, Douglas S, et al. Pleurocidin-family cationic antimicrobial peptides are cytolytic for breast carcinoma cells and prevent growth of tumor xenografts. Breast Cancer Res 2011;13:R102.
[10] Lim KJ, Sung BH, Shin JR, Lee YW, Kim DJ, et al. A cancer specific cell-penetrating peptide, BR2, for the efficient delivery of an scFv into cancer cells. PLoS One 2013; 8, e66084.
[11] Yang L, Cui Y, Shen J, Lin F, Wang X, et al. Antitumor activity of SA12, a novel peptide, on SKBr-3 breast cancer cells via the mitochondrial apoptosis pathway. Drug Des Devel Ther 2015;9:1319–30.
[12] Tyagi A, Kapoor P, Kumar R, Chaudhary K, Gautam A, et al. In silico models for designing and discovering novel anticancer peptides. Sci Rep 2013;3.
[13] Vijayakumar S, Ptv L. ACPP: a web server for prediction and design of anti-cancer peptides. Int J Pept Res Ther 2015;21:99–106.
[14] Zhong J, Wang W, Yang X, Yan X, Liu R. A novel cysteine-rich antimicrobial peptide from the mucus of the snail of Achatina fulica. Peptides 2013;39:1–5.
[15] Hancock WS. Handbook of HPLC for the separation of amino acids, peptides and proteins, Vols. I and II. Boca Raton, FL: CRC Press; 1984(512 pp.).
[16] Freshney RI. Culture of animal cells. New York: Wiley-Liss; 1994.
[17] Lowry OH, Rosebrough NJ, Farr AL, Randall RJ. Protein measurement with the Folin phenol reagent. J Biol Chem 1951;193:265–75.
[18] Perkins DN, Pappin DJC, Creasy DM, Cottrell JS. Probability-based protein identification by searching sequence databases using mass spectrometry data. Electrophoresis 1999;20:3551–67.
[19] Tyagi A, Tuknait A, Anand P, Gupta S, Sharma M, Mathur D, et al. CancerPPD: a database of anticancer peptides and proteins. Nucleic Acids Res 2015;43:D837–43.
[20] Gupta S, Kapoor P, Chaudhary K, Gautam A, Kumar R, et al. In silico approach for predicting toxicity of peptides and proteins. PLoS One 2013;8, e73957.
[21] Gautam A, Chaudhary K, Kumar R, Sharma A, Kapoor P, et al. In silico approaches for designing highly effective cell penetrating peptides. J Transl Med 2013;11:74.
[22] Thévenet P, Shen Y, Maupetit J, Guyon F, Derreumaux P, et al. PEP-FOLD: an updated de novo structure prediction server for both linear and disulfide bonded cyclic peptides. Nucleic Acids Res 2012;40:W288–93.
[23] Amin S, Prentis P, Gilding E, Pavasovic A. Assembly and annotation of a non-model gastropod (Nerita melanotragus) transcriptome: a comparison of de novo assemblers. BMC Res Notes 2014;7:488.
[24] Wang Z, Wang G. APD: the antimicrobial peptide database. Nucleic Acids Res 2004; 32:D590–2.
[25] Huang Y, Feng Q, Yan Q, Hai X, Chen Y. Alpha-helical cationic anticancer peptides: a promising candidate for novel anticancer drugs. Mini Rev Med Chem 2015;15: 73–81.
[26] Smith AM, Robinson TM, Salt MD, Hamilton KS, Silvia BE, et al. Robust cross-links in molluscan adhesive gels: testing for contributions from hydrophobic and electrostatic interactions. Comp Biochem Physiol B Biochem Mol Biol 2009;152:110–7.
[27] Kanzawa N, Shintani S, Ohta K, Kitajima S, Ehara T, et al. Achacin induces cell death in HeLa cells through two different mechanisms. Arch Biochem Biophys 2004;422: 103–9.
[28] Thorpe GW, Reodica M, Davies MJ, Heeren G, Jarolim S, et al. Superoxide radicals have a protective role during H2O2 stress. Mol Biol Cell 2013;24:2876–84.
[29] Alarifi S. Assessment of MCF-7 cells as an in vitro model system for evaluation of chemical oxidative stressors. Afr J Biotechnol 2011;10:3872–9.
[30] Li H, Parisi MG, Parrinello N, Cammarata M, Roch P. Molluscan antimicrobial peptides, a review from activity-based evidences to computer-assisted sequences. Isj-Invertebr Surviv J 2011;8:85–97.
[31] Mulder KCL, Lima LA, Miranda VJ, Dias SC, Franco OL. Current scenario of peptide-based drugs: the key roles of cationic antitumor and antiviral peptides. Front Microbiol 2013;4:321.

18

Network analysis of human post-mortem microarrays reveals novel genes, microRNAs, and mechanistic scenarios of potential importance in fighting huntington's disease

Sreedevi Chandrasekaran *, Danail Bonchev

Center for the Study of Biological Complexity, Virginia Commonwealth University, Richmond, VA, USA

ARTICLE INFO

Keywords:
Huntington's disease
Computational molecular neurobiology
Transcriptome
Protein interaction network
miRNAs
Microarray analysis

ABSTRACT

Huntington's disease is a progressive neurodegenerative disorder characterized by motor disturbances, cognitive decline, and neuropsychiatric symptoms. In this study, we utilized network-based analysis in an attempt to explore and understand the underlying molecular mechanism and to identify critical molecular players of this disease condition. Using human post-mortem microarrays from three brain regions (cerebellum, frontal cortex and caudate nucleus) we selected in a four-step procedure a seed set of highly modulated genes. Several protein–protein interaction networks, as well as microRNA–mRNA networks were constructed for these gene sets with the Elsevier Pathway Studio software and its associated ResNet database. We applied a gene prioritizing procedure based on vital network topological measures, such as high node connectivity and centrality. Adding to these criteria the guilt-by-association rule and exploring their innate biomolecular functions, we propose 19 novel genes from the analyzed microarrays, from which *CEBPA, CDK*1, *CX3CL1, EGR1, E2F1, ERBB2, LRP1, HSP90AA*1 and *ZNF148* might be of particular interest for experimental validation. A possibility is discussed for dual-level gene regulation by both transcription factors and microRNAs in Huntington's disease mechanism. We propose several possible scenarios for experimental studies initiated via the extra-cellular ligands TGFB1, FGF2 and TNF aiming at restoring the cellular homeostasis in Huntington's disease.

1. Introduction

In 1872, a young American physician named George Huntington was the first to recognize a specific inherited neurodegenerative disorder. Later it was named after him as Huntington's disease (HD) and in early 1990s the mutant gene Huntingtin (HTT) was discovered to be the cause of the disease. Expansion of 36 or more CAG trinucleotide (polyQ) repeats in HTT gene is the hallmark characteristic of the disease. PolyQ expanded HTT is considered as a trigger of the neurodegeneration that eventually caused all the Huntington's disease symptoms. The disease is described by progressive motor, cognitive as well as emotional disturbances. The motor symptoms include chorea, dystonia, rigidity, postural instability, etc. Depression and personality changes are the major emotional disturbances part of the disorder. Like Alzheimer's disease (AD), short-term memory loss, confusion and disorientation are some of the cognitive issues found in HD patients.

In Huntington's disease, it was suspected that the neurodegeneration is selective for striatal GABAergic medium-sized spiny neurons.

These neurons project to substantia nigra and globus pallidus parts of the brain affecting primarily motor coordination [1–3]. Fig. 1 depicts HTT gene along with its interacting partners which trigger the striatal neuronal loss that eventually manifests all the Huntington's disease symptoms. Like the other neurodegenerative disorders Parkinson's disease and Alzheimer's disease, HD also showed protein misfolding, ubiquitin proteasome system deregulation, autophagy dysfunction, metabolic and mitochondrial dysfunction as well as oxidative stress, which over the years culminates into motor and cognitive disorders [4–7]. The advances in understanding the molecular pathogenesis of HD, and the multiple research approaches undertaken since the discovery of the HD gene in 1993, are reviewed by Zuccato et al. [8]. Mattson summarized the evidence for the role of apoptosis and related pathways of oxidative stress, perturbed calcium homeostasis, mitochondrial dysfunction and caspases activation in neurodegenerative diseases. His analysis involved the protecting survival signals, which suppress oxygen radicals and stabilize calcium homeostasis and mitochondrial function [9].

In addition to HTT, mutations in HDL3, JPH3 and PRNP genes were also related to Huntington's disease pathogenesis (OMIM database, retrieved on Dec. 17, 2012). Other genes such as CCKBR, cytochrome c and GAPDH were known to contribute to the diseased state. It was

* Corresponding author at: Center for the Study of Biological Complexity, VCU, Harris Hall Room No. 3132, 1015 Floyd Ave., Richmond, VA 23284, USA
E-mail address: sreedevi.c@gmail.com (S. Chandrasekaran).

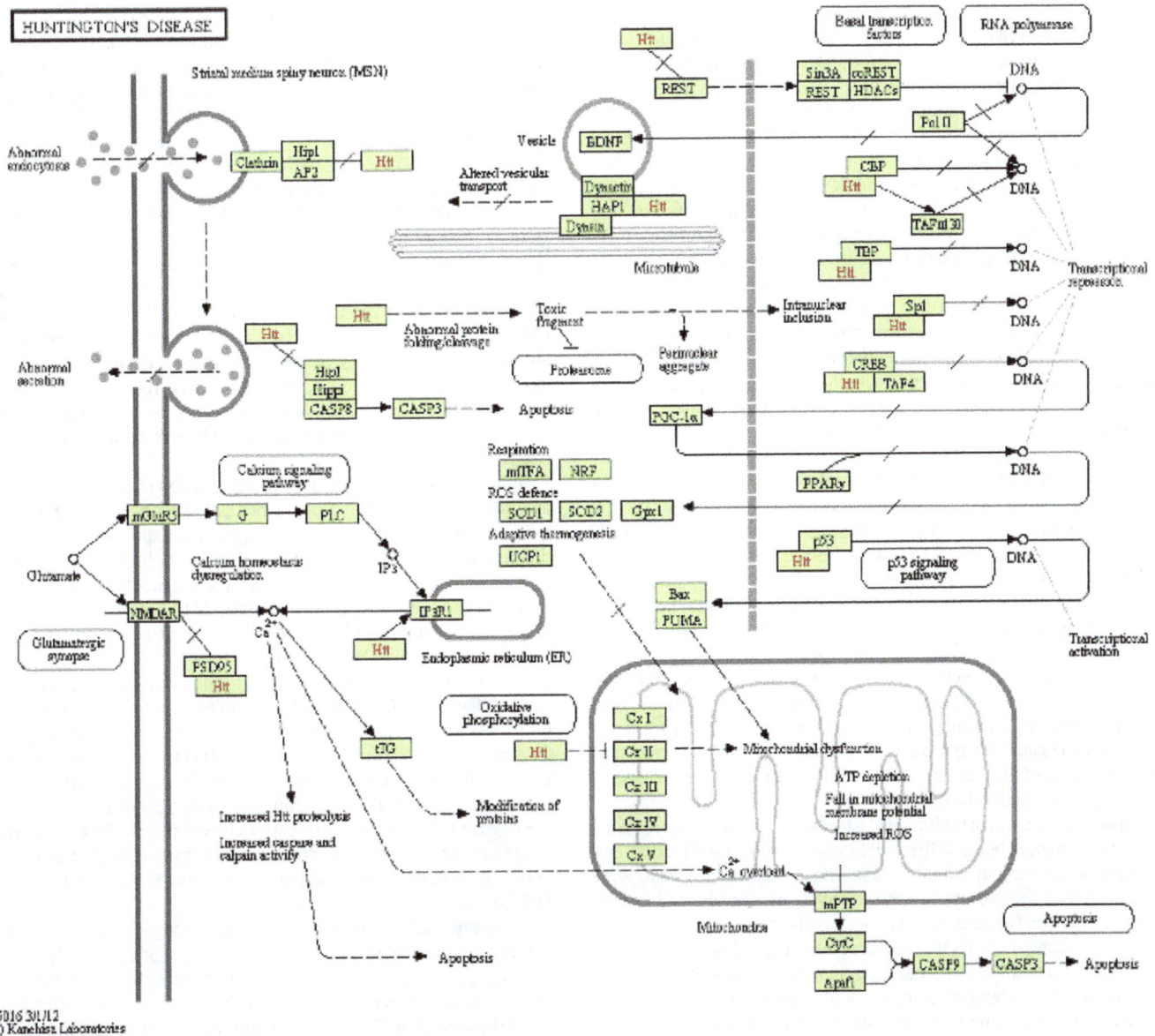

Fig. 1. Huntington's disease pathway from KEGG database. Biological processes and genes implicated in the Huntington's disease. Courtesy: Huntington's disease pathway from KEGG database, available at http://www.genome.jp/kegg/pathway/hsa/hsa05016.html retrieved on Apr. 3, 2013.

reported that there is a possible selective loss of Cholecystokinin receptors (CCKBR) containing neurons in cerebral cortex of Huntington's patients [10]. Cytochrome c release from mitochondria triggers the downstream caspase activation leading to apoptotic neuronal death in many neurodegenerative diseases. This kind of neuronal death plays a greater role at the end stage of HD [11]. Due to its selective binding to the CAG repeats in huntingtin gene, GAPDH activity was found reduced in HD brains, thereby reducing the cellular energy production [12,13].

The role of FOXO gene for preserving the normal neuronal capacity, a role that is suppressed by the Wnt receptor Ryk in HD was studied by Tourette et al. [14]. Such models also helped to identify genes that regulate the dysfunction of mutant polyglutamine neurons [15]. Brain-derived neurotrophic factor (BDNF) has been suggested to reduce amyloid-β neurotoxicity in Alzheimer's disease [16–18]. Some of the suggested beneficial pathways up-regulate the innate autophagy process and via increasing BDNF gene expression. Autophagy process protects against the toxic insults of mutant huntingtin proteins by enhancing its clearance from the cell. BDNF is necessary for the survival of striatal neurons in the brain and it promotes synaptic plasticity in addition to memory formation. It can also act as a neuromodulator affecting the pre-

synaptic release of neurotransmitters in central nervous system. Along with BDNF, genes like BCL2 and G-protein coupled receptors (GPCRs) could also be part of therapeutic measures in Huntington's disease. Human antioxidant defense proteins that were strongly induced in striatum, but also detectable in cortex, were identified as peroxiredoxins 1, 2, and 6, as well as glutathione peroxidases 1 and 6 [19].

Huntington's disease research benefited considerably from animal models (*C. elegans* and particularly mouse). These models expand the knowledge base of the disease, act as a valuable experimental tool for preclinical therapeutic trials, and provide biomarkers of disease progression which could be detectable in skin, muscle, blood or other peripherally accessible tissue [20,21]. This resulted in few valuable therapeutic measures to alleviate the disease symptoms. HD animal model data suggest a possible interaction between genetic and environmental factors [22]. Novel transcriptional changes in several genes that were involved in synaptic integrity and function were found in HD mice [23]. Another HD animal model study demonstrated that the mutant huntingtin directly or indirectly reduces the expression of a distinct set of genes involved in signaling pathways that were known to be critical for the functioning of striatal neurons [24]. When administered systemically or delivered via genetically-grafted cells, BDNF has shown to prevent striatal neurons from cell death in HD animal models [25]. FGF2 (fibroblast growth factor 2) was shown to improve motor performance and extend the lifespan by 20% by reducing the accumulation of polyQ aggregates in the brain [26, 27]. Zhang et al. [28] found that mouse caspase activation precedes pro-apoptotic changes in Bcl-2 family members. Understanding the chronology of apoptotic events provides important information for appropriate therapeutic targeting in this devastating and untreatable disease. Decreasing IRS2 (insulin receptor substrate 2) signaling could be part of a therapeutic approach to slow down the progression of HD [29]. Thier abundant presence in central nervous system as well as their complex interactions with many downstream targets have made GPCRs potential drug targets in many neurological diseases including Huntington's disease [30]. The transcription factor XBP1 deficiency was found to lead in animal studies to augmented expression of FOXO1, a key transcription factor regulating autophagy in neurons [31].

Recently, three papers used one of the two microarray datasets analyzed in our study (GSE3790 with Affymetrix GeneChip Human Genome HG-U133A). Neueder and Bates [32] uncovered previously unidentified transcription dysregulation in the HD cerebellum, and found that genes implicated in mitochondrial function, glycolysis, intracellular protein transport, proteasome and synaptic vesicles are commonly negatively correlated with HD in the cerebellum, frontal and caudate networks. Studying molecular mechanisms of HD Kalathur et al. [33] found indications for potential relevance of the cell cycle processes, RNA splicing, Wnt and ErbB signaling, and proposed a candidate set of 24 novel genetic modifiers. Alcaraz et al. [34] used the GSE3790 dataset in one of the case studies to prove the efficiency of their KeyPathwayMiner computational tool. An interesting moment in their analysis is that some of the proposed new HD-relevant genes (termed "exception" genes) are statistically insignificant. This approach (although not rigorously defined) parallels one part of the strategy of the present authors used in preceding articles [35–37] under the name "connecting" proteins (vide infra).

Since its discovery, a multitude of genetic and biomolecular research studies have contributed valuable information about the Huntington's disease underlying mechanism and the critical genes that were deregulated in the process. In this research work, we have utilized extensive network techniques to further expand the knowledge base of the Huntington's disease's biological mechanisms and the vital molecular players. Network-based analysis is a useful tool to understand and appreciate the underlying complexity of a system as a whole rather than a disconnected unit. This research work on Huntington's disease is part of a comprehensive study to understand the underlying common molecular mechanisms and genes involved in neurodegenerative diseases including Parkinson's disease and Alzheimer's disease. Our more important network-based findings about the latter two diseases can be found in Refs. [36,37].

2. Material and methods

The study rationale and workflow are similar for all three neurodegenerative disorders (NDDs) (Parkinson's, Alzheimer's and Huntington's diseases) that we chose to study in understanding the underlying common biomolecular mechanisms and genes. Detailed information about the study methodology is presented in our Parkinson's disease journal article [36]. Below, we describe the various steps involved in network-based analysis of Huntington's disease.

Microarray gene expression data is the fundamental means to carry out network-based analysis. Such datasets are available in major public data repositories like National Center of Biotechnology Information's (NCBI) Gene Expression Omnibus (GEO) and European Bioinformatics Institute's (EBI) ArrayExpress databases. For uniformity and cross-validation, Affymetrix microarray gene expression datasets were used for all three NDDs. For Huntington's disease, NCBI GEO's GSE3790 dataset [38] which contained post-mortem human brain tissue samples from both HD patients and controls (44 and 36 samples, respectively)

taken from three brain regions namely cerebellum, frontal cortex and caudate nucleus was used. For a detailed information about the samples, refer to Refs. [38,39]. The microarray expression dataset was then normalized using the Robust Multi-array Average (RMA) approach [40]. The differential gene expression changes were statistically evaluated by the empirical Bayes (eBayes) method from the limma Bioconductor package [41,42]. Probe-sets with p-values of <0.05 were considered to be significantly differentially expressed genes (SDEGs).

The GSE3790 microarray gene dataset had both Affymetrix GeneChip Human Genome HG-U133A and B. Following the above mentioned statistical plan, the two datasets were analyzed and the significantly differentially expressed genes lists called "seed genes" were generated. The lists generated from the GSE3790 dataset were denoted as CE, FL and CN, for the three types of brain tissue samples: cerebellum, frontal cortex and caudate nucleus, respectively. In addition, a "diagnosis" differential genes list was generated utilizing *all* the 201 microarray samples in which there were 87 controls and 114 HD cases. By including the diagnosis set, we were able to filter our SDEGs in a little more stringent and concise way.

A four-step strategy was applied for selection of "seed" genes in our analysis. It starts with the strongly modulated genes in each of the three types of brain tissues, then identifying the overlapping genes between different tissues from the *same microarray*, next eliminating the redundant genes from the overlap between the selected sets of the *two microarrays*. In GSE3790 HG-U133A, an overlap of 617 seed genes was found between the four sets of significantly differentially expressed genes (SDEGs), as shown in Fig. 2a. In a similar way, an overlap of 351 seed genes was found in the GSE3790 HG-U133B microarray gene expression dataset (Fig. 2b). Altogether, 925 seed genes were found after removing duplicates in GSE3790 U133A and B datasets. In order to partially compensate for not accounting for the multiple correlation, and still have a considerable number of overlapping genes, the fourth selection step included in the network evaluation only those SDEGs with the p-value lower than 0.01. Following this new cut-off criteria, 531 genes were treated as "seed genes" which were then subjected to comprehensive network analysis (See Supplementary Table S1 for the list of 531 SDEGs).

Pathway Studio 9.0 software package (http://www.elsevier.com/online-tools/pathway-studio) along with its proprietary molecular interaction database namely ResNet 9.0 (released October 15, 2011) was utilized to construct various networks such as *direct* interaction, *shortest-path* and *miRNA* regulation [43]. In addition, critical network topological characteristics such as node degree (local connectivity), closeness centrality (network monitoring) and betweenness centrality (traffic-influential) scores were also calculated using the Pajek software package [44,45]. Node degree is defined through the number of nearest neighbors in the network. The larger this number, the stronger the influence of this network node on its neighborhood in case of positive or negative modulation. Closeness centrality is defined as reciprocal of the sum of distances from a node to all other nodes in the network. High closeness centrality means more effective monitoring the network from a given node. Betweenness centrality measures another aspect of central location—the possibility to influence a larger portion of the network node–node communications. In short, these three measures of local, respectively global connectivity and centrality determine the speed with which harmful or/and beneficial signals will be transmitted through the entire system. Based on these prioritizing topological characteristics, as well as on their biological/molecular functions relevant for the neurodegenerative process, the seed genes were categorized as "already known HD-genes" and "genes of interest for HD". Such categorization was possible after careful review of various sources like Online Mendelian Inheritance in Man (OMIM) database (http://omim.org/), NCBI's PubMed database (http://www.ncbi.nlm.nih.gov/pubmed.com), MalaCards database (http://malacards.org/), and Google search for the latest publications (http://www.google.com). Later, the two categories were further divided in two subcategories, those found among the

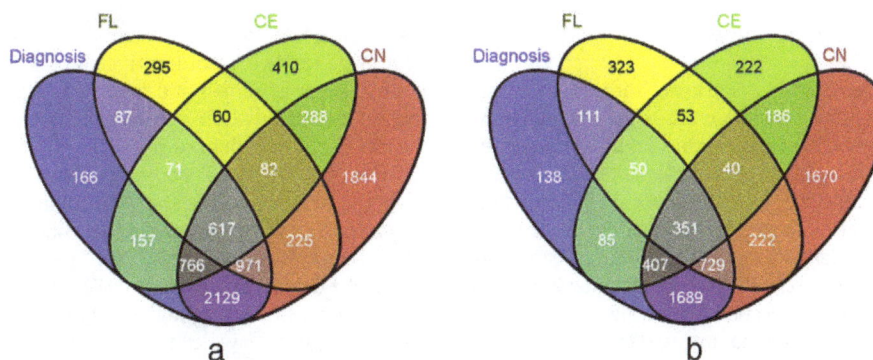

Fig. 2. (a) and (b). Four-set Venn diagram of the overlap of significantly differentially expressed genes (SDEGs) in (a) GSE3790 HG-U133A and (b) GSE3790 HG-U133B gene expression datasets. Courtesy: Oliveros, J.C. (2007–2015) Venny. An interactive tool for comparing lists with Venn's diagrams. http://bioinfogp.cnb.csic.es/tools/venny/index.html.

significantly differentially expression genes (SDEGs) and such emerging from the connecting proteins in *shortest-path* network.

Further the seed genes were subjected to Gene Ontology (GO) enrichment analysis using Database for Annotation, Visualization and Integrated Discovery (DAVID), a widely used Web-based application focusing on GO classification [46–48]. GO enrichment analysis provides biological functional interpretation of large lists of genes derived from genomic studies such as microarray, proteomics experiments, etc. We also used the core analysis in Ingenuity's IPA (Ingenuity Systems, http://www.ingenuity.com/) and Pathway Enrichment Analysis in Pathway Studio to explore the various canonical pathways that could be affected in Huntington's disease.

In conclusion, an *integrated* mechanistic disease network was constructed using the genes/proteins found in common in all enriched Kyoto Encyclopedia of Genes and Genomes (KEGG) pathways resulted from DAVID analysis [49]. Again, Pathway Studio software was used to construct the *direct interaction* network using the "mechanism genes" (genes found in common in all enriched KEGG pathways) to investigate the *integrated* Huntington's disease mechanism.

3. Results and discussion

Using the GSE3790 dataset, the original authors had performed gene set enrichment analysis using DAVID tools to identify biological processes and pathways significantly affected in HD [39]. One of the major conclusions of this study was that the differential gene expression in HD brains showed distinct regional pattern similar to an already known pattern of neuronal loss. The greatest number and magnitude of differentially expressed genes/proteins were detected in the caudate nucleus, followed by motor cortex, then in cerebellum tissue types.

Our statistical analysis with renormalized data revealed a similar gene expression difference in these neuronal tissue samples. In this study, we expanded upon their research work by subjecting the differentially expressed genes to various network techniques to explore the underlying cellular mechanisms and molecular players of Huntington's disease. We initiated our HD network analysis using 531 "seed genes".

3.1. Huntington's disease direct interaction network

Constructing networks using larger gene set such as 531 "seed genes" are beneficial in a couple of ways. First, one could have a broader view of the network neighborhood of any gene of interest along with all its complex interactions. Next, one could also shrink the network size in order to have a closer look on the proximity of those nodes that are critical. Fig. 3 shows the primary *direct* interaction (DI) network of the 531 Huntington's disease "seed genes". In this network, 224 of these genes *directly* interact with each other and it included such types of interactions like regulation, physical binding, co-expression, promoter binding, protein modification, molecular transport and direct regulation.

After exhausted literature review, we found in the direct interaction network 26 genes that have already been associated to Huntington's disease (some of them are discussed in Introduction). Another eight genes (CNTNAP1, CX3CL1, DPYSL5, FDFT1, FGFR1, FKBP5, RCAN2 and ZNF148) were identified as being of potential interest in Huntington's disease pathogenesis due to their specific molecular functions as detailed below. More details on how we conducted our literature search and gene classification are given in methods and data section above, as well as in our Parkinson's paper [36]. We constructed a *compact* direct interaction network using these 34 genes/proteins to understand their inter-connectivity, and to validate the eight genes of interest by the guilt-by-association rule. The network shown in Fig. 4 clearly reveals that except for FKBP5 and DPYSL5, all other genes/proteins of potential interest are connected to the HD-known ones.

Their innate physiological roles along with their vital network attributes, increases the chance of the eight candidate genes to be involved in the HD pathology, elucidating the neuroprotective mechanisms in Huntington's disease realm. For instance, fractalkine (CX3CL1) is a known Parkinson's disease gene where it exhibited neuro protective role against microglia activation as well as reduced motor coordination impairment [50,51]. Being a known neuroprotective agent for a similar neurodegeneration disease with movement disorder, it may have a potential therapeutic role in Huntington's disease domain too. As mentioned earlier, FGFs were proposed to improve the motor performance and to extend the lifespan in HD mouse model study. Being a receptor for fibroblast growth factors, FGFR1 could up-regulate FGF's beneficial activities in the cell. Studies have found that PPAR-γ together with PGC-1α (a transcriptional co-activator) is required for the regulation of mitochondrial biogenesis. PPAR-γ agonists are thought to be neuroprotective in amyotrophic lateral sclerosis (ALS) and HD [52]. Currently there is no cure available for HD patients. In a search for such a cure, zinc finger proteins were designed in such a way that these proteins were able to recognize and bind specifically with CAG repeats in mouse DNA [53]. The study reported that there was considerable reduction of the mutant huntingtin gene expression at both protein and mRNA levels (95% and 78% reduction, respectively). Many zinc finger proteins including ZBTB10, ZFP36L1 and ZNF148 were found significantly differentially expressed in the microarray dataset used in our study. These zinc finger proteins among which ZNF148 directly interacts with three known-HD genes (BCL2, CASP6 and IRS2) could emerge as a promising new gene therapy tool for Huntington's disease which could be extended and tested in human HD patients.

A number of the candidate genes shown in Fig. 4 are *direct* interacting partners of many already HD-associated genes. The CX3CL1 interacts with five already known-HD genes namely, BCL2, CLU, FGF2, GAPDH and PPARA. The modulation of these previously known-HD genes have been shown beneficial in reducing the disease pathogenesis. This makes this protein a significant player in the HD-related biomolecular mechanisms.

3.2. Huntington's disease shortest path network (SPNW)

Different from the previous *direct* interaction network type, *shortest-path* network help us to identify indirect protein–protein interactions that take place through intermediary nodes in the absence of direct relationship. With this in mind, we built the *shortest-path* network using only those seed genes which had at least 25 neighbors in Pathway Studio ResNet 9.0 database (in order to have concise and yet meaningful network). Two hundred fifty-eight out of 531 seed genes met this cut-off criteria and the resulted *shortest-path* network included 208 Pathway Studio software-added connecting genes. Following our methodology, we categorized these connecting genes into two groups. The genes that were already implicated in Huntington's disease belong to the known-HD genes group and those genes which could be of potential interest in HD due to their cellular functions were grouped separately. Table 1 shows the different categories and the number of genes in each.

We constructed a *compact* shortest-path network (CSPNW), using the 85 genes from Table 1 along with few additional connecting genes that were needed to have a unified well-connected network. The average node degree of this *compact* shortest-path network was 7.10. In this CSPNW, many of the known-HD genes such as AKT1, AR, BCL2, INSR and SP1 were among the top 25 nodes with node degree ≥10, as well as the top 25 with highest closeness (network monitors) and betweenness (traffic influential) centrality measures. Fig. 5 illustrates the interactions between the known and the genes of interest in the HD *compact* SPNW.

Interestingly, PRNP gene/protein was not statistically differentially expressed in our microarray dataset, but it emerged as connecting gene/protein in the *shortest-path* network. PRNP (prion protein) is a glycoprotein that tends to aggregate into rod-like structures causing neuronal cell death. Prion proteins have been associated with many neurodegenerative disorders including Huntington's, Creutzfeldt–Jakob diseases in human, and "mad cow" disease in cattle [54–56]. We found that PRNP was of importance for network topology as one of the top 15 nodes with highest visibility (closeness) and most influence (betweenness) in the *compact* shortest-path network.

In the next few paragraphs, we summarize the innate molecular characteristics of various genes that could be of potential interest in HD pathogenesis, in addition to their "guilt-by-association" relationship to some of the already implicated HD genes. Table 2 lists our proposed Huntington's disease candidate genes along with the number of known-HD genes to which they directly interact with (see Fig. 5).

From the *compact* shortest-path network, we found that EGR1 (early growth response 1) was the nearest interacting partner to an unusually high number (13) of previously known HD genes (AKT1, AR, BCL2, CLU, CTNNB1, CYCS, FGF2, MAOB, MMP9, SOD1, SP1, TGFB1 and TP53), which according to the "guilt-of-association" rule makes it gene of considerable interest for the HD-disease mechanisms. BIM (BCL2-like 11 apoptosis facilitator) plays an important role in neuronal apoptosis, a hallmark feature of many neurological diseases including Alzheimer's and Parkinson's diseases. Previous research study had demonstrated that EGR1 directly transactivate BIM gene expression to promote neuronal apoptosis. EGR1/BIM pathway has been suggested as a pro-apoptotic mechanism in neurological diseases. Mithramycin A, a U.S. Food and Drug Administration clinically approved drug has been studied to improve motor symptoms and extend life span in a mouse model of Huntington's disease. This drug was suspected to exploit the EGR1/BIM pathway to promote neuroprotective mechanism in HD models and thus could be a promising drug for the treatment for the same [57,58].

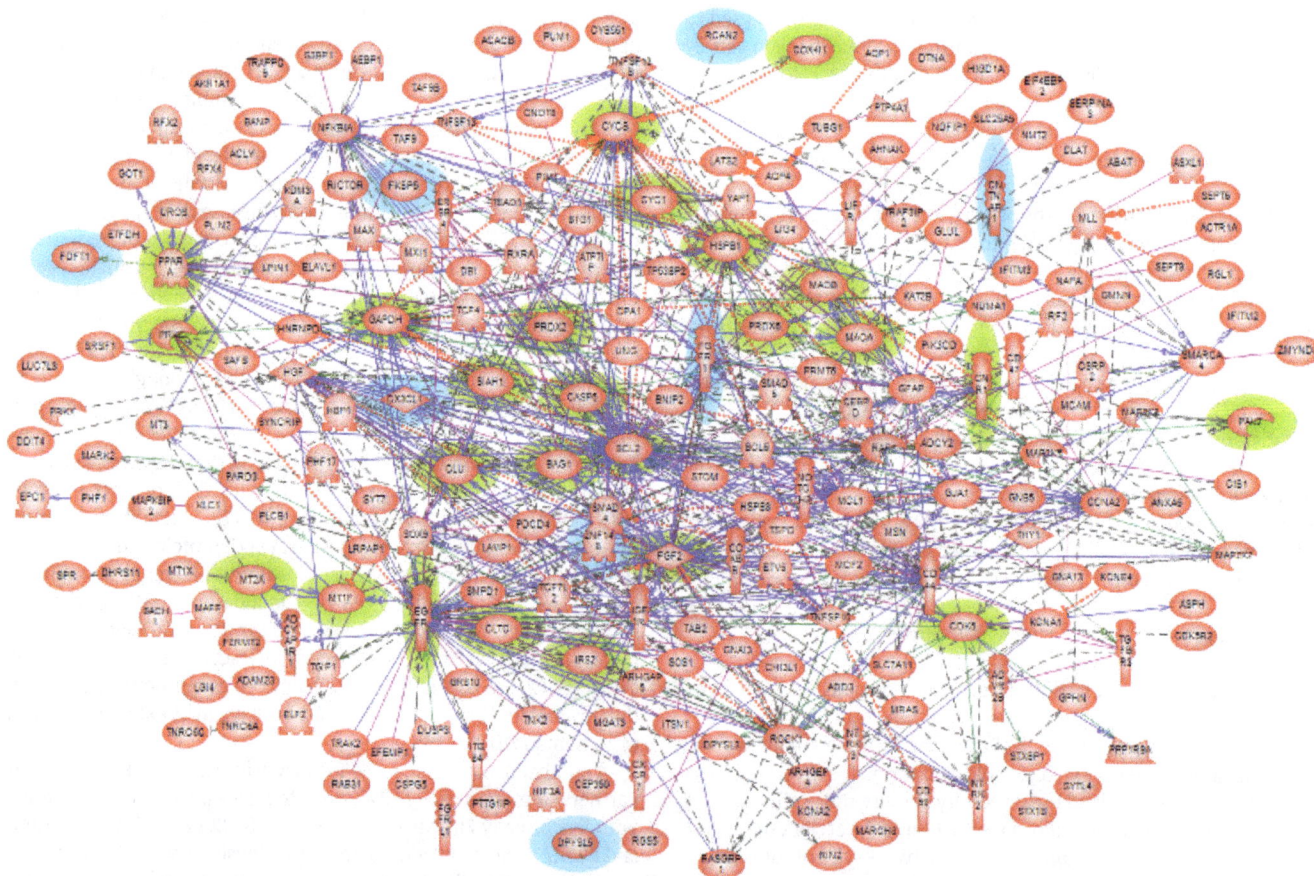

Fig. 3. Huntington's disease *direct* interaction network. The 26 genes/proteins implicated in HD pathology are highlighted in green and the eight genes/proteins of potential interest for that disease are highlighted in blue.

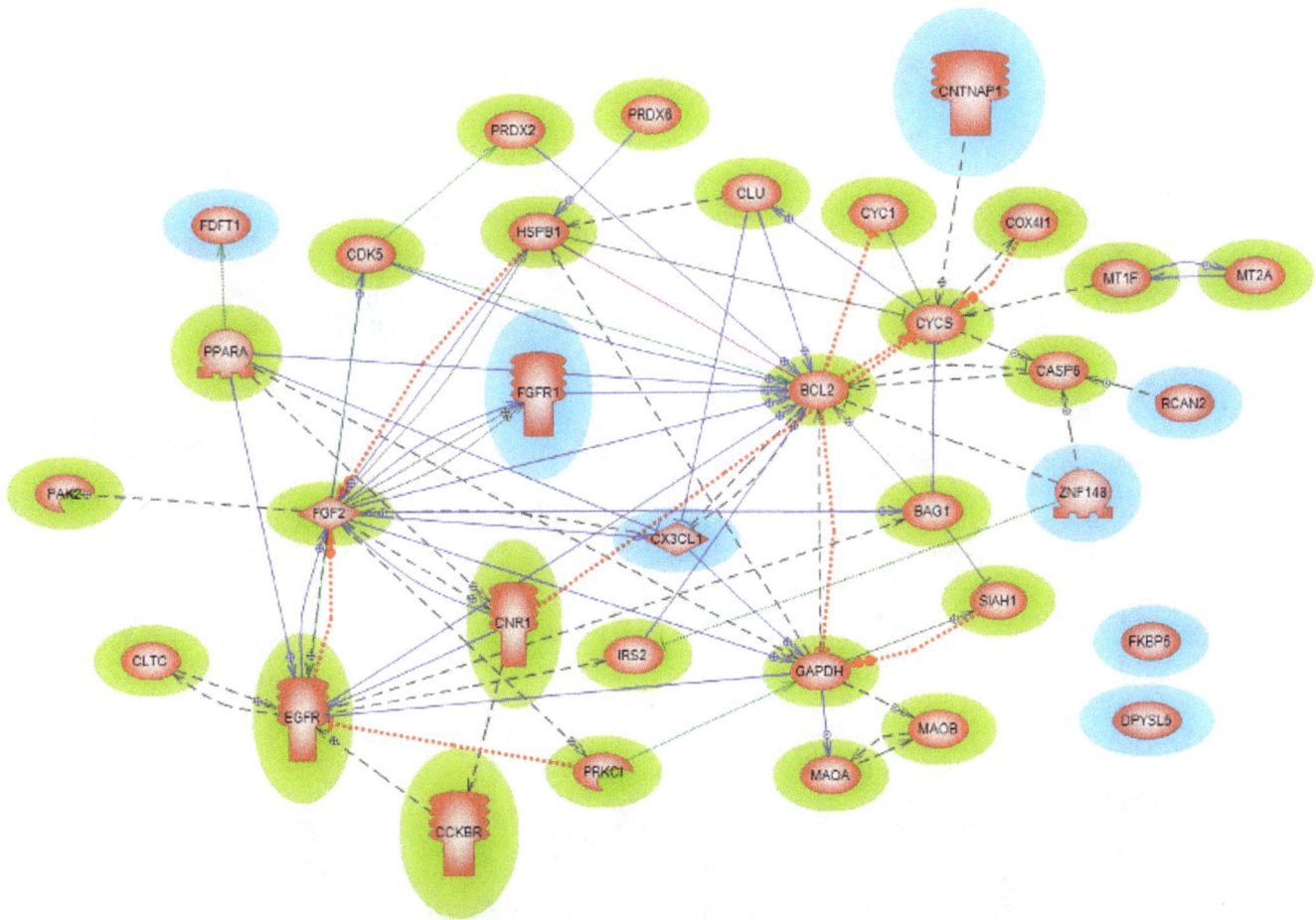

Fig. 4. Huntington's disease *compact* direct interaction network. The 26 genes/proteins implicated in HD pathology are highlighted in green and the eight genes/proteins of potential interest for that disease are highlighted in blue.

Our next HD candidate gene is CEBPA (CCAAT/enhancer binding protein, alpha). It has been shown to bind to the promoter and modulate the expression of the gene encoding for leptin, a protein that plays an important role in body weight homeostasis. Leptin receptors are found in various brain regions such as the hippocampus and cerebral cortex, and have known roles in neural development and neuroendocrine functions. Studies have indicated that leptin could be neuroprotective and thus enhance neuronal survival [59]. CEBPA, the promoter of leptin gene could also play a critical role in this neuroprotective mechanism. In the *compact* shortest-path network, CEBPA was the first-level interacting partners with nine known-HD genes.

Another recommendation for HD candidate gene is CDK1 (cyclin-dependent kinase 1).The abnormal activation of CDK1 is likely to be involved in the neuronal cell loss in neurodegenerative diseases including Alzheimer's disease and HIV [60]. Earlier, CDK5 was suspected to contribute to the deleterious protein accumulation in Alzheimer's disease [61–63]. In the *compact* SP network, CDK1 directly interacts with eight known-HD genes. CDK1 is a part of the kinase family that is actively contributing to the neurodegeneration process in similar disease conditions; it could have potential role in HD neurodegeneration mechanism too.

Thus, undeniably, these candidate genes should be further investigated for their molecular role in HD. However, we expect many of these novel genes to surpass the experimental verification due to their "guilt-by-association" with previously established Huntington's disease genes.

Apart from novel connecting genes, it is valuable to note that genes like EGFR, ESR1, HSBP1 and MAPT were also included in the *compact* SP network as connecting genes. They are previously known contributors as well as therapeutic agents in neurodegenerative disorders. Hyperphosphorylated tau (MAPT) is the major component of the neurofibrillary tangles, one of the hallmarks of neurodegenerative diseases [62, 64–66]. Similarly, huntingtin gene, which mutates strongly in HD was suggested to be indirectly associated with EGFR, thus deregulating the downstream actions of EGFR leading to cell death [67]. Considering the treatment measures, HSBP1 and ESR1 are suggested to offer such mechanisms. In general, heat shock proteins (HSBP1) are evaluated as therapeutic targets in mitigating or preventing protein aggregate formations [68]. One of the major conclusions of an animal HD model study was that the female sex hormone, estrogen (ESR1) could be a target for neuroprotective therapy aiming at postponing the onset and reducing the severity of HD. A similar pattern of late onset was also shown in a human HD study [69,70]. Moreover, these four connecting genes were among the top 25 nodes with highest connectivity (degree >10) as well as one of the top 25 nodes with highest visibility as measured by the closeness centrality scores. Except HSBP1, the other three genes were also among the top 25 nodes with highest accessibility to other nodes in the network as determined by the betweenness centrality in the *compact* SP network.

Table 1

Summary of genes of interest and genes already known in Huntington's disease.

Different categories	No. of genes	Node color code in figures
Genes of interest from SDEGs	13	blue
Known HD genes from SDEGs	25	green
Genes of interest in SPNW connecting nodes	23	orange
Known HD genes in SPNW connecting nodes	24	red

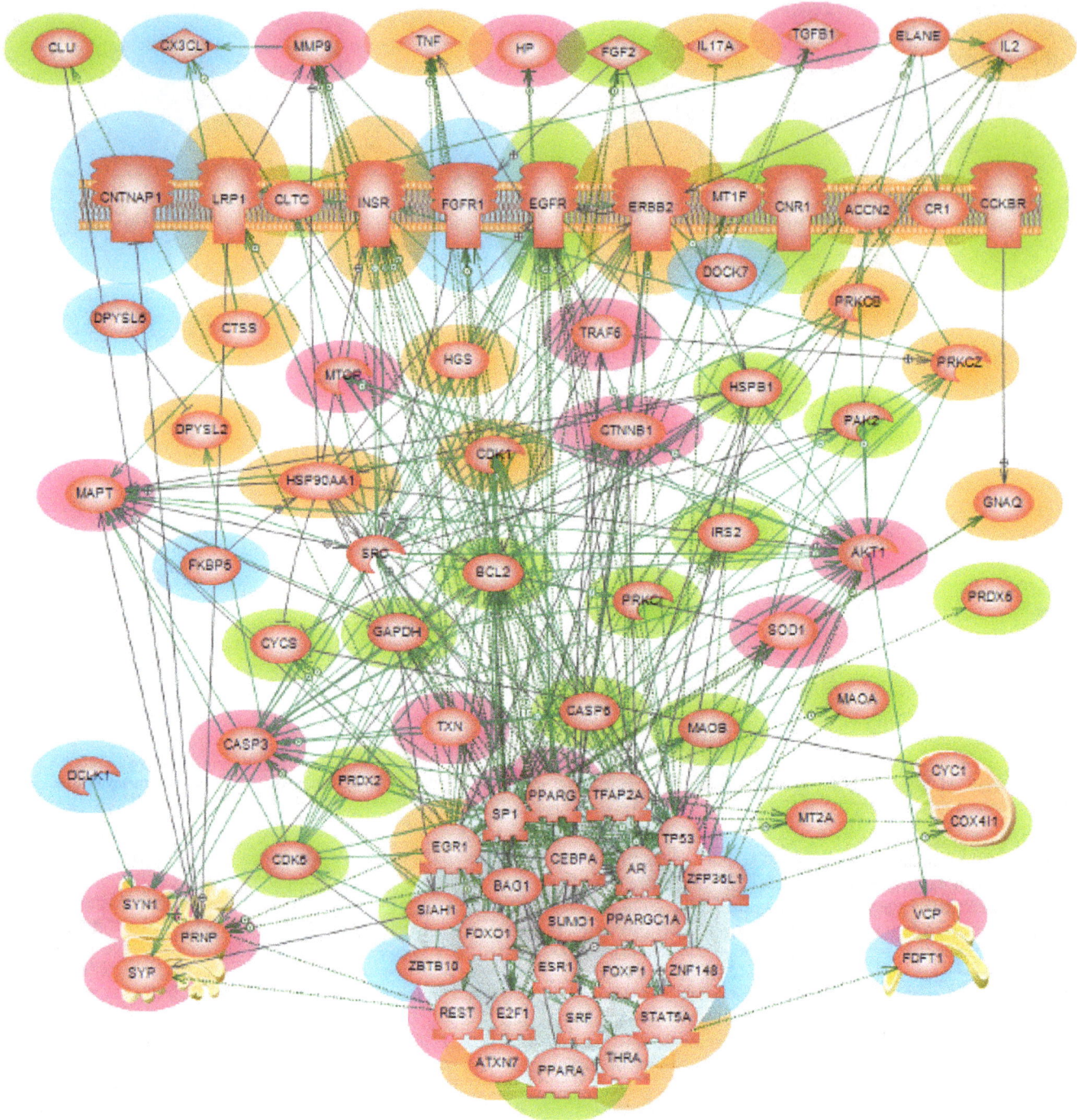

Fig. 5. Huntington's disease *compact* shortest path network. The genes/proteins implicated in HD pathology are highlighted in green and red. The genes/proteins of potential interest are highlighted in blue and orange (see Table 1. for details). Genes/proteins causing neuronal loss are highlighted in purple and those that help in neuronal survival are in yellow.

Neueder and Bates [32] employed weighted correlation network analysis to a number of post-mortem brain tissues in human, as well as in mouse samples. They identified extensive transcriptional dysregulation in the cerebellum of HD patients, similar to that observed in the frontal cortex and caudate nucleus. A common signature of gene expression changes in all three brain tissues networks of HD patients has been proposed. Our study generally confirms the disruption of a number of biological processes and first of all that of mitochondrial functions. However, our unweighted correlation network analysis did not show HD importance of some of the genes proposed in their work, e.g., E2F, NRF1, SF1, E4F1, and ELK1 were excluded from our "seed genes" list, due to insufficient statistical significance ($p > 0.05$).

Kalathur et al. [33] used the same GSE3970 database, along with a couple of other HD mouse model datasets. There is a considerable overlap of the enriched KEGG pathways and GO categories with our study. 20 out of 26 Kalahari's selected genes were significantly expressed in our analysis ($p < 0.05$), all of them expressed in caudate nucleus region of the brain. However, due to the more stringent p-value cut off criterion in our work ($p < 0.01$), only one gene (CDK5R2) out of these 20 was included in our network analysis. We added also MAP3K5 ($p < 0.05$), as the only gene with more than 25 neighbors.

There is no overlap of genes or biological pathways reported in the study of Alcaraz et al. [34] with those identified in our work. Out of the 11 new genes proposed by these authors seven genes (CTNNB1,

Table 2
Genes of interest for Huntington's disease identified by "guilt-by-association" with the known HD-related genes.

Genes of interest	Interacts with no. of known HD genes
EGR1	13
CEBPA	9
CDK1	8
HSP90AA1	7
PRKCZ	6
E2F1, STAT5A	5
SRF	4
ERBB2, FGFR1, IL2, INSR, LRP1, PRKCB, TNF, ZNF148	3
CX3CL1	2
CNTNAP1, RCAN2	1

GNAQ, GRB2, OPTN, TP53, UBE2K and YWHAB) have *p*-value of <0.01 but were not significantly expressed in all tissue types so they were not included into our 531 SDEGs.

Continuing with our network analysis, we subjected the genes in the *compact* shortest-path network to DAVID analysis to identify various enriched biological processes and pathways in Huntington's disease. Table 3 lists some of the Gene Ontology categories/subcategories related to nervous system and functions that were statistically significantly enriched in HD (with Benjamini-Hochberg multiple correction). DAVID analysis uncovered biological processes involving in oxidative stress, reactive oxygen species, deregulation in inflammatory response, steroid hormone receptor signaling, lipid binding and insulin receptor signaling pathways to be significantly affected in Huntington's disease, some of which were mentioned above. Other neurodegenerative signaling pathway including Alzheimer's and ALS were also considerably affected in Huntington's disease which reinforce our view for similar underlying molecular pattern in all these diseases. In the next section, we will provide detailed information about our proposed model for Huntington's disease mechanism based on the various biological pathways that were affected in this disease.

3.3. Integrated huntington's disease mechanism

In addition to various biological processes, DAVID analysis also recommended several KEGG pathways to be significantly (*p*-values of <0.05 with Benjamini-Hochberg multiple correction) affected in Huntington's disease of which we selected 10 pathways for further evaluation. These pathways were selected (listed in Table 4) on the basis of previous implications in Huntington's disease research work. Either the entire pathway or many important players of the pathways were found deregulated in HD pathogenesis [71–77].

Enriched KEGG pathways belong to endocrine system, cell communication, cell growth and death, signal transduction, neurodegenerative diseases, and endocrine and metabolic diseases classification. We used the 10 pathways (see Table 4) to search for any underlying molecular mechanism that could either cause or mitigate Huntington's disease pathology. To accomplish this task, we constructed an *integrated* HD mechanism network using the 41 genes found in common in all the 10 enriched KEGG pathways (see Fig. 6).

We then performed a Huntington's disease literature search to classify these 41 genes into two groups, namely, genes that aid in the neuronal survival or cause loss. Twenty-three out of 41 were implicated in neuronal loss and the remaining 18 genes were related to neuronal survival. This classification is depicted in Fig. 6 where genes are highlighted in purple and yellow, respectively. Similar to our *integrated* neurodegenerative disease mechanism networks of Parkinson's and Alzheimer's disease [36,37], we found a pattern of three extra-cellular ligands (FGF2, TNF, and TGFB1) initiating various downstream signaling cascades in the integrated Huntington's disease mechanism network as well.

As explained earlier, FGF2s are pursued as promising drug targets for its neuroprotective and neuroproliferative roles. TNF (tumor necrosis factor) and TGFB1 (transforming growth factor, beta 1) belong to inflammatory cytokine family which is involved in the regulation of a wide variety of biological processes including cell proliferation, differentiation, adhesion, apoptosis, lipid metabolism, and coagulation. Neuroinflammation has been implicated in many neurological disorders

Table 3
Gene set DAVID enrichment analysis of Huntington's disease *compact* shortest-path network.

Term	Gene count	Fold enrichment	Benjamini
GO:0048666 ~ neuron development	14	6.50	5.99E − 06
GO:0030182 ~ neuron differentiation	15	5.39	1.44E − 05
GO:0046324 ~ regulation of glucose import	6	28.60	4.33E − 05
GO:0043005 ~ neuron projection	13	6.07	1.37E − 04
GO:0050994 ~ regulation of lipid catabolic process	5	30.25	3.09E − 04
GO:0010001 ~ glial cell differentiation	6	17.81	3.09E − 04
GO:0043523 ~ regulation of neuron apoptosis	7	12.23	3.22E − 04
GO:0031175 ~ neuron projection development	10	6.14	4.40E − 04
GO:0043627 ~ response to estrogen stimulus	7	10.49	6.42E − 04
GO:0042063 ~ gliogenesis	6	14.52	6.76E − 04
GO:0048812 ~ neuron projection morphogenesis	9	6.65	6.99E − 04
GO:0006979 ~ response to oxidative stress	8	7.67	8.94E − 04
GO:0001836 ~ release of cytochrome c from mitochondria	4	29.96	0.003
GO:0048667 ~ cell morphogenesis involved in neuron differentiation	8	6.02	0.003
GO:0030425 ~ dendrite	8	7.84	0.004
GO:0030424 ~ axon	8	8.04	0.004
GO:0007568 ~ aging	6	8.58	0.006
GO:0008289 ~ lipid binding	11	3.82	0.009
GO:0000302 ~ response to reactive oxygen species	5	10.49	0.010
GO:0050727 ~ regulation of inflammatory response	5	10.35	0.011
GO:0007409 ~ axonogenesis	7	5.71	0.011
GO:0008286 ~ insulin receptor signaling pathway	4	17.01	0.013
GO:0050804 ~ regulation of synaptic transmission	6	6.94	0.013
GO:0031644 ~ regulation of neurological system process	6	6.17	0.020
GO:0016192 ~ vesicle-mediated transport	11	3.00	0.023
GO:0050767 ~ regulation of neurogenesis	6	5.69	0.027
GO:0030518 ~ steroid hormone receptor signaling pathway	4	10.85	0.037
GO:0006874 ~ cellular calcium ion homeostasis	6	5.16	0.038
GO:0030136 ~ clathrin-coated vesicle	6	7.26	0.039
GO:0055114 ~ oxidation reduction	11	2.71	0.042
GO:0045121 ~ membrane raft	6	6.70	0.045

Table 4
Enriched KEGG pathways in Huntington's disease resulted from DAVID analysis.

Term	Gene count	Fold enrichment	Benjamini	Genes
hsa05016:Huntington's disease	12	5.14	4.40E − 04	CASP3, GNAQ, SP1, CYCS, PPARG, CYC1, TP53, COX4I1, REST, CLTC, SOD1, PPARGC1A
hsa04010:MAPK signaling pathway	14	4.04	5.86E − 04	EGFR, FGFR1, TNF, TP53, SRF, TGFB1, PRKCB, AKT1, CASP3, PAK2, MAPT, HSPB1, TRAF6, FGF2
hsa04012:ErbB signaling pathway	8	7.08	0.002	EGFR, AKT1, PAK2, ERBB2, STAT5A, MTOR, SRC, PRKCB
hsa05010:Alzheimer's disease	10	4.73	0.003	CASP3, TNF, LRP1, GNAQ, MAPT, CYCS, CYC1, COX4I1, GAPDH, CDK5
hsa05014:Amyotrophic lateral sclerosis (ALS)	6	8.72	0.006	CASP3, TNF, BCL2, CYCS, TP53, SOD1
hsa04910:Insulin signaling pathway	8	4.57	0.011	AKT1, PRKCZ, IRS2, PRKCI, FOXO1, MTOR, INSR, PPARGC1A
hsa04920:Adipocytokine signaling pathway	6	6.90	0.012	AKT1, PPARA, IRS2, TNF, MTOR, PPARGC1A
hsa04930:Type II diabetes mellitus	5	8.20	0.015	PRKCZ, IRS2, TNF, MTOR, INSR
hsa04520:Adherens junction	6	6.00	0.016	EGFR, FGFR1, ERBB2, INSR, SRC, CTNNB1
hsa04115:p53 signaling pathway	5	5.67	0.049	CDK1, CASP3, CYCS, TP53, SIAH1

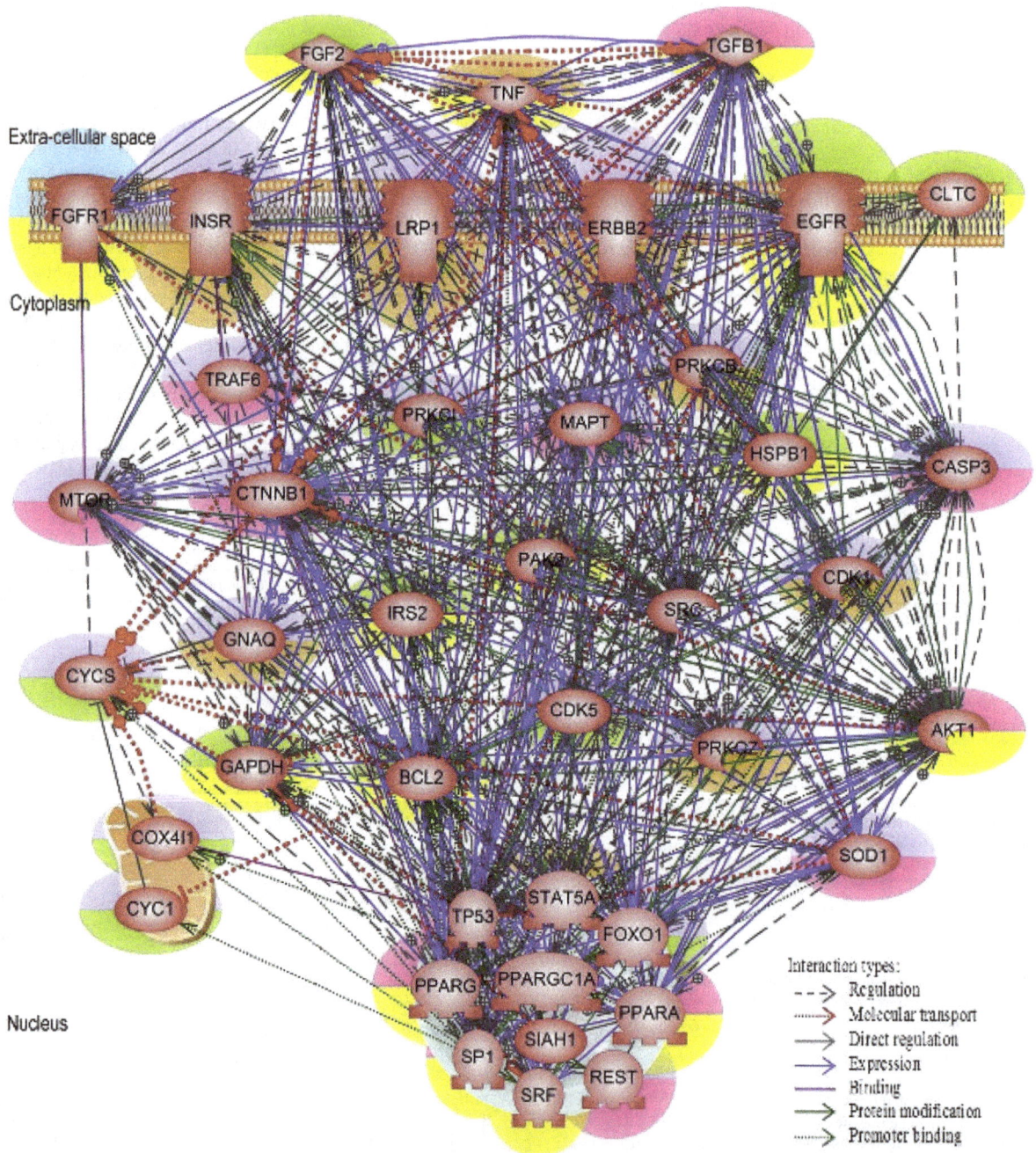

Fig. 6. *Integrated* Huntington's disease mechanism. The 41 genes/proteins are found in common in all 10 enriched KEGG pathways. Genes/proteins implicated in HD pathology are highlighted in green/red and the genes/proteins of potential interest are highlighted in blue/orange. Genes/proteins causing neuronal loss are highlighted in purple and those that help in neuronal survival are in yellow.

including Huntington's disease. In general, cytokines are required for normal functioning of cells. However, the formation of protein aggregates inside the cell triggers inflammatory mechanism which leads to increased cytokine activities thereby causing chronic cell stress [78,79]. A delicate balance has to be maintained in order to sustain homeostasis within cell. On examining the *integrated* network, we propose that the hemostasis in Huntington's disease environment could be restored by regulating the three extra-cellular ligands FGF2, TGFB1 and TNF thereby controlling their downstream signaling cascades of various target genes expression. On the other hand, even though the *integrated* HD mechanism network size was relatively small, due to the high interconnectedness of all the nodes it was difficult to suggest specific intracellular pathway(s) to implement in detail the proposed strategy for neuronal restoration in Huntington's disease via the three ligands.

Once again we went back to HD research literature looking for some molecular mechanisms and/or therapeutic pathways that are currently being utilized in this field. A recent review article by Zuccato et al. in 2010 [8] described the past achievements, the current status along with suspected disease mechanisms and therapeutic measures available in Huntington's disease realm. Several research works suggest few important players in HD whose regulation could promote neurogenesis. Under normal physiological conditions, HTT interacts with many genes/ proteins including BDNF, MTOR and REST to promote the survival of striatal neurons, the ones that are subjected to cell death in Huntington's disease. Interaction between BDNF (brain-derived neurotrophic factor) and HTT is important for the survival of striatal neurons as well as promoting synapse formation. In addition, HTT binds and sequesters mechanistic target of rapamycin (MTOR) inside the cytoplasm inhibiting MTOR's downstream regulation. In general, MTORs are negative regulators of autophagy. Autophagy is an essential, homeostatic process by which cells break down their own components. They are the debris clearance machineries in the cell that is required to protect against infections, autoimmune and inflammatory diseases [80]. Likewise, REST (RE1-silencing transcription factor) and HTT interaction is also important in HD pathogenesis. HTT binds with REST to maintain low levels of REST gene expression inside the cytoplasm thereby not affecting the transcription of BDNF gene.

In Huntington's disease, mutation in HTT causes protein aggregation formations which were not properly cleared from the cell thus disrupting the normal functioning of the stratial neurons. Due to transcription suppression by REST, the BDNF level was found reduced in neurodegenerative diseases including Alzheimer's, Parkinson's and Huntington's diseases [81,82]. Mutant HTTs were found inducing neuronal death via distinct but complementary pathways including deregulation of apoptosis and/or autophagy, altered transcription, metabolism and cellular stress responses. Currently one of the therapeutic measures suggested in Huntington's disease domain is clearing the HTT protein aggregates from the cell through the induction of autophagy by the MTOR inhibitor rapamycin. Another indicated treatment is through increasing the beneficial BDNF gene expression [8]. Animal HD model studies have shown that the use of rapamycin (MTOR inhibitor) improved striatal neuron survival and motor performance. However, due to deleterious side effects of rapamycin, it was not recommended for use as an exclusive drug in HD treatment. A combinatorial strategy with rapamycin or other drugs promoting autophagy has been suggested as relevant treatment for HD and other related diseases [83].

Following these HD literature suggested treatment ideas, we modified our *integrated* Huntington's disease mechanism network to include only those nodes (13 genes: AKT1, BCL2, GAPDH, EGFR, FGF2, FGFR1, INSR, MTOR, PPARGC1A, REST, SP1, TGFB1 and TNF) that might play a critical role in both inhibiting MTOR and improving BDNF gene expression. BDNF was added to the reduced network, as was done with HTT. The resulted *enriched* integrated HD mechanism network is shown in Fig. 7.

From this *enriched* network we propose two pathways through which homeostasis in HD could be restored by initiating the downstream signaling cascade of various target genes expression via primarily

through TGFB1, one of the three extra-cellular ligands. Our first proposal includes a two-step process of MTOR inhibition. Step 1: TGFB1 activates PPARGC1A gene expression in the nucleus, which in turn increases GAPDH gene/protein activity. Step 2: Up-regulation of GAPDH inhibits MTOR gene expression activity. Once MTOR is inhibited, autophagy mechanism could be boosted up in the cell. As soon as autophagy process is reestablished, HTT protein aggregates will be effectively cleared from the cell thus leading to neuronal survival.

Our second restoration pathway recommendation is via both FGF2 and TGFB1 ligand activation of EGFR receptors thereby initiating several downstream target genes expression. Among those upregulated genes, AKT1 (v-akt murine thymoma viral oncogene homolog 1) is a vital downstream target for EGFR and has been shown to be a critical mediator of neuronal survival. AKT1 has been suggested to be a promising therapeutic target to promote cell survival [84]. Apart from AKT1, EGFR interacts with SP1 (Sp1 transcription factor) which is involved in many cellular processes, including cell differentiation, cell growth, apoptosis, immune responses, response to DNA damage, and chromatin remodeling. SP1 fine-tunes the transcription of many genes including BCL2 and REST. BDNF transcription could be increased via maintaining tight regulation between REST and SP1. In addition, SP1 could also upregulate BCL2 gene expression, promoting anti-apoptosis. Thus, eventually striatal neurons could be protected by promoting BDNF activity, as well as by reducing the apoptotic process in the cell. Additionally, EGFR also promotes GAPDH gene expression eventually aiding in neuronal survival as detailed in our earlier pathway proposal.

From network analysis stand-point, the two proposed homeostasis restoration pathways show promising measures towards treatment plans in Huntington's disease. As a first step towards translating our proposed therapeutic networks into real world applications, such complex multi-player interconnected pathways could be evaluated using advanced dynamic modeling tools such as cellular automata [85].

3.4. Huntington's disease microRNA regulatory network

MicroRNAs perform important role in delivering post-transcriptional regulation of gene expression. Previous studies have found such microRNAs in this disease paradigm [86–88]. In order to identify the microRNAs and their potential targets in Huntington's disease domain, we constructed a *microRNA* regulatory network (MRN) using the 514 "seed genes". Before proceeding with the MRN construction, we first identified the microRNAs that could target our seed genes. This was accomplished using the *shortest-path* network option in Pathway Studio software where we subjected all the 514 HD "seed genes" to only microRNA interactions type. We found 132 microRNAs to target our HD genes. In order to obtain the microRNA–target gene interactions, we constructed a direct interaction network using the 132 microRNAs and the 514 "seed genes". [Note: The figure with the microRNA regulatory network containing over 1000 nodes is not shown, due to its extreme complexity].

The average node degree of the microRNA regulatory network was 4.3. Being the node with highest degree in the network, miR-9 was observed to target 35 genes. In addition, miR-9 was the node with highest closeness and betweenness centrality scores. Finding it in our microRNA regulatory network was exciting for a couple of reasons. First, miR-9 was previously known in Huntington's disease mechanism, as well as found to target REST (RE1-silencing transcription factor), one of the important players of HD pathogenesis. Secondly, miR-9 regulation has already been identified and found reduced in both Alzheimer's and Huntington's disease brains [81,82,89,90], providing thus another evidence for the conjectured unified underlined mechanism of the neurodegenerative diseases. Apart from miR-9 regulation, the network included miR-132 and miR-29a/b1 miRNAs, both already associated with HD pathogenesis [91,92].

The next top five microRNAs found in the regulatory network were miR-124, miR-135a, miR-141, miR-182 and miR-19a. All these

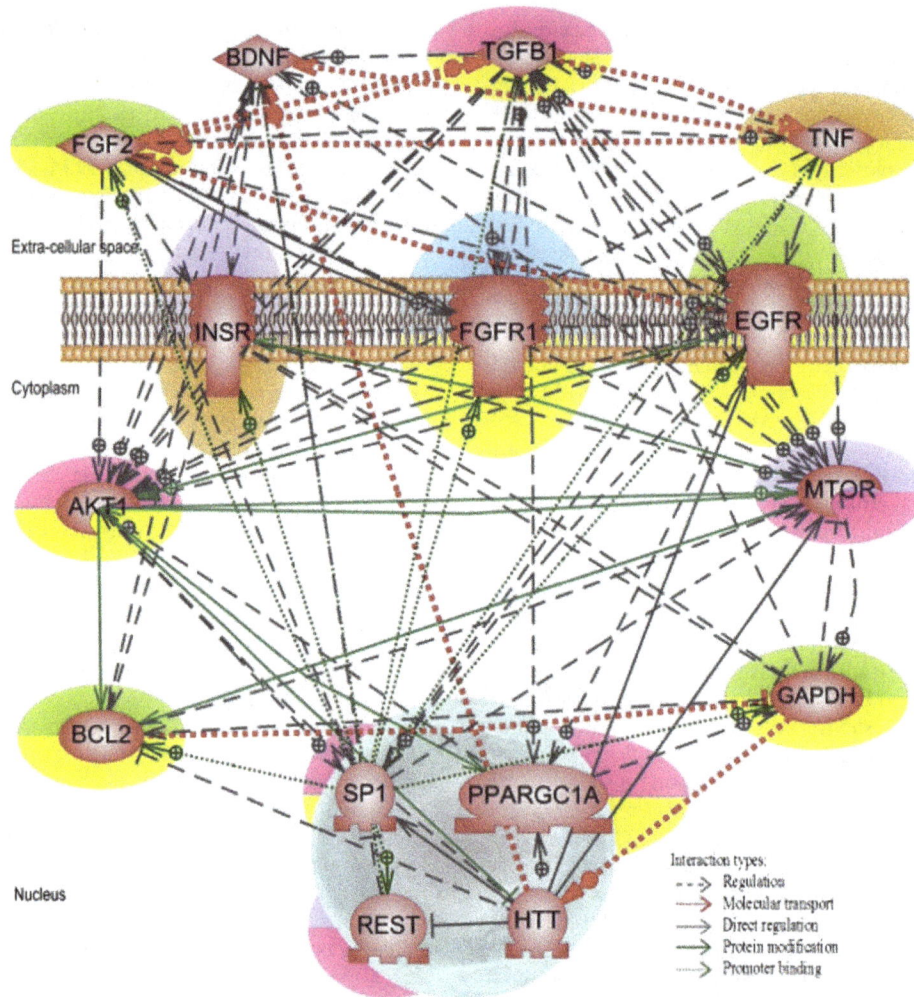

Fig. 7. Enriched integrated Huntington's disease mechanism. The 15 genes/proteins that were suggested to play a major role in HD treatment. Genes/proteins implicated in HD pathology are highlighted in green/red and the genes/proteins of potential interest are highlighted in blue/orange. Genes/proteins causing neuronal loss are highlighted in purple and those that help in neuronal survival are in yellow.

microRNAs were among the top 25 nodes with highest degree (node degree ≥ 12), and the top 25 nodes with highest closeness and betweenness centrality scores. miR-124 is one of the most abundantly expressed miRNAs in the nervous system, being widely expressed in neurons in the brain, retina, and spinal cord. It has been implicated in the modulation of neurite outgrowth, as well as cytoskeleton formation [93]. There have been no indications so far for involvement in neurodegenerative processes of miR-135a, known to target genes involved in blood pressure regulation [94]. Similarly, miR-141 and miR-182 were known to be involved only in DNA methylation and cancer metastasis, respectively [95,96]. miR-19 and miR-21 have been found to target PTEN, a gene/protein found localized in the neurofibrillary tangles (NFTs) and senile plaques in Alzheimer's disease brains [97,98]. MicroRNA regulatory network also uncovered that many of these top regulating microRNAs modulate several known-HD genes such as CNR1, FOXP1, GAPDH, and

IRS2. We suggest that the following nine microRNAs namely, miR-135 A1, miR-141, miR-153-1, miR-15 A, miR-16-1, miR-182, miR-19 A, miR-27 A and miR-96 could be of potential interest in HD. However, our *microRNA* regulatory analysis should be offered with some caution, because miRNAs that are enriched in the CNS are more likely to regulate targets enriched in the CNS. In addition, currently a high percentage of miRNA–target interactions in Pathway Studio ResNet 9.0 database are based on predictions, as verified from the references given in this database. Hence, further experimental verification is recommended.

Table 5 shows the genes of interest in the HD microRNA regulatory network and the number of microRNAs that are targeting each gene. In this network, DOCK7 (dedicator of cytokinesis 7) was the gene with the highest number of microRNA regulations, being regulated by seven members of miR-181 and miR-30 families. DOCK7 gene encodes for guanine nucleotide exchange factor (GEF) protein that plays a major role in axon formation and neuronal polarization. In general, GEFs are critical mediators of Rho GTPase activation by stimulating the exchange of GDP for GTP. Under normal physiological conditions, Rho GTPases act as molecular switches in intracellular signaling pathways and have many downstream targets. Mutations in GEFs and deregulated Rho GTPase signaling have been implicated in ALS, a debilitating motor neuron disease caused by neuronal degeneration. Based on its molecular function and its association with similar neurodegeneration disease, DOCK7 could be of potential interest in Huntington's disease mechanism as well.

Table 5
Genes of interest determined from Huntington's disease microRNA regulatory network.

Genes of interest	Target by no. of miRNAs
DOCK7, ZBTB10	7
DPYSL5, ZNF148	6
DCLK1, OSBPL11, RCAN2, ZFP36L1	5
CNTNAP1	4
FGFR1, FKBP5	2
CX3CL1, FDFT1	1

Moving on with other genes of interest in the microRNA regulatory network, zinc finger proteins (ZBTB10, ZNF148, and ZFP36L1) were highly targeted by multiple microRNAs including miR-20a and miR-29b1, known microRNAs in Alzheimer's and Huntington's diseases, respectively [99]. As reported in the previous section, zinc finger proteins are demonstrated to be a promising new gene therapy tool for Huntington's disease. Such a therapy could be enhanced by these microRNA regulations.

Additional to *microRNA* regulation, the network also included 43 genes that code for transcription factors. These significantly differentially expressed TFs indicate a possible integrated gene expression regulation mechanism in Huntington's disease. Like we noticed for the other two neurodegenerative disorders, there is a likelihood of dual-level gene expression regulation also occurring in HD paradigm. Thus, Huntington's disease should be considered as another complex disease system that involves highly interconnected molecular players and multi-level regulation.

4. Conclusions

Proceeding from 514 well selected "seed genes" that were significantly differentially expressed in HD postmortem samples we constructed several types of intracellular networks. Our network analysis was based on two basic principles. The guilt-by-association rule gives preference to genes being surrounded in the network predominantly by genes implicated with Huntington's disease. The best example of this strategy is our candidate gene *EGR*1 which interacts with 13 already known HD genes. Our second prioritizing rule was based on identifying *critical nodes in network topology*, i.e., nodes with high connectivity, visibility and traffic influence, as characterized by the node degree, closeness centrality and betweenness centrality. Thus, five of our novel candidate genes: *EGR1, CDK1, CEBPA, E2F1* and *INSR* were among the top 25 most highly connected, visible and influential genes. Moreover, some well-known Huntington's disease genes like *PRNP* which have not been significantly expressed in the post-mortem microarrays and was not included in our "seed gene" list, re-emerged as one of the top 15 critical nodes with highest visibility and most influence on the interaction traffic.

Overall, our network analysis prioritized 19 novel genes and nine miRNAs with pivotal positions in the HD-related networks built. Taking also in consideration their intrinsic molecular functions, we propose these genes and miRNAs as novel candidates for the analysis of Huntington's disease pathogenesis and survival, and briefly discuss some of those. *CEBPA* was among the best connected with known-HD genes. As promoter of leptin gene which enhances neuronal survival CEBPA could also play a critical role in the neuroprotective mechanism. Another well-connected gene is *CDK*1, a part of the kinase family that is actively contributing to the neurodegeneration process in similar disease conditions. The early growth response gene 1 (*EGR1*), which plays a role in memory formation, has a record-high connectivity to 13 HD-related genes and may also be considered as strong candidate for experimental confirmation of its role in the Huntington's disease. *E2F1* mediates neuronal death via activation of its transcriptional targets. *ERBB2* is of interest as a link between molecular pathways underlying neurodegeneration. *LRP1* seems to be of importance in elucidating the connection between cholesterol homeostasis and pathophysiology of HD. *HSP90AA*1 as a member of heat shock proteins, might be critically involved in the progression of HD. As a member of the zinc finger proteins family *ZNF*148 could be of interest in regulating the level of the mutant Huntingtin protein. Their critical role in the biological pathways that was significantly affected in Huntington's disease, as well as being directly associated with many known-HD genes, increases the probability that these proposed candidate genes could play a major part in the HD pathogenesis.

Through our *microRNA* regulatory network, we suggest that nine microRNAs could be potential regulators and drug targets in HD. However, the central role in our network is played by two well-known ones, miR9 and miR124. We plan to investigate the possibility for dual-level gene regulation by both microRNAs and transcription factors in Huntington's disease mechanism. Our future work on the role of miRNAs in HD pathogenesis and molecular mechanisms for fighting the disease will account for the limitations of the microarray analysis. The latter is capable of measuring the status of known transcripts only, and expression of low-abundance mRNAs is often not detected by the hybridization-based approach, thus opening the field for the more sensitive RNA-seq analysis, a revolutionary tool for transcriptomics and neurodegenerative diseases [100,101].

Analyzing our *integrated* network, we propose plans for several beneficial pathways of modulations of HD-related molecular factors, initiated via the extra-cellular ligands TGFB1, FGF2 and TNF. Restoring the normal homeostasis in Huntington's disease seems possible; one such plan is to up-regulate the innate autophagy process by inhibiting MTOR activity within the cell. Another plan aims to promote striatal neuron survival via increasing BDNF gene expression.

References

[1] Albin RL, Reiner A, Anderson KD, Penney JB, Young AB. Striatal and nigral neuron subpopulations in rigid Huntington's disease: implications for the functional anatomy of chorea and rigidity-akinesia. Ann Neurol 1990;27:357–65. http://dx.doi.org/10.1002/ana.410270403.
[2] Bates G. Huntingtin aggregation and toxicity in Huntington's disease. Lancet 2003; 361:1642–4. http://dx.doi.org/10.1016/S0140-6736(03)13304-1.
[3] Reiner A, Dragatsis I, Dietrich P. Genetics and neuropathology of Huntington's disease. Int Rev Neurobiol 2011;98:325–72. http://dx.doi.org/10.1016/B978-0-12-381328-2.00014-6.
[4] Davies JE, Sarkar S, Rubinsztein DC. The ubiquitin proteasome system in Huntington's disease and the spinocerebellar ataxias. BMC Biochem 2007; 8(Suppl. 1):S2. http://dx.doi.org/10.1186/1471-2091-8-S1-S2.
[5] Gil JM, Rego AC. Mechanisms of neurodegeneration in Huntington's disease. Eur J Neurosci 2008;27:2803–20. http://dx.doi.org/10.1111/j.1460-9568.2008.06310.x.
[6] Jellinger KA. Basic mechanisms of neurodegeneration: a critical update. J Cell Mol Med 2010;14:457–87. http://dx.doi.org/10.1111/j.1582-4934.2010.01010.x.
[7] Martinez-Vicente M, Talloczy Z, Wong E, Tang G, Koga H, Kaushik S, et al. Cargo recognition failure is responsible for inefficient autophagy in Huntington's disease. Nat Neurosci 2010;13:567–76. http://dx.doi.org/10.1038/nn.2528.
[8] Zuccato C, Valenza M, Cattaneo E. Molecular mechanisms and potential therapeutical targets in Huntington's disease. Physiol Rev 2010;90:905–81. http://dx.doi.org/10.1152/physrev.00041.2009.
[9] Mattson MP. Apoptosis in neurodegenerative disorders. Nat Rev Mol Cell Biol 2000; 1:120–9. http://dx.doi.org/10.1038/35040009.
[10] Hays SE, Goodwin FK, Paul SM. Cholecystokinin receptors are decreased in basal ganglia and cerebral cortex of Huntington's disease. Brain Res 1981;225:452–6.
[11] Kiechle T, Dedeoglu A, Kubilus J, Kowall NW, Beal MF, Friedlander RM, et al. Cytochrome C and caspase-9 expression in Huntington's disease. Neuromolecular Med 2002;1:183–95. http://dx.doi.org/10.1385/NMM:1:3:183.
[12] Burke JR, Enghild JJ, Martin ME, Jou YS, Myers RM, Roses AD, et al. Huntingtin and DRPLA proteins selectively interact with the enzyme GAPDH. Nat Med 1996;2: 347–50.
[13] Mazzola JL, Sirover MA. Reduction of glyceraldehyde-3-phosphate dehydrogenase activity in Alzheimer's disease and in Huntington's disease fibroblasts. J Neurochem 2001;76:442–9.
[14] Tourette C, Farina F, Vazquez-Manrique RP, Orfila A-M, Voisin J, Hernandez S, et al. The wnt receptor ryk reduces neuronal and cell survival capacity by repressing FOXO activity during the early phases of mutant huntingtin pathogenicity. PLoS Biol 2014;12, e1001895. http://dx.doi.org/10.1371/journal.pbio.1001895.
[15] Lejeune F-X, Mesrob L, Parmentier F, Bicep C, Vazquez-Manrique RP, JA Parker, et al. Large-scale functional RNAi screen in C. elegans identifies genes that regulate the dysfunction of mutant polyglutamine neurons. BMC Genomics 2012;13:91. http://dx.doi.org/10.1186/1471-2164-13-91.
[16] Ferrer I, Goutan E, Marín C, Rey MJ, Ribalta T. Brain-derived neurotrophic factor in Huntington disease. Brain Res 2000;866:257–61.
[17] Ross CA, Tabrizi SJ. Huntington's disease: from molecular pathogenesis to clinical treatment. Lancet Neurol 2011;10:83–98. http://dx.doi.org/10.1016/S1474-4422(10)70245-3.
[18] Tapia-Arancibia L, Aliaga E, Silhol M, Arancibia S. New insights into brain BDNF function in normal aging and Alzheimer disease. Brain Res Rev 2008;59:201–20. http://dx.doi.org/10.1016/j.brainresrev.2008.07.007.
[19] Sorolla MA, Reverter-Branchat G, Tamarit J, Ferrer I, Ros J, Cabiscol E. Proteomic and oxidative stress analysis in human brain samples of Huntington disease. Free Radic Biol Med 2008;45:667–78. http://dx.doi.org/10.1016/j.freeradbiomed.2008.05.014.
[20] Kuhn A, Goldstein DR, Hodges A, Strand AD, Sengstag T, Kooperberg C, et al. Mutant huntingtin's effects on striatal gene expression in mice recapitulate changes observed in human Huntington's disease brain and do not differ with mutant

huntingtin length or wild-type huntingtin dosage. Hum Mol Genet 2007;16: 1845–61. http://dx.doi.org/10.1093/hmg/ddm133.

[21] Luthi-Carter R, Hanson SA, Strand AD, Bergstrom DA, Chun W, Peters NL, et al. Dysregulation of gene expression in the R6/2 model of polyglutamine disease: parallel changes in muscle and brain. Hum Mol Genet 2002;11:1911–26.

[22] Benn CL, Luthi-Carter R, Kuhn A, Sadri-Vakili G, Blankson KL, Dalai SC, et al. Environmental enrichment reduces neuronal intranuclear inclusion load but has no effect on messenger RNA expression in a mouse model of Huntington disease. J Neuropathol Exp Neurol 2010;69:817–27. http://dx.doi.org/10.1097/NEN.0b013e3181ea167f.

[23] Becanovic K, Pouladi MA, Lim RS, Kuhn A, Pavlidis P, Luthi-Carter R, et al. Transcriptional changes in Huntington disease identified using genome-wide expression profiling and cross-platform analysis. Hum Mol Genet 2010;19:1438–52. http://dx.doi.org/10.1093/hmg/ddq018.

[24] Luthi-Carter R, Strand A, Peters NL, Solano SM, Hollingsworth ZR, Menon AS, et al. Decreased expression of striatal signaling genes in a mouse model of Huntington's disease. Hum Mol Genet 2000;9:1259–71.

[25] Spires TL, Grote HE, Varshney NK, Cordery PM, van Dellen A, Blakemore C, et al. Environmental enrichment rescues protein deficits in a mouse model of Huntington's disease, indicating a possible disease mechanism. J Neurosci 2004;24:2270–6. http://dx.doi.org/10.1523/JNEUROSCI.1658-03.2004.

[26] Jin K, LaFevre-Bernt M, Sun Y, Chen S, Gafni J, Crippen D, et al. FGF-2 promotes neurogenesis and neuroprotection and prolongs survival in a transgenic mouse model of Huntington's disease. Proc Natl Acad Sci U S A 2005;102:18189–94. http://dx.doi.org/10.1073/pnas.0506375102.

[27] La Spada AR. Huntington's disease and neurogenesis: FGF-2 to the rescue? Proc Natl Acad Sci U S A 2005;102:17889–90. http://dx.doi.org/10.1073/pnas.0509222102.

[28] Zhang Y, Ona VO, Li M, Drozda M, Dubois-Dauphin M, Przedborski S, et al. Sequential activation of individual caspases, and of alterations in Bcl-2 proapoptotic signals in a mouse model of Huntington's disease. J Neurochem 2003;87:1184–92.

[29] Sadagurski M, Cheng Z, Rozzo A, Palazzolo I, Kelley GR, Dong X, et al. IRS2 increases mitochondrial dysfunction and oxidative stress in a mouse model of Huntington disease. J Clin Invest 2011;121:4070–81. http://dx.doi.org/10.1172/JCI46305.

[30] Dowie MJ, Scotter EL, Molinari E, Glass M. The therapeutic potential of G-protein coupled receptors in Huntington's disease. Pharmacol Ther 2010;128:305–23. http://dx.doi.org/10.1016/j.pharmthera.2010.07.008.

[31] Vidal RL, Figueroa A, Court FA, Thielen P, Molina C, Wirth C, et al. Targeting the UPR transcription factor XBP1 protects against Huntington's disease through the regulation of FoxO1 and autophagy. Hum Mol Genet 2012;21:2245–62. http://dx.doi.org/10.1093/hmg/dds040.

[32] Neueder A, Bates GP. A common gene expression signature in Huntington's disease patient brain regions. BMC Med Genomics 2014;7:60. http://dx.doi.org/10.1186/s12920-014-0060-2.

[33] Kalathur RKR, Hernández-Prieto MA, Futschik ME. Huntington's disease and its therapeutic target genes: a global functional profile based on the HD Research Crossroads database. BMC Neurol 2012;12:47. http://dx.doi.org/10.1186/1471-2377-12-47.

[34] Alcaraz N, Friedrich T, Kötzing T, Krohmer A, Müller J, Pauling J, et al. Efficient key pathway mining: combining networks and OMICS data. Integr Biol 2012;4:756–64. http://dx.doi.org/10.1039/c2ib00133k.

[35] Chandrasekaran S, Bonchev DG. A network view on schizophrenia related genes. Netw Biol 2012;2:16–25.

[36] Chandrasekaran S, Bonchev D. A network view on Parkinson's disease. Comput Struct Biotechnol J 2013;7, e201304004. http://dx.doi.org/10.5936/csbj.201304004.

[37] Chandrasekaran S, Bonchev D. Network topology analysis of post-mortem brain microarrays identifies more Alzheimer's related genes and MicroRNAs and points to novel routes for fighting with the disease. PLoS One 2016;11, e0144052. http://dx.doi.org/10.1371/journal.pone.0144052.

[38] NCBI GEO Dataset: GSE3790 n.d. http://www.ncbi.nlm.nih.gov/geo/query/acc.cgi?acc=GSE3790.

[39] Hodges A, Strand AD, Aragaki AK, Kuhn A, Sengstag T, Hughes G, et al. Regional and cellular gene expression changes in human Huntington's disease brain. Hum Mol Genet 2006;15:965–77. http://dx.doi.org/10.1093/hmg/ddl013.

[40] Irizarry RA, Hobbs B, Collin F, Beazer-Barclay YD, Antonellis KJ, Scherf U, et al. Exploration, normalization, and summaries of high density oligonucleotide array probe level data. Biostatistics 2003;4:249–64. http://dx.doi.org/10.1093/biostatistics/4.2.249.

[41] Gentleman RC, Carey VJ, Bates DM, Bolstad B, Dettling M, Dudoit S, et al. Bioconductor: open software development for computational biology and bioinformatics. Genome Biol 2004;5:R80. http://dx.doi.org/10.1186/gb-2004-5-10-r80.

[42] Smyth GK. Linear models and empirical bayes methods for assessing differential expression in microarray experiments. Stat Appl Genet Mol Biol 2004;3. http://dx.doi.org/10.2202/1544-6115.1027 Article3.

[43] Nikitin A, Egorov S, Daraselia N, Mazo I. Pathway studio—the analysis and navigation of molecular networks. Bioinformatics 2003;19:2155–7.

[44] Batagelj V, Mrvar A. Pajek - program for large network analysis. Connections 1998; 21:1–11.

[45] Estrada E. The structure of complex networks. Hardback: Oxford University Press; 2011.

[46] Glynn Dennis J, Dennis G, Sherman BT, Hosack DA, Yang J, Gao W, et al. DAVID: database for annotation, visualization, and integrated discovery. Genome Biol 2003;4:P3.

[47] Huang DW, Sherman BT, Lempicki RA. Systematic and integrative analysis of large gene lists using DAVID bioinformatics resources. Nat Protoc 2009;4:44–57. http://dx.doi.org/10.1038/nprot.2008.211.

[48] Huang DW, Sherman BT, Lempicki RA. Bioinformatics enrichment tools: paths toward the comprehensive functional analysis of large gene lists. Nucleic Acids Res 2009;37:1–13. http://dx.doi.org/10.1093/nar/gkn923.

[49] Kanehisa M, Goto S. KEGG: Kyoto encyclopedia of genes and genomes. Nucleic Acids Res 2000;28:27–30.

[50] Pabon MM, Bachstetter AD, Hudson CE, Gemma C, Bickford PC. CX3CL1 reduces neurotoxicity and microglial activation in a rat model of Parkinson's disease. J Neuroinflammation 2011;8:9. http://dx.doi.org/10.1186/1742-2094-8-9.

[51] Morganti JM, Nash KR, Grimmig BA, Ranjit S, Small B, Bickford PC, et al. The soluble isoform of CX3CL1 is necessary for neuroprotection in a mouse model of Parkinson's disease. J Neurosci 2012;32:14592–601. http://dx.doi.org/10.1523/JNEUROSCI.0539-12.2012.

[52] Xiang Z, Valenza M, Cui L, Leoni V, Jeong H-K, Brilli E, et al. Peroxisome-proliferator-activated receptor gamma coactivator 1 α contributes to dysmyelination in experimental models of Huntington's disease. J Neurosci 2011;31:9544–53. http://dx.doi.org/10.1523/JNEUROSCI.1291-11.2011.

[53] Garriga-Canut M, Agustín-Pavón C, Herrmann F, Sánchez A, Dierssen M, Fillat C, et al. Synthetic zinc finger repressors reduce mutant huntingtin expression in the brain of R6/2 mice. Proc Natl Acad Sci U S A 2012;109:E3136–45. http://dx.doi.org/10.1073/pnas.1206506109.

[54] Imran M, Mahmood S. An overview of animal prion diseases. Virol J 2011;8:493. http://dx.doi.org/10.1186/1743-422X-8-493.

[55] Moore RC, Xiang F, Monaghan J, Han D, Zhang Z, Edström L, et al. Huntington disease phenocopy is a familial prion disease. Am J Hum Genet 2001;69:1385–8. http://dx.doi.org/10.1086/324414.

[56] Wagner W, Reuter A, Hüller P, Löwer J, Wessler S. Peroxiredoxin 6 promotes upregulation of the prion protein (PrP) in neuronal cells of prion-infected mice. Cell Commun Signal 2012;10:38. http://dx.doi.org/10.1186/1478-811X-10-38.

[57] Ferrante RJ, Ryu H, Kubilus JK, D'Mello S, Sugars KL, Lee J, et al. Chemotherapy for the brain: the antitumor antibiotic mithramycin prolongs survival in a mouse model of Huntington's disease. J Neurosci 2004;24:10335–42. http://dx.doi.org/10.1523/JNEUROSCI.2599-04.2004.

[58] Xie B, Wang C, Zheng Z, Song B, Ma C, Thiel G, et al. Egr-1 transactivates bim gene expression to promote neuronal apoptosis. J Neurosci 2011;31:5032–44. http://dx.doi.org/10.1523/JNEUROSCI.5504-10.2011.

[59] Tang BL. Leptin as a neuroprotective agent. Biochem Biophys Res Commun 2008; 368:181–5. http://dx.doi.org/10.1016/j.bbrc.2008.01.063.

[60] Castedo M, Perfettini J-L, Roumier T, Kroemer G. Cyclin-dependent kinase-1: linking apoptosis to cell cycle and mitotic catastrophe. Cell Death Differ 2002;9: 1287–93. http://dx.doi.org/10.1038/sj.cdd.4401130.

[61] Iijima K, Ando K, Takeda S, Satoh Y, Seki T, Itohara S, et al. Neuron-specific phosphorylation of Alzheimer's beta-amyloid precursor protein by cyclin-dependent kinase 5. J Neurochem 2000;75:1085–91.

[62] Crews L, Masliah E. Molecular mechanisms of neurodegeneration in Alzheimer's disease. Hum Mol Genet 2010;19:R12–20. http://dx.doi.org/10.1093/hmg/ddq160.

[63] Mateo I, Vázquez-Higuera JL, Sánchez-Juan P, Rodríguez-Rodríguez E, Infante J, García-Gorostiaga I, et al. Epistasis between tau phosphorylation regulating genes (CDK5R1 and GSK-3beta) and Alzheimer's disease risk. Acta Neurol Scand 2009; 120:130–3. http://dx.doi.org/10.1111/j.1600-0404.2008.01128.x.

[64] Jellinger KA. Recent advances in our understanding of neurodegeneration. J Neural Transm 2009;116:1111–62. http://dx.doi.org/10.1007/s00702-009-0240-y.

[65] Pittman AM, Fung H-C, de Silva R. Untangling the tau gene association with neurodegenerative disorders. Hum Mol Genet 2006;15. http://dx.doi.org/10.1093/hmg/ddl190 (Spec No:R188–95).

[66] Lei P, Ayton S, Finkelstein DI, Adlard PA, Masters CL, Bush AI. Tau protein: relevance to Parkinson's disease. Int J Biochem Cell Biol 2010;42:1775–8. http://dx.doi.org/10.1016/j.biocel.2010.07.016.

[67] Liu YF, Deth RC, Devys D. SH3 domain-dependent association of huntingtin with epidermal growth factor receptor signaling complexes. J Biol Chem 1997;272: 8121–4.

[68] Mymrikov EV, Seit-Nebi AS, Gusev NB. Large potentials of small heat shock proteins. Physiol Rev 2011;91:1123–59. http://dx.doi.org/10.1152/physrev.00023.2010.

[69] Bode FJ, Stephan M, Suhling H, Pabst R, Straub RH, Raber KA, et al. Sex differences in a transgenic rat model of Huntington's disease: decreased 17beta-estradiol levels correlate with reduced numbers of DARPP32 + neurons in males. Hum Mol Genet 2008;17:2595–609. http://dx.doi.org/10.1093/hmg/ddn159.

[70] Roos RA, Vegter-van der Vlis M, Hermans J, Elshove HM, Moll AC, van de Kamp JJ, et al. Age at onset in Huntington's disease: effect of line of inheritance and patient's sex. J Med Genet 1991;28:515–9. http://dx.doi.org/10.1136/jmg.28.8.515.

[71] Andreassen OA, Dedeoglu A, Stanojevic V, Hughes DB, Browne SE, Leech CA, et al. Huntington's disease of the endocrine pancreas: insulin deficiency and diabetes mellitus due to impaired insulin gene expression. Neurobiol Dis 2002; 11:410–24.

[72] Bae B-I, Xu H, Igarashi S, Fujimuro M, Agrawal N, Taya Y, et al. p53 mediates cellular dysfunction and behavioral abnormalities in Huntington's disease. Neuron 2005; 47:29–41. http://dx.doi.org/10.1016/j.neuron.2005.06.005.

[73] Apostol BL, Illes K, Pallos J, Bodai L, Wu J, Strand A, et al. Mutant huntingtin alters MAPK signaling pathways in PC12 and striatal cells: ERK1/2 protects against mutant huntingtin-associated toxicity. Hum Mol Genet 2006;15:273–85. http://dx.doi.org/10.1093/hmg/ddi443.

[74] Lalić NM, Marić J, Svetel M, Jotić A, Stefanova E, Lalić K, et al. Glucose homeostasis in Huntington disease: abnormalities in insulin sensitivity and early-phase insulin secretion. Arch Neurol 2008;65:476–80. http://dx.doi.org/10.1001/archneur.65.4.476.

[75] Phan J, Hickey MA, Zhang P, Chesselet M-F, Reue K. Adipose tissue dysfunction tracks disease progression in two Huntington's disease mouse models. Hum Mol Genet 2009;18:1006–16. http://dx.doi.org/10.1093/hmg/ddn428.

[76] Reis SA, Thompson MN, Lee J-M, Fossale E, Kim H-H, Liao JK, et al. Striatal neurons expressing full-length mutant huntingtin exhibit decreased N-cadherin and altered neuritogenesis. Hum Mol Genet 2011;20:2344–55. http://dx.doi.org/10.1093/hmg/ddr127.

[77] Moreira Sousa C, McGuire JR, Thion MS, Gentien D, de la Grange P, Tezenas du Montcel S, et al. The Huntington disease protein accelerates breast tumour development and metastasis through ErbB2/HER2 signalling. EMBO Mol Med 2013;5:309–25. http://dx.doi.org/10.1002/emmm.201201546.

[78] Battaglia G, Cannella M, Riozzi B, Orobello S, Maat-Schieman ML, Aronica E, et al. Platform presentation—early defect of transforming growth factor β1 formation in Huntington's disease. Neurotherapeutics 2010;7:135–6. http://dx.doi.org/10.1016/j.nurt.2009.09.025.

[79] Möller T. Neuroinflammation in Huntington's disease. J Neural Transm 2010;117:1001–8. http://dx.doi.org/10.1007/s00702-010-0430-7.

[80] Levine B, Mizushima N, Virgin HW. Autophagy in immunity and inflammation. Nature 2011;469:323–35. http://dx.doi.org/10.1038/nature09782.

[81] Zuccato C, Ciammola A, Rigamonti D, Leavitt BR, Goffredo D, Conti L, et al. Loss of huntingtin-mediated BDNF gene transcription in Huntington's disease. Science 2001;293:493–8. http://dx.doi.org/10.1126/science.1059581.

[82] Zuccato C, Cattaneo E. Role of brain-derived neurotrophic factor in Huntington's disease. Prog Neurobiol 2007;81:294–330. http://dx.doi.org/10.1016/j.pneurobio.2007.01.003.

[83] Ravikumar B, Vacher C, Berger Z, Davies JE, Luo S, Oroz LG, et al. Inhibition of mTOR induces autophagy and reduces toxicity of polyglutamine expansions in fly and mouse models of Huntington disease. Nat Genet 2004;36:585–95. http://dx.doi.org/10.1038/ng1362.

[84] Dudek H. Regulation of neuronal survival by the serine–threonine protein kinase Akt. Science 1997;275(80):661–5. http://dx.doi.org/10.1126/science.275.5300.661.

[85] Ermentrout GB, Edelstein-Keshet L. Cellular automata approaches to biological modeling. J Theor Biol 1993;160:97–133. http://dx.doi.org/10.1006/jtbi.1993.1007.

[86] Kocerha J, Xu Y, Prucha MS, Zhao D. Chan AWS. microRNA-128a dysregulation in transgenic Huntington's disease monkeys. Mol Brain 2014;7:46. http://dx.doi.org/10.1186/1756-6606-7-46.

[87] Roshan R, Ghosh T, Scaria V, Pillai B. MicroRNAs: novel therapeutic targets in neurodegenerative diseases. Drug Discov Today 2009;14:1123–9. http://dx.doi.org/10.1016/j.drudis.2009.09.009.

[88] Liu T, Im W, Mook-Jung I, Kim M. MicroRNA-124 slows down the progression of Huntington's disease by promoting neurogenesis in the striatum. Neural Regen Res 2015;10:786–91. http://dx.doi.org/10.4103/1673-5374.156978.

[89] Lukiw WJ. Micro-RNA speciation in fetal, adult and Alzheimer's disease hippocampus. Neuroreport 2007;18:297–300. http://dx.doi.org/10.1097/WNR.0b013e3280148e8b.

[90] Packer AN, Xing Y, Harper SQ, Jones L, Davidson BL. The bifunctional microRNA miR-9/miR-9* regulates REST and CoREST and is downregulated in Huntington's disease. J Neurosci 2008;28:14341–6. http://dx.doi.org/10.1523/JNEUROSCI.2390-08.2008.

[91] Junn E, Mouradian MM. MicroRNAs in neurodegenerative diseases and their therapeutic potential. Pharmacol Ther 2012;133:142–50. http://dx.doi.org/10.1016/j.pharmthera.2011.10.002.

[92] Lau P, de Strooper B. Dysregulated microRNAs in neurodegenerative disorders. Semin Cell Dev Biol 2010;21:768–73. http://dx.doi.org/10.1016/j.semcdb.2010.01.009.

[93] Yu J-Y, Chung K-H, Deo M, Thompson RC, Turner DL. MicroRNA miR-124 regulates neurite outgrowth during neuronal differentiation. Exp Cell Res 2008;314:2618–33. http://dx.doi.org/10.1016/j.yexcr.2008.06.002.

[94] Söber S, Laan M, Annilo T. MicroRNAs miR-124 and miR-135a are potential regulators of the mineralocorticoid receptor gene (NR3C2) expression. Biochem Biophys Res Commun 2010;391:727–32. http://dx.doi.org/10.1016/j.bbrc.2009.11.128.

[95] Segura MF, Hanniford D, Menendez S, Reavie L, Zou X, Alvarez-Diaz S, et al. Aberrant miR-182 expression promotes melanoma metastasis by repressing FOXO3 and microphthalmia-associated transcription factor. Proc Natl Acad Sci U S A 2009;106:1814–9. http://dx.doi.org/10.1073/pnas.0808263106.

[96] Vrba L, Jensen TJ, Garbe JC, Heimark RL, Cress AE, Dickinson S, et al. Role for DNA methylation in the regulation of miR-200c and miR-141 expression in normal and cancer cells. PLoS One 2010;5, e8697. http://dx.doi.org/10.1371/journal.pone.0008697.

[97] Pezzolesi MG, Platzer P, Waite KA, Eng C. Differential expression of PTEN-targeting microRNAs miR-19a and miR-21 in Cowden syndrome. Am J Hum Genet 2008;82:1141–9. http://dx.doi.org/10.1016/j.ajhg.2008.04.005.

[98] Sonoda Y, Mukai H, Matsuo K, Takahashi M, Ono Y, Maeda K, et al. Accumulation of tumor-suppressor PTEN in Alzheimer neurofibrillary tangles. Neurosci Lett 2010;471:20–4. http://dx.doi.org/10.1016/j.neulet.2009.12.078.

[99] K-C. Sonntag. MicroRNAs and deregulated gene expression networks in neurodegeneration. Brain Res 2010;1338:48–57. http://dx.doi.org/10.1016/j.brainres.2010.03.106.

[100] Labadorf A, Hoss AG, Lagomarsino V, Latourelle JC, Hadzi TC, Bregu J, et al. RNA Sequence analysis of Human Huntington disease brain reveals an extensive increase in inflammatory and developmental Gene Expression. PLoS One 2015;10, e0143563. http://dx.doi.org/10.1371/journal.pone.0143563.

[101] Crotti A, Benner C, Kerman BE, Gosselin D, Lagier-Tourenne C, Zuccato C, et al. Mutant huntingtin promotes autonomous microglia activation via myeloid lineage-determining factors. Nat Neurosci 2014;17:513–21. http://dx.doi.org/10.1038/nn.3668.

19

A Gene Module-Based eQTL Analysis Prioritizing Disease Genes and Pathways in Kidney Cancer

Mary Qu Yang [a,b,*], Dan Li [a,b], William Yang [c], Yifan Zhang [a,b], Jun Liu [d], Weida Tong [e]

[a] *Joint Bioinformatics Graduate Program, Department of Information Science, George W. Donaghey College of Engineering and Information Technology, University of Arkansas at Little Rock, USA*
[b] *University of Arkansas for Medical Sciences, 2801 S. University Ave, Little Rock, AR 72204, USA*
[c] *School of Computer Science, Carnegie Mellon University, 5000 Forbes Ave, Pittsburgh, PA 15213, USA*
[d] *Department of Statistics, Harvard University, Cambridge, MA 02138, USA*
[e] *Divisions of Bioinformatics and Biostatistics, National Center for Toxicological Research, US Food and Drug Administration, 3900 NCTR Road, Jefferson, AR 72079, USA*

ARTICLE INFO

Keywords:
ccRCC
Causative mutation
Pathways
Protein-protein interaction
Gene module
eQTL

ABSTRACT

Clear cell renal cell carcinoma (ccRCC) is the most common and most aggressive form of renal cell cancer (RCC). The incidence of RCC has increased steadily in recent years. The pathogenesis of renal cell cancer remains poorly understood. Many of the tumor suppressor genes, oncogenes, and dysregulated pathways in ccRCC need to be revealed for improvement of the overall clinical outlook of the disease. Here, we developed a systems biology approach to prioritize the somatic mutated genes that lead to dysregulation of pathways in ccRCC. The method integrated multi-layer information to infer causative mutations and disease genes. First, we identified differential gene modules in ccRCC by coupling transcriptome and protein-protein interactions. Each of these modules consisted of interacting genes that were involved in similar biological processes and their combined expression alterations were significantly associated with disease type. Then, subsequent gene module-based eQTL analysis revealed somatic mutated genes that had driven the expression alterations of differential gene modules. Our study yielded a list of candidate disease genes, including several known ccRCC causative genes such as *BAP1* and *PBRM1*, as well as novel genes such as *NOD2*, *RRM1*, *CSRNP1*, *SLC4A2*, *TTLL1* and *CNTN1*. The differential gene modules and their driver genes revealed by our study provided a new perspective for understanding the molecular mechanisms underlying the disease. Moreover, we validated the results in independent ccRCC patient datasets. Our study provided a new method for prioritizing disease genes and pathways.

1. Introduction

Kidney cancer is the sixth most common form of cancer for men and the tenth most common form of cancer for women. In 2016, over 63,000 newly diagnosed cases and 14,400 kidney cancer deaths were reported in the United States [1]. The vast majority of kidney cancers are renal cell carcinomas (RCC), among which nearly 75% are clear cell renal cell carcinomas (ccRCC) [2]. Despite recent advances, metastatic RCC remains largely an incurable disease [3,4]. Patients with this disease often have no apparent symptoms or laboratory abnormalities in the early stages. The incidence of ccRCC has been rising steadily in recent years due to the prevalence of adverse lifestyle changes and exposure to toxins such as smoke [5].

ccRCC is characterized by the presence of *VHL* gene mutation in most cases [6]. However, the loss of *VHL* alone is not sufficient for tumor initiation and survival, and a fraction of ccRCCs contain wild-type *VHL* genes, suggesting additional genetic alterations are required in the course of tumor development. Recent large-scale sequencing studies of ccRCC, including TCGA (The Cancer Genome Atlas) project have discovered several new and prevalent genomic mutations such as *PBRM1* and *BAP1* [7–9]. Despite these findings, the mortality rate of ccRCC has not significantly decreased, indicating that the genetic basis of the disease occurrence and development remains to be elucidated. Additionally, previous studies have shown that ccRCC is a highly heterogeneous disease [10,11], creating the need to identify new disease genes and pathways.

The expression quantitative trait loci (eQTL) analysis has been used to identify single-nucleotide polymorphisms (SNPs) that are significantly associated with gene expressions [12–14]. Most eQTL analysis performed testing on transcript-SNP pairs to identify genetic mutations that significantly affected individual gene expression. Here, we presented a gene module-based eQTL method to identify the somatic mutations

---

Abbreviations: RCC, Renal cell cancer; ccRCC, Clear cell renal cell carcinoma; eQTL, Expression quantitative trait loci; SVM, Support vector machine; TCGA, The Cancer Genome Atlas; KEGG, Kyoto Encyclopedia of Genes and Genomes; DEG, Differentially expressed gene; DGM, Differential gene module; AUC, Area Under Curve; ROC, Receiver Operating Characteristic.

* Corresponding author at: Joint Bioinformatics Graduate Program, Department of Information Science, George W. Donaghey College of Engineering and Information Technology, University of Arkansas at Little Rock, USA.

E-mail address: mqyang@ualr.edu (M.Q. Yang).

that are associated with gene clusters, which potentially function in the same pathway. We first identified differentially expressed gene modules (DGMs). The DGMs are comprised of a set of interacting genes based on protein-protein interactions and expression profile. The Gene Ontology analysis suggested that majority DGMs contained genes involved in the same biological processes. Additionally, the genes inside the same DGM tended to be co-expressed. Hence, these gene modules most likely contained genes function together in the disease-affected pathway. Disease genes are not always differentially expressed. The integration of gene expressions and protein interactions empower the discovery of disease genes, as disease genes without significant expression alterations could be revealed by DGMs through interacting with the differentially expressed genes in the gene modules. The subsequent eQTL analysis further established the linkages of somatic mutations with the DGMs. Collectively, the DGMs and their associated genetic mutations lead to the identification of novel disease genes and pathways. Moreover, we examined the DGMs on four independent ccRCC patient cohorts. The results showed DGMs accurately classified the tissue types blindly.

2. Results

An interacting pathway regulates the expression of a group of genes that often perform certain functions together. When a pathway is perturbed by genetic mutations, then expression levels of interacting genes associated with the pathway can be altered accordingly and can further contribute to malignant transformation. By integrating gene expression and protein-protein interactions here, we developed a new method of identifying gene clusters in the pathways impacted by the disease. Then, we performed an eQTL analysis to infer potential driver mutations and disease affected pathways. The procedure of our study was illustrated in Fig. 1.

2.1. Differentially Expressed Gene Modules Identification

The RNA-Seq expression profile of 19,768 protein-coding genes was obtained from TCGA 539 ccRCC and 72 paired normal tissue samples. After filtering out the genes with very low expression levels (Methods), a total of 16,343 genes remained for the subsequent analysis. Then, we coupled gene expression and protein-protein using a network approach to systematically reveal gene modules that were differentially expressed

in ccRCC. At first, each individual gene was employed as the seed of a module, and new genes were added to the module in an iterative manner. At each step, all genes that interacted with any gene member of the module were assessed using an activity score. A higher activity score suggested the expression level of the corresponding module was more likely associated with the tissue phenotype (Methods, Fig. 1). Hence, the gene that maximized the activity score was selected and added to the module. After a gene module was built, we applied three statistical tests to evaluate the significance of the module compared to background. The three tests included permuting tissue phenotype, randomizing genes in the module, and randomizing genes in the module with the same seed protein, respectively (Methods). Finally, we identified 1066 significant gene modules with activity scores equal or larger than 0.34 (P-value < 0.001 in all three statistical tests, Fig. 1). We referred to these gene modules as differential gene modules (DGMs).

2.2. Performance Evaluation Classification Based on Differential Gene Modules

The DGMs represented gene clusters that were significantly associated with tissue phenotypes. Thus, we hypothesized that the expression levels of DGMs can be utilized as features to distinguish ccRCC tissue from normal tissue samples. We examined the hypothesis using the TCGA-ccRCC dataset and three independent ccRCC patient datasets obtained from GEO (Methods, Table 1) [4,9]. The TCGA dataset contained an imbalance between ccRCC and normal samples (539 ccRCC versus 72 normal samples), whereas the other three data sets contained more balanced samples (Table 1).

The differentially expressed genes (DEG) based evaluation was performed for comparison as well. The TCGA expression profile was generated using RNA-Seq data, whereas the expression profiles of the other three independent ccRCC patient cohorts were produced using Microarray data (Methods). We used edgeR for the TCGA RNA-Seq dataset, and t-test followed by multiple-test correction for microarray datasets to perform differential expression analyses (Methods, Supp. Fig. 1).

We conducted hierarchical clustering analysis, using DGMs and DEGs, respectively, on the four ccRCC datasets including the TCGA dataset, GSE36895, GSE40435 and GSE46699. For GSE36895 and GSE40435, both using DGMs and DEGs yielded distinctive tumor and normal tissue clusters with perfect homogeneity (Table 1). However, for the

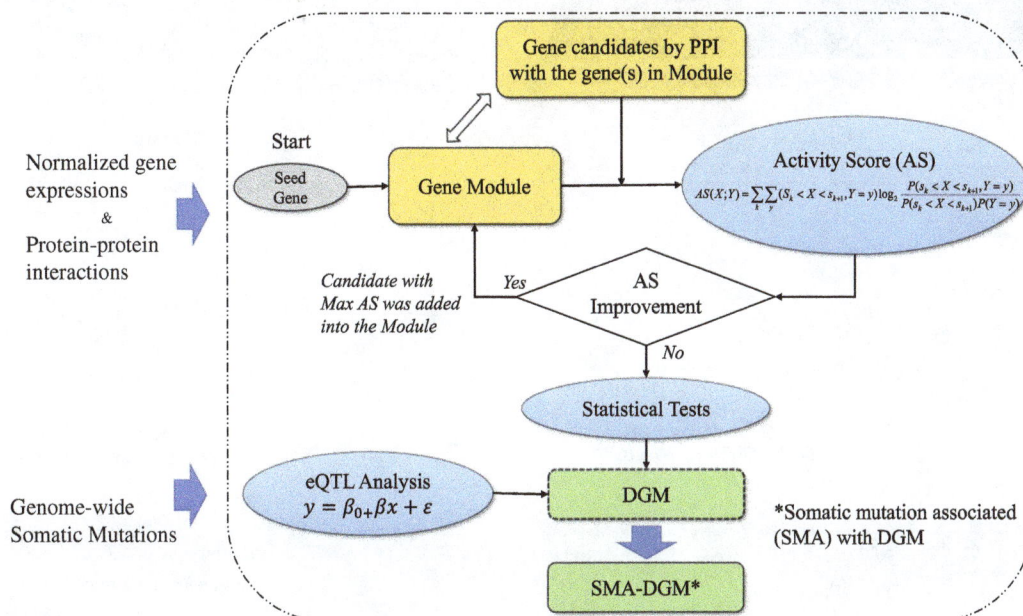

Fig. 1. The procedure of our study. After the differentially expressed gene modules were identified by coupling PPI with gene expression, somatic mutations were linked with the DGMs using eQTL analysis. Here, SMA-DGM refers to somatic mutations associated the DGMs.

Table 1
The performance of DGM and DEG based hierarchy clustering and SVM classifiers on the TCGA ccRCC patient group and three independent ccRCC datasets.

ccRCC patient cohorts	Normal	Tumor	Misclustered tissue samples		AUC of the classifiers	
			DGM-based	DEG-based	DGM-based	DEG-based
TCGA-ccRCC	72	539	2	4	**0.942**	**0.767**
GSE36895	23	29	0	0	0.923	1.0
GSE46699	63	67	9	15	0.953	0.949
GSE40435	101	101	0	0	0.956	0.997

The DGM based classifier significantly outperformed the DGE based classifier by 22.8%((0.942 -0.767) / 0.767, Table 1 bold number) on the TCGA-ccRCC which is an imbalanced data set (72 normal vs 539 tumor samples).

TCGA data and GSE46699, the clusters yielded by DGMs tended to be more homogeneous as compared to the clusters generated by the DEGs. Four tumor tissue samples were misclustered using DEGs (Fig. 2A top panel, Table 1), whereas the number of misclustered tumor tissues was reduced to two using DGMs for the TCGA data (Fig. 2A bottom panel, Table 1). For the GSE46699, using DGM resulted in 9 misclustered tumor samples (Fig. 2B top panel, Table 1), whereas using DGE yielded 15 misclustered ccRCC samples (Fig. 2B bottom, Table 1).

Moreover, we built SVM-based (Support Vector Machine) classifiers to predict tissue type using the expression levels of DGMs and DEGs genes as input features, respectively. The area under curve (AUC) of the receiver operating characteristic (ROC) curve generated by three-fold validation was measured for classification performance assessment. The AUC of the classifier using DGM as features is 0.942, which is significantly higher than 0.767 for the classifier using DEG as features, in predicting TCGA dataset tissue types (Fig. 2C left panel). However,

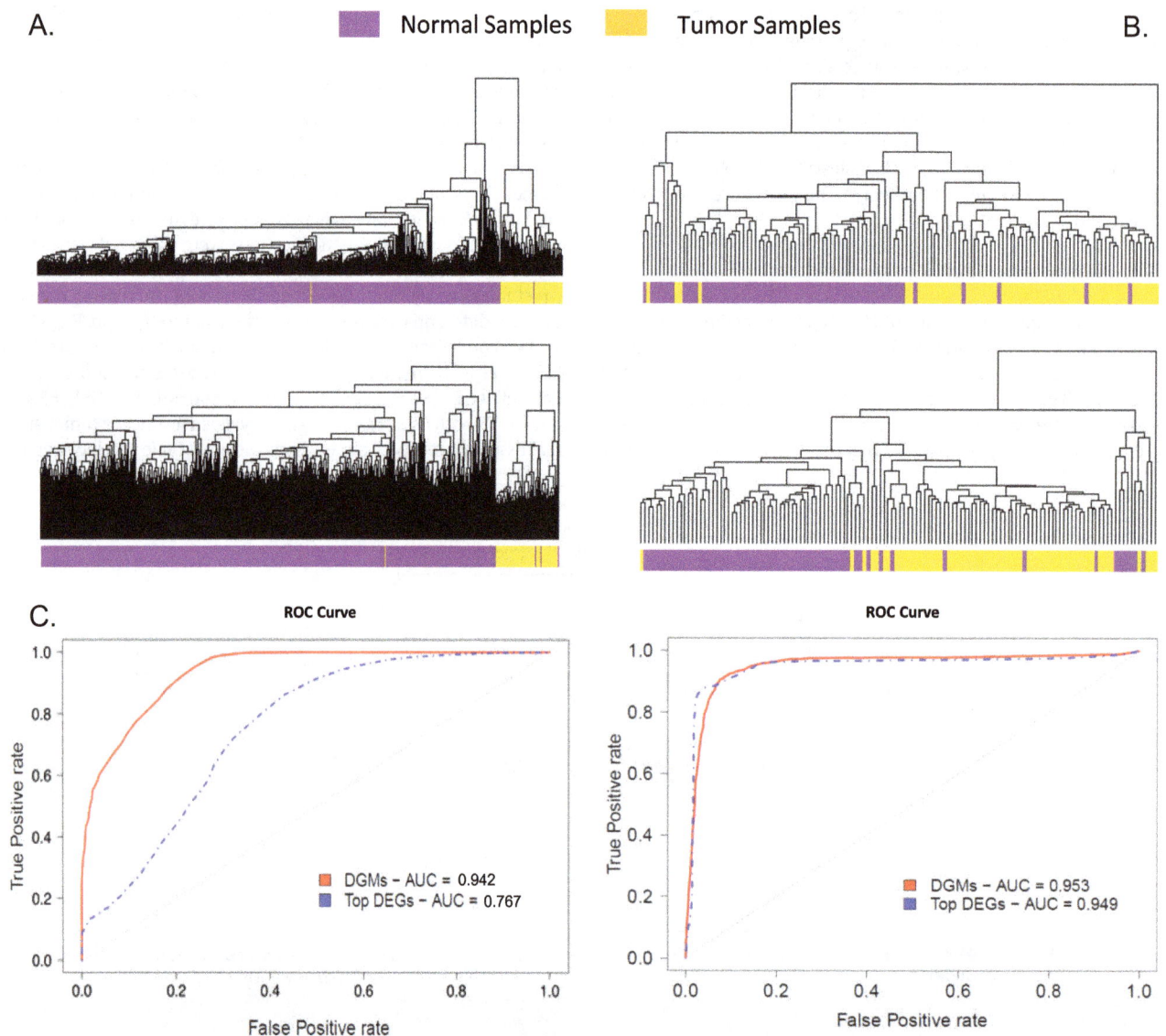

Fig. 2. The performance comparisons of clustering and classification based on DGMs and DEGs. (A) Hierarchical clustering of TCGA 539 ccRCC and 72 normal tissues based on the expression of DGMs (top) and DEGs (bottom). (B) Hierarchical clustering of an independent ccRCC (GSE46699) tumor and normal tissues based on DGMs (top) and DEGs (bottom). (C) The ROC curves of the classifiers using the expression of DGM and DEG for the TCGA dataset locate at left panel, for GSE46699 locate at right panel.

additional classifications on three independent ccRCC datasets, which included fewer but more balanced tissue samples than the TCGA dataset, showed that the DEG-based classifiers performed slightly better or very similar compared to the performance of DGM based classifiers (Table 1, Fig. 2C right panel). Nevertheless, the DGM yielded good performance more robustly in both classification and clustering and suggests that genes in the differential modules were significantly associated with ccRCC; they are coordinately expressed; and, they likely function together in the disease pathways.

2.3. Functional Assessment of the Gene Modules

We conducted Gene Ontology (GO) analysis on individual DGMs. The GO terms enrichment of were assessed by hypergeometric test ($P < 0.01$). Given that the median size of the gene modules is 6, we found that 90.4%, 65.4% and 39.9% (963/1065, 696/1065, and 425/1065) of these modules contained at least two, three, and four genes that participated in the same significantly enriched biological process, respectively. In contrast, none of the random modules that had the same topology and size as the DEMs contained more than one gene in the same biological process.

Additionally, our expression analysis showed that the majority of the genes in the DGMs appeared to be co-expressed (74.5%, 793/1065). Thus, the significant modules more likely consisted of genes functioning together in the disease-related pathways.

We found a total of 22 enriched biological process terms that were significantly associated with at least 18.9% (201/1065) of the DGMs (Supp. Table 1), including several known cancer-related biological processes. For instance, 209 gene modules were prevalent in the neurotrophin Tropomyosin Receptor Kinase (TRK) receptor signaling pathway, a pathway involving malignant gliomas [15].

Moreover, we identified 26 Kyoto Encyclopedia of Genes and Genomes (KEGG) pathways that were significantly enriched in at least 8.5% (90/1065) of the gene modules (Supp. Table 2). Thyroid cancer pathway, a top-affected pathway in our list, was significantly associated with 27.9% (297/1065) of gene modules ($P < 0.05$, hypergeometric test). It has been reported that ccRCC is most frequent of origin of thyroid metastases and represents 12 to 34% of all secondary thyroid tumors [16–18]. 18.4% (196/1065) modules included genes that are significantly prevalent in fatty acid degradation pathways. Cellular proliferation requires fatty acids for synthesis of membranes and signaling molecules. Dysregulation of cellular proliferation is associated with the occurrence of cancer.

2.4. Gene Module-based eQTL Analysis

We performed eQTL analysis on differential gene modules. The mutation of *VHL*, a known ccRCC causative gene, was found to be significantly associated with multiple DGMs (FDR < 0.03, Fig. 3A and B). These modules were enriched of genes in MAPK signaling pathway, apoptosis, pathways in cancer ($P < 0.03$, hypergeometric test).

To further prioritize the most significant somatic mutations that were associated with the differential gene modules, we employed FDR < 0.0001 as the cutoff in the eQTL analysis. The number of somatic mutations that were significantly associated with each differential module was assessed. Overall, we found 780 of 1065 modules were significantly associated with at least one somatic mutation. The median number of associated somatic mutations with DGMs is 8 (Supp. Fig. 2). 188 modules were significantly associated with five or less somatic mutations (Table 2). Some mutated genes influenced many DGMs.

BAP1 and *PBPM1* mutations significantly impacted 42.5% (80 of 188) and 18.1% (34/188) of the DGMs (Table 2). *BAP1* loss has been reported to define a new class of ccRCC and acts as a tumor suppressor [4]. In addition, ccRCC patients with *BAP1* somatic mutations had poor 5-year survival rates ($P < 0.014$, Fig. 4A). *PBRM1* encodes a protein that changes chromatin structure and influences p53 transcriptional activity. The previous study suggested that PBRM1 protein is regulated by p53-induced protein degradation in renal cell carcinomas [19]. Interestingly,

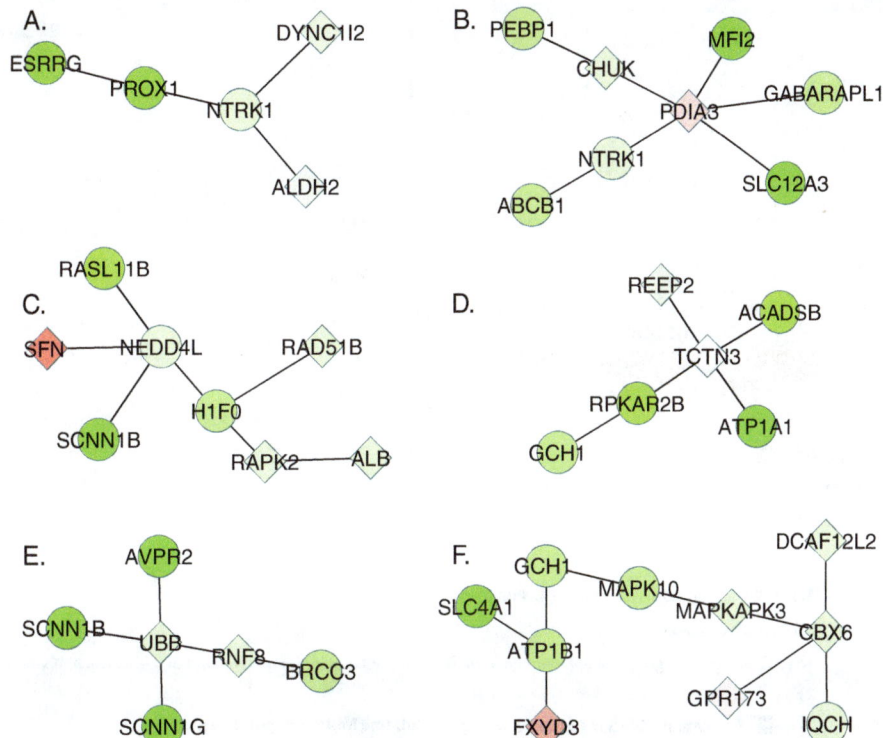

Fig. 3. The examples of differentially expressed gene modules. The genes colored in red were up-regulated, whereas genes colored in green were down-regulated in ccRCC. The intensity of the color is proportioned to log2 fold-change of the gene expression. Circle nodes refer to the expression levels of genes that were significantly changed, whereas diamond nodes refer to the genes without significantly altered expression levels.

Table 2
A total of 188 DGMs were significantly associated with five or less mutated genes (FDR < 0.0001).

Num. of mutated gene(s) associated with each DGM	Num. of associated DGMs	Mutated genes associated with the DGMs[*]
1	116	**BAP1(40)**, NOD2(18), **RRM1(13)**, CSRNP1(8), **PBRM1(8),** CNTN1(5), PGM5(3), RBM27(2), FAM19A1(2), KCNT2(2), PHKB(2), ZNF624(2), MARCH1(1), CACNA1E(1), IPO4(1), PLXNA2(1), SLC12A9(1), SLC15A4(1), SSRP1(1), TACC1(1), UBN2(1), ZNF844(1)
2	47	**BAP1(23), PBRM1(13), RRM1(11)**, NOD2(9), SLC4A2(9), TTLL1(9), CSRNP1(2), PHKB(2), March1(1), ATG3(1), ATG4C(1), CD180(1), CNTN1(1), EPHA1(1), EVPL(1), LAMB4(1), PARD6A(1), RBM26(1), SETD2(1), SPTBN1(1), TMEM17(1), ZNF711(1)
3	16	**BAP1(12), PBRM1(8),** CNTN1(4), CSRNP1(4), PHKB(3), SLC4A2(3), TTLL1(3), CHD8(2), FAM19A1(2), NOD2(2), CACNA1E(1), FRS2(1), **RRM1(1),** SPTBN1(1), UBA7(1)
4	4	**PBRM1(5),** **BAP1(4),** CAST(1), CD200R1(1), COL14A1(1), GRM3(1), NCOA5(1), PCOLCE2(1), ZNF572(1)
5	5	MEIS3(3), MOB3B(3), R3HCC1(3), SHISA5(3), WEE1(3), CAST(2), CD200R1(2), COL14A1(2), PCOLCE2(2), **BAP1(1), RRM1(1)**

[*] The number in the parentheses after a gene symbol represents the number of the DGMs that were linked with this mutated gene.

the mutations of somatic mutations of *BAP1* and *PBRM1* tend to be mutually exclusive ($P < 0.03$, Fisher Exact Test, Fig. 4C). *RRM1*, a target of five Food and Drug Administration (FDA) approved cancer drugs, were significantly associated with 13.8% (26/188) of differential modules. *RRM1* is involved in carcinogenesis, tumor progression, and induces metastasis suppression though PTEN-regulated pathways [20,21].

The other somatic mutated genes that significantly affected at least 5.3% of differential gene modules included *NOD2, RRM1, CSRNP1, SLC4A2, TTLL1* and *CNTN1*. The genetic alterations of these eight genes were significantly associated with poor survival rates of ccRCC patients ($P < 0.054$, Supp. Fig. 3). However, the mutations only presented in 2.6% (11/421) TCGA ccRCC patients. The association of the genetic mutations and survival rage need to be interpreted with caution. Presently, the

functional roles of eight genes in ccRCC have not yet been studied. Collectively, our results indicated that the gene module-based eQTL analysis yielded a list of putative disease genes, including known ccRCC genes such as *VHL, BAP1,* and *PBRM1,* as well as novel disease genes.

3. Methods

3.1. The Whole-exome and Transcriptome Data of ccRCC Patients

The whole-exome sequencing and RNA sequencing data were obtained from the TCGA data portal. The tumor and paired normal tissue samples were collected from newly diagnosed ccRCC patients having no prior treatment for this disease, including chemotherapy or

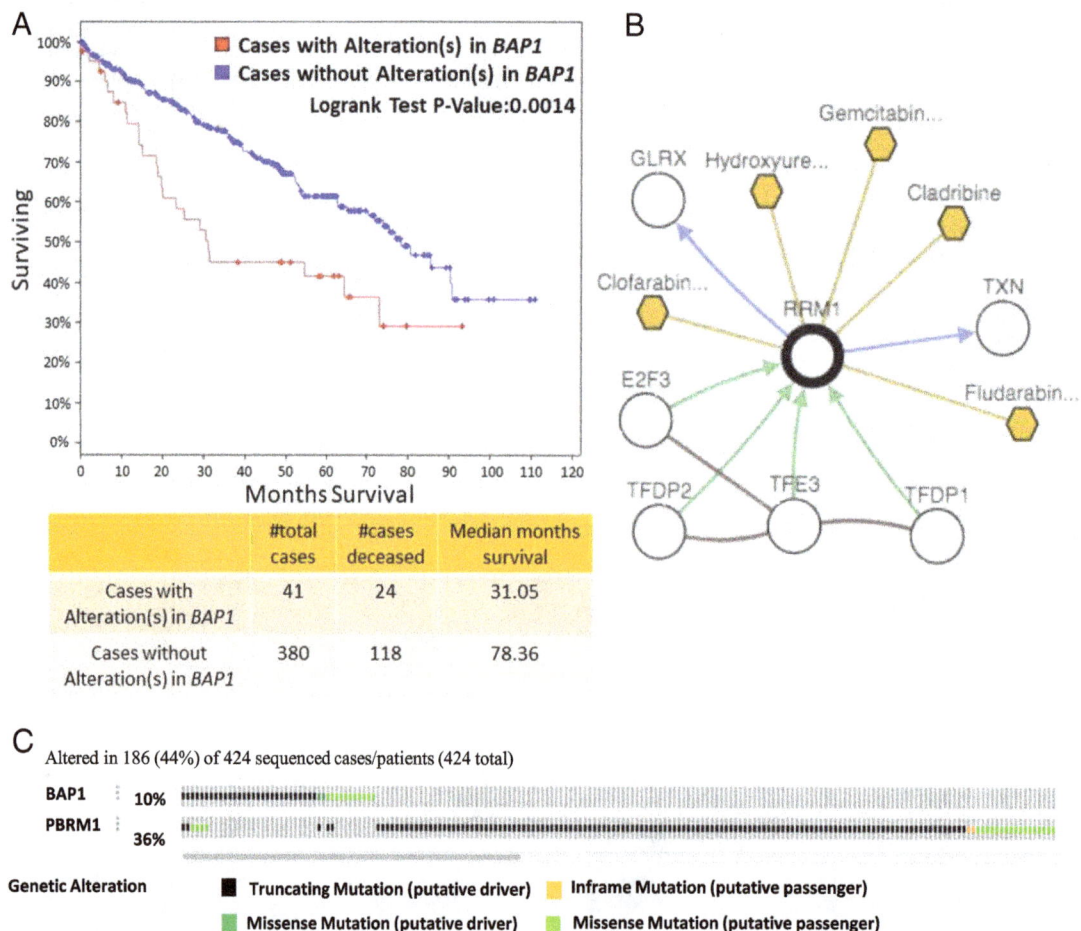

Fig. 4. The analysis of the genes harbored significant somatic mutations and were associated with the DGMs. (A) The ccRCC patients with *BAP1* somatic mutations had poor five-year survival rate. (B) Five FDA-approved cancer drugs (colored in gold) target at RRM1. (C) The mutation of BAP1 and PBRM1 tend to mutual exclusively at $P < 0.03$.

radiotherapy. The sequencing data were generated using Illumina Hiseq 2000 pair-end sequencing. The ccRCC patients consisted of 65% male, 35% females, and represented 93% Caucasian, 3.4% African American/ Black, and 1.6% Asian. The median age of patients at diagnosis is 60.9 years [9]. The whole-exome sequencing reads from 417 of paired tumor and normal tissue samples were aligned to human reference genome using Blat-like Fast Accurate Search Tool (BFAST) [22]. Then, genome-wide somatic mutations were detected using the MuTect algorithm [23]. The RNA sequencing data was generated from 539 tumor and 72 matched normal tissue samples. After poor quality reads were removed using the srf2fastq tool (Staden package), the RNA sequencing reads were aligned to reference transcript database using BWA algorithm [24]. The genes that have the mean Fragments Per Kilobase of transcript per Million mapped reads (FPKM) values < 0.1 in tumor samples as well as normal samples were removed from the expression profile. The differential expressed genes were detected using edgeR package in R [25]. The output of edgeR includes fold change and false discovery rates for individual genes. The gene that satisfied two criteria simultaneously, FC > 2 and FDR < 0.01, were considered as differentially expressed genes.

We also attained three independent ccRCC microarray datasets from the Gene Expression Omnibus database (GEO) for validation. The first dataset (GSE36895) contained 23 normal and 29 tumor tissue samples of a group of ccRCC patients [4]. The second dataset (GSE46699) contained 65 normal and 65 ccRCC paired samples of individual patients [26]. The third dataset (GSE40435) contained 101 normal and 101 paired ccRCC tissue samples [27]. The microarray data was generated using Affymetrix Human Genome U133 Plus 2.0 arrays. We applied t-test followed by Benjamini-Hochberg multiple test correction to identify the differentially expressed genes in ccRCC.

3.2. Protein-protein Interactions

Protein-protein interaction (PPI) data were obtained by combining five public PPI databases: intAct, MINT, BioGrid, DIP, and Reactome. The intAct database is quite comprehensive, containing information from 12 databases such as MINT, UniPort, mpidb, etc. Only human PPIs were selected for our study. For instance, we used "homo_sapiens.mitab.interactions" to attain human PPI in humans in the Reactome database. After filtering the redundancy of the union set of all PPIs, we attained a total of unique 440,747 PPIs. Then, we performed two expression correlation assessments using the expression profile of TCGA ccRCC tumor and normal tissue samples. The first correlation assessment is simply the correlation of the expression levels of the two genes across all tissue samples, while the second is the difference in the expression level correlations of the two genes between tumor and normal tissues. The PPI pairs that rank simultaneously below 5% in both correlation assessments were removed. A total of 319,291 PPIs were retained for differential gene module construction.

3.3. Gene Modules Construction

The expression levels of individual genes were normalized across samples by Z-score transformation ($\mu = 0$). A gene module started with a seed gene. Then, more genes were added interactively into the module based on PPIs and mutual information assessment. According known PPIs at each step, all genes interacting with any gene members in the current modules were evaluated by mutation information. Mutation information measures the degree to which two random variables are independent. When a random variable X is independent of another random variable Y, the resulting mutation information is 0. Here, we tested for whether the expression levels of gene modules (X) are associated with tissue types (Y) (ccRCC versus normal). The candidate gene that maximized the mutual information was selected. Here, we referred to the value of mutual information as the activity score. As X is a discrete variable, we discretized normalized expression level (Z) by

dividing the range of Z into equally spaced bins defined by split points s_k, resulting in the following expression for the activity score calculation:

$$AS(X;Y) = \sum_k \sum_y P(s_k < X < s_{k+1}, Y = y) \log_2 \frac{P(s_k < X < s_{k+1}, Y = y)}{P(s_k < X < s_{k+1})P(Y = y)}$$

At each iterative step, we calculated the improvement of the activity score. The searching procedure was terminated if there is no further improvement by adding new genes into the gene module. Then, we performed three statistical tests to assess the significance of all gene modules. We permuted the tissue phenotype 1000 times to obtain the null distribution in order to test the hypothesis that the gene module is significantly associated with tissue phenotypes. Then, we constructed the other two null distributions by randomly selecting the same number of genes as the gene module retaining seed genes, both with and without seed genes, 1000 times to test the hypothesis that the gene modules are significantly different from the background.

3.4. A SVM-based Classifier

A SVM R package "e1071" based on widely used "libsvm" was applied to build the classifier. We adopted "sigmoid" as the kernel function, and default values for all other parameters. The normalized average expression levels of genes in DGMs were used as features for the differential gene module-based classifier.

3.5. eQTL Analysis

We used Matrix eQTL to assess the association of somatic mutations and differential gene modules [28]. The linear regression model was adopted in the eQTL analysis. If any gene member in the module had significant associations with the point somatic mutations (FDR < 0.0001), we considered the mutations to be associated with the gene module.

4. Discussion

The molecular pathogenesis of many cancer types, including ccRCC, is poorly understood, and can be partially attributed to a limited understanding about comprehensive causative genes and pathways that govern disease initiation and development. The method we developed in this study included two stages: constructing differentially expressed-gene modules and identifying causative mutated genes associated with gene modules. The results yielded by both steps can lead to an expansion of current ccRCC genes and pathways sets.

The differential gene modules were built by coupling known PPIs and expression profiles of ccRCC patients. The number of PPIs has been increasing exponentially in recent years. On the contrary, databases of pathways remain incomplete and largely generic. At present, the majority of the pathways represent summaries of the most conserved components of such pathways and not necessarily what really occurs in each individual case. In addition, pathways can change between tissues, cell types, individuals, and species [29]. Our method offered a way to dynamically discover the ways in which gene clusters function together in the disease state. Gene ontology and pathways enrichment analysis suggested differential gene modules presented a set of genes that function together in the same biological process related to the diseases. Given that the median size of gene modules is six, 90.4% and 45% of gene modules contained at least two genes were significantly associated with biological process and KEGG pathways, respectively. In contrast, none of the random gene modules having the same topology and size as the differential gene modules had two genes associated with the same biological process or pathways. Thus, our results have lead to the discovery pathways involved in ccRCC.

We used normalized expression levels of gene modules as input features to build a SVM-based classifier for predicating tissue types. The classifier achieved over 0.97 AUC in classifying over 600 TCGA ccRCC tissue samples, compared to 0.77 for a SVM classifier using the expression levels of individual genes as input features. The differential gene modules-based classifier achieved over 0.92 AUC for prediction three independent ccRCC patient cohorts (GSE36895, GSE46699, and GSE40435). Thus, the DGMs could be used as molecular signatures to infer tissue phenotypes.

The eQTL has often been applied on transcript-SNP pairs. Here, we implemented the eQTL mapping to the differential gene modules. As the gene modules represent gene clusters in the same pathway, the significant somatic mutations can be linked directly to the disease-affected pathways and suggested potential association between mutations and the pathways. The known ccRCC genes, *BAP1* and *PBRM1*, were revealed by our study. The mutations of *BAP1* and *PRMB1* were the most frequently associated with DGMs (Table 2). The *BAP1* encodes a protein called ubiquitin carboxyl-terminal hydrolase BRCA1-associated protein 1 (BAP1). The BAP1 is associated with multi-protein complex, which regulated several crucial cellular pathways including cell cycle, cell death, the DNA damage response and gluconeogenesis [30]. BAP1 is inactive in 15% of ccRCCs and the loss of BAP1 has defined a new class of ccRCC [31]. The germline mutation of *BAP1* has been associated with high risk of neoplasms [32]. *PRBM1* (Polybromo 1), a SWI/SNF chromatin remodeling complex gene, is frequently mutated in ccRCC [33].

A set of new genes *NOD2, RRM1, CSRNP1, SLC4A2, TTLL1* and *CNTN1*, as well as their associated gene modules, were identified. The mutation of *NOD2* can lead to impaired activation of NFKB in vitro [34] and has associated with colorectal, ovarian and breast cancer [35–37]. *RRM1* is reported as metastasis suppressor gene by inducing expression of PTEN [38]. Currently, five FDA approved cancer drug target at *RRM1* (Fig. 3). *CSRNP1* involves in apoptotic process and may play a role in apoptosis [39]. *SLC4A2* encodes anion exchanger 2 (AE2) and AE2 has been associated with multiple cancer types [40]. CNTN1, a protein encoded by *CNTN1*, promoted lung cancer invasive and metastasis [41]. Despite that these genes have been linked to the tumorgenesis of various cancer, their roles in ccRCC have not been extensively studied yet. The genes and their associated DGMs can offer guidance to perform experiments to further validate their functional roles in ccRCC. Thus, our two-stage method provides a new way for identifying new disease genes and their affect pathways.

To date, PPI databases may still contain false positives, e.g., bias in the PPI experiments (some proteins have been studied more than others). Our co-expression assessments may help to reduce the negative effect. Additionally, our results suggested that eQTL analysis could prioritize disease candidate genes, however, true associations may be overlooked and further experimental validation may be needed. The eQTL analysis was based on transcription level, which are quantitative traits relying on accurate and precise measurement of gene expression. Additionally, the sample size may limit the sensitivity for identifying true associations. Similar to the GWAS study, increasing sample size will lead to more association discovery. On the other hand, eQTL analysis may introduce false positives. The significant expression difference between normal and tumor could also attribute to the other somatic alterations such as copy number variations and methylation events. Nevertheless, our study yielded novel candidate genes for further experimental validation, which could potentially advance our understanding of ccRCC.

5. Conclusions

Our method integrated whole-exome sequencing data, transcriptome, and PPIs to identify disease genes. These genes harbored somatic mutations that significantly impacted the expression alteration of differential gene modules. The differential gene modules were shown to function in the biological process and their expression levels can be used as molecular signatures to predict unknown tissue types. Our results confirmed several known ccRCC causative as well as novel genes involved in diseases.

Competing Interests

The authors declare that they have no competing interests.

Acknowledgements

This study was supported in part by United States National Institutes of Health (NIH) Academic Research Enhancement Award 1R15GM114739 and National Institute of General Medical Sciences (NIH/NIGMS) 5P20GM103429, United States Food and Drug Administration (FDA) HHSF223201510172C through Arkansas Research Alliance (ARA) BAA-15-00121 and Arkansas Science and Technology Authority (ASTA) Basic Science Research 15-B-23 and 15-B-38.

References

[1] Siegel RL, Miller KD, Jemal A. Cancer statistics, 2016. CA Cancer J Clin Jan-Feb, 2016; 66(1):7–30.
[2] Baldewijns MM, van Vlodrop IJ, Schouten LJ, Soetekouw PM, de Bruine AP, van Engeland M. Genetics and epigenetics of renal cell cancer. Biochim Biophys Acta Apr, 2008;1785(2):133–55.
[3] Brugarolas J. Renal-cell carcinoma—molecular pathways and therapies. N Engl J Med Jan 11, 2007;356(2):185–7.
[4] Pena-Llopis S, Vega-Rubin-de-Celis S, Liao A, Leng N, Pavia-Jimenez A, Wang S, et al. BAP1 loss defines a new class of renal cell carcinoma. Nat Genet Jun 10, 2012;44(7): 751–9.
[5] Weiss RH, Lin PY. Kidney cancer: identification of novel targets for therapy. Kidney Int Jan, 2006;69(2):224–32.
[6] Rini BI, Campbell SC, Escudier B. Renal cell carcinoma. Lancet Mar 28, 2009; 373(9669):1119–32.
[7] Varela I, Tarpey P, Raine K, Huang D, Ong CK, Stephens P, et al. Exome sequencing identifies frequent mutation of the SWI/SNF complex gene PBRM1 in renal carcinoma. Nature Jan 27, 2011;469(7331):539–42.
[8] Dalgliesh GL, Furge K, Greenman C, Chen L, Bignell G, Butler A, et al. Systematic sequencing of renal carcinoma reveals inactivation of histone modifying genes. Nature Jan 21, 2010;463(7279):360–3.
[9] Comprehensive molecular characterization of clear cell renal cell carcinoma, Nature Jul 04, 2013;499(7456):43–9.
[10] Tomaszewski JJ, Uzzo RG, Smaldone MC. Heterogeneity and renal mass biopsy: a review of its role and reliability. Cancer Biol Med Sep, 2014;11(3):162–72.
[11] Yang W, Yoshigoe K, Qin X, Liu JS, Yang JY, Niemierko A, et al. Identification of genes and pathways involved in kidney renal clear cell carcinoma. BMC Bioinf 2014; 15(Suppl. 17):S2.
[12] Littlejohn MD, Tiplady K, Fink TA, Lehnert K, Lopdell T, Johnson T, et al. Sequence-based association analysis reveals an MGST1 eQTL with pleiotropic effects on bovine milk composition. Sci Rep May 05, 2016;6:25376.
[13] Deelen P, Zhernakova DV, de Haan M, van der Sijde M, Bonder MJ, Karjalainen J, et al. Calling genotypes from public RNA-sequencing data enables identification of genetic variants that affect gene-expression levels. Genome Med 2015;7(1):30.
[14] Breitling R, Li Y, Tesson BM, Fu J, Wu C, Wiltshire T, et al. Genetical genomics: spotlight on QTL hotspots. PLoS Genet Oct, 2008;4(10):e1000232.
[15] Lawn S, Krishna N, Pisklakova A, Qu X, Fenstermacher DA, Fournier M, et al. Neurotrophin signaling via TrkB and TrkC receptors promotes the growth of brain tumor-initiating cells. J Biol Chem Feb 06, 2015;290(6):3814–24.
[16] Medas F, Calo PG, Lai ML, Tuveri M, Pisano G, Nicolosi A. Renal cell carcinoma metastasis to thyroid tumor: a case report and review of the literature. J Med Case Reports Dec 10, 2013;7:265.
[17] Heffess CS, Wenig BM, Thompson LD. Metastatic renal cell carcinoma to the thyroid gland: a clinicopathologic study of 36 cases. Cancer Nov 01, 2002;95(9):1869–78.
[18] Zamarron C, Abdulkader I, Areses MC, Garcia-Paz V, Leon L, Cameselle-Teijeiro J. Metastases of renal cell carcinoma to the thyroid gland with synchronous benign and malignant follicular cell-derived neoplasms. Case Rep Oncol Med 2013;2013: 485025.
[19] Macher-Goeppinger S, Keith M, Tagscherer KE, Singer S, Winkler J, Hofmann TG, et al. PBRM1 (BAF180) protein is functionally regulated by p53-induced protein degradation in renal cell carcinomas. J Pathol Dec, 2015;237(4):460–71.
[20] Zheng Z, Chen T, Li X, Haura E, Sharma A, Bepler G. DNA synthesis and repair genes RRM1 and ERCC1 in lung cancer. N Engl J Med Feb 22, 2007;356(8):800–8.
[21] Gautam A, Li ZR, Bepler G. RRM1-induced metastasis suppression through PTEN-regulated pathways. Oncogene Apr 10, 2003;22(14):2135–42.

[22] Homer N, Merriman B, Nelson SF. BFAST: an alignment tool for large scale genome resequencing. PLoS One Nov 11, 2009;4(11):e7767.

[23] Cibulskis K, Lawrence MS, Carter SL, Sivachenko A, Jaffe D, Sougnez C, et al. Sensitive detection of somatic point mutations in impure and heterogeneous cancer samples. Nat Biotechnol 2013, Mar;31(3):213–9.

[24] Li H, Durbin R. Fast and accurate short read alignment with burrows-wheeler transform. Bioinformatics Jul 15, 2009;25(14):1754–60.

[25] Robinson MD, McCarthy DJ, Smyth GK. edgeR: a Bioconductor package for differential expression analysis of digital gene expression data. Bioinformatics 2010, Jan 01; 26(1):139–40.

[26] Ji ZC, Ji HK. TSCAN: pseudo-time reconstruction and evaluation in single-cell RNA-Seq analysis. Nucleic Acids Res 2016;44(13):e117 [PMC. Web 16 Sept. 2017].

[27] Wozniak MB, Calvez-Kelm FL, Abedi-Ardekani B, Byrnes G, Durand G, Carreira C, et al. Integrative genome-wide gene expression profiling of clear cell renal cell carcinoma in Czech Republic and in the United States. In: Hoque Mohammad O, editor. PLoS ONE, Vol. 8.3. ; 2013. p. e57886 [PMC. Web. 16 Sept. 2017].

[28] Shabalin AA. Matrix eQTL: ultra fast eQTL analysis via large matrix operations. Bioinformatics May 15, 2012;28(10):1353–8.

[29] Werner T. Next generation sequencing in functional genomics. Brief Bioinform Sep, 2010;11(5):499–511.

[30] Carbone M, Yang H, Pass HI, Krausz T, Testa JR, Gaudino G. BAP1 and cancer. Nat Rev Cancer 2013;13(3):153–9.

[31] Peña-Llopis S, Samuel, Vega-Rubín-de-Celis S, Liao A, Leng N, Pavía-Jiménez A, et al. BAP1 loss defines a new class of renal cell carcinoma. Nat Genet 2012;44(7):751–9.

[32] Carbone M, Ferris LK, Baumann F, Napolitano A, Lum CA, Erin G, et al. BAP1 cancer syndrome: malignant mesothelioma, uveal and cutaneous melanoma, and MBAITs. J Transl Med 2012;10:179.

[33] Varela I, Tarpey P, Raine K, Huang D, Ong CK, Stephens P, et al. Exome sequencing identifies frequent mutation of the SWI/SNF complex gene PBRM1 in renal carcinoma. Nature 2011:539–42.

[34] Folwaczny M, Glas J, Török H-P, Mauermann D, Folwaczny C. The 3020insC mutation of the NOD2/CARD15 gene in patients with periodontal disease. Eur J Oral Sci 2004: 316–9.

[35] Kurzawski G, Suchy J, Kładny J, Grabowska E, Mierzejewski M, Jakubowska A, et al. The NOD2 3020insC mutation and the risk of colorectal cancer. Cancer Res 2004: 1604–6.

[36] Magnowski P, Medrek K, Magnowska M, Stawicka M, Kedzia H, Górski B, et al. The 3020insC NOD2 gene mutation in patients with ovarian cancer. Ginekol Pol 2008: 544–9.

[37] Huzarski T, Lener M, Domagała W, Gronwald J, Byrski T, Kurzawski G, et al. The 3020insC allele of NOD2 predisposes to early-onset breast cancer. Breast Cancer Res Treat 2011:539–42.

[38] Gautam A, Li Z-R, Bepler G. RRM1-induced metastasis suppression through PTEN-regulated pathways. Oncogene 2003:2135–42.

[39] Ishiguro H, Tsunoda T, Tanaka T, Fujii Y, Nakamura Y, Furukawa Y. Identification of AXUD1, a novel human gene induced by AXIN1 and its reduced expression in human carcinomas of the lung, liver, colon and kidney. Oncogene 2001:5062–6.

[40] Zhang L-J, Lu R, Song Y-N, Zhu JY, Xia W, Zhang M, et al. Knockdown of anion exchanger 2 suppressed the growth of ovarian cancer cells via mTOR/p70S6K1 signaling. Sci Rep 2017;7(1):6362.

[41] J-L Su, Yang C-Y, Shih J-Y, Wei L-H, Hsieh C-Y, Jeng Y-M, et al. Knockdown of contactin-1 expression suppresses invasion and metastasis of lung adenocarcinoma. Cancer Res 2006:2553–61.

20

Temperature dependent dynamics of DegP-trimer: A molecular dynamics study

Nivedita Rai, Amutha Ramaswamy *

Centre for Bioinformatics, School of Life Sciences, Pondicherry University, Puducherry 605014, India

ARTICLE INFO

Keywords:
DegP-trimer
PDZ1
PDZ2
LA loop
Molecular dynamics simulation
Principal component analysis

ABSTRACT

DegP is a heat shock protein from high temperature requirement protease A family, which reacts to the environmental stress conditions in an ATP independent way. The objective of the present analysis emerged from the temperature dependent functional diversity of DegP between chaperonic and protease activities at temperatures below and above 28 °C, respectively. DegP is a multimeric protein and the minimal functional unit, DegP-trimer, is of great importance in understanding the DegP pathway. The structural aspects of DegP-trimer with respect to temperature variation have been studied using molecular dynamics simulations (for 100 ns) and principal component analysis to highlight the temperature dependent dynamics facilitating its functional diversity. The DegP-trimer revealed a pronounced dynamics at both 280 and 320 K, when compared to the dynamics observed at 300 K. The LA loop is identified as the highly flexible region during dynamics and at extreme temperatures, the residues 46–80 of LA loop express a flip towards right (at 280) and left (at 320 K) with respect to the fixed β-sheet connecting the LA loop of protease for which Phe46 acts as one of the key residues. Such dynamics of LA loop facilitates inter-monomeric interaction with the PDZ1 domain of the neighbouring monomer and explains its active participation when DegP exists as trimer. Hence, the LA loop mediated dynamics of DegP-trimer is expected to provide further insight into the temperature dependent dynamics of DegP towards the understanding of its assembly and functional diversity in the presence of substrate.

1. Introduction

Every living organism has special mechanisms to respond against a variety of stress conditions such as temperature, pH, and oxidative stress. The heat shock proteins are a group of evolutionarily conserved proteins which respond to stress conditions by protein folding/degradation in the presence of ATP [1,2]. HtrA (high temperature requirement protease A) is one such heat shock protein, which fulfils these roles without consuming ATP. It is a widely conserved protein lying on the extra-cytosolic compartment of prokaryotes as well as eukaryotes. There are various homologues of HtrA found in *Escherichia coli* such as DegP, DegQ and DegS [3,4]. DegP is an essential protein of *E. coli* which behaves as molecular chaperone and protease under stress condition in a temperature dependent manner for cell survival. It is experimentally reported that below 28 °C, DegP behaves as chaperone, but efficiently degrades unfolded protein above 28 °C [5,6].

DegP shares a common structural architecture with all the HtrA proteins which comprises serine protease and PDZ (*P*ostsynaptic density

protein/*D*lg1/*Z*o1) domains. A mature DegP monomer has 448 amino acid residues in which residues 1–259 form N-terminal serine protease domain and two PDZ domains at the C-terminal (namely PDZ1 and PDZ2) are formed by residues 260–358 and 359–448, respectively [7]. The protease domain also contains (i) catalytic triad formed by Ser210, His105 and Asp135 residues and (ii) activation loops: LD (168–175), L1 (205–209), L2 (227–238), L3 (185–198) and LA (36–81) [8,9]. The LA loop of DegP contains a disulphide bridge between residues Cys57 and Cys69, which makes the structure comparatively compact and reduction of this S–S bond leads to autocatalysis of DegP [10,11]. The N-terminal protease domain is not only required for the protease activity but also plays a vital role in chaperonic activity, whereas the C-terminal domain involves in substrate interaction and DegP oligomerization [12]. The protease domain is connected with PDZ1 domain by a flexible loop region where Arg262 and Gly263 act as a hinge and play a key role in changing the orientation of PDZ1 domain. The PDZ2 domain is connected to PDZ1 via flexible linker residues Ser357–Ser364. Both these linkers are essential for protease activity of hexameric DegP [13].

DegP is a multimeric protein that exists in active and inactive forms [14]. The minimal functional unit of DegP is a trimer which is capable of performing both protease and chaperonic activities [15]. This DegP

* Corresponding author.
E-mail addresses: amutha_ramu@yahoo.com, ramutha@bicpu.edu.in
(A. Ramaswamy).

trimer (hereafter referred as DegP-trimer) is stabilized by the interaction between the protease domains of adjacent monomers. DegP exists in an inactive state when two trimeric units form a hexamer. In this hexameric form, the LA loop of each monomers in one trimeric subunit interacts with L1, L2 and LD loops of other trimer and blocks the catalytic site due to which, the hexameric unit adopts resting/inactive state. In this resting/inactive state, DegP exists either in open or closed form [16]. In the open form, the two trimeric units are connected only via the LA loops and provide a wide lateral passage for the substrate binding. Whereas, in the closed form, in addition to the LA loop interaction (as observed in open form), there is cross-talk between the PDZ domains of the two trimeric units, i.e., the PDZ1 of one trimeric unit interacts with the PDZ2 of other trimeric subunit [17–19]. As a consequence of these domain rearrangements, the proteolytic site is completely shielded from the solvent [13]. The protease and PDZ1 domains govern the protease activity while the protease domain alone is sufficient to express chaperonic activity [15]. The PDZ1 domain of DegP contains a deep substrate binding hydrophobic cleft referred as E–L–G–I pocket (similar to G–L–G–F motif of serine protease family) formed by the carboxylate binding loop residues from 264 to 267 [20]. When substrate binds to the inactive hexamer, the hydrophobic cleft of PDZ1 domain recognizes the C-terminal residues of the substrate and places it into the inner cavity formed by the protease domain [21–23]. This binding signal is then transmitted to the activation loops (L1, L2 and LA) via the sensory L3 loop and mediates the activation of hexamer by dissociating into transient trimeric units. Subsequently, the trimers assemble into catalytically active higher order oligomers such as 12-mer and 24-mer. These oligomers are reverted to the inactive hexameric form after the completion of its function [24].

Molecular dynamics simulation methods have been successfully applied in exploring the structural and functional aspects of various bio-molecules including conformational transition/diversity in nanosecond timescale [25–28]. Various studies on temperature dependent simulations have demonstrated its ability to explore the functional diversity of biomolecules with respect to temperature [29–31]. Principal component analysis (PCA) is a useful technique in analysing the trajectories from molecular dynamics simulations. PCA transforms a number of possibly correlated variables into a smaller set of linear variables and this reduction in dimension gives low frequency independent subspace, called as essential subspace, in which the functionally relevant motions occur [32,33]. The total mobility of a system is described by the sum of eigenvalues, in which the first few eigenvectors with higher percentage of variance describe the collective dynamics of the molecule. Hence, the PCA of molecular dynamics trajectory effectively differentiates the low frequency collective motions and high frequency localized motions [34,35]. Recently, the allosteric regulation of HtrA2 by PDZ domain is well documented using molecular dynamics simulations [36]. In this present work, the structural dynamics of DegP-trimer is studied at various temperatures like 280, 300 and 320 K using molecular dynamics simulations. The present study is a first report explaining the structural dynamics of DegP-trimer with special emphasis on the effect of temperature using computational methods.

2. Materials and methods

2.1. Modelling of DegP-trimer

The crystallographic structure of DegP deposited in the Protein Data Bank (PDB ID: 1KY9) was used for the present study. DegP contains 448 amino acid residues, out of which, the structure of residues forming regulatory loop regions (Asp52–Gly78 and Ser188–Tyr195) and other residues like Gly370–Ala374 and Met447–Gln448 are not reported. Structure of the LA loop region (Asp52–Gly78) was predicted using ab-initio structure prediction tool Quark [37] and the other missing residues were modelled using SPDBV v4.0.1 software [38]. The monomeric coordinates were replicated using VMD

software according to the symmetry reported in crystallization studies to model the structure of DegP-trimer [39]. This modelled and validated DegP-trimer was used for further molecular dynamics simulation studies.

2.2. Molecular dynamics simulations

MD simulations of DegP-trimer at various temperatures (280, 300 and 320 K) were performed in GROMACS 4.5.5 for a period of 100 ns using GROMOS43a1 force field [40]. The structure of DegP-trimer was solvated using SPC/E water model extending 12 Å from the extents of DegP and was neutralized by replacing six water molecules with Cl^- ions. The system was initially relaxed using the steepest descent algorithm followed by conjugate gradient algorithm and the minimization was automatically truncated when the force is lesser than 1000.0 kJ/mol/nm. The long range electrostatic interactions were evaluated using the Particle Mesh Ewald method [41] with a cut-off of 10 Å. All non-bonded interactions were treated by the Lennard–Jones interaction with a cut-off of 10 Å. The bonds involving hydrogen atoms were constrained using LINCS algorithm [42]. Constant temperature was maintained using V-rescale thermostat for NVT ensemble with a coupling constant of 0.1 ps [43]. The Parrinello–Rahman Barostat for NPT ensemble was used at a constant pressure of 1 bar, coupling constant of 2.0 ps and compressibility of $4.5e-5$/bar [44]. The entire system was equilibrated using NVT and NPT ensembles for 500 and 100 ps, respectively. Finally, the system was simulated without any constraints using NPT ensemble for a period of 100 ns to understand the temperature dependent structural dynamics of DegP-trimer.

2.2.1. Cross correlation analysis

The cooperative domain motions of DegP-trimer were analysed using the correlation matrix R_{ij} generated for the Cα-atoms (N = 1314) ranging between −1 and 1 [45,46]. The extent of displacement between the residues is calculated by correlation coefficient of each pairs of Cα atoms i and j as given below.

$$R_{ij} = \frac{\langle \Delta r_i \cdot \Delta r_j \rangle}{\sqrt{\langle \Delta r_i^2 \rangle \langle \Delta r_j^2 \rangle}}$$

where, Δr_i and Δr_j are the displacements from the mean position of ith and jth of Cα atoms and are averaged over the entire trajectory. Cross-correlation matrix is generated by using cpptraj module of Amber Tools 13 [47].

2.2.2. Principal component analysis

PCA is one of the convenient methods to examine the structural evolutions explored by a MD trajectory. PCA is based on the covariance matrix, which captures the degree of co-linearity of atomic motions describing the internal dynamics. The principal components (PCs) are obtained by the orthogonal transformation of protein Cartesian coordinates [48]. The covariance matrix (C_{ij}) is calculated from the mass-weighted Cartesian coordinates (i and j) of the N-particle system sampled over the simulated trajectory and is given by

$$C_{ij} = \langle (x_i - \langle x_i \rangle)(x_j - \langle x_j \rangle) \rangle.$$

The covariance matrix C_{ij} is diagonalized by an orthogonal coordinate transformation matrix R to get the diagonal matrix with eigenvalues λ

$$R^T C_{ij} R = I\lambda.$$

a

b

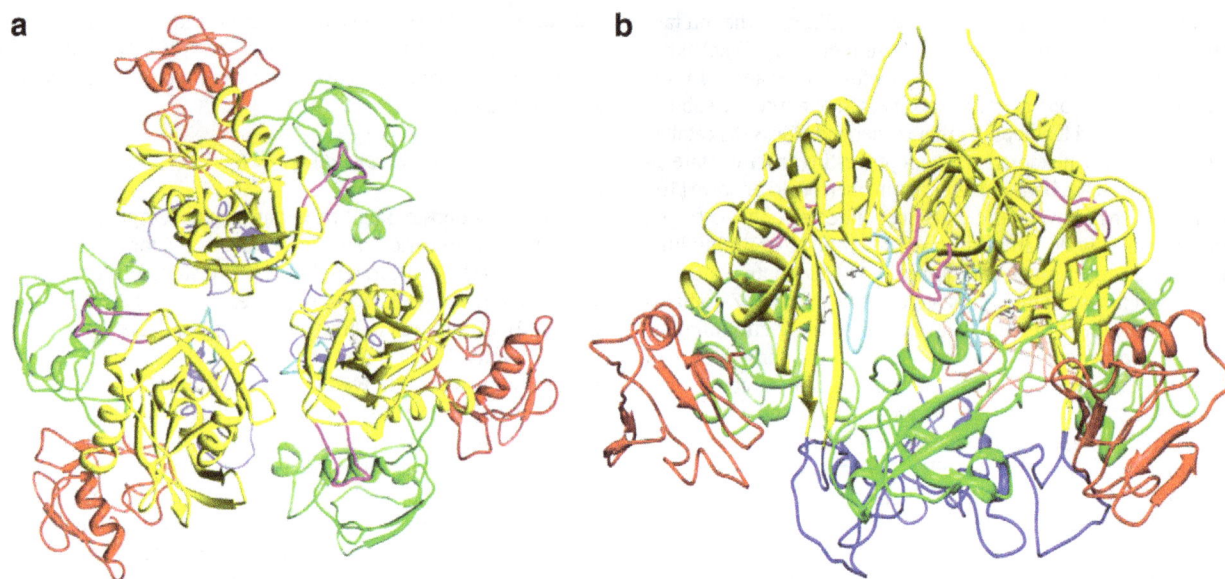

Fig. 1. The structure of modelled DegP-trimer in top (a) and side (b) views. The LA, L2 and L3 loops are coloured as blue, cyan and magenta, respectively and the rest of protease domain is coloured as yellow. The domain PDZ1 is in green colour and the PDZ2 is coloured red.

Here, R represents the eigenvectors or principal modes and I is an identity matrix of dimension 3N and λ represents eigenvalues. Each eigenvalue is associated to an eigenvector, which gives the direction of the new coordinate. The projection of eigenvectors with respect to these eigenvalues gives the PCs (p_i, where $i = 1, ..., 3N$).

$$p = R^T(x - \langle x \rangle)$$

Here, the eigenvalue λ denotes the mean-square fluctuation in the direction of respective principal mode for the DegP-trimer having 1314 Cα-atoms with 3942 eigenvectors.

3. Results and discussion

The structure of DegP-trimer was modelled and validated using the Ramachandran plot in which 84.2% residues come under the most favoured regions and 13.8% residues are present in additionally favoured regions. The modelled structure of DegP-trimer is shown in Fig. 1. All atom simulations performed on the modelled DegP-trimer for a period of 100 ns were analysed systematically. It is reported that DegP expresses chaperonic activity at temperature below 28 °C and involves in protease activity above 28 °C [49]. In line with this report, the structural dynamics of DegP-trimer was studied at various temperatures (280, 300 and 320 K) to elucidate the temperature dependent structural aspects behind the functional diversity of DegP.

The root mean square deviation (RMSD) calculated for the backbone of DegP-trimer during the simulation of 100 ns is shown in Fig. 2(a). In the simulations performed at 280, 300 and 320 K, DegP-trimer expressed stable dynamics after 20 ns and hence, the analyses of DegP-trimer have been performed on the trajectories observed between 20 and 100 ns. DegP-trimer at 280 K expresses stable dynamics with RMSD about 0.75 nm, whereas, at both 300 and 320 K, it expressed comparatively increased dynamics with RMSD closer to 1.15 nm. The RMSD of individual monomers of DegP was examined further to analyse such higher RMSD. Supplementary Fig. 1 clearly explains the RMSD of protease domain (with and without LA loop region), PDZ1 and PDZ2 domains as well. The analysis revealed that the domains of DegP monomers are highly stable during dynamics and the observed higher RMSD is a consequence of LA loop dynamics as well as inter-monomeric interactions.

The compactness of DegP-trimeric assembly is monitored (Fig. 2(b)) by calculating the radius of gyration (Rg). The DegP-trimer at 280 K is highly stable with a Rg value of 3.38 nm and is assembled more closely at both 300 and 320 K due to which a decrease in Rg value (3.17 and 3.07 nm, respectively) was observed. All these observations ensure a temperature dependent change in the configuration of DegP-trimer.

a

b

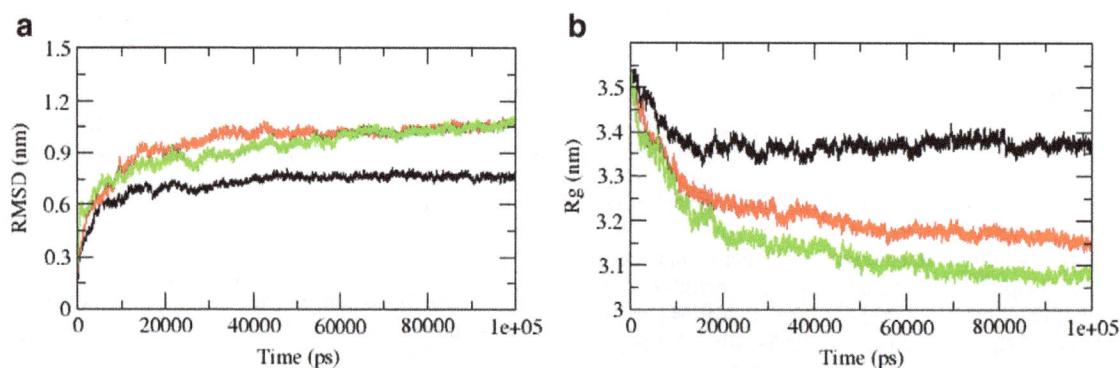

Fig. 2. The variation in backbone RMSD (a) and radius of gyration of Cα-atoms (b) of DegP-trimer observed during dynamics at 280 (black lines), 300 (red lines) and 320 K (green lines), respectively.

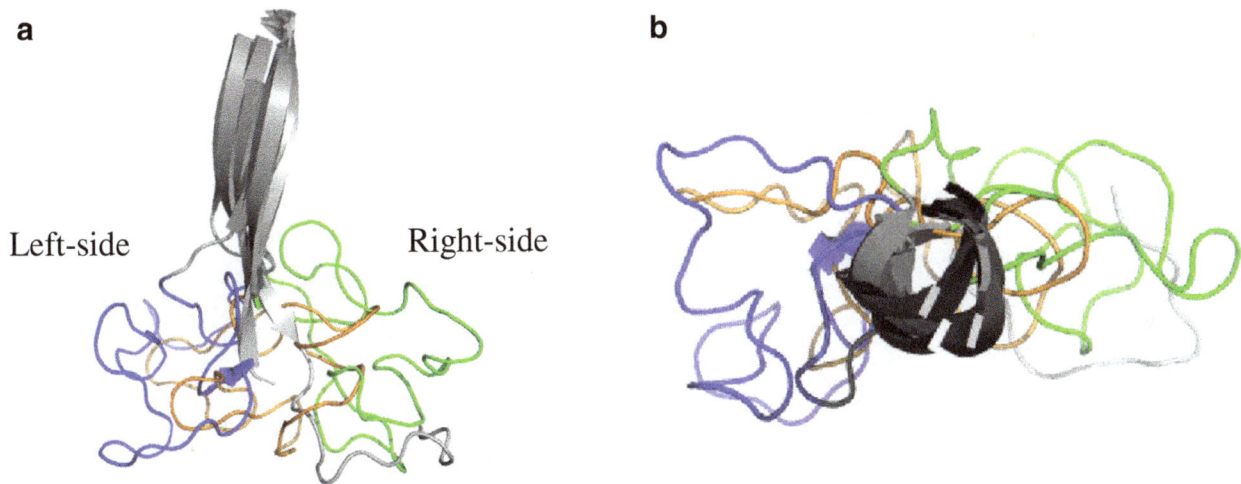

Fig. 3. Temperature dependent dynamics of LA loop displayed in side view (a) and top view (b) of monomer 1 simulated at 280, 300 and 320 K along with the cryo-EM (PDB ID: 2ZLE) structure and are coloured as green, orange, blue and grey, respectively.

3.1. Intra- and inter-monomeric interactions in DegP-trimer

The stability of DegP-trimer, governed by the cooperative and anti-cooperative intra- as well as inter-monomeric motions, is studied by plotting the cross-correlation matrices. Fig. 4 shows the correlated motions observed in DegP-trimer at various temperatures such as 280 (a), 300 (b) and 320 K (c), in which each monomer is highlighted by a box. The positive regions indicate cooperative motions between residues, where the strength of correlation increases from yellow to red (scaled between 0.1 and 1.0). The negative regions coloured in blue (scaled between -0.6 and -0.1) are defined by the anti-cooperative domain motions. The uncorrelated regions separating both correlated and anti-correlated are coloured in cyan (from -0.1 to 0.1).

DegP-trimer expresses strong intra- as well as inter-monomeric motions. The strong cooperative motions observed in DegP-trimer at 280 K are between: (i) LA loop of monomer 1 and PDZ1 (from Arg262 to Ala322 except Gln292 to Ala302) of monomer 2, and (ii) LA loop of monomer 2 with both LA loop and PDZ1 of monomer 3. In DegP trimer, the protease domain in all three monomers shares a strong cooperative motion among themselves. The LA loop of monomer 1 is anti-cooperative with the PDZ2 of monomer 2. The PDZ2 of monomer 2 shares a strong anti-cooperative motion with the PDZ1 of monomer 3. In general, it is also observed that all three monomers express a similar pattern of correlated and anti-correlated inter- as well as intra-domain motions at 280 K.

At 300 K, the prominent anti- cooperative motion of LA loop with respect to its protease domains is not expressive as observed at 280 K. Similarly, the anti-cooperative motion of PDZ2 of monomer 1 against the protease and PDZ1 of monomer 2 is also not expressed at 300 K. DegP-trimer ensures an inter-monomeric cooperative interaction of the LA loop of monomer 1 with the PDZ1 of monomer 2. In addition, the protease and PDZ2 (which moves cooperatively) of monomer 2 share a cooperative motion with the protease domain of monomer 1. The PDZ2 of monomer 2 moves together with both PDZ1 and PDZ2 domains of monomer 3. The amplitude of correlation is significantly diminished at 300 K when compared to the observed correlation at 280 K.

The anti-cooperative motions are marginally more pronounced when DegP-trimer is heated to 320 K. The monomer 1 involves in a strong intra-domain interactions, except the LA loop, which significantly expresses an anti-correlated motion with the rest of its domains. The protease (except LA loop) and PDZ1 of monomer 1 express a significant anti-cooperative motion with both PDZ1 and PDZ2 domains of monomers 2 and 3. These strong anti-cooperative motions promote the LA loop to adopt a new conformation during dynamics. The cooperative motion of both protease and PDZ1 domains at 320 K shows their role in protease activity, which is not observed with the simulation at 280 K.

The observed coordinated motions of LA loop with various parts of DegP and its monomeric units emphasize its active participation for the functional dynamics of DegP-trimer. The coordinated motions

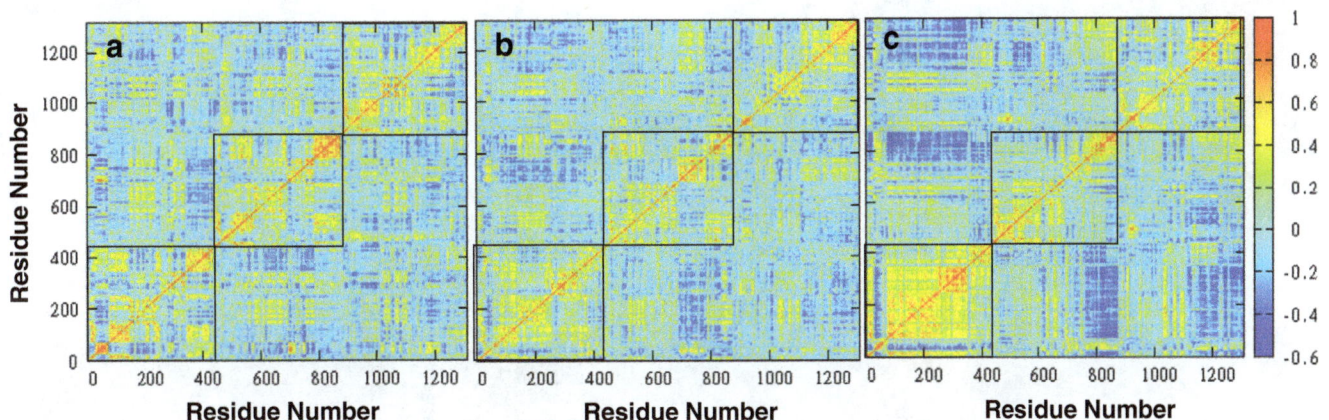

Fig. 4. The correlative motions between the Cα-atom of residues of DegP-trimer observed during the last 80 ns of simulation performed at 280 (a), 300 (b) and 320 K (c), respectively. Each monomeric unit is outlined by a box. The amplitude of correlation varies from blue (anti-cooperative region) to red (highly cooperative region) via the uncorrelated region coloured as cyan.

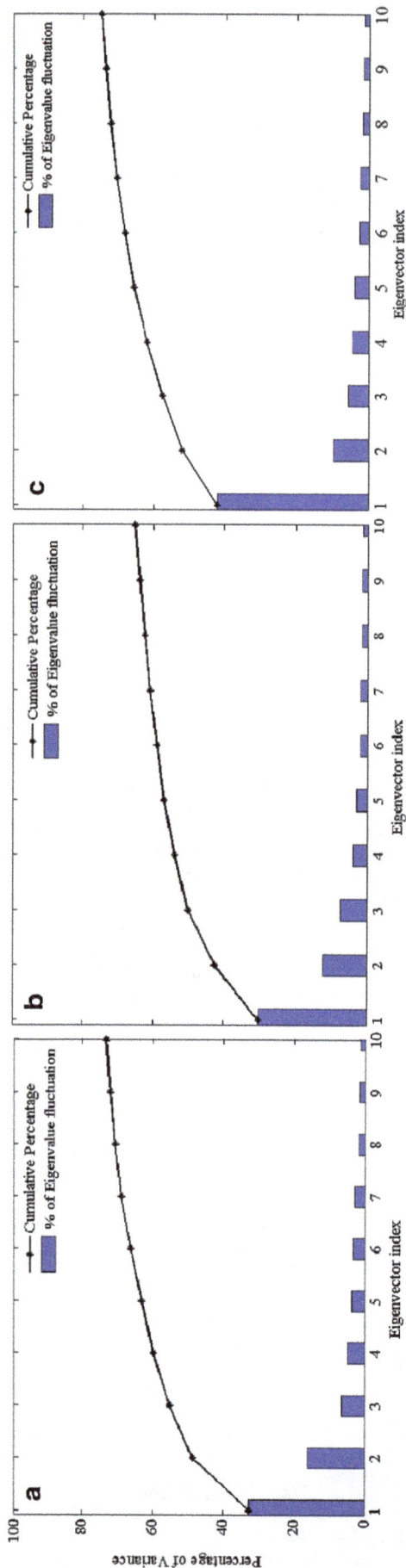

Fig. 5. Percentage (blue bars) and cumulative percentage (black line) of variance for the first 10 eigenvalues obtained from the covariance matrix of Cα atoms observed during the last 80 ns simulations performed at 280 (a), 300 (b) and 320 K (c), respectively.

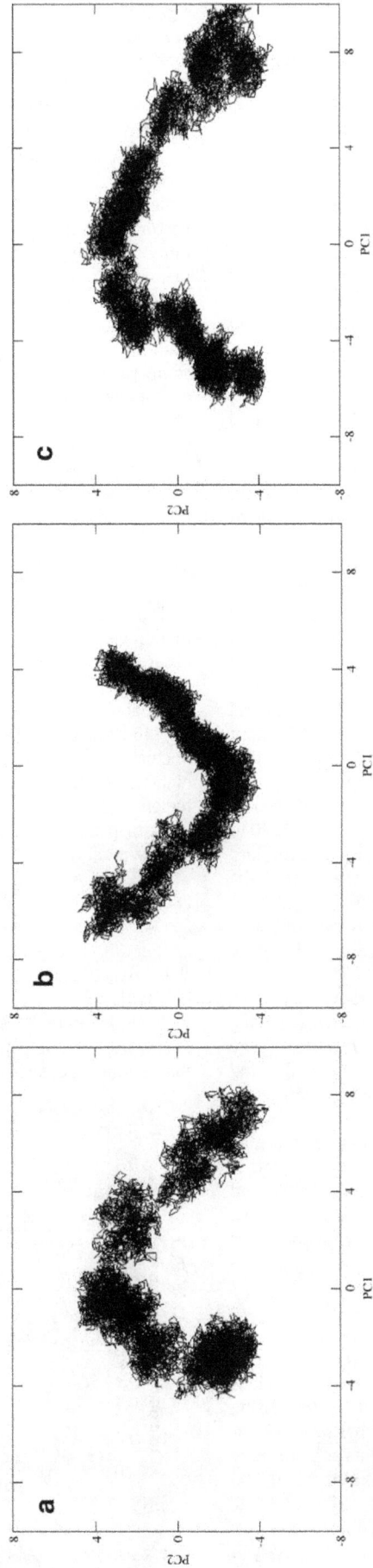

Fig. 6. Projection of Cα-atom trajectories on the first two eigenvectors (in nm) of DegP-trimer simulated at 280 (a), 300 (b) and 320 K (c), respectively.

observed between protease domains also ensure their role in maintaining the assembly of DegP-trimer. All these observations evidence temperature dependent domain motions between the monomeric units of DegP-trimer.

3.2. Dynamics of LA loop

The LA loop, which is an important regulatory element of DegP-trimer, expresses dynamic behaviour and effectively participates in both intra- and inter-monomeric domain motions. The LA loop is highly networked by ~22 intra-LA H-bonds and is consistent irrespective of the temperature during simulation. The H-bonds between LA loop and other interacting domains of the monomeric units were also examined. At room temperature, the LA loop is networked with more number of inter-H-bonds (i.e., 8 ± 2) and is comparatively free at 280 and 320 K with lesser inter-H-bonds (3 ± 1 and 4 ± 1, respectively) due to its flipping motion from the protease core.

Analysis on the temperature dependent dynamics of LA loop revealed variable conformations with respect to temperature and is shown in Fig. 3. During simulations, the LA loop residues Thr36–Thr39, Arg41–Arg44, Gln48, and Gly72–Gln81 form instantaneous H-bond with the neighbouring domains of the same monomer. Simulations at 320 K revealed a transformation of this highly flexible LA loop from the initial conformation to a new conformation, i.e., at 320 K, the region of LA loop encompassed by residues 46–80 is completely flipped off from its initial position and reoriented towards its PDZ2 domain and the PDZ1 domain of the neighbouring monomeric unit. Precisely, this flip is towards the left side (indicated in Fig. 3) with respect to the fixed β-sheets that connect the LA loop of protease domain. To identify the residues responsible for this transformation at 320 K, the dihedral angles (φ and ψ) of residues Phe46–Phe50 and Gly78–Glr82 (those forming the neck of the flipped region) were monitored. It is clear that the dihedral angle of these residues shows significant variation at 320 K, i.e., the φ (ψ) angle of residues Phe46, Gln80 and Gln81 varies in the range -129 to $-179°$ (-139 to -104), -176 to -159 (40 to 176) and -177 to -166 (-179 to -180), respectively. The residues 47–49 modulate their dihedral angles according to the variation in the dihedrals observed for Phe46 to facilitate the transition.

At 280 K, there is no such flipping of LA loop as observed at 320 K. The LA loop expresses stable dynamics around the initial resting position and communicates interactions with its PDZ1 domain (green ribbon in Fig. 3). This conformation of LA loop is closer to the conformations observed from the cryo-EM method (PDB ID: 2ZLE), when the DegP exists in chaperonic conformation (grey ribbon in Fig. 3). This observation evidences its ability in adopting similar conformation (favouring chaperonic activity) even when DegP exists as trimer.

Similar to the moderate interactions revealed by the correlation map (at 300 K), the LA loop is not involved in any conformational flip but is stabilized in between the two conformations observed at 280 and 320 K. Overall, the present analysis discloses a temperature sensitive dynamics of LA loop and promotes further insight into the dynamics of DegP-trimer.

3.3. Principal component analysis

The last 80 ns of the simulated trajectories of DegP-trimers was used to perform PCA. Fig. 5 shows the plot of the first ten eigenvalues resolved from the covariance matrix of fluctuations in decreasing order against the eigen-indices for DegP-trimer. Comparison of eigenvalues indicates that the first ten principal components account more than 65% of the global motion (73.27, 65.22 and 74.81% at 280, 300 and 320 K, respectively). It is observed that PC1 contributes 32.46, 30.36 and 42.26% for the motion, whereas the PC2 accounts 16.1, 12.24 and 9.74% for the motion at 280, 300 and 320 K, respectively. It is also visible that both magnitude and percentage of contribution of

eigenvalues at 280 and 320 K for PC1 are higher than that of simulation at 300 K and evidences the presence of a single major essential subspace for the structural dynamics of DegP-trimer at lower and higher temperatures. In addition, it is clear that the first eigenvalue contribute significantly to the global motions when compared to the rest of eigenvalues that diminishes quickly as they correspond to the localized fluctuations.

In PCA, the direction of motion is extracted by projecting the trajectory observed between 20 and 100 ns over the desired eigenvector to understand the dynamics of protein in the direction of respective eigenvector. Fig. 6 depicts the projection of first two PCs to display the motions of DegP-trimer in a 2-dimensional subspace. It provides a measure of mobility where each cluster represents the tertiary conformational changes observed along the trajectory. At 280 K, the DegP-trimer spans in a larger conformational subspace, while at 300 K, it spans in a comparatively smaller with lesser amplitude and reveals a comparatively confined motion of DegP-trimer. At 300 K, the direction of motion is exactly reverse when compared to the direction of motion observed at 280 and 320 K. When temperature increases to 320 K, the DegP-trimer adopts similar pattern of motion as observed at 280 K, but with increased amplitude in the direction of PC1. These plots ensure that DegP-trimer exists in a structurally active state at higher and lower temperatures than the room temperature.

3.4. Residue fluctuations in PC1

Fig. 7 shows the displacement of residues corresponding to the motion described by first PC at 280, 300 and 320 K, respectively. It is observed that all the monomers show variation in their amplitude of motion. DegP possesses many dynamic regions such as LA, L2 and L3 loops, and PDZ1 and PDZ2 domains. Among these regions, the LA loop (Phe46 to Gly65) of DegP expresses higher fluctuation. Such higher fluctuation of LA loop explains its temperature dependent mobility. At higher temperature (i.e., 320 K), the L3 loop, which acts as a signalling region, expresses higher fluctuation in all the monomers and indicates its participation during dynamics. The PDZ1 domain motion is comparatively moderate at 320 K and few residues like Gly370, Asp382, Gln398, and Asp440 of PDZ2 domain show higher fluctuations. At 280 K, apart from LA loops, few more residues like Ser118, Gly315, Leu342 and Asp440 express fluctuation. The residues of sensory loop L3 (Ser188, Gly189 and Glu193) that transmits the substrate binding signal to activation domain are highly triggered by high temperature [16,50,51].

In DegP-trimer, the secondary structures like β-barrel and β-sheets are involved in stable dynamics throughout the simulation. In general, the residue fluctuations at 280 and 320 K are highly pronounced

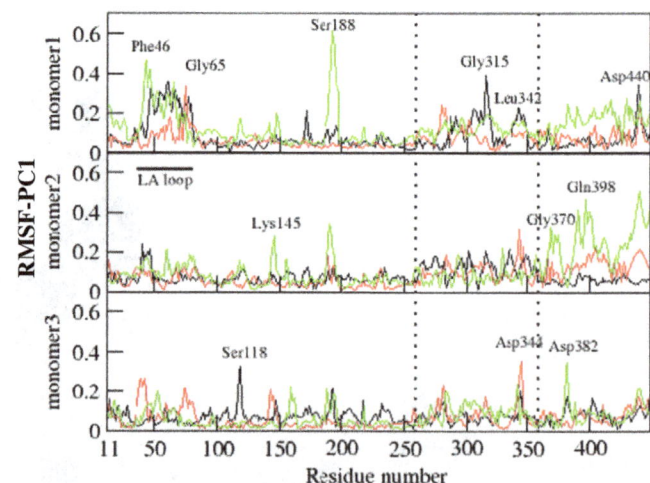

Fig. 7. Residue displacements (in nm) in the subspace spanned by the first eigenvector of DegP-trimer at 280 K (black), 300 K (red) and 320 K (green).

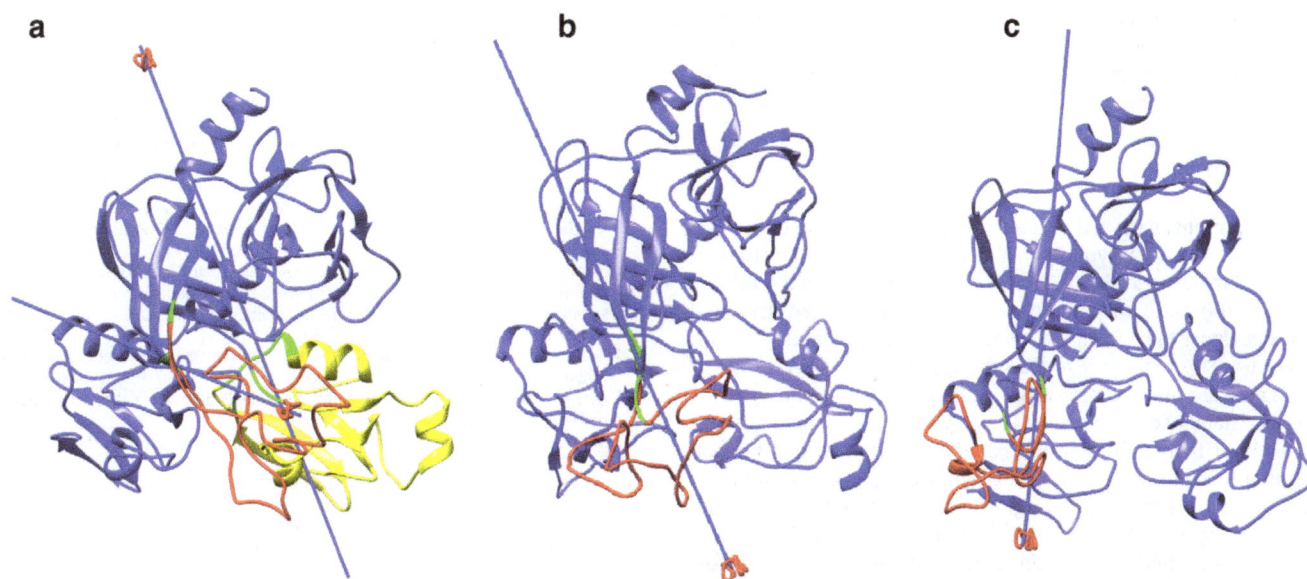

Fig. 8. Domain motions described by the monomer from DegP-trimer at 280 (a), 300 (b) and 320 K (c) temperatures respectively. The fixed and moving domains are coloured in blue, red, and yellow respectively. The axis of rotation of moving domain with respect to the fixed domain is also depicted.

when compared to that observed at 300 K and explain the dual nature of DegP-trimer at extreme (higher and lower) temperatures and the passive role of DegP-trimer at room temperature.

3.5. Domain motions in DegP-trimer

The domain motions have also been analysed using DynDom programme [52] to quantify the translational cum rotational dynamics of individual domains (Fig. 8 and Table 1). At 280 K, the residues Ala16–Ser34 and Met85–Gln259 of protease, Val260–Gly263, Thr330–Val333, Leu354–Ser358 of PDZ1 and Gln359–Leu446 of PDZ2 domains are defined as the fixed domain with respect to which the translational cum rotational motions of the moving domains are explained. DegP possesses two moving domains defined by (i) Thr36–Phe81 of LA loop including Thr35, Gln82–Phe84 of protease domain (referred as moving domain 1) and (ii) Glu264–Gly329 and Val333–Glu353 of PDZ1 domain (referred as moving domain 2) for which the identified hinge residues are (i) Ser34–Thr35, and Phe84–Met85 and (ii) Gly263–Glu264, Gly329–Val333, and Glu353–Leu354, respectively. Both of the moving domains 1 and 2 express rotational (of about ~57°) cum translational motion (of 0.3 and 2.1 Å, respectively) with respect to the fixed domain. It is also observed that, the substrate binding hydrophobic loop (E–L–G–I motif) is identified as the hinge region for these motions. Both LA loop and PDZ1 moving domains express a closure of 41.3 and 24.9%, respectively with respect to the fixed domain. As the degree of closure measures the closeness of respective moving domains towards the fixed

domain, the observed 41.3% closure of LA loop towards the PDZ1 domain indicates that at lower temperature (i.e. at 280 K), the LA loop moves closure to the chaperonic conformation.

At 300 K, the residues Thr36–Gln80 are observed as moving domain with respect to the fixed domain formed by Pro17–Gln259 of protease, Val260–Ser358 of PDZ1 and Gln359–Thr442 of PDZ2 domains, for which the residues of LA loop (Ser34–Thr35 and Gln80–Gln81) act as hinge. The moving LA loop expresses a translational (−4.6 Å) cum rotational (58.8°) motion with 66.9% of closure property.

At 320 K, the only identified moving domain (Thr36–Gly78 of LA loop) expressed rotational (82.4°) cum translational (−3.7 Å) motion with 61.3% of closure with respect to the fixed domain. The residues Pro13–Thr35 and Gly79–Gln259 of protease along with both PDZ domains form the fixed domain and the residues Thr36–Gly78 serve as the hinge residues. From the DynDom analysis of DegP-trimer at 280, 300 and 320 K, it is clear that DegP at 300 K exists in a comparatively intermediate state, while at lower and higher temperatures, DegP adopts two different conformations, which discriminate the dual function of DegP. The axis of rotation also supports the flip of LA loop to both the right and left sides of the anti-parallel beta sheet for the functional diversity of DegP-trimer.

It is also hypothesized that the signals are transmitted consecutively rather than simultaneously; an asymmetry is induced in the structure [51]. Our present analysis also reveals a structural asymmetry during dynamics, i.e. high amplitude motion is expressed by one LA loop and the interacting L3 loop, when compared to the dynamics of

Table 1
Relative motions of the moving domains with respect to the fixed domain.

Relative domain motions	280 K		300 K	320 K
Fixed domain	Protease: 16–34, 85–259		Protease: 17–34, 81–259	Protease: 13–35, 79–259
	PDZ1: 260–263, 330–333, 354–358		PDZ1: 260–358	PDZ1: 260–358
	PDZ2: 359–446		PDZ2: 359–442	PDZ2: 359–446
Moving domain 1	Protease: 35, 82–84		LA loop: 36–80	LA loop: 36–78
	LA loop: 36–81			
Moving domain 2	PDZ1: 264–329, 333–353		Nil	Nil
Angle of rotation (°)	LA loop terminal	PDZ1	58.8	82.4
	57.7	56.0		
Translation along axis (Å)	0.3	2.1	−4.6	−3.7
Closure (%)	41.3	24.9	66.9	61.3
Bending residue	34–35, 84–85	263–264, 329–333, 353–354	34–35, 80–81	35–36, 78–79

LA and L3 loops in other monomeric units. In general, all these observations reveal a dynamics structural state of DegP-trimer like other oligomer and also a temperature dependent dynamics of LA loop, which plays a key role in determining the function of DegP.

4. Conclusion

In this present analysis, the structure of DegP-trimer was built based on the reported crystal symmetry. Molecular dynamics simulations at various temperatures 280, 300 and 320 K were performed on DegP-trimer for 100 ns, to understand the temperature induced structural dynamics. Under the influence of temperature, both molecular dynamics as well as PCA analyses revealed a stable dynamics of DegP-trimer by relaxing the tertiary structures while maintaining the secondary structures. The stability of DegP-trimer is highly mediated by both intra- as well as inter-monomeric motions. Specifically, the LA and L3 loop regions and PDZ1 domain contribute significantly to the dynamics of DegP. At room temperature, DegP-trimer exists in a passive state with less dynamic property. Simulations at 280 and 320 K expressed pronounced dynamics of DegP-trimer. At 280 K, the LA adopts a conformation closer to the conformation observed when DegP functions as chaperone, whereas the LA loop at 320 K flips with respect to Phe46 in a direction opposite to the flip at 280 K and such dynamics might coordinate the signalling between the neighbouring monomers. It is reported that the phenylalanine residues present at the core region of LA loop play a key role in promoting the flexible dynamics of LA loop according to the temperature shift [53]. Our observation of Phe46 as one of the key residues in promoting the conformational changes of LA loop also reinforces its role in dynamics. Even though, all the three monomers are involved in the functional dynamics of DegP-trimer, only one monomer takes the lead in expressing high amplitude motions of LA loop in association with temperature shift.

In general, it is interesting to observe the temperature induced high amplitude dynamics of DegP-trimer (via loop regions and PDZ1 domain) at both lower and higher temperatures. The overall observations signify the role of flexible LA loop over the structural dynamics of DegP-trimer which might promote further understanding on the temperature dependent functions of DegP. Various intriguing aspects on DegP mechanism like fate of substrate and transition between different oligomeric states are yet to be explored to further understand the function of DegP. In line with these observations, the structural dynamics of DegP-hexamer in the presence of substrate is in progress to understand the allosteric activation mechanism of DegP from the structural point of view.

Acknowledgement

Amutha Ramaswamy acknowledges the Science and Engineering Research Board, India for providing computational facilities under Fast Track Research Project for Young Scientists Scheme. Molecular graphic images and movies were generated using the UCSF Chimera package from the University of California, San Francisco.

References

[1] Hightower LE. Stress proteins in biology and medicine. In: Morimoto Richard I, Tissieres Alfred, Georgopoulos Costa, editors. Cold Spring Harbor, NY: Cold Spring Harbor Laboratory; 1990 [x, 450 pp., illus. $97. Cold Spring Harbor Monograph Series 19. Science 249: 572–573].

[2] Lindquist S, Craig EA. The heat-shock proteins. Annu Rev Genet 1988;22:631–77.

[3] Kim DY, Kim KK. Structure and function of HtrA family proteins, the key players in protein quality control. J Biochem Mol Biol 2005;38:266–74.

[4] Pallen MJ, Wren BW. The HtrA family of serine proteases. Mol Microbiol 1997;26:209–21.

[5] Clausen T, Southan C, Ehrmann M. The HtrA family of proteases: implications for protein composition and cell fate. Mol Cell 2002;10:443–55.

[6] Spiess C, Beil A, Ehrmann M. A temperature-dependent switch from chaperone to protease in a widely conserved heat shock protein. Cell 1999;97:339–47.

[7] Krojer T, Garrido-Franco M, Huber R, Ehrmann M, Clausen T. Crystal structure of DegP (HtrA) reveals a new protease-chaperone machine. Nature 2002;416:455–9.

[8] Krojer T, Sawa J, Schafer E, Saibil HR, Ehrmann M, et al. Structural basis for the regulated protease and chaperone function of DegP. Nature 2008;453:885–90.

[9] Sobiecka-Szkatula A, Gieldon A, Scire A, Tanfani F, Figaj D, Denkiewicz M, et al. The role of the L2 loop in the regulation and maintaining the proteolytic activity of HtrA (DegP) protein from Escherichia coli. Arch Biochem Biophys 2010;500:123–30.

[10] Jomaa A, Iwanczyk J, Tran J, Ortega J. Characterization of the autocleavage process of the Escherichia coli HtrA protein: implications for its physiological role. J Bacteriol 2009;191:1924–32.

[11] Skorko-Glonek J, Zurawa D, Tanfani F, Scire A, Wawrzynow A, Narkiewicz J, et al. The N-terminal region of HtrA heat shock protease from Escherichia coli is essential for stabilization of HtrA primary structure and maintaining of its oligomeric structure. Biochim Biophys Acta 2003;1649:171–82.

[12] Iwanczyk J, Damjanovic D, Kooistra J, Leong V, Jomaa A, Ghirlando R, et al. Role of the PDZ domains in Escherichia coli DegP protein. J Bacteriol 2007;189:3176–86.

[13] Ortega J, Iwanczyk J, Jomaa A. Escherichia coli DegP: a structure-driven functional model. J Bacteriol 2009;191:4705–13.

[14] Maurizi MR. Love it or cleave it: tough choices in protein quality control. Nat Struct Biol 2002;9:410–2.

[15] Jomaa A, Damjanovic D, Leong V, Ghirlando R, Iwanczyk J, Ortega J, et al. The inner cavity of Escherichia coli DegP protein is not essential for molecular chaperone and proteolytic activity. J Bacteriol 2007;189:706–16.

[16] Krojer T, Sawa J, Huber R, Clausen T. HtrA proteases have a conserved activation mechanism that can be triggered by distinct molecular cues. Nat Struct Mol Biol 2010;17:844–52.

[17] Figaj D, Gieldon A, Polit A, Sobiecka-Szkatula A, Koper T, Denkiewicz M, et al. The LA loop as an important regulatory element of the HtrA (DegP) protease from Escherichia coli: structural and functional studies. J Biol Chem 2014;289:15880–93.

[18] Sassoon N, Arie JP, Betton JM. PDZ domains determine the native oligomeric structure of the DegP (HtrA) protease. Mol Microbiol 1999;33:583–9.

[19] Subrini O, Betton JM. Assemblies of DegP underlie its dual chaperone and protease function. FEMS Microbiol Lett 2009;296:143–8.

[20] Jones CH, Dexter P, Evans AK, Liu C, Hultgren SJ, Hruby DE, et al. Escherichia coli DegP protease cleaves between paired hydrophobic residues in a natural substrate: the PapA pilin. J Bacteriol 2002;184:5762–71.

[21] Jiang J, Zhang X, Chen Y, Wu Y, Zhou ZH, Chang Z, et al. Activation of DegP chaperone-protease via formation of large cage-like oligomers upon binding to substrate proteins. Proc Natl Acad Sci U S A 2008;105:11939–44.

[22] Meltzer M, Hasenbein S, Mamant N, Merdanovic M, Poepsel S, Hauske P, et al. Structure, function and regulation of the conserved serine proteases DegP and DegS of Escherichia coli. Res Microbiol 2009;160:660–6.

[23] Meltzer M, Hasenbein S, Hauske P, Kucz N, Merdanovic M, Grau S, et al. Allosteric activation of HtrA protease DegP by stress signals during bacterial protein quality control. Angew Chem Int Ed Engl 2008;47:1332–4.

[24] Merdanovic M, Mamant N, Meltzer M, Poepsel S, Auckenthaler A, Melgaard R, et al. Determinants of structural and functional plasticity of a widely conserved protease chaperone complex. Nat Struct Mol Biol 2010;17:837–43.

[25] Dror RO, Dirks RM, Grossman JP, Xu H, Shaw DE. Biomolecular simulation: a computational microscope for molecular biology. Annu Rev Biophys 2012;41:429–52.

[26] Karplus M. Molecular dynamics simulations of biomolecules. Acc Chem Res 2002;35:321–3.

[27] Ramaswamy A, Froeyen M, Herdewijn P, Ceulemans A. Helical structure of xylose-DNA. J Am Chem Soc 2009;132:587–95.

[28] Sangeetha B, Muthukumaran R, Amutha R. The dynamics of interconverting D- and E-forms of the HIV-1 integrase N-terminal domain. Eur Biophys J 2014;43:485–98.

[29] Maisuradze GG, Zhou R, Liwo A, Xiao Y, Scheraga HA. Effects of mutation, truncation, and temperature on the folding kinetics of a WW domain. J Mol Biol 2012;420:350–65.

[30] Tang L, Liu H. A comparative molecular dynamics study of thermophilic and mesophilic ribonuclease HI enzymes. J Biomol Struct Dyn 2007;24:379–92.

[31] Vijayakumar S, Vishveshwara S, Ravishanker G, Beveridge DL. Differential stability of beta-sheets and alpha-helices in beta-lactamase: a high temperature molecular dynamics study of unfolding intermediates. Biophys J 1993;65:2304–12.

[32] Amadei A, Linssen AB, Berendsen HJ. Essential dynamics of proteins. Proteins 1993;17:412–25.

[33] Levy RM, Srinivasan AR, Olson WK, McCammon JA. Quasi-harmonic method for studying very low frequency modes in proteins. Biopolymers 1984;23:1099–112.

[34] Hayward S, Go N. Collective variable description of native protein dynamics. Annu Rev Phys Chem 1995;46:223–50.

[35] Hess B. Similarities between principal components of protein dynamics and random diffusion. Phys Rev E 2000;62:8438–48.

[36] Bejugam PR, Kuppili RR, Singh N, Gadewal N, Chaganti LK, Sastry GM, et al. Allosteric regulation of serine protease HtrA2 through novel non-canonical substrate binding pocket. PLoS ONE 2013;8:e55416.

[37] Xu D, Zhang Y. Ab initio protein structure assembly using continuous structure fragments and optimized knowledge-based force field. Proteins 2012;80:1715–35.

[38] Guex N, Peitsch MC. SWISS-MODEL and the Swiss-PdbViewer: an environment for comparative protein modeling. Electrophoresis 1997;18:2714–23.

[39] Humphrey W, Dalke A, Schulten K. VMD: visual molecular dynamics. J Mol Graph 1996;14:33–8 [27–38].

[40] Scott WRP, Haunenberger PH, Tironi IG, Mark AE, Billeter SR, Fennen J, et al. The GROMOS biomolecular simulation program package. J Phys Chem A 1999;103:3596–607.

[41] Darden TYD, Pedersen L. Particle mesh Ewald: an Nlog(N) method for Ewald sums in large systems. J Chem Phys 1993;98:10089–93.

[42] Hess B, Bekker H, Berendsen HJC, Fraaije JGEM. LINCS: a linear constraint solver for molecular simulations. J Comput Chem 1997;18:1463–72.

[43] Bussi G, Donadio D, Parrinello M. Canonical sampling through velocity rescaling. J Chem Phys 2007;126:014101.

[44] Parrinello M, Rahman A. Polymorphic transitions in single crystals: a new molecular dynamics method. J Appl Phys 1981;52:7182–90.

[45] Hunenberger PH, Mark AE, van Gunsteren WF. Fluctuation and cross-correlation analysis of protein motions observed in nanosecond molecular dynamics simulations. J Mol Biol 1995;252:492–503.

[46] Ichiye T, Karplus M. Collective motions in proteins: a covariance analysis of atomic fluctuations in molecular dynamics and normal mode simulations. Proteins 1991; 11:205–17.

[47] Roe DR, Cheatham TE. PTRAJ and CPPTRAJ: software for processing and analysis of molecular dynamics trajectory data. J Chem Theory Comput 2013;9:3084–95.

[48] Hess B. Similarities between principal components of protein dynamics and random diffusion. Phys Rev E Stat Phys Plasmas Fluids Relat Interdiscip Topics 2000;62: 8438–48.

[49] Lipinska B, Zylicz M, Georgopoulos C. The HtrA (DegP) protein, essential for *Escherichia coli* survival at high temperatures, is an endopeptidase. J Bacteriol 1990;172:1791–7.

[50] Clausen T, Kaiser M, Huber R, Ehrmann M. HTRA proteases: regulated proteolysis in protein quality control. Nat Rev Mol Cell Biol 2011;12:152–62.

[51] Thompson NJ, Merdanovic M, Ehrmann M, van Duijn E, Heck AJ. Substrate occupancy at the onset of oligomeric transitions of DegP. Structure 2014;22:281–90.

[52] Poornam GP, Matsumoto A, Ishida H, Hayward S. A method for the analysis of domain movements in large biomolecular complexes. Proteins 2009;76:201–12.

[53] Sobiecka-Szkatula A, Polit A, Scire A, Gieldon A, Tanfani F, Szkarlat Z, et al. Temperature-induced conformational changes within the regulatory loops L1–L2–LA of the HtrA heat-shock protease from *Escherichia coli*. Biochim Biophys Acta (BBA) Proteins Proteom 2009;1794:1573–82.

21

A New Paradigm for Known Metabolite Identification in Metabonomics/Metabolomics: Metabolite Identification Efficiency

Jeremy R. Everett *

Medway Metabonomics Research Group, University of Greenwich, Chatham Maritime, Kent ME4 4TB, United Kingdom

A R T I C L E   I N F O

Keywords:
Metabolite identification efficiency (MIE)
Metabolite identification carbon efficiency
(MICE)
Metabonomics
Metabolomics
NMR spectroscopy
Molecular spectroscopic information

A B S T R A C T

A new paradigm is proposed for assessing confidence in the identification of known metabolites in metabonomics studies using NMR spectroscopy approaches. This new paradigm is based upon the analysis of the amount of metabolite identification information retrieved from NMR spectra relative to the molecular size of the metabolite. Several new indices are proposed including: metabolite identification efficiency (MIE) and metabolite identification carbon efficiency (MICE), both of which can be easily calculated. These indices, together with some guidelines, can be used to provide a better indication of known metabolite identification confidence in metabonomics studies than existing methods. Since known metabolite identification in untargeted metabonomics studies is one of the key bottlenecks facing the science currently, it is hoped that these concepts based on molecular spectroscopic informatics, will find utility in the field.

1. Introduction

Metabonomics is defined as 'The study of the metabolic response of organisms to disease, environmental change or genetic modification' [1] and has emerged as a leading technology in a number of fields, including biology and medicine [2], with new areas emerging recently, such as pharmacometabonomics for personalised medicine [3–5]. The alternative term metabolomics [6] was defined a little later as a 'comprehensive analysis in which all the metabolites of a biological system are identified and quantified'. The two terms are now used interchangeably but in this article we will refer to the original term throughout.

Metabonomics studies are typically conducted with either nuclear magnetic resonance (NMR) spectroscopy or a hyphenated mass spectrometry (MS) technology, such as liquid chromatography–MS (LC–MS) [7], to acquire information on the identities and quantities of metabolites in the particular samples of interest. The studies are conducted either in a targeted fashion, where a pre-defined set of metabolites are measured, or in an untargeted fashion, where no preconceptions of the metabolites of importance are imposed. The choice of analytical technology used often depends upon the particular study requirements.

In this article, the focus will be on the use of NMR spectroscopy rather than MS, although the two technologies are quite complementary and it is often advantageous to use them together in concert.

Metabonomics/metabolomics studies have a number of important elements including:

1. definition of study aims e.g., understanding the metabolic consequences of disease progression in a particular group of patients
2. ethical approval
3. sample collection and storage
4. sample preparation
5. NMR data acquisition
6. quality control of the acquired data to ensure adequate signal-to-noise, lineshape and resolution
7. spectroscopic data pre-processing steps such as zero-filling, apodisation, Fourier transform, phasing and baseline correction
8. statistical data pre-processing steps such as peak alignment, scaling and normalisation of the data
9. statistical analysis of the data to interrogate differences between groups of subjects e.g., healthy volunteers vs patients with disease
10. identification of metabolites responsible for any inter-group differences discovered in the study
11. rationalisation of the role of the discriminating metabolites in terms of physiological and biochemical changes in the subject.

Many of the 11 steps above have been subject to rigorous study and guidelines have emerged for several areas including NMR-based sample preparation, data acquisition, data pre-processing and statistical analysis of the data, especially by multivariate methods [8–15]. However, the critical step for many untargeted metabonomics studies is the

E-mail address: j.r.everett@greenwich.ac.uk.

identification of the metabolites that are responsible for discriminating between different groups of subjects in the study: step 10. This remains problematical for both MS [16,17] and NMR spectroscopy [14,18–24] and is a significant bottleneck for the development of the science.

The issue with metabolite identification was nicely illustrated by Wishart who contrasted the 4 different bases in the human genome, and the 20 natural amino acids in the human proteome, with the thousands of different metabolites in the human metabolome: this is the cause of the issue [25].

For studies where many samples are available, statistical methods of metabolite identification, such as STOCSY and variants thereof, are powerful tools that can be used for metabolite and biomarker identification [26,27]. In genome wide association studies on metabonomics data, the pathway information that can be gleaned can also be used to help identify key metabolites, including by metabomatching [28,29].

Metabolite identification by NMR spectroscopy has recently been significantly facilitated by the development of spectral databases of metabolites [30], such as the Human Metabolome Database (HMDB) [31], the BioMagResBank (BMRB) [32] and the Birmingham Metabolite Library (BML) [33]. These libraries not only store information on the NMR spectra of a vast array of metabolites, which helps metabolite identification, but more powerfully, some also allow downloading of the original NMR free induction decay data from the databases, to facilitate comparison of the spectral features of authentic metabolites with those of unidentified metabolites in users' biological samples.

Some progress has been made towards the automated identification of metabolites but these methods are not yet at the stage that they can be routinely used to identify more than a fraction of the metabolites in complex biofluids such as urine. The Birmingham Metabolite Library (BML) provides a facility for the matching of experimental 2D ^{1}H J-resolved spectra with those of reference metabolites stored in a database, which is a good approach, but is limited by the low number of metabolites in that database [33]. Approaches such as MetaboHunter have been applied to the identification of mixtures of standard compounds but not to a biofluid [21]. An approach based on 1D ^{1}H NMR profiles, has had success in identifying metabolites in human serum and cerebrospinal fluid, but was less successful in identifying metabolites in urine due to spectral complexity and the lack of a complete reference set [34]. A different 1D ^{1}H NMR approach based on extraction of relevant variables for analysis (ERVA) has been applied to simulated mixtures and to the analysis of tomato extracts, but again relies on the availability of authentic spectra of the metabolites and fails for compounds that have only a single peak in their 1D ^{1}H NMR spectrum [24]. Thus, at the present time, the only robust way to identify known metabolites in the 1D ^{1}H NMR spectra of complex biofluids such as urine is by manual analysis by an expert NMR spectroscopist.

The metabolite identification issue is in two distinct categories: first the *structure elucidation* of truly novel metabolites, not previously reported, and secondly, the *structure confirmation* or *structure identification* of previously reported or known metabolites. This simple language and description is consistent with decades of molecular structure elucidation literature, and is preferable to the more complex and confusing labelling of metabolites as 'unknown unknowns' or 'known unknowns' that has emerged more recently [25,35]. For the structure elucidation of truly novel metabolites, there is a consensus that the same rigorous processes used in the natural product field should be adopted in metabolite identification. This would usually involve extraction and purification of the novel metabolite, followed by full structure elucidation by ultraviolet, infrared and NMR spectroscopies in concert with MS [36,37].

However, the process of structure confirmation of known metabolites remains an issue, due to differences in approaches across the metabonomics/metabolomics community. In order to address the metabolite identification issue, the Metabolomics Standards Initiative

(MSI) [38] set up a Chemical Analysis Working Group (CAWG) which proposed a 4-level classification system (Table 1) for the structure confirmation of known metabolites in 2007 [36].

In the seven years since these proposals were made, they have not been widely adopted by the community [39]. There are two basic problems with the original proposals: firstly, the requirement of comparison of experimental data for known metabolites to an authentic reference standard in the lab, is often too strict and not always appropriate for an NMR-based study, and secondly, the system is too coarse and does not define closely enough the confidence achieved in the metabolite identification. Recently, new proposals emerged to update the 4-level system with either: (i) addition of sub-levels to grade confidence better, (ii) an alternative quantitative identification points scoring system or (iii) quantitative enhancement of the current 4-level system to indicate confidence [40]. An overlapping subset of the same authors also proposed quantitative and alphanumeric metabolite identification metrics [41]. The quantitative scoring proposal in the latter publication contains a mixture of excellent metabolite identification criteria with precision e.g., accurate mass of parent ion (<5 ppm) and processes such as having a COSY NMR, with no precision or scoring for matching. It was commented that it is difficult to see how scoring for matching of metabolite data to standards could be achieved [41]. A call to the community was made for engagement with this important problem [40].

This paper responds to those calls. A new approach to the understanding of the NMR spectroscopic information theoretically embedded in metabolites is put forward, and compared with the data that is actually obtained in the course of metabonomics experiments. Conclusions and proposals are arrived at in terms of a different approach to metabolite identification confidence, which should be applicable in spirit to any other analytical technology, in addition to NMR spectroscopy.

2. Material and methods

2.1. Subjects, sample preparation and NMR spectroscopy

The 75 metabolites included in this study were identified manually from the proton NMR spectra of the urine from a C57BL/6 mouse at 30 weeks of age, and the urine of a diabetic patient on an exercise study at La Sapienza University, Rome. Both studies were ethically approved [42].

The diabetic urine sample was prepared by mixing urine (630 μl) with phosphate buffer (70 μl of an 81:19 (v/v) mixture of 1.0 M K_2HPO4 and 1.0 M NaH_2PO4 pH 7.4). After standing at room temperature for 10 min, the sample was centrifuged at 13,000 g for five minutes at 4 °C to enable separation of clear supernatant (600 μl) from any particulate matter. The supernatant was mixed with a solution of the chemical shift reference material, sodium 3-(trimethylsilyl) propionate-2, 2, 3, 3-d4 (TSP) in D_2O (60 μl), to give a final TSP concentration of 0.18 mM.

The mouse urine sample was prepared by mixing pooled urine (500 μl) from a single C57BL/6 mouse with phosphate buffer (150 μl of a 81:19 (v/v) mixture of 0.6 M K_2HPO4 and NaH_2PO4 in 100% ^2H_2O, pH 7.4, containing 0.5 mM TSP as a reference and 9 mM sodium azide). The sodium azide was added to prevent bacterial growth in the urine sample.

All NMR experiments were conducted on a Bruker Avance spectrometer operating at 600.44 MHz for ^{1}H NMR, at ambient temperature, in 5 mm NMR tubes (508-UP-7). All chemical shifts are on the δ_H or δ_C scales relative to TSP at 0.

The identification of the metabolites used a combination of standard 1D and 2D NMR methods, including J-resolved (JRES), COSY, TOCSY, HSQC and HMBC experiments. The 1D ^{1}H NMR experiments used the 1D NOESY presaturation pulse sequence, noesypr1d. Free induction decays were collected into 65,536 data points with 256 scans and 4 dummy scans and a spectral width of 12,019.2 Hz. The resulting spectra were zero-filled to 131,072 or 262,144 points, baseline corrected automatically, phase corrected automatically (with manual override, as required) and apodised for resolution enhancement using Gaussian multiplication. The detailed parameters for the acquisition of the 2D NMR spectra are given in Supplementary Table 1.

2.2. Theoretical analysis of the NMR spectroscopic information content of the 75 metabolites

All 75 metabolites were characterised by their common names, IUPAC name, HMDB code, SMILES string, InChi code and InChi key (see Table 4 and Supplementary Data). 14 Molecular features were analysed manually for each of the 75 metabolites represented in this study (Table 2):

The following rules were applied to this feature analysis:

1. Only non-exchanging protons were included in the analysis of the number of proton chemical shifts present in the metabolites, and this included non-exchanging (on the NMR timescale) amides but not hydroxyl, amine or acid protons
2. The number of multiplicities is simply the sum total of the number of singlets, doublets, doublet of doublets etc. contained in a metabolite: for example, if a metabolite has one singlet and two doublet proton signals, the multiplicity count for that metabolite is three
3. The total number of coupling constants was calculated for all possible 2- and 3-bond proton-to-proton couplings involving non-exchanging protons
4. COSY cross-peaks between two protons were only counted once: therefore the number of COSY peaks must equal the number of coupling constants: long-range COSY connectivities were not counted
5. All HSQC cross-peaks including those from non-equivalent methylene protons on the same carbon were counted. However, symmetrically-equivalent HSQC or HMBC cross-peaks, such as those that occur in succinic acid for example, were counted only once: the analysis reflects the number of peaks that can be seen in the spectra.
6. The count of theoretical ^{1}H, ^{13}C HMBC NMR cross-peaks includes all possible 2- and 3-bond carbon-to-proton couplings, including those between pseudo-equivalent groups e.g., the methyl groups in trimethylamine, as these are real and provide useful information for the identification of small metabolites
7. The second-order flag was only set in cases where the presence of magnetically non-equivalent but chemically equivalent protons would give rise to additional transitions in the spectra, not merely for cases where the signals have intensity distortions. The flag is set to 1 if there are ≥1 of these second order features and 0 otherwise.
8. For sugars, the count of features is applied to *both* anomers.

Thirteen parameters, A to M, were then calculated from the 14 features (Table 3): see also the Supplementary Data.

Table 1
The four levels of known metabolite identification from the CAWG 2007 [36].

Level 1	*Identified Compound*: A minimum of two independent and orthogonal data (such as retention time and mass spectrum) compared directly relative to an authentic reference standard
Level 2	*Putatively Annotated Compound*: Compound identified by analysis of spectral data and/or similarity to data in a public database but without direct comparison to a reference standard as for Level 1
Level 3	*Putatively Characterised Compound Class*: unidentified per se but the data available allows the metabolite to be placed in a compound class
Level 4	*Unknown Compound*: unidentified or unclassified but characterised by spectral data

Table 2
The 14 molecular and spectroscopic features calculated for the 75 metabolites.

1. Number of hydrogen atoms	2. Number of carbon atoms	3. Number of oxygen atoms
4. Number of nitrogen atoms	5. Number of sulphur atoms	6. Nominal mass in Da
7. Number of chiral centres	8. Number of ^{1}H NMR chemical shifts	9. Number of multiplicities
10. Number of 2- or 3-bond H, H coupling constants	11. Second order flag = 0 or 1	12. Number of 2D ^{1}H COSY cross-peaks
13. Number of 2D ^{1}H, ^{13}C HSQC cross-peaks	14. Number of 2D ^{1}H, ^{13}C HMBC cross-peaks	

2.3. Analysis of the NMR spectroscopic data and metabolite identification

All spectral processing was conducted in MNova version 9.0.0-12821 (Mestrelab Research S.L.).

Analysis of the spectroscopic information content of the urinary metabolites was conducted manually and captured and further analysed in Excel for Mac 2011 version 14.4.6 (Microsoft Corporation). Student t-tests were run in Excel using 2-tailed, unpaired calculations to determine the statistical significance of differences in values between groups of data. A p value of <0.05 was used as the cut-off for statistical significance [43]. IUPAC names, SMILES strings and InChi codes for the metabolites were downloaded from either the Human Metabolome Database [31] or from ChemSpider (Royal Society of Chemistry). Although 2D ^{1}H JRES NMR gives no new information (except for 2nd order systems and the discrimination of homonuclear and heteronuclear coupling), it was used to assist with the analysis of the 1D ^{1}H NMR spectra, and some coupling and multiplet information was abstracted from 2D ^{1}H JRES NMR rather than the 1D ^{1}H NMR spectra if appropriate. Similarly, TOCSY data was occasionally used to assist spectral analysis in crowded regions, although it theoretically provides no new information over COSY in the absence of spectral crowding.

Table 3
Metabolite identification parameters calculated for the 75 metabolites.

Parameter	Calculation
A. Total number of heavy atoms	Sum of features 2 to 5 in Table 2
B. Total number of spectroscopic information bits available from 1D ^{1}H NMR	Sum of features 8 to 11
C. Total number of spectroscopic information bits available from 1D ^{1}H and 2D ^{1}H COSY NMR	Sum of features 8 to 12
D. Total number of spectroscopic information bits available from 1D ^{1}H and 2D ^{1}H COSY and HSQC NMR	Sum of features 8 to 13
E. Total number of spectroscopic information bits available from 1D ^{1}H and 2D ^{1}H COSY, HSQC and HMBC NMR	Sum of features 8 to 14
F. Theoretical metabolite identification carbon efficiency (MICE) for 1D ^{1}H NMR	(Sum of features 8 to 11)/number of carbon atoms
G. Theoretical metabolite identification carbon efficiency (MICE) for 1D ^{1}H and 2D ^{1}H COSY NMR	(Sum of features 8 to 12)/number of carbon atoms
H. Theoretical metabolite identification carbon efficiency (MICE) for 1D ^{1}H and 2D ^{1}H COSY and HSQC NMR	(Sum of features 8 to 13)/number of carbon atoms
I. Theoretical metabolite identification carbon efficiency (MICE) for 1D ^{1}H and 2D ^{1}H COSY, HSQC and HMBC NMR	(Sum of features 8 to 14)/number of carbon atoms
J. Theoretical metabolite identification efficiency (MIE) for 1D ^{1}H NMR	(Sum of features 8 to 11)/number of heavy atoms
K. Theoretical metabolite identification efficiency (MIE) for 1D ^{1}H and 2D ^{1}H COSY NMR	(Sum of features 8 to 12)/number of heavy atoms
L. Theoretical metabolite identification efficiency (MIE) for 1D ^{1}H and 2D ^{1}H COSY and HSQC NMR	(Sum of features 8 to 13)/number of heavy atoms
M. Theoretical metabolite identification efficiency (MIE) for 1D ^{1}H and 2D ^{1}H COSY, HSQC and HMBC NMR	(Sum of features 8 to 14)/number of heavy atoms

Table 4

The 75 metabolites identified by NMR spectroscopy in recent metabonomics studies on human and mouse urine.

Metabolite class	Common name	IUPAC name
Carboxylic acids	Formic acid	Formic acid
	Acetic acid	Acetic acid
	Propionic acid	Propanoic acid
	Butyric acid	Butanoic acid
	Isobutyric acid	2-Methylpropanoic acid
	Isovaleric acid	2-Methylbutanoic acid
	Ketoleucine	4-Methyl-2-oxopentanoic acid
	Benzoic acid	benzoic acid
	Phenylacetic acid	2-Phenylacetic acid
	Para-hydroxy-phenylacetic acid	2-(4-Hydroxyphenyl)acetic acid
	Hydrocinnamic acid	3-Phenylpropanoic acid
Hydroxycarboxylic acids	Glycolic acid	2-Hydroxyacetic acid
	Lactic acid	(2S)-2-hydroxypropanoic acid
	2-Hydroxyisobutyric acid	2-Hydroxy-2-methylpropanoic acid
	3-Hydroxyisobutyric acid	(2S)-3-hydroxy-2-methylpropanoic acid
Dicarboxylic acids	Succinic acid	Butanedioic acid
	L-Malic acid	(2S)-2-hydroxybutanedioic acid
	Tartaric acid	(2R,3R)-2,3-Dihydroxybutanedioic acid
	Methylsuccinic acid	2-Methylbutanedioic acid
	Glutaric acid	Pentanedioic acid
	2-Hydroxyglutaric acid	(2S)-2-hydroxypentanedioic acid
	2-Ketoglutaric acid	2-Oxopentanedioic acid
	2-Isopropylmalic acid	(2S)-2-hydroxy-2-(propan-2-yl)butanedioic acid
Tricarboxylic acid	Citric acid	2-Hydroxypropane-1,2,3-tricarboxylic acid
	Isocitric acid	1-Hydroxypropane-1,2,3-tricarboxylic acid
	cis-Aconitic acid	(1Z)-Prop-1-ene-1,2,3-tricarboxylic acid
	Trans-aconitic acid	(1E)-Prop-1-ene-1,2,3-tricarboxylic acid
Small alcohols	Ethanol	Ethanol
	Chiral 2, 3-butanediol	(2R,3R)-butane-2,3-diol or (2S,3S)-butane-2,3-diol
	Meso-2, 3-butanediol	(2R,3S)-2,3-butanediol
Ketones	Butanone	Butan-2-one
	Acetoin	3-Hydroxybutan-2-one
Sugars and sugar acids	D-Xylose	(3R,4S,5R)-oxane-2,3,4,5-tetrol
	L-Fucose	(3S,4R,5S,6S)-6-methyloxane-2,3,4,5-tetrol
	D-Glucose	(3R,4S,5S,6R)-6-(hydroxymethyl)oxane-2,3,4,5-tetrol
	Mannitol	(2R,3R,4R,5R)-hexane-1,2,3,4,5,6-hexol
	D-Glucaric acid	(2R,3S,4S,5S)-2,3,4,5-tetrahydroxyhexanedioic acid
	D-Glucuronic acid	(2S,3S,4S,5R,6S)-3,4,5,6-tetrahydroxyoxane-2-carboxylic acid
	Para-cresol glucuronide	(2S,3S,4S,5R,6S)-3,4,5-trihydroxy-6-(4-methylphenoxy)oxane-2-carboxylic acid
Amines	Methylamine	Methanamine
	Dimethylamine	Dimethylamine
	Trimethylamine	Trimethylamine
	Trimethylamine N-oxide	N,N-dimethylmethanamine oxide
	Ethanolamine	2-Aminoethan-1-ol
	Choline	(2-Hydroxyethyl)trimethylazanium
	3-Methylhistamine	2-(1-Methyl-1H-imidazol-5-yl)ethan-1-amine
	Hypotaurine	2-Aminoethane-1-sulfinic acid
	Taurine	2-Aminoethane-1-sulfonic acid
	3-Indoxyl sulphate	1H-indol-3-yloxidanesulfonic acid
	Putrescine	Butane-1,4-diamine
	Creatinine	2-Imino-1-methylimidazolidin-4-one
	Creatine	2-(1-Methylcarbamimidamido)acetic acid
	L-Carnitine	(3R)-3-hydroxy-4-(trimethylazaniumyl)butanoate
Amino acids and amides	Glycine	2-Aminoacetic acid
	N-methylglycine, sarcosine	2-(Methylamino)acetic acid
	Dimethylglycine	2-(Dimethylamino)acetic acid
	N,N,N-trimethylglycine, betaine	2-(Trimethylazaniumyl)acetate
	N-acetylglycine	2-Acetamidoacetic acid
	N-propionylglycine	2-Propanamidoacetic acid
	N-butyrylglycine	2-Butanamidoacetic acid
	N-isovalerylglycine	2-(3-Methylbutanamido)acetic acid
	Hippuric acid, benzoylglycine	2-(Phenylformamido)acetic acid
	Phenylacetylglycine	2-(2-Phenylacetamido)acetic acid
	Guanidoacetic acid	2-Carbamimidamidoacetic acid
	Ureidopropionic acid	3-(Carbamoylamino)propanoic acid
	L-Alanine	(2S)-2-aminopropanoic acid
	Beta-alanine	3-Aminopropanoic acid
	Pyroglutamic acid	(2S)-5-oxopyrrolidine-2-carboxylic acid
	L-Histidine	(2S)-2-amino-3-(1H-imidazol-4-yl)propanoic acid
	1-Methylhistidine	(2S)-2-amino-3-(1-methyl-1H-imidazol-4-yl)propanoic acid

Table 4 (*continued*)

Metabolite class	Common name	IUPAC name
Amino acids and amides	Allantoin	(2,5-Dioxoimidazolidin-4-yl)urea
	Trigonelline	1-Methylpyridin-1-ium-3-carboxylate
	1-Methylnicotinamide	3-Carbamoyl-1-methylpyridin-1-ium
	Cytosine	6-Amino-1,2-dihydropyrimidin-2-one
Other metabolites	Para-cresol sulphate	(4-Methylphenyl)oxidanesulfonic acid

Features numbered 8 to 14 in Section 2.2 above were then analysed in the *actual* 1D and 2D NMR spectra of the urines of the mouse and diabetic patient, as features 8′ to 14′ respectively. For example, the total number of ^{1}H NMR chemical shifts (8′) actually observed for each metabolite (as opposed to the theoretical number calculated in Section 2.2 above) was measured. Parameters equivalent to B to M in Section 2.2 above were then calculated from features 8′ to 14′ to give the values for B′ to M′ respectively, for comparison with the theoretical values. For instance, the *actual* metabolite identification carbon efficiency (MICE) for 1D ^{1}H NMR level data is parameter F′. The full spreadsheet containing all these data is available as Supplementary Data.

3. Results and discussion

3.1. The identification of 75 human and mouse urinary metabolites

The manual analysis of a range of 1D and 2D NMR spectra of mouse and human urines from two recent studies [42] had resulted in the identification of a total of 75 metabolites. These metabolites were identified on the basis of NMR spectral analysis and comparison of the spectral data of the metabolites with that available for standard reference metabolites in the Human Metabolome Database [31], the BioMagResBank (BMRB) [32] and the Birmingham Metabolite Library (BML) [33]. The exact methodology for the analysis will not be detailed here but typically involved: (i) comprehensive comparison of the 2D ^{1}H, ^{13}C HSQC data with reference data in the HMDB and (ii) further interrogation of the data using all available resolution-enhanced 1D ^{1}H and 2D ^{1}H JRES, COSY, TOCSY and HMBC data.

As an example, the alpha and beta anomers of L-fucose (6-deoxy-L-galactose), a methyl sugar, were identified in the mouse urine. The process of this identification is described here. The 600 MHz 1D ^{1}H NMR spectrum of the mouse urine is shown in Fig. 1. Hundreds of signals are seen for dozens of metabolites. The identification of the known metabolite L-fucose commenced by matching the cross-peak at 1.25, 18.47 in the HSQC spectrum (Supplementary Fig. 1) to the methyl group of the beta anomer of L-fucose by an HMDB 2D HSQC search. The database gives figures of 1.26, 18.3 for the beta anomer of L-fucose. In confirmation, the $^3J_{H,H}$ coupling constant of the doublet at ca 1.25 in the mouse urine was measured as 6.5 Hz, in accordance with the HMDB figure. Naturally, if the beta anomer of L-fucose is present, then the alpha anomer must also be detected, as they are in dynamic equilibrium, although it is expected to be present at lower levels.

No signals for the alpha anomer were clearly visible in the 1D ^{1}H NMR spectrum, but the HSQC spectrum displayed a cross-peak at 1.22, 18.4, which corresponded well with the HMDB data for the authentic reference material (1.20, 18.3). The $^3J_{H,H}$ coupling constant between the alpha methyl and H5 was 6.6 Hz, in good agreement with HMDB (6.7 Hz). This was measured in the 2D J-resolved ^{1}H NMR spectrum

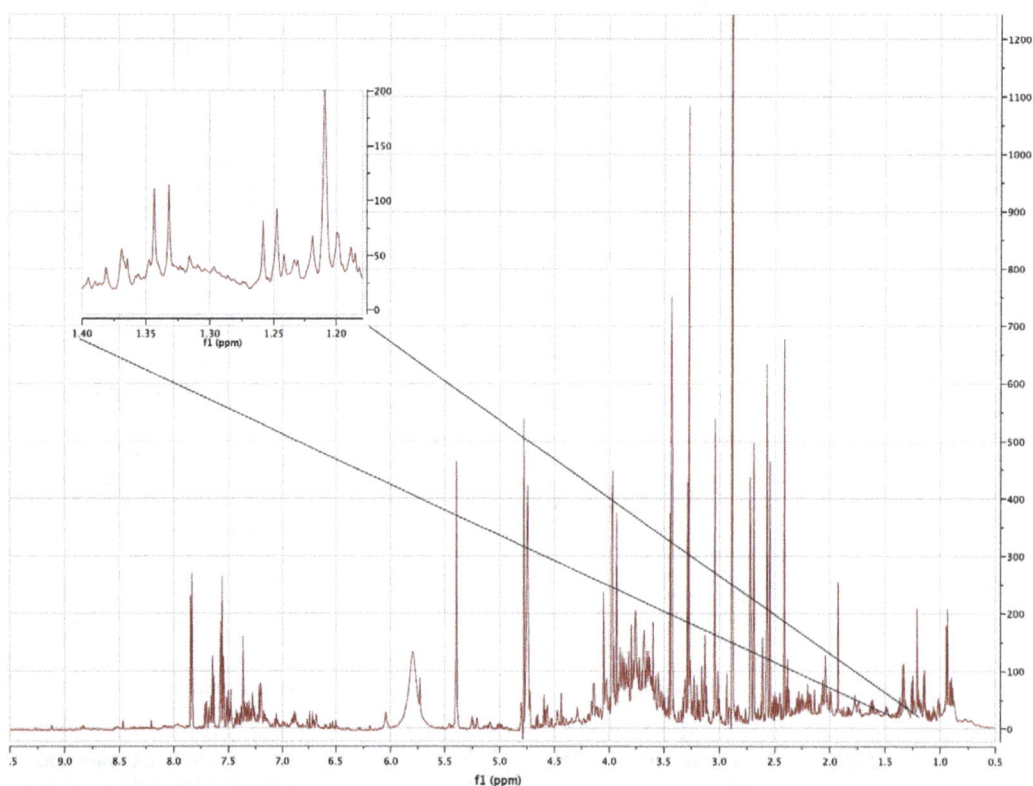

Fig. 1. The 600 MHz ^{1}H NMR spectrum of the urine from a C57BL/6 mouse and an expansion in the region of the methyl signals from lactic acid and the two anomers of L-fucose. The spectrum is moderately resolution-enhanced by Gaussian multiplication.

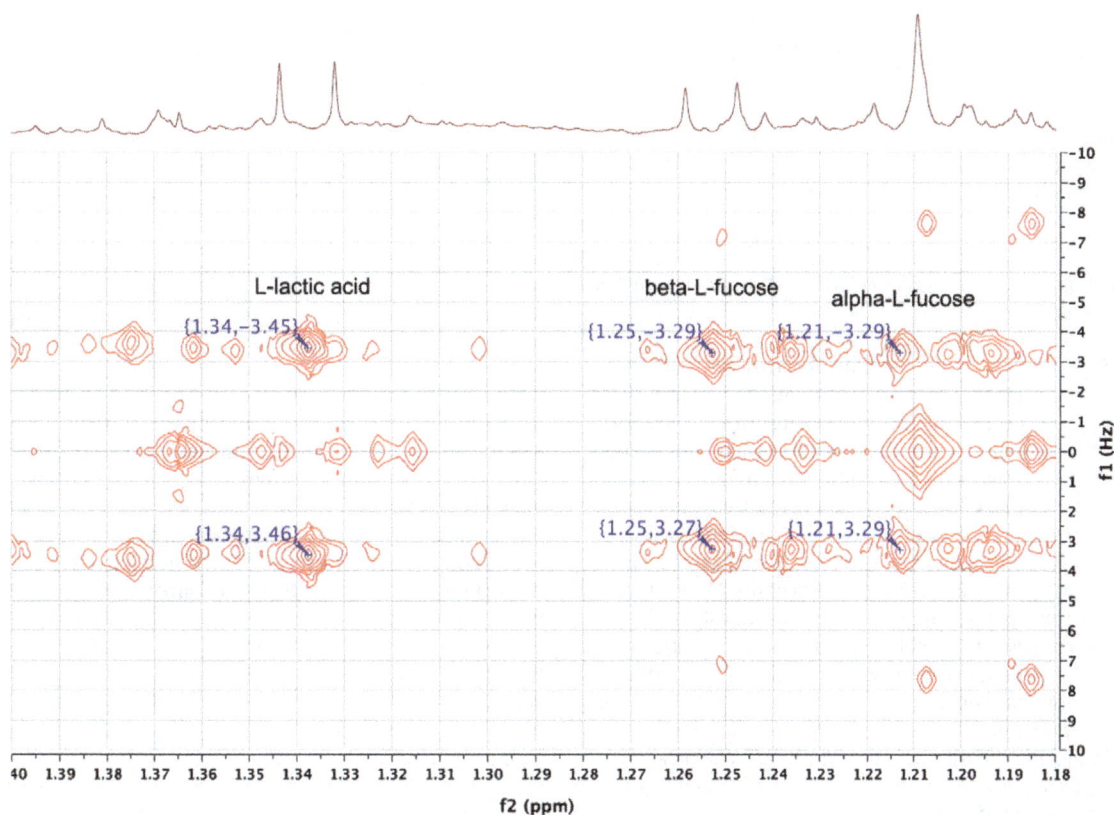

Fig. 2. An expansion of the 600 MHz 2D ^{1}H J-resolved NMR spectrum of the urine from a C57BL/6 mouse in the region of the methyl signals from lactic acid and the two anomers of L-fucose, underneath the corresponding region of the 1D ^{1}H NMR spectrum.

(Fig. 2), where the hidden alpha anomer signal is revealed by the spreading-out of the overlapped metabolite signals across a second dimension.

Further confirmation that these signals belonged to L-fucose came from a 2D ^{1}H COSY NMR spectrum (Supplementary Fig. 2), which showed that the methyl doublets resonating at ca 1.25 and at ca 1.21 in the mouse urine spectrum were connected to protons resonating at 3.80 and 4.20 respectively, exactly as expected for the beta and alpha anomers of L-fucose according to HMDB00174, which gives 3.80 and 4.18 respectively. A 2D ^{1}H, ^{13}C HMBC NMR spectrum also showed that the methyl protons at ca 1.25 connected to a carbon resonating at 73.7, which is a good match for C5 in beta-L-fucose (HMDB00174 gives 73.5).

In addition to these data, signals for the alpha and beta anomeric protons of L-fucose were detected at 5.22 (doublet (d), ca 4.0 Hz, COSY to H-2 at ca 3.78, HSQC to 95.5) and 4.57 (d, ca 7.8 Hz, COSY to 3.46, HSQC to 99.5) respectively (data not shown). Interestingly, the latter COSY revealed that there had been a data misinterpretation in HMDB (HMDB00174, accessed 29th November 2014) as the resonance for H-2 beta is given as 3.64 instead of 3.46, even though the coupling data matches the signal reported in HMDB at 3.46 and not at 3.64. HMDB00174 gives 5.19 (d, 3.9 Hz), 95.1 and 4.54 (d, 7.9 Hz), 99.0 for the anomeric protons and carbons of the alpha and beta anomers respectively.

It seemed that the metabolite whose signals were observed in the mouse urine was definitely L-fucose. However, according to the MSI guidelines, the identification could only be classified as a putative annotation, as the comparisons were made relative to the data in the HMDB, rather than to an authentic reference standard (Table 1). Indeed, all 75 metabolites identified in the studies of the mouse and human urine (Table 4) could only be described as putatively annotated by these rules. This seemed inappropriate and unsatisfactory, as the confidence in the identification of the vast majority of these metabolites was very

high. It seems that the MSI 4-Level system is too conservative for metabolite identification based on NMR spectroscopic data, which in comparison to chromatographic retention time data, or electrospray MS signal intensity data, for example, is more predictable and precise.

In order to explore what information had been acquired and how it compared with what was theoretically available, an analysis of the spectroscopic information present in the 75 metabolites identified in the two metabonomics studies was undertaken.

The 75 metabolites from the two studies were combined to provide a realistic representation of the range of metabolites that a typical metabonomics study by high field NMR might identify. Analysis of these molecules showed that they had a molecular weight range of 31 to 284 Da (nominal mass) with an average of 126.7 ± 46.6 Da. The number of carbon atoms ranged from 1 to 13 with an average of 4.9 ± 2.2 (standard deviations). See the Supplementary Data for more information.

The subsequent analysis was completed in three parts: (i) an analysis of the information content of 1D and 2D ^{1}H NMR spectra; (ii) an analysis of the NMR spectroscopic features theoretically present in the 75 metabolites and (iii) a comparison of the features theoretically present in the metabolites with those actually found in the course of the metabonomics studies. The aim of these analyses was to determine how much structural information was present in the metabonomics data and therefore how much confidence could be ascribed to metabolite identification. This analysis proved to be both informative and thought provoking.

3.2. The Information content of NMR spectra in the context of metabonomics experiments

NMR spectroscopy provides a surprisingly rich quantity of information on the molecules under study. The following list of 11 NMR spectral features is not exhaustive but includes those that are useful for the purposes of metabolic profiling, and is focused on ^{1}H NMR-detected experiments: (1) chemical shifts, (2) signal multiplicities, (3) coupling constants, (4) 1st or 2nd order signal nature, (5) signal half-bandwidth, (6) signal integral, (7) COSY cross peaks, (8) HSQC cross-peaks, (9) HMBC cross-peaks, (10) TOCSY cross-peaks and (11) signal rate of change [44]. A detailed analysis of these features is provided in Supplementary Table 2.

3.3. Analysis of the theoretical 1D and 2D ^{1}H NMR spectroscopic information content of metabolites

Of the 11 features of 1D or 2D ^{1}H NMR spectroscopy outlined in Section 3.2 above, the analysis here focused on just 7: chemical shifts, multiplicities, coupling constants, 2nd order nature and COSY, HSQC and HMBC cross-peaks, for further study, as these are of most importance for metabolite identification by NMR spectroscopy. A manual analysis of the *number* of *each* of these 7 features expected to occur in *each* of the 75 metabolites was conducted (see Supplementary Data).

Table 5 shows the number of bits of spectroscopic information theoretically present in the 1D or 2D ^{1}H NMR spectra of the 75 metabolites, for a range of different approaches to metabolite identification. The first would involve just the use of 1D ^{1}H NMR; the second, the additional use of COSY, the third the additional use of HSQC and finally, the additional use of HMBC information. For example, the number of bits of spectroscopic information in a metabolite for an approach based on just 1D ^{1}H NMR would include the total number of ^{1}H NMR chemical shifts, multiplicities and coupling constants in a metabolite, plus a flag for a second order spin system, if present.

Thus, as expected, the amount of spectroscopic information available to assist with metabolite identification increases in going from approaches based solely on 1D ^{1}H NMR, to those involving significant utilisation of 2D NMR methods. The distribution of the data across the 75 metabolites is informative (Fig. 3). *The bits of spectroscopic information can be considered to be bits of metabolite identification information, each of importance to the valid identification of metabolites.* What is immediately apparent is that even with a simple 1D ^{1}H NMR approach, some metabolites contain a surprisingly large number of bits of information that can be used to identify them: up to 42 bits in one metabolite in this set.

It is important to understand how the metabolite identification information content of the metabolites varies with their structures. Fig. 4 shows the variation in the number of bits of metabolite identification information against the number of carbon atoms in the molecule. An approximately linear relationship is observed, apart from three clear outliers (filled diamonds in Fig. 4) due to the sugars xylose, fucose and glucose in the set. Removal of the three outliers improves the linear correlation to an R^2 of 0.47, with the equation $y = 1.74x - 0.48$.

The analysis was then developed using a concept from drug discovery. In 2004, Alex, Groom and Hopkins introduced the concept of ligand efficiency as a tool to assist lead and drug discovery [45,46]. The essence of this approach is to calculate the binding energy of ligands *per heavy atom in the molecule*, in order to drive drug discovery projects towards molecules that have the highest binding energy with the lowest

molecular weight. A corresponding approach to metabolite identification analysis would use the concept of metabolite identification efficiency (MIE). In contrast to ligand efficiency (LE) where the total molecular weight is of importance, in MIE, the number of carbon atoms in the metabolite is also of importance, as the carbon atoms carry the vast majority of the non-exchangeable hydrogen atoms observed in ^{1}H NMR experiments. We thus introduce the concept of metabolite identification efficiency in two forms:

MIE = number of bits of metabolite identification information/number of heavy atoms in metabolite

MICE = number of bits of metabolite identification information/number of carbons in metabolite

where MICE is the Metabolite Identification Carbon Efficiency. Like MIE, MICE can be calculated separately for each metabolite according to the approach taken to the analysis of the metabonomics data, be that solely based on 1D ^{1}H NMR, or involving significant utilisation of 2D NMR methods (Fig. 5).

It is clear that the MICE for metabolites varies broadly and that the use of additional 2D technologies including COSY, HSQC and HMBC can significantly boost the theoretical amount of metabolite identification information per carbon atom in the metabolite. The theoretical MICE values range from an average of 1.8 ± 1.3 for 1D ^{1}H NMR alone, to 2.2 ± 1.7 for approaches that include COSY data, to 2.9 ± 2.1 for approaches that also include COSY and HSQC and to 4.6 ± 3.3 bits per carbon atom for approaches that include COSY, HSQC and HMBC (standard deviations).

For comparison, the theoretical MIE values range from an average of 1.0 ± 0.7 for 1D ^{1}H NMR alone, to 1.2 ± 0.9 for approaches that also include COSY data, to 1.6 ± 1.1 for approaches that also include COSY and HSQC and to 2.6 ± 1.8 bits per heavy atom (standard deviations) for approaches that also include HMBC. The MIE values are naturally lower as the number of heavy atoms is approximately double the number of carbon atoms across this set of 75 metabolites.

The amount of metabolite identification information per heavy atom or per carbon atom in the metabolites is quite high and gives a perspective on what can be achieved via modern, high field NMR spectroscopy approaches to metabonomics. This efficiency-based approach is critical in understanding how much metabolite identification information is being obtained relative to the molecular size of the metabolite.

The theoretical MIE and MICE values also varied significantly according to the type of metabolites under study. The metabolites were sorted between those containing 1 to 5 chiral centres (n = 24) and those containing no chiral centres (n = 51). Table 6 shows the differences between the chiral and non-chiral metabolites in terms of their theoretical number of bits of metabolite identification information at the level of 1D ^{1}H and 2D COSY and HSQC NMR data plus the corresponding MIE and MICE values.

Table 6 clearly shows that: (i) the information content of the chiral metabolites is significantly greater than that of the non-chiral metabolites (p = 0.0007), and also that (ii) the information density per heavy atom (MIE, p = 0.0022) or per carbon atom (MICE, p = 0.0014, all from two-tailed, unpaired student t-tests) is also significantly higher for chiral metabolites. In all cases the p values from the student *t*-test

Table 5
The number of bits of spectroscopic information per metabolite *theoretically contained* in the group of 75 metabolites, from four NMR-based metabonomics approaches: each bit corresponds to a bit of metabolite identification information.

Feature/methodology	1D ^{1}H NMR	1D ^{1}H NMR plus 2D COSY	1D ^{1}H NMR plus 2D COSY and HSQC	1D ^{1}H NMR plus 2D COSY, HSQC and HMBC
Minimum number of bits	2	2	3	3
Maximum number of bits	42	56	70	106
Average number of bits	9.2	11.3	14.7	24.3
Median number of bits	7	8	11	16
Standard deviation	7.9	10.6	13.1	21.8

theoretical number of bits of metabolite ID information from NMR

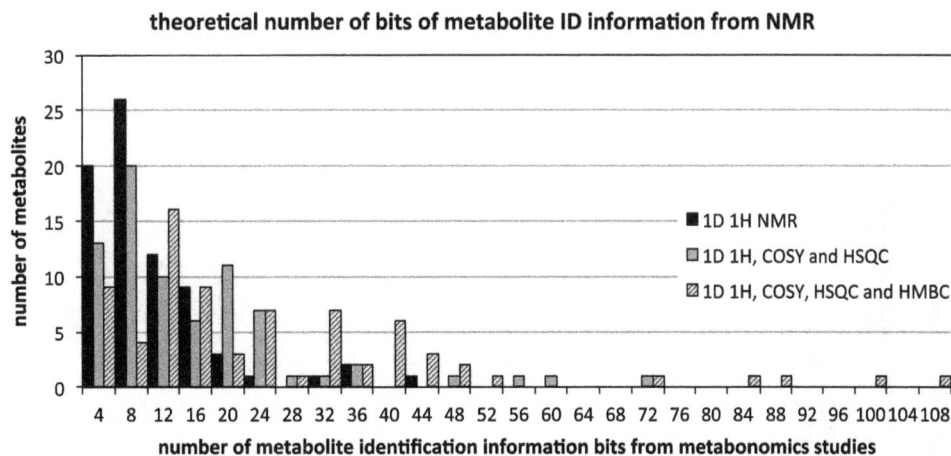

Fig. 3. The distribution of the theoretical number of bits of metabolite identification (ID) information available from three different NMR approaches across the 75 metabolites. The number of bits is calculated in bins ranging from 0 to 4, 5 to 8 etc. up to 105 to 108 bits. Each bit represents a ^{1}H NMR chemical shift, multiplicity, coupling constant, 2nd order flag, COSY cross-peak, HSQC cross peak or HMBC cross peak, that theoretically should be observed for the metabolite in question. Data for approaches using ^{1}H plus COSY data not shown for clarity of presentation.

are less than 0.05, the cut-off for statistical significance of the differences in the values, with 95% confidence. The reason for these significant differences is principally the raising of the chemical shift degeneracy for methylene protons in the environment of a chiral centre: this significantly increases the number of metabolite identification information bits in a metabolite.

A similar theoretical analysis using an NMR approach including 1D ^{1}H NMR, COSY and HSQC data was conducted of differences between the classes of metabolites in Table 4. This demonstrated that the MIE values for the tricarboxylic acids (0.8 ± 0.5, n = 4) are significantly lower than the corresponding values for the cluster formed of the small alcohols and ketones (1.9 ± 0.5, n = 5, grouped together) with a p value of 0.012. In addition the group of sugars and sugar acids have an MIE value (3.6 ± 2.1) that is significantly greater than those of all other groups (additional values are: 1.5 ± 0.4, 1.3 ± 1.1, 1.4 ± 1.0, 1.4 ± 0.6 and 1.4 ± 0.7 for the carboxylic acids, n = 11, the hydroxycarboxylic acids, n = 4, the dicarboxylic acids, n = 8, the amines, n = 14 and the amino acids and amides, n = 21 respectively) with p values all ≤0.039, apart from the cluster formed of the small alcohols and ketones (p = 0.079).

The corresponding theoretical MICE analysis (at the level of 1D ^{1}H NMR, COSY and HSQC data) showed that the values of the sugars and sugar acids (7.1 ± 4.1) are significantly greater than those of all other groups, with p values ranging from 0.012 to 0.031. The MICE values for the other groups are 2.3 ± 0.6, 2.5 ± 1.9, 2.7 ± 1.9, 1.7 ± 1.1, 2.8 ± 0.7, 2.5 ± 1.1 and 2.6 ± 1.2 for the carboxylic acids, the hydroxycarboxylic acids, the dicarboxylic acids, the tricarboxylic acids, the small alcohols and ketones, the amines, and the amino acids and amides, respectively. No other groups showed significantly different MICE values in pairwise comparisons. The non-significance (MIE) vs the significance (MICE) in the differences between the values for the sugars and sugar acids and the small alcohols and ketones, reflects the fact that the carbon to oxygen ratio is at least 2 to 1 for the alcohols and ketones whereas it is ca 1:1 for most of the sugars and sugar acids. This has the effect of scaling down the MIE values for the sugars and sugar acids and making the difference between their average values and those of the small alcohols and ketones non-significant.

It is worth noting that a third approach, different from either the MIE or MICE approaches is possible. This third approach involves simply counting the number of bits of metabolite identification information

plot of total theoretical number of metabolite ID information bits from 1D 1H NMR against number of carbon atoms in a metabolite

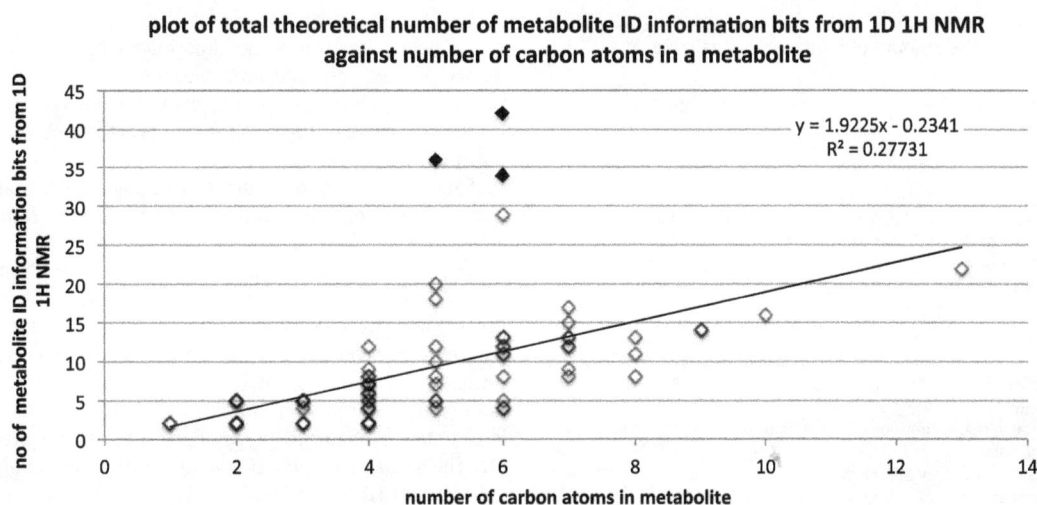

Fig. 4. The number of metabolite identification (ID) information bits theoretically available from 1D ^{1}H NMR plotted against the number of carbon atoms for all 75 metabolites. Three outliers due to xylose, fucose and glucose are highlighted with filled, as opposed to open diamonds.

theoretical metabolite identification carbon efficiency (MICE) for all 75 metabolites

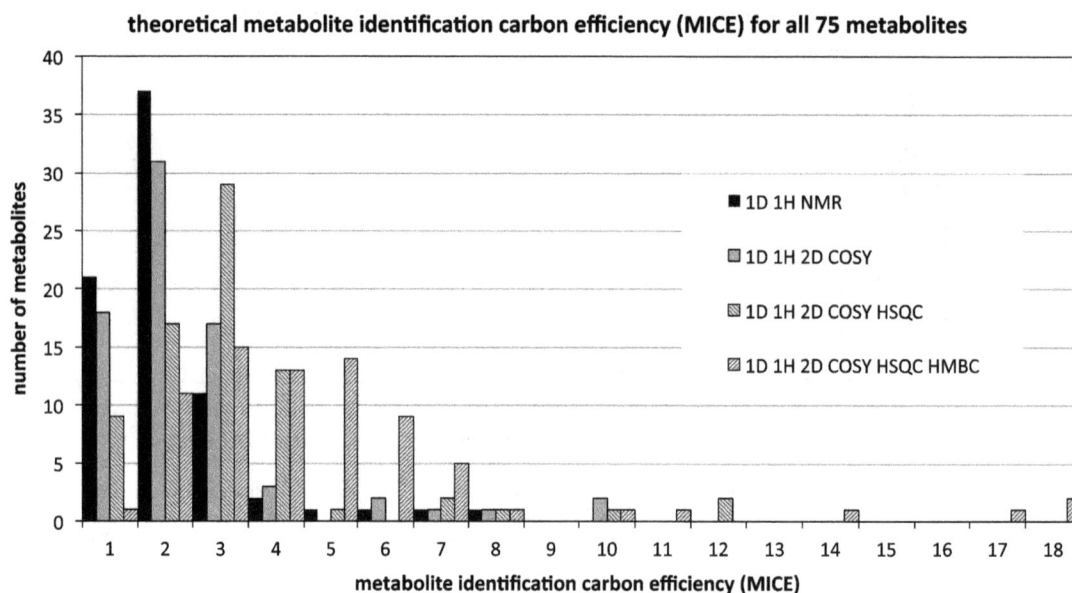

Fig. 5. The theoretical metabolite identification carbon efficiency (MICE) for all 75 metabolites and for four separate metabolic profiling approaches: 1D ^{1}H NMR alone, 1D ^{1}H plus COSY, 1D ^{1}H plus COSY and HSQC data and 1D ^{1}H plus COSY, HSQC and HMBC data. The histogram shows the number of metabolites for each approach with MICE values in bins of 0 to 1, >1 to 2, >2 to 3 etc. up to >17 to18.

theoretically present in each of the metabolites, with each level of NMR approach, from 1D ^{1}H NMR alone, up to the combined usage of 1D ^{1}H NMR together with COSY, HSQC and HMBC data. The number of theoretical metabolite ID information bits can then be compared with the actual number of bits experimentally observed to give a metabolite identification hydrogen fraction (MIHF). This analysis is conducted in Section 3.4 below.

3.4. An analysis of the actual 1D and 2D ^{1}H NMR spectroscopic information content of metabolites, retrieved from analysis of biofluid NMR spectra

Theoretical analyses are all well and good but a key question is how much metabolite identification information is *actually retrieved* in typical metabonomics experiments. Issues such as relatively low abundance of a particular metabolite and/or spectral crowding in some chemical shift regions will reduce the actual amount of metabolite identification information retrieved for metabolites, relative to the theoretical maximal amount. In addition, the small size and lack of hydrogen atoms in some metabolites limit the amount of information available.

Table 6
A theoretical analysis of the total number of metabolite identification information bits, metabolite identification efficiency (MIE) and metabolite identification carbon efficiency (MICE) for chiral (24) vs non-chiral (n = 51) metabolites in this study (all analyses at the level of data from 1D ^{1}H and 2D COSY and HSQC NMR.

Feature/parameter	Average value	Standard deviation
Total number of metabolite identification information bits for chiral metabolites	24.58	17.98
Total number of metabolite identification information bits for non-chiral metabolites	9.98	6.03
Metabolite identification efficiency MIE, chiral	2.36	1.52
Metabolite identification efficiency MIE, non-chiral	1.27	0.56
Metabolite identification carbon efficiency (MICE), chiral	4.45	3.03
Metabolite identification carbon efficiency (MICE), non-chiral	2.19	0.88

Table 7 lists the information obtained using four different levels of NMR spectroscopy for the 75 metabolites studied in this work. A direct comparison with Table 5 will illustrate that there is a significant drop in the amount of information obtained from the analysis of the experimental NMR spectra, compared with that which is theoretically available. Table 8 provides another view of the data, providing the total number of bits of metabolite identification information actually obtained in four different modes of NMR-based metabonomics versus the bits of information theoretically available.

The drop off in metabolite identification information observed relative to that theoretically available is particularly steep for the HMBC data. Only 82 bits of information out of a possible total of 725 bits were obtained across all 75 metabolites from HMBC experiments. This is unsurprising given the difficulty in acquiring HMBC data on low abundance metabolites in biofluids with good sensitivity in a reasonable period of time. However, for HSQC, an encouraging 129 bits of information were obtained from a theoretical maximum of 250 across the 75 metabolites.

Supplementary Fig. 3 shows a histogram comparing the actual number of metabolite identification information bits retrieved in the experiments reported here compared with the amount theoretically available, for an approach combining information from 1D ^{1}H NMR, COSY and HSQC experiments. The clustering of the actual information retrieved to lower bin sizes is clear.

Finally, Fig. 6 shows the actual metabolite identification carbon efficiency (MICE) obtained in the experiments with four different NMR approaches.

The data in Fig. 6 can be directly compared with that in Fig. 5. It is clear that the actual, experimental MICE values for metabolites vary broadly and that the use of additional 2D technologies including COSY and HSQC does boost the actual amount of metabolite identification information per carbon atom in the metabolite. However, in these experiments, the additional information from HMBC did not augment the information available to anywhere near the extent theoretically possible. The actual average MICE values over all metabolites range from 1.3 ± 0.8 for 1D ^{1}H NMR alone, to 1.6 ± 1.0 for approaches that also include COSY, to 1.9 ± 1.1 for approaches that also include COSY and HSQC and to 2.2 ± 1.1 bits per carbon atom (standard deviations) for approaches that also include HMBC. The corresponding actual MIE

Table 7
The bits of metabolite identification information per metabolite *actually obtained* from four NMR-based metabonomics approaches in the group of 75 metabolites.

Feature/methodology	1D ^{1}H NMR	1D ^{1}H NMR plus 2D COSY	1D ^{1}H NMR plus 2D COSY and HSQC	1D ^{1}H NMR plus 2D COSY, HSQC and HMBC
Minimum number of bits	2	2	2	2
Maximum number of bits	22	28	31	31
Average number of bits	6.2	7.5	9.2	10.3
Median number of bits	5	6	8	9
Standard deviation	4.5	5.8	6.5	6.8

values (Fig. 7) averaged over all 75 metabolites are: 0.7 ± 0.4 for 1D ^{1}H NMR alone, 0.9 ± 0.5 for approaches that also include COSY, 1.1 ± 0.6 for approaches that also include COSY and HSQC and 1.2 ± 0.6 bits per heavy atom for approaches that also include COSY, HSQC and HMBC (standard deviations).

As mentioned in Section 3.3 above, another approach to take to the question of metabolite identification confidence would be to compare simply the number of metabolite identification information bits obtained experimentally, with the number of bits theoretically present in each metabolite to arrive at a metabolite identification hydrogen fraction (MIHF) as defined below:

MIHF = NMIIo/NMIIt
NMIIo = Number of bits of Metabolite Identification Information
    actually **o**bserved
NMIIt = Number of bits of Metabolite Identification Information
    theoretically present

MIHF can be calculated for single metabolites, sub-groups of metabolites or an entire collection. This analysis is also illuminating (Fig. 8).

It is striking that 42 out of 75 metabolites (56%) have MIHF values of >0.9 for a simple 1D ^{1}H NMR approach to metabolite identification, indicating that the majority of metabolites studied here are displaying >90% of the available 1D ^{1}H NMR information bits. This % drops off as the NMR approach includes the use of more and more 2D NMR methods and is lowest for the approach combining 1D ^{1}H NMR with 2D COSY, HSQC and HMBC approaches. This is due to the difficulty of observing all HMBC cross-peaks for metabolites present in a biofluid at relatively low concentrations.

As discussed above, the MIHF values clustered significantly at the high end of the range of possible values and provided a less good discrimination between metabolites than the corresponding MIE or MICE values. In addition, the MIHF values can seem misleadingly low for chiral metabolites, where there is typically more information than is required for confident metabolite identification, due to the raising of the degeneracy of methylene proton signals. For instance, a comparison of achiral, 2-hydroxyisobutyric acid (2-hydroxy-2-methylpropanoic acid, HMDB00729) with its chiral isomer, 3-hydroxyisobutyric acid ((2S)-3-hydroxy-2-methylpropanoic acid, HMDB00023) shows that the former has a total of just 2 bits of spectroscopic information at the level of 1D ^{1}H NMR information bits, whereas the latter has 12! Finally, it is also a concern that it may be easier for a small metabolite with a low number of signals to get a very high MIHF score, compared with a more complex metabolite with more signals. Consequently, the MIE and MICE

measures of confidence in metabolite identification were used in the rest of this analysis in preference to the MIHF.

3.5. How much NMR information is enough for confident metabolite identification?

This is the key question. The Metabolomics Standards Initiative (MSI) approach differentiates between the situation where: (i) the experimental metabonomics data is compared with an authentic reference standard (Level 1, Identified Metabolite) and (ii) where comparison is made to the literature or a public domain database such as the HMDB (Level 2, Putatively Annotated Metabolite): see Table 1. On the basis of the analysis of the NMR-derived data in this study, that differentiation is not appropriate and it is perfectly possible to confidently identify known metabolites based on reference to the literature or the public databases. The guidelines to enable this are proposed to be as follows:

1. experimental metabolite identification carbon efficiency (MICE) ideally ≥ 1 and/or metabolite identification efficiency (MIE) > 0.5, [these are guidelines, not rigid cut-offs, based on the experience with the metabolites in this study and are for MICE and MIE values with NMR approaches based on 1D ^{1}H plus 2D COSY and HSQC data]
2. the fit of the experimental data to reference data should be precise, generally within ± 0.03 ppm for 1D ^{1}H and ± 0.5 ppm for ^{13}C NMR shifts and ± 0.2 Hz for homonuclear proton couplings: values outside these limits need explanation: in addition, the reference database entries should be double-checked for consistency with other literature values and general accuracy and self-consistency, including by downloading of actual free induction decay data e.g., from the HMDB [31]
3. the NMR spectral data should provide 'coverage' of all parts of the molecule: for example, for para-cresol glucuronide, a molecule with two distinct parts, it is important to have NMR data from both the cresol and glucuronide parts for good confidence
4. the signal-to-noise ratio and the resolution (actual and digital) in the spectra should be sufficient to measure the signal features with confidence, with high resolution having the added benefit of enabling the observation of long-range, homonuclear ^{1}H $-^1$H and two-bond ^{1}H $-^{14}$N couplings that can be diagnostic for certain metabolites
5. care needs to be applied in the assignments of signals in regions of the ^{1}H NMR spectrum that are crowded with signals from other metabolites, as the possibility of miss-assignment is higher in these regions: high spectral and digital resolution is

Table 8
A comparison of the total amount of metabolite identification information actually obtained versus that theoretically available from four NMR-based metabonomics approaches across the group of 75 metabolites as a whole.

Feature/methodology	1D ^{1}H NMR	1D ^{1}H NMR plus 2D COSY	1D ^{1}H NMR plus 2D COSY and HSQC	1D ^{1}H NMR plus 2D COSY, HSQC and HMBC
Theoretical total number of metabolite identification bits available	688	849	1099	1824
Actual total number of metabolite identification bits observed	467	560	689	771

actual metabolite identification carbon efficiency (MICE) for all 75 metabolites

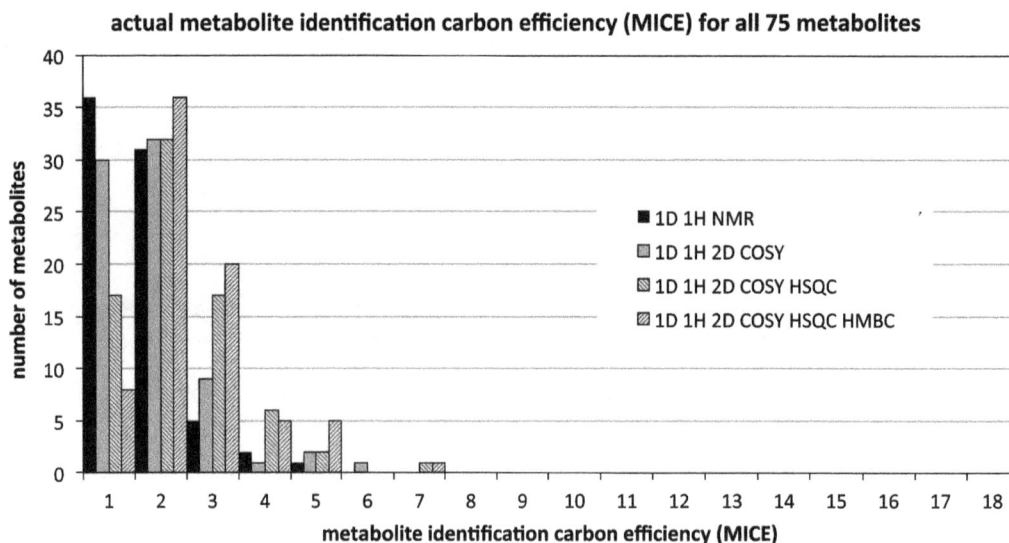

Fig. 6. The actual experimental metabolite identification carbon efficiency (MICE) for all 75 metabolites and for four separate metabolic profiling approaches: 1D ^{1}H NMR alone, 1D ^{1}H plus COSY, 1D ^{1}H plus COSY and HSQC data and 1D ^{1}H plus COSY, HSQC and HMBC data. The histogram shows the number of metabolites for each approach with MICE values in bins of 0 to 1, >1 to 2, >2 to 3 etc. up to >17 to18.

even more critical, as is the ability to correlate the correct signals together: TOCSY and J-resolved spectra can be enabling here

6. HSQC data is extremely important in resolving metabolite identification issues, as the chemical shift sensitivity of ^{13}C NMR is ca 20× that of ^{1}H NMR and it provides a superb orthogonal data source, as recommended by MSI: reliance solely on 1D ^{1}H NMR data will lead to confident assignments of major metabolites but will struggle with the confident identification of less prominent metabolites in crowded spectral regions

7. even though HMBC provided only a small proportion of the metabolite identification bits that were theoretically possible in these experiments, it is sometimes the only way to categorically identify metabolites. HMBC is extremely valuable for defining inter-atomic connectivities to quaternary carbon atoms, as

well as through quaternary carbon atoms and heteroatoms, and should be used as much as possible.

The example of the identification of L-fucose is a good one, if on the extreme end of proving a point. A total of 15 bits of 1D ^{1}H NMR information were discovered in the experimental data. This figure increased to 20, 24 and 25 bits of information if COSY, COSY plus HSQC or COSY plus HSQC and HMBC data respectively, were considered in addition to the 1D ^{1}H NMR data. The experimental MICE values were 2.5, 3.3, 4.0 and 4.2 for the 1D ^{1}H NMR, 1D ^{1}H plus 2D COSY, 1D ^{1}H plus 2D COSY and HSQC, and 1D ^{1}H plus 2D COSY, HSQC and HMBC data approaches respectively. The corresponding MIE values were: 1.4, 1.8, 2.2 and 2.3 respectively. All the experimental bits of metabolite identification information were in good agreement with those reported for authentic L-fucose, HMDB00174, in the HMDB (see Section 3.1 above).

actual metabolite identification efficiency (MIE) for all 75 metabolites

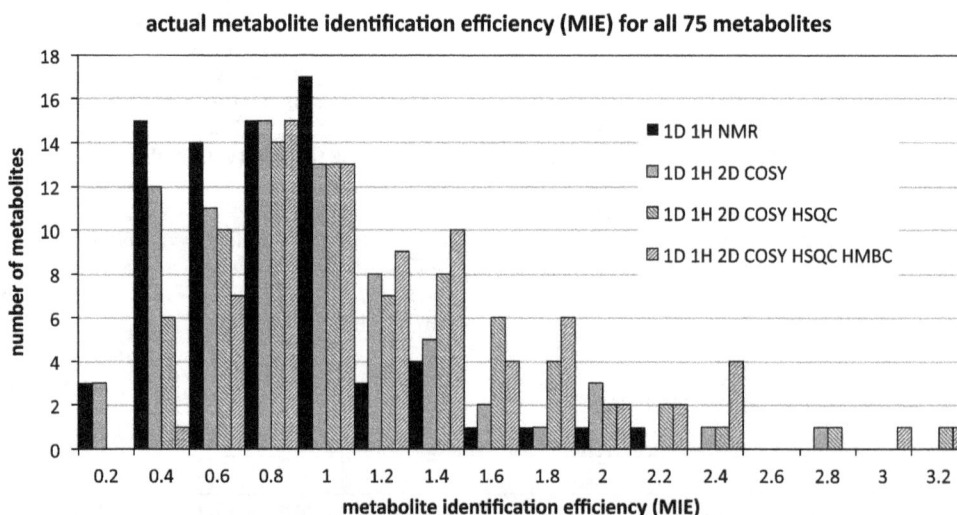

Fig. 7. The actual experimental metabolite identification efficiency (MIE) for all 75 metabolites and for four separate metabolic profiling approaches: 1D ^{1}H NMR alone, 1D ^{1}H plus COSY, 1D ^{1}H plus COSY and HSQC data and 1D ^{1}H plus COSY, HSQC and HMBC data. The histogram shows the number of metabolites for each approach with MIE values in bins of 0 to 0.2, >0.2 to 0.4, >0.4 to 0.6 etc. up to >3.0 to 3.2.

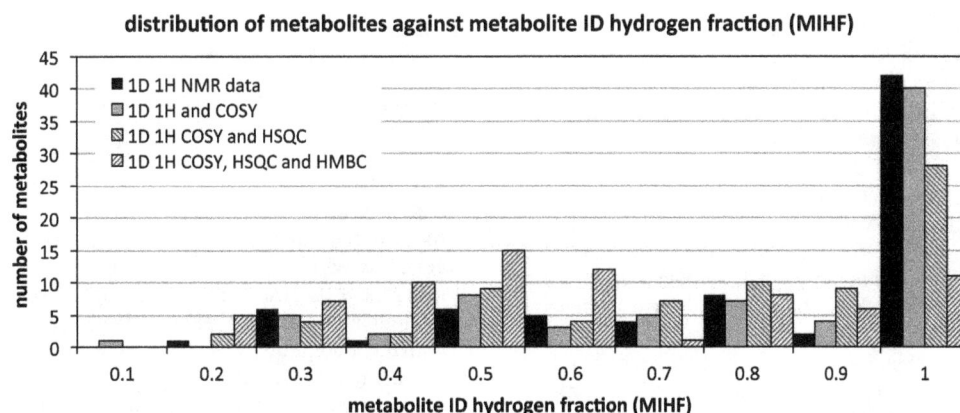

Fig. 8. A histogram of the number of metabolites in the collection of 75 metabolites analysed here against the metabolite identification hydrogen fraction (MIHF) in buckets of 0.1 from 0 to 1. The analysis is shown for four separate NMR approaches to metabolite identification: use of 1D ^{1}H NMR data alone and the additional uses of COSY, HSQC and HMBC data.

These figures indicate great confidence in the metabolite identification and no need for any further direct comparisons with an actual sample of authentic L-fucose as recommended in the original MSI publication [36].

A metabolite with an MICE value just under average for approaches based on 1D ^{1}H plus 2D COSY and HSQC data is ketoleucine (HMDB00695). This is a more normal example of a metabolite that was identified in the mouse urine. The methyl groups were observed as a doublet at 0.941 (d, 6.6 Hz), 24.5 with a COSY to 2.098 (triplet of septets), and the latter signal had a COSY to 2.618 (d, 7.0 Hz). The identification of three chemical shifts, three multiplicities, two coupling constants, two COSY and one HSQC cross-peaks gave a total of 11 bits of information. The corresponding data for HMDB00695 was 2.60 (d, 7.0), 50.8; 2.09, 26.7 and 0.93 (d, 6.7 Hz), 24.4 and is an excellent match to the experimental data. Ketoleucine has 6 carbon atoms, so the MICE value is 11/6 = 1.8, just under the average MICE value of 1.9 bits per carbon for all the metabolites in this study, and at this level. Ketoleucine, a metabolite with a below average MICE value is considered confidently identified.

3.6. How confident can we be in metabolite identification with MIE < 0.5 or MICE < 1 (using 1D ^{1}H plus COSY and HSQC NMR data)?

This analysis will be exemplified for NMR approaches that use 1D ^{1}H plus COSY and HSQC NMR data, as this is routine in metabonomics/metabolomics studies. Of the 75 metabolites in this study, five have a theoretical MIE of <0.5 and four of these five have a theoretical MICE of <1 based on combined 1D ^{1}H plus COSY and HSQC NMR data (See Supplementary Data). These metabolites are: 2-hydroxyisobutyric acid, succinic acid, tartaric acid, allantoin and guanidoacetic acid. All five metabolites have just a single singlet in their 1D ^{1}H NMR spectrum,

severely limiting the amount of NMR information that can be obtained. In practice, the relatively distinctive chemical shifts of the first four, the availability of HSQC information for all five and HMBC information for all except 2-hydroxyisobutyric acid, means that their identification is unambiguous (see Supplementary Information). However, in these cases, where the MIE < 0.5 and/or MICE is < 1, it is critical to have orthogonal confirmation of metabolite identities via HSQC/HMBC data, as achieved in the experiments reported here, and all five metabolites are considered confidently identified. The actual experimental MICE, MIE values were: 2-hydroxyisobutyric acid (0.8, 0.4), succinic acid (0.8, 0.4), tartaric acid (0.8, 0.3), allantoin (0.8, 0.3) and guanidoacetic acid (1.0, 0.4), all being identical to the maximum theoretical values in this case.

In addition, 13 of the 75 metabolites studied have *actual* MIE scores of <0.5 and/or actual MICE scores of <1.0 based on the combined experimental 1D ^{1}H plus COSY and HSQC NMR data (See Supplementary Data). These 13 naturally include the five metabolites analysed above. Table 9 extracts the data that was available for the MICE scores of the 8 additional metabolites from the Supplementary materials.

So, for these metabolites, how confident is their identification based on the information given in Table 9? For phenylacetic acid, three additional HMBC connectivities were observed from the acid, and ipso and ortho aromatic carbons to the methylene protons, which also had a long-range, 4-bond COSY to the ortho aromatic protons, so this identification is considered confident. For methylsuccinic acid, no additional information was available and therefore this metabolite should be described as putatively annotated, to keep consistency with the MSI nomenclature. For trans-aconitic acid, in addition to the 1D ^{1}H NMR chemical shift, multiplicity and HSQC information, a long-range, 4-bond COSY was observed between the olefin and methylene protons,

Table 9

the actual NMR-based metabolic identification information available from 1D ^{1}H plus COSY and HSQC NMR experiments on eight metabolites with MICE scores of <1.0.

Common name	Number of carbon atoms	Number of 1D 1H δH	Number of mult.	Number of nJHH	Actual 2nd order flag	Number of COSY cross-peaks	Number of HSQC peaks	Actual total info 1D 1H, COSY & HSQC	Actual MICE based on 1D 1H COSY HSQC
Phenylacetic acid	8	1	1	0	1	0	4	7	0.9
Methylsuccinic acid	5	1	1	1	0	1	0	4	0.8
Trans-aconitic acid	6	2	2	0	0	0	1	5	0.8
Choline	5	1	1	0	0	0	1	3	0.6
L-Carnitine	7	1	1	0	0	0	1	3	0.4
Dimethylglycine	4	1	1	0	0	0	1	3	0.8
N,N,N-trimethylglycine, betaine	5	1	1	0	0	0	1	3	0.6
N-propionylglycine	5	1	1	1	0	1	0	4	0.8

both of which displayed characteristic 0.8 Hz couplings on resolution enhancement of the spectra. Reprocessing the reference NMR data in HMDB (HMDB00958) with resolution enhancement reveals couplings of ca 0.7 and 0.8 Hz on the methylene and olefin signals respectively, in agreement, so this metabolite is considered confidently identified. For choline, in addition to the 1D ^{1}H NMR chemical shift, multiplicity and HSQC information, cross-methyl, and N–CH$_2$ carbon to methyl proton HMBC peaks were observed. Remarkably, due to the quasi-symmetrical environment of the quadrupolar nitrogen-14 atom, resolution enhancement of the methyl signal revealed a small $^2J_{NH}$ of ca 0.6 Hz (triplet 1:1:1) which is diagnostic and also present on reprocessing the HMDB reference spectrum with resolution enhancement (HMDB00097). This identification is thus considered confident. For L-carnitine, in addition to the proton chemical shift and multiplicity of the methyl group, an HSQC cross-peak to the methyl carbon was observed, together with HMBC cross-peaks from the methyl carbon and the N–CH$_2$ carbon to the methyl protons. This metabolite is considered putatively annotated however, as none of the metabolite identification information covers the carboxylic acid portion of the molecule (see Guideline 3 in Section 3.5 above). For dimethylglycine, in addition to the 1D ^{1}H NMR chemical shift, multiplicity and HSQC information, cross-methyl and N–CH$_2$ carbon to methyl HMBC peaks were observed, confirming the identification of this metabolite. For betaine, in addition to the 1D ^{1}H NMR chemical shift, multiplicity and HSQC information, cross-methyl, and N–CH$_2$ carbon to methyl proton HMBC peaks were observed, confirming the identification. For N-propionylglycine however, no further information was available and thus, this metabolite should be described as putatively annotated also.

In summary for 13 metabolites with *actual* MIE scores of <0.5 and/or actual MICE scores of <1.0 based on the combined experimental 1D ^{1}H plus COSY and HSQC NMR data, a total of three metabolites were classed as putatively annotated: the rest were confidently identified. Thus, even with relatively low MIE or MICE scores, it is still possible to confidently identify a very large number of metabolites, *as long as additional, high quality 1D and 2D NMR data is available.*

4. Conclusions

This work represents a novel, more quantitative approach to the issue of confidence in metabolite identification. The spectroscopic information content of the 1D and 2D ^{1}H NMR spectra of metabolites has been investigated from a metabolite identification perspective for the first time. New theoretical and experimental measures of metabolite identification efficiency have been delineated: the metabolite identification efficiency (MIE), the metabolite identification carbon efficiency (MICE) and the metabolite identification hydrogen fraction (MIHF). These are expected to be useful in helping to establish the confidence of metabolite identifications in future metabonomics/metabolomics studies.

The main recommendation emerging from this work is that the requirement for comparison with an authentic reference standard for confident metabolite identification is unnecessary for NMR-based metabolite identifications as long as the 7 recommendations for metabolite identification confidence below are acted upon (see also Section 3.5). Metabolites can be confidently identified by comparison with data in online databases such as HMDB [31]. Examples have been given of confident identifications of metabolites with high, average and relatively low MIE/MICE values, using data at the level of 1D ^{1}H plus 2D COSY and HSQC data. Metabolites with low MIE/MICE values will need corroboration with other data. In the case of experiments run at the level of 1D ^{1}H plus 2D COSY and HSQC data, this may be HMBC or long-range coupling data, for example.

The 7 recommendations for confident identification of known metabolites based on comparison with the NMR spectra of those

metabolites in reference databases such as HMDB that emerged from this work (see Section 3.4 above for more details) are:

1. the experimental metabolite identification carbon efficiency (MICE) obtained in the experiments ideally should be ≥1, or the metabolite identification efficiency (MIE) > 0.5: these are guidelines, not absolute numbers, and are for approaches using 1D ^{1}H plus 2D COSY and HSQC data
2. the fit of the experimental data to reference data should be precise, generally within ±0.03 ppm for ^{1}H, and ±0.5 ppm for ^{13}C NMR shifts and ±0.2 Hz for proton couplings: the database entries should be double-checked for self-consistency, accuracy and agreement with other literature, including by downloading of actual free induction decay data (HMDB)
3. the NMR spectral data should provide 'coverage' of all parts of the molecule
4. the signal-to-noise ratio and the resolution (actual and digital) in the spectra should be sufficient to measure the signal features with confidence
5. care should be applied when assigning signals in crowded spectral regions
6. HSQC data is important in metabolite identification, as it provides an excellent orthogonal data source via the ^{13}C NMR chemical shift
7. HMBC data should be used wherever possible to corroborate identifications.

A further recommendation from this work is that metabonomics/metabolomics researchers publish more detail on the spectroscopic data on which they are basing their metabolite identifications. This additional information could include the MIE or MICE values for each of the metabolites identified. Confidence in metabolite identification is critical for any subsequent biochemical or biological interpretation of the data.

Thus, in summary, as long as the 7 recommendations above are acted upon, confident identifications of *known metabolites* can be made by reference to on-line databases such as the Human Metabolome Database (HMDB). Out of 75 known metabolites studied in this work, it is asserted that 72 of 75 (96%) are confidently identified and only 3 metabolites (4%) fell into the putatively annotated category.

One of the reasons for being less conservative in the identification of known metabolites using NMR spectroscopic methods is that NMR technology is stable and reproducible. Having an excellent resource like the HMDB [31] available, that not only provides access to *information* on the NMR spectra of the metabolites in both 1D and 2D forms, but also enables access to the raw free induction decay data, is equivalent in many cases to having access to an authentic reference standard for direct comparisons. However, as always, the researcher needs to double-check all database entries for coherence and accuracy: errors in the databases do occur.

It is hoped that this work provides a new paradigm for NMR-based metabolite identification of *known metabolites*. It is expected that other researchers will investigate and test the methodology and no doubt develop it further. However, it is hoped and expected that the provision of a metabolite identification confidence index such as MIE or MICE will help solve the current issue of confidence in metabolite identification. Finally, it should be noted that the ideas herein are equally applicable to mass spectrometry, and to other analytical techniques, and should have broad utility.

Acknowledgements

I would like to gratefully acknowledge the following people: Professor Liz Shephard and Ms. Flora Scott for a collaboration on mouse phenotypes; Professor Stefano Balducci for a collaboration on the effects of exercise on diabetic patients; Professors Jeremy Nicholson, John Lindon and Elaine Holmes for long-term collaborations on metabonomics and pharmacometabonomics, and for access to the 600 MHz NMR facilities; Dr. Anthony Dona for assistance with the 600 NMR spectrometer and

Ms Dorsa Varshavi for assistance with the sample preparation and spectral analysis.

Appendix A

A.1. Glossary of terms

Term	Meaning
1D	One-dimensional
2D	Two-dimensional
CAWG	Chemical Analysis Working Group
CE–MS	Capillary electrophoresis mass spectrometry
COSY	COrrelated SpectroscopY
δ_H	Hydrogen-1 or proton NMR chemical shift
δ_C	Carbon-13 NMR chemical shift
GC–MS	Gas chromatography mass spectrometry
HMBC	Heteronuclear multiple bond correlation spectroscopy
HMDB	Human Metabolome Database
HSQC	Heteronuclear single quantum correlation spectroscopy
ID	Identification
$^{3}J_{H,H}$	Three-bond spin–spin coupling between two hydrogens etc
JRES	J-resolved spectroscopy
LC–MS	Liquid chromatography mass spectrometry
MIE	Metabolite identification efficiency
MICE	Metabolite identification carbon efficiency
MIHF	Metabolite Identification Hydrogen Fraction
MS	Mass spectrometry
MSI	Metabolomics Standards Initiative
NOESY	Nuclear Overhauser spectroscopy
NMR	Nuclear magnetic resonance
TOCSY	TOtal Correlation SpectroscopY
TSP	Sodium 3-(trimethylsilyl) propionate-2, 2, 3, 3-d4
UPLC–MS	Ultra-performance liquid chromatography mass spectrometry

References

[1] Lindon J, Nicholson J, Holmes E, Everett J. Metabonomics: metabolic processes studied by NMR spectroscopy of biofluids. Concepts Magn Reson 2000;12(5):289–320.

[2] Lindon JC, Nicholson JK, Holmes E. The handbook of metabonomics and metabolomics. Amsterdam; Oxford: Elsevier; 2007.

[3] Everett JR, Loo RL, Pullen FS. Pharmacometabonomics and personalized medicine. Ann Clin Biochem 2013;50(6):523–45.

[4] Clayton TA, Lindon JC, Cloarec O, Antti H, Charuel C, Hanton G, et al. Pharmacometabonomic phenotyping and personalized drug treatment. Nature 2006; 440(7087):1073–7.

[5] Clayton TA, Baker D, Lindon JC, Everett JR, Nicholson JK. Pharmacometabonomic identification of a significant host-microbiome metabolic interaction affecting human drug metabolism. Proc Natl Acad Sci U S A 2009;106(34):14728–33.

[6] Fiehn O. Metabolomics—the link between genotypes and phenotypes. Plant Mol Biol 2002;48(1–2):155–71.

[7] Jankevics A, Merlo ME, de Vries M, Vonk RJ, Takano E, Breitling R. Separating the wheat from the chaff: a prioritisation pipeline for the analysis of metabolomics datasets. Metabolomics 2012;8(1):S29–36.

[8] Beckonert O, Keun HC, Ebbels TMD, Bundy J, Holmes E, Lindon JC, et al. Metabolic profiling, metabolomic and metabonomic procedures for NMR spectroscopy of urine, plasma, serum and tissue extracts. Nat Protoc 2007;2(11):2692–703.

[9] Craig A, Cloarec O, Holmes E, Nicholson JK, Lindon JC. Scaling and normalization effects in NMR spectroscopic metabonomic data sets. Anal Chem 2006;78(7):2262–7.

[10] Bylesjo M, Rantalainen M, Cloarec O, Nicholson JK, Holmes E, Trygg J. OPLS discriminant analysis: combining the strengths of PLS-DA and SIMCA classification. J Chemometr 2006;20(8–10):341–51.

[11] Dona AC, Jimenez B, Schaefer H, Humpfer E, Spraul M, Lewis MR, et al. Precision high-throughput proton nmr spectroscopy of human urine, serum, and plasma for large-scale metabolic phenotyping. Anal Chem 2014;86(19):9887–94.

[12] Posma JM, Garcia-Perez I, De Iorio M, Lindon JC, Elliott P, Holmes E, et al. Subset optimization by reference matching (STORM): an optimized statistical approach for recovery of metabolic biomarker structural information from H-1 NMR spectra of biofluids. Anal Chem 2012;84(24):10694–701.

[13] Bouatra S, Aziat F, Mandal R, Chi Guo A, Wilson MR, Knox C, et al. The human urine metabolome. PLoS ONE 2013;8(9).

[14] Emwas A-HM, Salek RM, Griffin JL, Merzaban J. NMR-based metabolomics in human disease diagnosis: applications, limitations, and recommendations. Metabolomics 2013;9(5):1048–72.

[15] Claridge T, High-Resolution NMR. Techniques in organic chemistry. Oxford, UK: Elsevier; 2009.

[16] Dunn WB, Erban A, Weber RJM, Creek DJ, Brown M, Breitling R, et al. Mass appeal: metabolite identification in mass spectrometry-focused untargeted metabolomics. Metabolomics 2013;9(1):S44–66.

[17] Watson DG. A rough guide to metabolite identification using high resolution liquid chromatography mass spectrometry in metabolomic profiling in metazoans. Comput Struct Biotechnol J 2013;4 [e201301005-e201301005].

[18] Fonville JM, Maher AD, Coen M, Holmes E, Lindon JC, Nicholson JK. Evaluation of full-resolution J-resolved 1H NMR projections of biofluids for metabonomics information retrieval and biomarker identification. Anal Chem 2010;82(5):1811–21.

[19] Ludwig C, Viant MR. Two-dimensional J-resolved NMR spectroscopy: review of a key methodology in the metabolomics toolbox. Phytochem Anal 2010;21(1).

[20] Cui Q, Lewis IA, Hegeman AD, Anderson ME, Li J, Schulte CF, et al. Metabolite identification via the Madison metabolomics consortium database. Nat Biotechnol 2008; 26(2):162–4.

[21] Tulpan D, Leger S, Belliveau L, Culf A, Cuperlovic-Culf M. MetaboHunter: an automatic approach for identification of metabolites from H-1-NMR spectra of complex mixtures. BMC Bioinformatics 2011;12.

[22] van der Hooft JJJ, de Vos RCH, Ridder L, Vervoort J, Bino RJ. Structural elucidation of low abundant metabolites in complex sample matrices. Metabolomics 2013;9(5): 1009–18.

[23] van der Hooft JJJ, Mihaleva V, de Vos RCH, Bino RJ, Vervoort J. A strategy for fast structural elucidation of metabolites in small volume plant extracts using automated MS-guided LC-MS-SPE-NMR. Magn Reson Chem 2011;49:S55–60.

[24] Jacob D, Deborde C, Moing A. An efficient spectra processing method for metabolite identification from H-1-NMR metabolomics data. Anal Bioanal Chem 2013;405(15): 5049–61.

[25] Wishart DS. Advances in metabolite identification. Bioanalysis 2011;3(15):1769–82.

[26] Sands CJ, Coen M, Ebbels TMD, Holmes E, Lindon JC, Nicholson JK. Data-driven approach for metabolite relationship recovery in biological H-1 NMR data sets using iterative statistical total correlation spectroscopy. Anal Chem 2011;83(6):2075–82.

[27] Robinette SL, Lindon JC, Nicholson JK. Statistical spectroscopic tools for biomarker discovery and systems medicine. Anal Chem 2013;85(11):5297–303.

[28] Krumsiek J, Suhre K, Evans AM, Mitchell MW, Mohney RP, Milburn MV, et al. Mining the unknown: a systems approach to metabolite identification combining genetic and metabolic information. PLoS Genet 2012;8(10).

[29] Rueedi R, Ledda M, Nicholls AW, Salek RM, Marques-Vidal P, Morya E, et al. Genome-wide association study of metabolic traits reveals novel gene-metabolite-disease links. PLoS Genet 2014;10(2) [e1004132-e1004132].

[30] Go EP. Database resources in metabolomics: an overview. J Neuroimmune Pharmacol 2010;5(1):18–30.

[31] Wishart DS, Jewison T, Guo AC, Wilson M, Knox C, Liu Y, et al. HMDB 3.0—the human metabolome database in 2013. Nucleic Acids Res 2013;41(D1):D801–7.

[32] Ulrich EL, Akutsu H, Doreleijers JF, Harano Y, Ioannidis YE, Lin J, et al. BioMagResBank. Nucleic Acids Res 2008:D402–8.

[33] Ludwig C, Easton JM, Lodi A, Tiziani S, Manzoor SE, Southam AD, et al. Birmingham Metabolite Library: a publicly accessible database of 1-D H-1 and 2-D H-1 J-resolved NMR spectra of authentic metabolite standards (BML-NMR). Metabolomics 2012; 8(1):8–18.

[34] Mercier P, Lewis MJ, Chang D, Baker D, Wishart DS. Towards automatic metabolomic profiling of high-resolution one-dimensional proton NMR spectra. J Biomol NMR 2011;49(3–4):307–23.

[35] Wishart DS. Computational strategies for metabolite identification in metabolomics. Bioanalysis 2009;1(9).

[36] Sumner LW, Amberg A, Barrett D, Beale MH, Beger R, Daykin CA, et al. Proposed minimum reporting standards for chemical analysis. Metabolomics 2007;3(3):211–21.

[37] Banks R, Blanchflower S, Everett J, Manger B, Reading C. Novel anthelmintic metabolites from an Aspergillus species; the aspergillimides. J Antibiot (Tokyo) 1997; 50(10):840–6.

[38] Fiehn O, Robertson D, Griffin J, van der Werf M, Nikolau B, Morrison N, et al. The metabolomics standards initiative (MSI). Metabolomics 2007;3(3):175–8.

[39] Salek RM, Steinbeck C, Viant MR, Goodacre R, Dunn WB. The role of reporting standards for metabolite annotation and identification in metabolomic studies. GigaScience 2013;2(1):13-13.

[40] Creek DJ, Dunn WB, Fiehn O, Griffin JL, Hall RD, Lei Z, et al. Metabolite identification: are you sure? And how do your peers gauge your confidence? Metabolomics 2014; 10(3):350-3.

[41] Sumner L, Lei Z, Nikolau B, Saito K, Roessner U, Trengove R. Proposed quantitative and alphanumeric metabolite identification metrics. Metabolomics 2014;10(6): 1047–9.

[42] Everett JR. Unpublished. (2014).

[43] Bowers D. Medical statistics from scratch: an introduction for health professionals. Chichester: Wiley; 2008.

[44] Tomlins A, Foxall P, Lynch M, Parkinson J, Everett J, Nicholson J. High resolution (1)H NMR spectroscopic studies on dynamic biochemical processes in incubated human seminal fluid samples. Biochim Biophys Acta Gen Subj 1998;1379(3):367–80.

[45] Hopkins AL, Groom CR, Alex A. Ligand efficiency: a useful metric for lead selection. Drug Discov Today 2004;9(10):430–1.

[46] Hopkins AL, Keserue GM, Leeson PD, Rees DC, Reynolds CH. The role of ligand efficiency metrics in drug discovery. Nat Rev Drug Discov 2014;13(2):105–21.

Permissions

All chapters in this book were first published in CSBJ, by Elsevier; hereby published with permission under the Creative Commons Attribution License or equivalent. Every chapter published in this book has been scrutinized by our experts. Their significance has been extensively debated. The topics covered herein carry significant findings which will fuel the growth of the discipline. They may even be implemented as practical applications or may be referred to as a beginning point for another development.

The contributors of this book come from diverse backgrounds, making this book a truly international effort. This book will bring forth new frontiers with its revolutionizing research information and detailed analysis of the nascent developments around the world.

We would like to thank all the contributing authors for lending their expertise to make the book truly unique. They have played a crucial role in the development of this book. Without their invaluable contributions this book wouldn't have been possible. They have made vital efforts to compile up to date information on the varied aspects of this subject to make this book a valuable addition to the collection of many professionals and students.

This book was conceptualized with the vision of imparting up-to-date information and advanced data in this field. To ensure the same, a matchless editorial board was set up. Every individual on the board went through rigorous rounds of assessment to prove their worth. After which they invested a large part of their time researching and compiling the most relevant data for our readers.

The editorial board has been involved in producing this book since its inception. They have spent rigorous hours researching and exploring the diverse topics which have resulted in the successful publishing of this book. They have passed on their knowledge of decades through this book. To expedite this challenging task, the publisher supported the team at every step. A small team of assistant editors was also appointed to further simplify the editing procedure and attain best results for the readers.

Apart from the editorial board, the designing team has also invested a significant amount of their time in understanding the subject and creating the most relevant covers. They scrutinized every image to scout for the most suitable representation of the subject and create an appropriate cover for the book.

The publishing team has been an ardent support to the editorial, designing and production team. Their endless efforts to recruit the best for this project, has resulted in the accomplishment of this book. They are a veteran in the field of academics and their pool of knowledge is as vast as their experience in printing. Their expertise and guidance has proved useful at every step. Their uncompromising quality standards have made this book an exceptional effort. Their encouragement from time to time has been an inspiration for everyone.

The publisher and the editorial board hope that this book will prove to be a valuable piece of knowledge for researchers, students, practitioners and scholars across the globe.

List of Contributors

Bashar Ibrahim
Bio System Analysis Group, Friedrich-Schiller-University Jena, and Jena Centre for Bioinformatics (JCB), 07743 Jena, Germany Umm Al-Qura University, 1109 Makkah, Saudi Arabia
Al-Qunfudah Center for Scientific Research (QCSR), 21912 Al-Qunfudah, Saudi Arabia

Adam Elhofy
Essential Pharmaceuticals, Ewing, New Jersey, United States

Raunaq Malhotra and Manjari Jha
The School of Electrical Engineering and Computer Science, The Pennsylvania State University, University Park, PA, 16802, USA

Mary Poss
Department of Biology, The Pennsylvania State University, University Park, PA 16802, USA

Raj Acharya
School of Informatics and Computing, Indiana University, Bloomington, IN 47405, USA

Pía Francesca Loren Reyes, Tom Michoel, Anagha Joshi and Guillaume Devailly
The Roslin Institute, The University of Edinburgh, Easter Bush, Midlothian, EH25 9RG, Scotland, UK

Hervé Seligmann
Aix-Marseille Univ, Unité de Recherche sur les Maladies Infectieuses et Tropicales Emergentes, UM 63, CNRS UMR7278, IRD 198, INSERM U1095, Institut Hospitalo-Universitaire Méditerranée -Infection, Marseille, Postal code 13385, France
Dept. Ecol Evol Behav, Alexander Silberman Inst Life Sci, The Hebrew University of Jerusalem, IL-91904 Jerusalem, Israel

Ganesh Warthi
Aix-Marseille Univ, Unité de Recherche sur les Maladies Infectieuses et Tropicales Emergentes, UM 63, CNRS UMR7278, IRD 198, INSERM U1095, Institut Hospitalo-Universitaire Méditerranée -Infection, Marseille, Postal code 13385, France

Ruibang Luo
Department of Computer Science, Johns Hopkins University, United States
Center for Computational Biology, McKusick-Nathans Institute of Genetic Medicine, Johns Hopkins University School of Medicine, United States

Fritz J. Sedlazeck and Charlotte A. Darby
Department of Computer Science, Johns Hopkins University, United States

Stephen M. Kelly
Center for Health Informatics and Bioinformatics, New York University School of Medicine, United States

Michael C. Schatz
Department of Computer Science, Johns Hopkins University, United States
Center for Computational Biology, McKusick-Nathans Institute of Genetic Medicine, Johns Hopkins University School of Medicine, United States
Simons Center for Quantitative Biology, Cold Spring Harbor Laboratory, United States

Ashok Palaniappan
Dept of Biotechnology, Sri Venkateswara College of Engineering, Post Bag No. 3, Pennalur, Sriperumbudur 602117, India

Eric Jakobsson
University of Illinois at Urbana–Champaign, IL 61820, USA

Seanna Hewitt, Benjamin Kilian, Richard Sharpe and Amit Dhingra
Molecular Plant Sciences Graduate Program, Washington State University, Pullman, WA 99164, United States
Department of Horticulture, Washington State University, Pullman, WA 99164-6414, United States

Tyson Koepke
Molecular Plant Sciences Graduate Program, Washington State University, Pullman, WA 99164, United States
Department of Horticulture, Washington State University, Pullman, WA 99164-6414, United States
Phytelligence Inc., 1615 NE Eastgate Blvd #3, Pullman,WA 99163

Ramyya Hari
Department of Horticulture, Washington State University, Pullman, WA 99164-6414, United States

Alexey V. Sulimov, Danil C. Kutov, Ekaterina V. Katkova and Vladimir B. Sulimov
Dimonta, Ltd, Nagornaya Street 15, Bldg. 8, Moscow 117186, Russia
Research Computer Center, Moscow State University, Leninskie Gory 1, Bldg. 4, Moscow 119992, Russia

Igor V. Oferkin
Dimonta, Ltd, Nagornaya Street 15, Bldg. 8, Moscow 117186, Russia

Dmitry A. Zheltkov
Faculty of Computational Mathematics and Cybernetics of Lomonosov Moscow State University, Leninskie Gory 1, Bldg. 52, Moscow 119992, Russia

Eugene E. Tyrtyshnikov
Faculty of Computational Mathematics and Cybernetics of Lomonosov Moscow State University, Leninskie Gory 1, Bldg. 52, Moscow 119992, Russia
Institute of Numerical Mathematics of Russian Academy of Sciences, Gubkin Street 8, Moscow, 119333, Russia

Amornpan Klanchui
Biological Engineering Program, Faculty of Engineering, King Mongkut's University of Technology Thonburi, Bangkok 10140, Thailand

Supapon Cheevadhanarak
Division of Biotechnology, School of Bioresources and Technology, King Mongkut's University of Technology Thonburi, Bangkok 10150, Thailand

Peerada Prommeenate
Biochemical Engineering and Pilot Plant Research and Development (BEC) Unit, National Center for Genetic Engineering and Biotechnology, National Science and Technology Development Agency at King Mongkut's University of Technology Thonburi, Bangkok 10150, Thailand

Asawin Meechai
Department of Chemical Engineering, Faculty of Engineering, King Mongkut's University of Technology Thonburi, Bangkok 10140, Thailand

Tamsyn A. Hilder
School of Chemical and Physical Sciences, Victoria University of Wellington, Wellington 6040, New Zealand
Computational Biophysics Group, Research School of Biology, Canberra, ACT 0200, Australia

Justin M. Hodgkiss
School of Chemical and Physical Sciences, Victoria University of Wellington, Wellington 6040, New Zealand
The MacDiarmid Institute of Advanced Materials and Nanotechnology, New Zealand

Paul W. Bible and Lai Wei
State Key Laboratory of Ophthalmology, Zhongshan Ophthalmic Center, Sun Yat-sen University, Guangzhou 510060, China

Hong-Wei Sun
Biodata Mining and Discovery Section, Office of Science and Technology, Intramural Research Program, National Institute of Arthritis and Musculoskeletal and Skin Diseases, Bethesda, Maryland

Maria I. Morasso
Laboratory of Skin Biology, Intramural Research Program, National Institute of Arthritis and Musculoskeletal and Skin Diseases, Bethesda, Maryland

Rasiah Loganantharaj
Laboratory of Bioinformatics, Center for Advanced Computer Studies, University of Louisiana at Lafayette, Lafayette, Louisiana

Jorge Duitama
Agrobiodiversity Research Area, International Center for Tropical Agriculture (CIAT), Cali, Colombia
Systems and Computing Engineering Department, Universidad de los Andes, Bogotá, Colombia

Lina Kafuri and Daniel Tello
Plant Breeding and Genetics Laboratory, Joint FAO/IAEA Division, International Atomic Energy Agency, Seibersdorf, Austria
Department of Biological Sciences, School of Natural Sciences, Universidad Icesi, Cali, Colombia

Ana María Leiva, Ericson Aranzales and Hernán Ceballos
Agrobiodiversity Research Area, International Center for Tropical Agriculture (CIAT), Cali, Colombia

Bernhard Hofinger, Sneha Datta and Bradley Till
Plant Breeding and Genetics Laboratory, Joint FAO/IAEA Division, International Atomic Energy Agency, Seibersdorf, Austria

Zaida Lentini
Department of Biological Sciences, School of Natural Sciences, Universidad Icesi, Cali, Colombia

Seyed Morteza Najibi
Department of Statistics, College of Sciences, Shiraz University, Shiraz, Iran

Mehdi Maadooliat
Department of Mathematics, Statistics and Computer Science, Marquette University, WI 53201-1881, USA
Center for Human Genetics, Marshfield Clinic Research Institute, Marshfield, WI 54449, USA

Lan Zhou and Jianhua Z. Huang
Department of Statistics, Texas A&M University, TX 77843-3143, USA

Xin Gao
Computational Bioscience Research Center (CBRC), Computer, Electrical and Mathematical Sciences and Engineering Division, King Abdullah University of Science and Technology (KAUST), Thuwal 23955-6900, Saudi Arabia

Jincheol Park
Department of Statistics, Keimyung University, South Korea

Cenny Taslim
Ohio State University Medical Center, USA

Shili Lin
Department of Statistics, State University, USA
Department of Statistics, The Ohio State University, 1958 Neil Avenue, Columbus, OH 43210-1247, USA

Raffaele Fronza and Manfred Schmidt
Department of Translational Oncology, National Center for Tumor Diseases and German Cancer Research Center, Im Neuenheimer Feld 581, 69120 Heidelberg, Germany

Alessandro Vasciaveo
Department of Translational Oncology, National Center for Tumor Diseases and German Cancer Research Center, Im Neuenheimer Feld 581, 69120 Heidelberg, Germany
Department of Control and Computer Engineering, Politecnico di Torino, Corso Duca degli Abruzzi 24, 10129 Torino, Italy

Alfredo Benso
Department of Control and Computer Engineering, Politecnico di Torino, Corso Duca degli Abruzzi 24, 10129 Torino, Italy

Teerasak E-kobon
Department of Genetics, Faculty of Science, Kasetsart University, Bangkok 10900, Thailand

Pennapa Thongararm and Pramote Chumnanpuen
Department of Zoology, Faculty of Science, Kasetsart University, Bangkok 10900, Thailand

Sittiruk Roytrakul
National Center for Genetic Engineering and Biotechnology, Thailand Science Park, Pathum Thani 12120, Thailand

Ladda Meesuk
Faculty of Dentistry, Thammasat University, Pathum Thani 12120, Thailand

Sreedevi Chandrasekaran and Danail Bonchev
Center for the Study of Biological Complexity, Virginia Commonwealth University, Richmond, VA, USA

Mary Qu Yang, Dan Li and Yifan Zhang
Joint Bioinformatics Graduate Program, Department of Information Science, George W. Donaghey College of Engineering and Information Technology, University of Arkansas at Little Rock, USA
University of Arkansas for Medical Sciences, 2801 S. University Ave, Little Rock, AR 72204, USA

William Yang
School of Computer Science, Carnegie Mellon University, 5000 Forbes Ave, Pittsburgh, PA 15213, USA

Jun Liu
Department of Statistics, Harvard University, Cambridge, MA 02138, USA

Weida Tong
Divisions of Bioinformatics and Biostatistics, National Center for Toxicological Research, US Food and Drug Administration, 3900 NCTR Road, Jefferson, AR 72079, USA

Nivedita Raib and Amutha Ramaswamy
Centre for Bioinformatics, School of Life Sciences, Pondicherry University, Puducherry 605014, India

Jeremy R. Everett
Medway Metabonomics Research Group, University of Greenwich, Chatham Maritime, Kent ME4 4TB, United Kingdom

Index

www.ingramcontent.com/pod-product-compliance
Lightning Source LLC
Chambersburg PA
CBHW080703200326
41458CB00013B/4945